Spatial Linear Models for Environmental Data

Many applied researchers equate spatial statistics with prediction or mapping, but this book naturally extends linear models, which includes regression and ANOVA as pillars of applied statistics, to achieve a more comprehensive treatment of the analysis of spatially autocorrelated data. *Spatial Linear Models for Environmental Data*, aimed at students and professionals with a master's level training in statistics, presents a unique, applied, and thorough treatment of spatial linear models within a statistics framework. Two subfields, one called geostatistics and the other called areal or lattice models, are extensively covered. Zimmerman and Ver Hoef present topics clearly, using many examples and simulation studies to illustrate ideas. By mimicking their examples and R code, readers will be able to fit spatial linear models to their data and draw proper scientific conclusions.

Topics covered include

- Exploratory methods for spatial data including outlier detection, (semi)variograms, Moran's I, and Geary's c
- Ordinary and generalized least squares regression methods and their application to spatial data
- Suitable parametric models for the mean and covariance structure of geostatistical and areal data
- Model-fitting, including inference methods for explanatory variables and likelihood-based methods for covariance parameters
- Practical use of spatial linear models including prediction (kriging), spatial sampling, and spatial design of experiments for solving real-world problems

All concepts are introduced in a natural order and illustrated throughout the book using four datasets. All analyses, tables, and figures are completely reproducible using open-source R code provided at a GitHub site. Exercises are given at the end of each chapter, with full solutions provided on an instructor's FTP site supplied by the publisher.

Dale L. Zimmerman is Professor of Statistics at the University of Iowa, and **Jay M. Ver Hoef** is Senior Scientist and Statistician, Alaska Fisheries Science Center, NOAA Fisheries. Both are Fellows of the American Statistical Association and recipients of that association's Section for Statistics and the Environment Distinguished Achievement Award.

Spatial Linear Models for Environmental Data

Dale L. Zimmerman and Jay M. Ver Hoef

CRC Press
Taylor & Francis Group
Boca Raton London New York

CRC Press is an imprint of the
Taylor & Francis Group, an **informa** business

A CHAPMAN & HALL BOOK

Designed cover image: © Jay M. Ver Hoef
Back Cover Image: Photo Credit ©Dave Withrow, NOAA Fisheries, MMPA Permit No. 782-1765

First edition published 2024
by CRC Press
2385 NW Executive Center Drive, Suite 320, Boca Raton FL 33431

and by CRC Press
4 Park Square, Milton Park, Abingdon, Oxon, OX14 4RN

CRC Press is an imprint of Taylor & Francis Group, LLC

ISBN: 978-0-367-18334-9 (hbk)
ISBN: 978-1-032-72527-7 (pbk)
ISBN: 978-0-429-06087-8 (ebk)

DOI: 10.1201/9780429060878

Typeset in CMR10 font
by KnowledgeWorks Global Ltd.

Publisher's note: This book has been prepared from camera-ready copy provided by the authors.

Access the [Instructor and Student Resources/Support Material]: https://www.routledge.com/9780367183349

Dedication

To my grandchildren Dean, Ivy, and Milo, and their future
sibling(s) and cousins — DZ
To my wonderful family Mary, Lander, and Mariah — JVH

Contents

Preface

Two main branches of spatial statistics are the subjects of this book: geostatistics and areal (sometimes called lattice) data analysis. Geostatistics is the branch that deals with observations of variables that exist more-or-less continuously in space, such as soil moisture, atmospheric ozone levels, etc., while in areal data analysis each observation is taken over a bounded region in space, e.g., the number of moose in each of several contiguous management units, or the percent tree canopy within pixels as determined by satellite imagery. These two types of spatial data occur in a wide variety of environmental sciences including geology, ecology, hydrology, and atmospheric science. Although geostatistical data and areal data differ in their support, with the consequence that statistical methods for their analysis traditionally have been regarded as distinct, we see those methods as being united, for data measured on a continuous scale, by the common thread of a linear model. The only real difference between them is how the information on spatial location is used to construct the covariance matrix of the model. The primary objective of our book is to present statistical methods for geostatistical and areal data from the unifying perspective of a spatial linear model. This perspective leads naturally to the inclusion or expansion of some topics not typically presented in spatial statistics books (e.g., spatial confounding and case-deletion diagnostics for spatial regression), but admittedly also imposes limitations (e.g., modeling of continuous-scale data only). There do not seem to be any other books that present methods for analyzing geostatistical and areal data from a pure linear models perspective. Another unique feature of this book is its extensive use of examples, both real and toy, and simulations to gain understanding of various models and methods.

The book has two intended uses. The first is as a textbook for an MS-level course on spatial statistics or spatial modeling. The first author has used the book, in the form of a continually evolving set of "lecture notes," for this purpose—specifically for a one-semester, three-credit course in spatial statistics at the University of Iowa, which he has taught roughly a dozen times. The exercises provided at the ends of most chapters complement this usage of the book. However, because the book does not include models and statistical methods for spatial point pattern data, he supplements this book with another that covers those topics, such as *Spatial Point Patterns: Methodology and Applications with R* by A. Baddeley, E. Rubak, and R. Turner, 2016, CRC Press. The second intended use is as a reference book for applied scientists in various environmental fields who have had good training in statistics. Whether using the book as a text or a reference, readers should have had an undergraduate-level course in mathematical statistics and some previous training in regression modeling at the upper undergraduate or Master's level, using vector and matrix notation. Background in linear algebra and linear model theory is not assumed, but would be an asset. The only theorems in the book give some relevant linear algebra results and are placed in an appendix, without proofs.

Analyses of spatial datasets from four environmental studies are presented throughout the book. An R data package consisting of the datasets can be found at https://github.com/jayverhoef/ZVHdata, along with installation instructions. R scripts for the analyses of these data can be found at https://github.com/jayverhoef/ZVHrcode. It is our goal that everything in this book be reproducible, so there are R scripts for all computations and figures at the same github site. Many of these scripts pertain to the analyses of the four datasets, but others correspond to toy examples and simulation studies that are likewise woven into the presentation. The R scripts are organized by Chapter, Section, and Subsection.

The book is organized as follows. Chapter 1 introduces the four spatial datasets, then briefly describes the general form of a spatial linear model. Owing to the major role played by the covariance matrix of a spatial linear model, a general review of covariances and related quantities is presented in Chapter 2, followed by an overview of various spatial properties that they may satisfy. Chapter 3 presents variography and other exploratory data analytic methods useful for formulating a spatial general linear model appropriate for a given spatial dataset. Chapter 4 considers the provisional estimation of the mean structure of a spatial linear model by the method of ordinary least squares. This method generally does not yield optimal inference for spatial linear models because it ignores possible heteroscedasticity and residual spatial correlation. However, the residuals from the ordinary least squares fit allow for additional model formulation, using methods also described in Chapter 4. Generalized least squares for a spatial Aitken model is the topic of Chapter 5. An entire chapter is devoted to this topic in spite of the inadequacy of the spatial Aitken model in practice because it turns out that a reasonable procedure for estimating the mean structure of a spatial general linear model is to apply generalized least squares to a spatial Aitken model obtained by substituting estimates of the variance-covariance parameters for their true values. Chapters 6 and 7 describe parametric models for the covariance structure of spatial general linear models for geostatistical data and areal data, respectively. The former include compactly supported models such as the spherical, and globally supported models such as the Matérn family, and the latter include simultaneous and conditional autoregressive models. Inferential methods for spatial general linear models are the topics of the next two chapters, with likelihood-based methods for estimating model parameters and comparing models considered in Chapter 8, and spatial prediction (kriging) of predictands featured in Chapter 9. Spatial sampling design and the design and analysis of spatial experiments based on a spatial general linear model are presented in Chapters 10 and 11, respectively. Finally, Chapter 12 briefly reviews extensions of many of the book's topics to non-Euclidean spatial domains (surfaces of spheres and river networks, in particular), space-time data, multivariate data, and discrete data.

We recognize that there are two major topics, in addition to spatial point pattern analysis, that are present in many books on spatial statistics but not included herein: methods for discrete spatial data, and Bayesian methods for spatial data of all types. We briefly review discrete data methods in Chapter 12, but we barely mention Bayesian methods anywhere. Nevertheless, since Bayesian methods still require a likelihood function, we believe that the models and likelihood-based methods featured in this book will be valuable to those practitioners who wish to perform a Bayesian analysis of their data.

A list of errata for the book is available from a link on the book's webpage, www.routledge.com/9780367183349. Also, solutions to most of the book's exercises are available to instructors from the instructor hub, for which there is a link at the same webpage.

We could not have written this book without the assistance of a number of people, for which we are most grateful. Among these are the many students who took the first author's spatial statistics course over the years. Each new cohort of students provided useful feedback, which improved the course notes as they evolved into the book. We are especially grateful to two of those students, Jun Tang and Ruida Song, for writing code for some of the examples, creating early versions of some figures, and preparing solutions to some of the exercises. While we were writing the book, Michael Dumelle, at the EPA Pacific Ecological Systems Division in Corvallis, and Matt Higham at the St. Lawrence

University, New York, in conjunction with the second author, developed the R package `spmodel`, which complements our book perfectly and was used for many examples and simulations. DZ thanks the University of Iowa for providing office space, materials, and released time from teaching for writing the book. JVH wants to thank the Marine Mammal Laboratory at the NOAA-NMFS Alaska Fisheries Science center for providing data, time and salary for working on this book, and in particular to John Bengtson and Peter Boveng for providing support and encouragement. Finally, we thank several anonymous reviewers for their comments and suggestions on earlier drafts of the book manuscript, and John Kimmel and Lara Spieker at Chapman & Hall/CRC Press for their guidance through the writing and publication process.

Dale L. Zimmerman
University of Iowa
Iowa City, IA
USA

Jay M. Ver Hoef
Marine Mammal Laboratory
Alaska Fisheries Science Center
NOAA Fisheries
Fairbanks, AK
USA

1

Introduction

1.1 Spatial data and predictands

This book considers models for, and statistical analysis of, **spatial data** that have basic form

$$\{(S_i, y_i) : i = 1, \ldots, n\},$$

where S_i is a "site" (spatial location) in a spatial domain $\mathcal{A}$ of interest and y_i is the observed value of a real-valued "response" variable y at S_i. Typically, the S_i's are distinct, i.e., there is only one observation taken at each site, but this is not strictly required. The spatial domain $\mathcal{A}$ is usually a region within $\mathbb{R}^d$, where $d = 1$, 2, or 3 and is often Euclidean, but again this is not strictly required; non-Euclidean spatial domains of considerable interest include the surface of a sphere (such as one that approximates the earth's surface) or a spatial network (such as a river network). It is assumed that the response variable is of either of continuous type or of discrete type with enough levels that models for a variable of continuous type may be applied with negligible effect. A vector $\mathbf{x}_i$ of "explanatory variables" (additional variables that may help to explain the response) may also be observed at each site, in which case the basic form of the data is expanded to $\{(S_i, y_i, \mathbf{x}_i) : i = 1, \ldots, n\}$. It is also possible that the response is multivariate rather than univariate, in which case the scalar y_i in these data representations is replaced with a vector $\mathbf{y}_i$ of response variables observed at S_i.

The term "site" in the preceding paragraph is intentionally imprecise, so as to encompass the two types of spatial data commonly known as **areal data** and **geostatistical data**. Areal data are observations taken at nonoverlapping, contiguous (or nearly so) subregions $\{A_i : i = 1, \ldots, n\}$ of $\mathcal{A}$. The subregions have well-defined boundaries, which may be regularly or irregularly shaped. Such data often arise by counting discrete events (e.g., disease cases) within administrative units (such as counties or wildlife management units), or by averaging a continuous variable (e.g., soil moisture) over well-defined areas (such as field plots or watersheds). Geostatistical data, like areal data, are taken at sites that are nonoverlapping subregions, but in contrast to areal data, the subregions are not contiguous and are so small relative to the spacing between them that nothing of consequence is lost by idealizing them as individual points $\{\mathbf{s}_i : i = 1, \ldots, n\}$, one representing each subregion. Here, the elements of the d-dimensional vector $\mathbf{s}_i$ represent the ith point's spatial coordinates in a coordinate system overlaid upon $\mathcal{A}$. Such data are often measurements of a continuous variable at a particular time or aggregated over time, such as the maximum ozone level recorded on a particular day at each monitoring site in a network of such sites distributed throughout a large city, or the total rainfall amounts at a network of rain gauges in a given year.

The following are introductions to four spatial datasets, two areal and two geostatistical, that will be featured throughout this book. For those readers who may not have read the Preface, we note again here that these data, as an R data package, can be found at https://github.com/jayverhoef/ZVHdata, along with installation instructions. R

DOI: 10.1201/9780429060878-1

FIGURE 1.1
Map of wet sulfate deposition data locations, where color classes indicate annual rates of deposition, in kg/ha. The legend shows the break points between color classes.

scripts for analyses of these data that are presented throughout the book can be found at https://github.com/jayverhoef/ZVHrcode. It is our goal that everything in this book is reproducible, so there are R scripts for all computations and figures at the same github site. The R scripts are organized by Chapter, Section, and Subsection. Many of the analyses feature the new R package spmodel (Dumelle *et al.*, 2023a), available on CRAN, which includes models for both geostatistical and areal data. Parameters are estimated using various approaches, but likelihood methods are featured, with additional modeling options that include anisotropy, random effects, partition factors, and methods for big data. Model-fit statistics are available, as well as prediction (kriging) at unobserved locations.

Example 1.1. Wet sulfate deposition

Figure 1.1 is a map showing data from our first example. The data are total wet sulfate (SO_4) deposition amounts (in kilograms per hectare) for 1987, at locations of 197 sites in the continental United States that were part of the National Atmospheric Deposition Program/National Trends Network (NADP/NTN) and met quality assurance criteria in that year. Deposition amounts of several ion species were recorded weekly at sites, but we focus only on sulfate and use the total annual deposition here. No explanatory variables were measured. Thus, in terms of the notation introduced earlier, $\mathcal{A}$ is the continental United States and S_i is the subregion occupied by the bucket that collects precipitation. It is obvious from the map that the distances separating the buckets are very large relative to the size of the buckets, so it is reasonable to regard the data as geostatistical with point sites $\mathbf{s}_1, \ldots, \mathbf{s}_n$ giving the (longitude, latitude) coordinates of each NADP/NTN site. Strictly, in this case $\mathcal{A}$ is not Euclidean; however, one might argue that the curvature of the earth over the continental United States is sufficiently small that a planar projection, such as the Albers equal area projection (Snyder, 1987) shown in Figure 1.1, adequately represents the spatial information in the data.

Some questions that an analysis of these data should attempt to address are as follows:

- Where in the United States is wet sulfate deposition greatest? Where is it so small as to be of no concern?

- Is the wet sulfate deposition surface of the United States "smooth" or do some sites have unusually high levels relative to nearby sites?

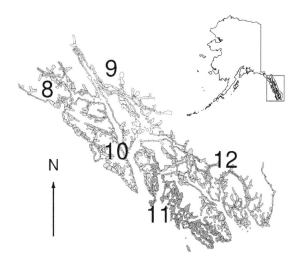

FIGURE 1.2
Polygons where harbor seals occur. The upper right corner shows a map of Alaska, USA, and the study area, known as Southeast Alaska, outlined in red. All polygons encapsulate coastal rocks along the mainland, or around islands, where seals hauled out and were counted. There are 12 seal stocks (genetically distinct populations) along the southern coast of Alaska, and stocks 8–12 occur in Southeast Alaska. The set of polygons forming the geographic range of each stock is labeled and shown as a distinct color.

- Is there spatial dependence in the data, and if so, at what scale does it exist?

- Does the answer to the previous question depend on direction, i.e., on the relative orientation of sites?

- Can we predict what the contemporaneous sulfate deposition was at an arbitrary unsampled location, or averaged over a selected subregion?

■

Example 1.2. Trends in harbor seal abundance

Our second example is a data set of trends in harbor seal abundance (Ver Hoef *et al.*, 2018b). Figure 1.2 is a map showing Southeast Alaska, USA, in the upper right, and the study area outlined in red. Harbor seals haul out on rocks along the coast, but the exact rocks that they use change somewhat from survey to survey. Hence, 463 sample units, as polygons, were created so that counts of harbor seals could be obtained from any rocks they might use within the polygons. The polygons encapsulate all known seal haulouts from more than 30 years of monitoring and, in contrast to the spatial locations for the wet sulfate deposition data, have substantial area and are mostly contiguous, so these data are areal

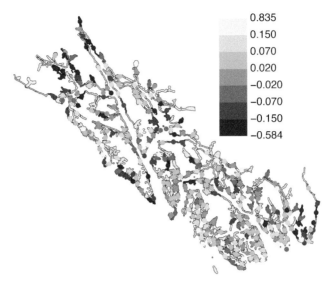

	0.835
	0.150
	0.070
	0.020
	−0.020
	−0.070
	−0.150
	−0.584

FIGURE 1.3

Map of raw harbor seal trends in Southeast Alaska. Colors show the magnitude of the estimated annual trends on the natural log scale. Each polygon is colored, but because some polygons are small and difficult to see, a circle was plotted over the centroid of each polygon and given the same color. Polygons colored grey have missing data.

rather than geostatistical. The harbor seal population in Alaska extends from the Aleutian Islands in the west to Southeast Alaska and has been divided into 12 stocks (genetically distinct populations). Five of the stocks, numbered 8–12, are shown in Figure 1.2. Seals were counted at various intensities (from 0 to 6 times per year per polygon) from aerial surveys over a 14-year period. Using counts as a provisional response variable, temporal trends for most polygons were estimated using Poisson regression (with year as the independent variable). To eliminate the effect of small sample sizes and imprecise estimates, polygons with fewer than two surveys over the 14-year period were eliminated, along with trends (linear on the log scale) that had estimated variances greater than 0.1. For demonstration purposes, we ignored the estimated variances and treated the estimated annual trends, on the natural log scale, as the final response variable. Those estimated trends are shown in Figure 1.3.

Some polygons had missing data, having fewer than two total surveys or an estimated variance greater than 0.1, resulting in 306 polygons with estimated trends and 157 that were missing. A common way to model spatial correlation for areal data such as these uses the concept of neighbors. When sample units are relatively large polygons, distance between them is not well defined, as distance could be between the closest two points from each polygon, the centroids of each polygon, etc. Using the concept of neighbors is a flexible way to form associations, such as among those that share a boundary or touch in some way (this can also be used in non-spatial situations such as social networks). Figure 1.4A shows neighbors that we defined for this study area, which are largely those that touch (with a few instances of an isolated polygon joined to its nearest neighbor). The definition of neighbor can be flexible, so in Figure 1.4B we included second-order neighbors, defined as neighbors of the neighbors found in Figure 1.4A. Figure 1.4B zooms into the area colored in gray in Figure 1.4A, as otherwise too much detail is lost. Figure 1.4C, which is a display of what might be regarded as fourth-order neighbors, includes the neighbors of neighbors found in Figure 1.4B. A more precise definition of fourth-order neighbors will be given later.

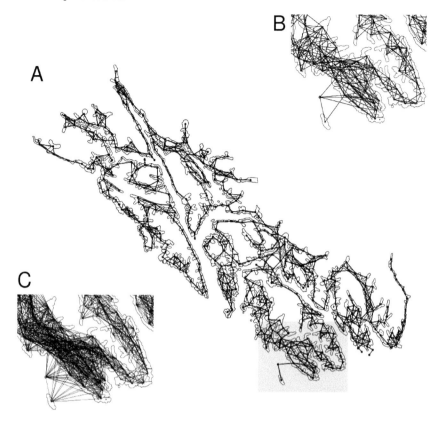

FIGURE 1.4
Map showing neighbors for harbor seal trends data. Lines connecting polygons indicate that they are neighbors. A) First-order neighbors. Generally, defining neighbors means that polygons share at least one point in common. In cases of isolated polygons, the nearest polygon, based on centroids, was used. B) Second-order neighbors. These include the neighbors of the neighbors for first-order polygons. C) Fourth-order neighbors. These include the neighbors of the neighbors for second-order polygons.

Some questions that an analysis of these data might attempt to address are as follows:

- Are there differences in trends among the five stocks?

- Is there spatial dependence in the data, after having estimated an effect for each stock? Another way to think of this is as "residual spatial correlation."

- If there is residual spatial correlation, must we account for it? That is, how does it affect our trend estimates?

- Which is a better model for residual spatial correlation, the first-order, second-order, or fourth-order neighborhood structure? How can we decide among them?

- Can we make predictions, along with uncertainty intervals, for polygons with missing data?

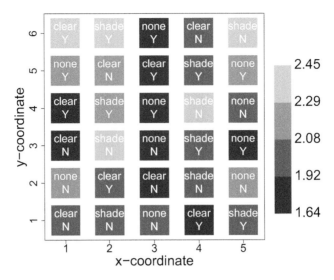

FIGURE 1.5
Designed experiment to determine how climate might affect forage biomass for caribou. There were two treatment factors: 1) a tarp over the plot, which was either clear, shade, or none, and 2) adding water, which was yes or no. The response (in percent nitrogen) to each treatment combination is colored, with the legend on the right giving the values for each color class. In the plots themselves, the upper label is the tarp treatment, and the lower label is the water treatment, where Y indicates water added, and N indicates no water added.

- Can we smooth over the observed data to reveal localized areas of increasing and decreasing trends?

■

Example 1.3. Caribou forage experiment

Global climate change is expected to have many consequences for ecosystems. Caribou are adapted to certain types of vegetation and may be limited by the amount and nutrition of their forage as climate changes. An experiment was designed in Alaska to determine what factors might affect plant nutrition in the summer range of caribou. Details of the study and data are given in Lenart *et al.* (2002). The experimental units were plots of dimensions 1.8 by 3.6 m, spaced 7.5 m apart horizontally and 9.8 m apart vertically, and arrayed in a 5-column × 6-row grid (Figure 1.5). Plots were covered with a clear tarp (to create increased temperature and less moisture) or a shade tarp (to create cooler and somewhat drier conditions, as the tarp allowed some light and moisture to pass through), or left uncovered. To separate the moisture effect, water was added, or not, to the plot as an additional main effect, creating six combinations of tarp and added moisture. In the terminology of classical designed experiments, this is a replicated two-factor design, and it can allow for interactions between the factors. (It is neither a completely randomized design nor a randomized block design, however. The exact nature of the design will be revealed

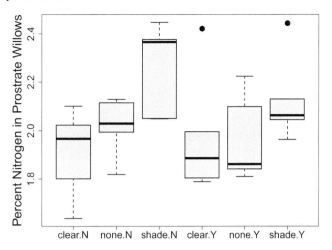

FIGURE 1.6
Boxplots of percent nitrogen in prostrate willows during the third clipping (peak biomass) of the second year of treatments in the caribou forage experiment. Below each boxplot, "clear," "none," and "shade" indicate the tarp treatment, followed by the water treatment after the dot, where "Y" indicates that water was added and "N" indicates that no water was added.

later.) The treatments were established in two years, 1994 and 1995. Each plot had eight subplots. One subplot was clipped during each of the green-up, late spring, peak biomass, and senescence time periods in each year, so all eight plots were eventually clipped. Plant material was separated into graminoids, forbs, and prostrate willows. Nitrogen is a critical component of caribou diet, and percent nitrogen was determined by laboratory analyses for each sample. We focus on percent nitrogen in prostrate willows in the third clipping (peak biomass) of the second year, to capture possible accumulated effects. These data are categorized into four color classes in Figure 1.5. Boxplots of the data for the six treatment combinations are shown in Figure 1.6.

Some questions that we might ask of these data, which clearly are areal, are as follows:

- Is there a significant difference in the response among the tarp treatments, and which one provided the highest response?

- Is there a significant difference in the response when water is added to the plots?

- Is there an interaction effect, where, for example, water added to clear tarps has a greater effect than water added to shade or no tarp?

- After accounting for the treatment effects, is there still spatial dependence in the data, and if so, must we account for it? That is, how does residual spatial correlation affect our estimates of the treatments?

- What should we consider as our neighborhood structure for residual spatial correlation? If we try several, how should we decide among them?

- While these data fit our conceptual framework for areal data, the nice regular grid also allows us to use distance among plots (centroid to centroid) to model spatial correlation (geostatistical models). How might that compare to using neighborhood models for areal data?

■

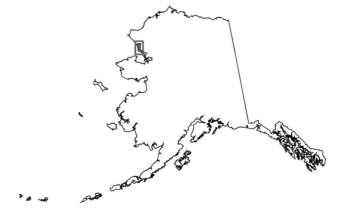

FIGURE 1.7
Map showing the location of Cape Krusenstern National Park in Alaska, outlined in red.

Example 1.4. Heavy metal concentrations in mosses

Figure 1.7 is a map showing the location of Cape Krusenstern National Park in Alaska, and Figure 1.8A is a map of the National Park itself, which shows the sampled locations. To the north and east of Cape Krustenstern is the Red Dog mine, which is the United States' largest source of zinc. Other heavy metals, such as lead and cadmium, are also mined there. A haul road from the Red Dog Mine traverses Cape Krusenstern to the coast of the Arctic Ocean. Trucks haul ore to the coast to be barged away during the short Alaskan summers. It is theorized that dust escapes into the environment from trucks. Mosses obtain much

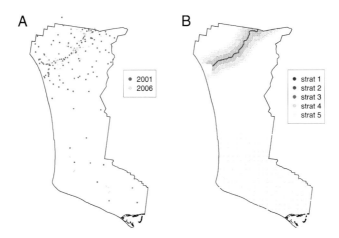

FIGURE 1.8
A) Map showing sample locations for two years (2001 and 2006) in and around Cape Krusenstern National Park. B) Prediction locations at various spatial densities, with higher densities near the road (Strata 1).

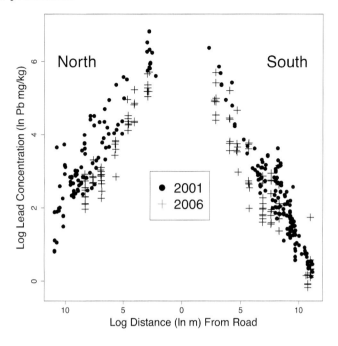

FIGURE 1.9
Log concentrations of lead in moss tissues as a function of both the side of the road (numbers to the left of 0 on the horizontal axis are north, while numbers to the right are south) and the log of the distance from the road. Data from both years, 2001 and 2006, are shown.

of their nutrients from the air, so they are ideal biomonitors for dust laden with heavy metals that mosses absorb. In 2001 (Hasselbach *et al.*, 2005) and again in 2006 (Neitlich *et al.*, 2017) mosses were sampled for heavy metals, with the sampling being more dense near the road (Figure 1.8A). Moss tissue samples were collected by snipping current annual growth, and then this was ground and homogenized before sending it out for laboratory analysis. Here, we just consider lead concentrations, although many other elements were analyzed. While we will have many of the same goals as for Example 1.3, here we have several potentially important explanatory variables, including distance from the haul road, side of the road (north or south), year of sample, elevation, slope, and aspect. Figure 1.9 shows higher concentrations of lead near the road, for both years 2001 and 2006, with possible differences in the relationship on the north side of the road versus the south side, and possible differences by year as well. The data are geostatistical, as the moss samples were very small and widely separated, relative to the size of the study region.

Some questions that might be posed for these data include:

- Is there a significant relationship between lead concentration and distance from road?

- Is there a significant difference in the relationship between lead concentration and distance from road on the north side of the road versus the south side?

- In response to the analysis of the 2001 data, trucks on the haul road began using hydraulic covers rather than tarps to minimize dust escaping while hauling the ore. Is there a significant difference in the relationship between lead concentration and distance from road in 2006 versus 2001? That is, did the use of hydraulic covers reduce environmental contamination?

- After accounting for the explanatory variables, is there still spatial dependence in the data, and if so, must we account for it? That is, how does residual spatial correlation affect our estimates of the relationships that we explored in the previous questions?

- Can we use explanatory variables and exploit the spatial correlation to predict lead concentrations at arbitrary unsampled locations, or obtain averaged values over selected subregions (given by the strata in Figure 1.8B)?

■

As can be seen from these examples, the specific questions associated with particular spatial datasets may vary. However, some general types of inquiries are common to many applications. For example, frequently one wants to understand how the response variable y varies over the spatial domain. Both large-scale variability (**spatial trend**) and small-scale variability (**spatial dependence** or **residual spatial correlation**) are of interest. Also, if explanatory variables are observed, one may want to know whether, and how, y is related to those variables after accounting for the spatial variation in y. Furthermore, it is often the case that one wants to predict the value of y, or perhaps the value of y minus the error made in its observation, at a set of **prediction sites** $\mathcal{P} \subset \mathcal{A}$. This problem is called **spatial prediction**, and the quantities to be predicted are called **spatial predictands**. Prediction sites, like data sites, may be either subregions or points, and they may be a subset or superset of the data sites or completely distinct from them. Furthermore, prediction sites need not have the same support as data sites, leading to what is sometimes called a change-of-support problem. For example, in a geostatistical setting where all the data sites are (idealized as) points, one may wish to predict the average of the response over a relatively large subregion. If $\mathcal{P}$ is either a finite partition of $\mathcal{A}$ or the set of all points in $\mathcal{A}$, then spatial prediction essentially amounts to producing a map of predictions over $\mathcal{A}$. In practice, it suffices to consider a finite (but possibly very large) collection of (prediction site, predictand) pairs, denoted as $\{(S_i, u_i) : i = n+1, \ldots, n+n_u\}$ where n_u is the number of such pairs. If the same explanatory variables observed at data sites are also observed at prediction sites, then the pairs may be expanded to triplets $\{(S_i, u_i, \mathbf{x}_i) : i = n+1, \ldots, n+n_u\}$, and the information in the additional $\mathbf{x}_i$'s may be used to possibly improve the predictions.

1.2 Statistical sampling framework

It is possible to address some questions about spatial data via nonstatistical methods, i.e., methods that are not based on any statistical sampling framework or model. For example, one common nonstatistical spatial prediction procedure predicts the value of a response at a site at which it is not observed by a weighted linear combination of the observed responses, where each weight is equal to the reciprocal of the distance from the observation site to the prediction site. This prediction approach, known as inverse distance weighting, generally yields sensible predictions; however, it does not yield an accompanying measure of the uncertainty associated with the prediction. To quantify that uncertainty, and more generally to make statistical inferences about spatial trend, spatial dependence, and spatial predictions, a statistical sampling framework must be specified for the responses and predictands. Within this framework, the responses $y_1, \ldots, y_n$ are regarded as observed values of random variables $Y_1 = Y(S_1), \ldots, Y_n = Y(S_n)$, and the predictands $u_{n+1}, \ldots, u_{n+n_u}$ are regarded as unobserved values of random variables $U_{n+1} = U(S_{n+1}), \ldots, U_{n+n_u} = U(S_{n+n_u})$. In the next section, we will introduce a class of models for these variables called **spatial linear models**. In the remainder of this section, we merely describe an appropriate sampling framework for these variables.

The classical random sampling framework, with its assumptions that the random variables are independent and identically distributed, is generally too restrictive in this context. Spatial data often adhere, to a large extent, to Tobler's first law of geography, which says that "everything is related to everything else, but near things are more related than distant things" (Tobler, 1970). In particular, spatial data generally are not identically distributed but instead vary across their domain in a manner that appears to include a smooth (but nonconstant) systematic component – what we referred to previously as a spatial trend. Furthermore, because some spatial processes and interactions may operate at relatively small scales, residuals from the trend corresponding to sites close to one another tend to be more similar than residuals corresponding to sites further apart – what we referred to previously as residual spatial correlation. Thus, the responses are not properly regarded as multiple independent observations from a univariate distribution, but rather as a single, vector-valued observation $\mathbf{y} = (y_1, y_2, \ldots, y_n)^T$ from an n-variate distribution, and any realistic statistical model for $\mathbf{y}$ should allow its elements to exhibit spatial trend and residual spatial correlation.

For some purposes, it may be useful to imagine the observed responses as a subset of a larger set of observed and "potential" responses. In the areal case, for example, for one reason or another the response may not be observed at every subregion of the partition of $\mathcal{A}$ and it may be desirable to predict the responses corresponding to those subregions where it was not observed. Alternatively, one may wish to predict the response corresponding to a subregion that lies within one, or overlaps with two or more, of the subregions. Regarding the observed response vector $\mathbf{y}$ as a subvector of a larger vector formed by appending the desired predictands to $\mathbf{y}$ may be convenient for solving these prediction problems. In the geostatistical case, a potential response exists, in theory at least, at every point $\mathbf{s}$ in $\mathcal{A}$, and for prediction of one or more of them it may be useful to regard $\mathbf{y}$ as the finite, observed portion of an infinite collection of observed and potential response variables $\{Y(\mathbf{s}) : \mathbf{s} \in \mathcal{A}\}$, henceforth referred to as a **geostatistical process**.

To perform spatial prediction statistically, we must regard the predictands $u_{n+1}, \ldots, u_{n+n_u}$, like the responses, as random variables. The same considerations that led us to regard the responses as comprising a single realization from a multivariate distribution also lead us to assume that the predictands are not independent and identically distributed but rather have a multivariate distribution exhibiting spatial trend and spatial dependence. Furthermore, because the prediction sites belong to the same spatial domain as the data sites, the spatial trend of, and spatial dependence among, the predictands is likely to be related to that of the responses. This suggests that we model $\mathbf{y}$ and $\mathbf{u} \equiv (u_{n+1}, \ldots, u_{n+n_u})^T$ jointly rather separately. In fact, in a geostatistical context, where the number of potential predictands is infinite, it may be convenient to specify a model for the joint distribution of the infinite collection of random variables $\{Y(\mathbf{s}) : \mathbf{s} \in \mathcal{A}\}$, or more generally for $\mathbf{y}$ and another related random process $\{U(\mathbf{s}) : \mathbf{s} \in \mathcal{A}\}$ of predictands.

For some purposes, such as point estimation of spatial trend parameters, it may be sufficient to specify the functional form of only some low-order moments (e.g., means, variances, and covariances) of the joint distribution of the responses and predictands; specification of the functional form of the distribution function is unnecessary. For obtaining classical parametric confidence intervals and prediction intervals or for performing statistical hypothesis tests, however, distributional assumptions may be necessary. When responses and predictands are continuous (rather than categorical or discrete), a very convenient distribution for these purposes is the multivariate normal distribution. Accordingly, we will often assume that the joint distribution of $\mathbf{y}$ and $\mathbf{u}$, possibly after a transformation and/or removal of outliers, is multivariate normal (or nearly so). In the geostatistical case, we will likewise often assume that $\{Y(\mathbf{s}) : \mathbf{s} \in \mathcal{A}\}$ and/or any other relevant process of predictands is a **Gaussian process** (an infinite collection of random variables whose finite-dimensional distributions are all multivariate normal).

Although we regard the responses and predictands as random variables, unless noted otherwise we take the explanatory variables to be nonrandom, as in classical linear models. Arguably, in many applications it may be just as reasonable to regard one or more of the explanatory variables as a random variable as it is to so regard the response. In such a situation, the model must be interpreted as a model for the conditional distribution of $\mathbf{y}$ and the predictands, given the observed values of the explanatory variables, and all inferences made on the basis of the model must be regarded as conditional (on the observed values of the explanatory variables) rather than unconditional. Furthermore, we assume throughout that the data sites $(S_1, \ldots, S_n)$ are selected independently of $\mathbf{y}$. That is, they are not preferentially sampled, using the terminology of Diggle *et al.* (2010); see also Section 10.5.

1.3   An introduction to spatial linear models

The first section of this chapter noted that spatial data are analyzed statistically in hopes of understanding mainly how the response varies over space, how explanatory variables affect the response after adjusting for its spatial variation, and how unobserved quantities (predictands) may be predicted from the observed responses. The second section indicated that these objectives are to be achieved within a statistical framework in which the data and predictands are regarded as a realization of a random vector (in the areal case) or a random process (in the geostatistical case). In this final section, we briefly introduce the class of models for the random vector of observed responses (in either case) that is the subject of this book: spatial linear models. More details and notation associated with spatial linear models, and extensions to a random process and/or to incorporate predictands, will be given in Chapters 2 and 9.

A **linear model** for a response variable Y postulates that the response is related to the observed values of p explanatory variables $X_1, X_2, \ldots, X_p$ via an equation

$$Y = X_1\beta_1 + X_2\beta_2 + \cdots + X_p\beta_p + e,$$

where $\beta_1, \ldots, \beta_p$ are unknown (and typically unconstrained) parameters called **coefficients** and e is a random variable called the **error**. The model further asserts that the error has mean zero and finite positive variance. The model is linear in two important ways. First, the **mean structure** of the model, i.e., $X_1\beta_1 + \cdots + X_p\beta_p$, is a linear function of the unknown parameters $\beta_1, \ldots, \beta_p$. Second, the error enters the model additively (rather than multiplicatively, say) and may itself be a linear combination of random variables.

When n observations $(y_1, \mathbf{x}_1), \ldots, (y_n, \mathbf{x}_n)$ are taken on the response and explanatory variables, the linear model as just defined implies that

$$y_i = \mathbf{x}_i^T \boldsymbol{\beta} + e_i \quad (i = 1, \ldots, n),$$

where $\boldsymbol{\beta} = (\beta_1, \beta_2, \ldots, \beta_p)^T$ and $e_1, \ldots, e_n$ are n possibly correlated random variables with zero means and finite positive variances. Using additional vector and matrix notation, this may also be written as

$$\mathbf{y} = \mathbf{X}\boldsymbol{\beta} + \mathbf{e}, \tag{1.1}$$

where

$$\mathbf{y} = \begin{pmatrix} y_1 \\ y_2 \\ \vdots \\ y_n \end{pmatrix}, \quad \mathbf{X} = \begin{pmatrix} \mathbf{x}_1^T \\ \mathbf{x}_2^T \\ \vdots \\ \mathbf{x}_n^T \end{pmatrix}, \quad \text{and} \quad \mathbf{e} = \begin{pmatrix} e_1 \\ e_2 \\ \vdots \\ e_n \end{pmatrix}.$$

Terminology associated with this notation is as follows: $\mathbf{y}$ is called the **response vector**, $\mathbf{X}$ is called the **model matrix**, and $\mathbf{e}$ is called the **error vector**. The variance-covariance matrix, or more simply the **covariance matrix** of $\mathbf{e}$, written as Var$(\mathbf{e})$ and denoted by $\boldsymbol{\Sigma} = (\sigma_{ij})$, is the $n \times n$ matrix whose ith main diagonal element, σ_{ii}, is the variance of e_i (and y_i), and whose (i,j)th off-diagonal element, σ_{ij} $(i \neq j)$, is the covariance between e_i and e_j (and, equivalently, between y_i and y_j). Note that here we have represented the covariance matrix by putting its (i,j)th element inside parentheses; this notational device will be used frequently throughout this book.

Linear models have been studied extensively by statisticians and used effectively by practitioners of virtually every scientific discipline. The models may be distinguished from one another in various ways, of which the most important for our purposes is in regard to their covariance structure, i.e., the form of $\boldsymbol{\Sigma}$. In a **Gauss-Markov** linear model, the errors have common variance (i.e., $\sigma_{ii} = \sigma^2$ for all i) and are uncorrelated (i.e., $\sigma_{ij} = 0$ for all $i \neq j$); equivalently, $\boldsymbol{\Sigma} = \sigma^2 \mathbf{I}_n$, where $\mathbf{I}_n$ is the $n \times n$ identity matrix (written merely as $\mathbf{I}$ when the dimension is unimportant or clear from the context). This is the class of linear models featured in most applied regression courses, and it is assumed throughout this book that the reader has had some previous exposure to it. Its covariance structure is extremely parsimonious and simple, involving only one unknown parameter (σ^2) in a multiplicative way. The classical approach to parameter estimation for this model, known as **ordinary least squares (OLS) estimation**, yields best (minimum variance) linear unbiased estimators of estimable linear functions of $\boldsymbol{\beta}$ and the best quadratic unbiased estimator of σ^2 when the Gauss-Markov model with the assumed model matrix $\mathbf{X}$ is the correct model. Ordinary least squares estimation and some associated inferential methods are presented in some detail in Chapter 4.

A Gauss-Markov linear model is appropriate for many, but by no means all, situations. For some types of data, a more flexible covariance structure than $\boldsymbol{\Sigma} = \sigma^2 \mathbf{I}$ is more appropriate, either because of *a priori* scientific considerations or because preliminary data analysis reveals evidence of **heteroscedasticity** (different error variances) or **autocorrelation** (correlation among the errors). Spatial data are such a type primarily because, as noted in the previous section, they tend to exhibit residual spatial correlation. A **spatial linear model** is defined in this book as a linear model for spatial data for which the elements of $\boldsymbol{\Sigma}$ are spatially structured functions of the data sites, where the meaning of "spatially structured" will become clear in the next chapter. Often, a spatial linear model also includes the spatial coordinates of the data sites or some function(s) of those coordinates among the explanatory variables in its mean structure, but that is not a defining characteristic of the model.

The simplest spatial linear model (from an inferential perspective) is an extension of the Gauss-Markov model that we call herein the **spatial Aitken model**. For this model,

$$\boldsymbol{\Sigma} = \sigma^2 \mathbf{R}, \tag{1.2}$$

where σ^2 is again an unknown positive parameter and $\mathbf{R}$ is a *known* (i.e., completely specified by the modeler) positive definite matrix (see below for what this means) whose elements are spatially structured functions of the data sites. We refer to $\mathbf{R}$ as the **scale-free covariance matrix** of the observations; in some settings, but not all, it is a correlation matrix. Most readers who have taken an applied regression course will have encountered (nonspatial)

Aitken models and the classical approach to the estimation of their parameters known as **generalized least squares (GLS) estimation**. GLS estimation yields best (minimum variance) linear unbiased estimators of estimable linear functions of $\boldsymbol{\beta}$ and the best quadratic unbiased estimator of σ^2 when the Aitken model with the assumed model matrix $\mathbf{X}$ and assumed scale-free covariance matrix $\mathbf{R}$ is the correct model. GLS estimation and attendant inferential methods are the topics of Chapter 5.

As a (toy) example of a spatial Aitken model, consider four observations located at the corners of the unit square in $\mathbb{R}^2$, ordered such that the first and last observations are located at opposite corners of the square. Suppose that the model for the observations is given by (1.1)-(1.2) with an arbitrary model matrix $\mathbf{X}$ and

$$\mathbf{R} = \begin{pmatrix} 1 & e^{-1} & e^{-1} & e^{-\sqrt{2}} \\ e^{-1} & 1 & e^{-\sqrt{2}} & e^{-1} \\ e^{-1} & e^{-\sqrt{2}} & 1 & e^{-1} \\ e^{-\sqrt{2}} & e^{-1} & e^{-1} & 1 \end{pmatrix}.$$

This scale-free covariance matrix is completely specified, and it is spatially structured in the sense that its (i,j)th element is a monotone decreasing function of the distance between data sites i and j. Furthermore, it has two properties that every covariance matrix must satisfy: it is **symmetric**, meaning that its (i,j)th element is equal to its (j,i)th element for all i and j; and it is **positive definite**, meaning that the variance of any linear combination of observations having this covariance matrix, except the trivial linear combination with all coefficients equal to zero, is positive. The symmetry of this $\mathbf{R}$ is obvious, but not its positive definiteness. Methods for determining whether a symmetric matrix is positive definite will be described in later chapters.

Unfortunately, the spatial Aitken model as it stands is inadequate for meeting the objectives of a spatial data analysis because it is unrealistically optimistic to expect a data analyst to correctly guess an appropriate fully specified scale-free covariance matrix $\mathbf{R}$ for the data (from among the class of all symmetric positive definite matrices). A more flexible spatial linear model that has a better chance of approximating the truth is a **spatial general linear model**, for which $\boldsymbol{\Sigma}$ is not fully specified but instead is given by a *known* spatially structured, matrix-valued parametric function $\boldsymbol{\Sigma}(\boldsymbol{\theta})$, where $\boldsymbol{\theta}$ is an *unknown* parameter vector. More precisely, the spatial general linear model is given by (1.1), with moment assumptions $\mathrm{E}(\mathbf{e}) = \mathbf{0}$ and $\mathrm{Var}(\mathbf{e}) \equiv \boldsymbol{\Sigma}(\boldsymbol{\theta})$ and with joint parameter space for $\boldsymbol{\beta}$ and $\boldsymbol{\theta}$ generally taken to be

$$\{(\boldsymbol{\beta}, \boldsymbol{\theta}) : \boldsymbol{\beta} \in \mathbb{R}^p, \boldsymbol{\theta} \in \Theta \subset \mathbb{R}^m\},$$

where Θ is the set of vectors $\boldsymbol{\theta}$ for which $\boldsymbol{\Sigma}(\boldsymbol{\theta})$ is symmetric and positive definite, or possibly some subset of that set. The matrix-valued functions $\boldsymbol{\Sigma}(\boldsymbol{\theta})$ commonly used in spatial general linear models depend on whether the data are geostatistical or areal, but have in common that they impart to $\boldsymbol{\Sigma}$ a spatial structure, including but not limited to a correlation structure that is reasonably consistent with Tobler's first law. Because the value of $\boldsymbol{\theta}$ is not specified, certain aspects of this structure, such as its strength, can be modeled flexibly. A spatial Aitken model is a special case of a spatial general linear model, with $\boldsymbol{\theta} \equiv \sigma^2$ and $\mathbf{R}$ not functionally dependent on any unknown parameters. More general spatial linear models allow the elements of $\mathbf{R}$ to depend functionally on unknown parameters. In practice, the number of unknown parameters in $\boldsymbol{\theta}$ is relatively small, usually two, three, or four; any more than that are difficult to estimate reliably.

As an example of a spatial general linear model, consider a spatial configuration of data sites exactly the same as that in the previous example of a spatial Aitken model, but suppose that the observations' covariance matrix is

$$
\boldsymbol{\Sigma}(\boldsymbol{\theta}) = \sigma^2
\begin{pmatrix}
1 & e^{-1/\alpha} & e^{-1/\alpha} & e^{-\sqrt{2}/\alpha} \\
e^{-1/\alpha} & 1 & e^{-\sqrt{2}/\alpha} & e^{-1/\alpha} \\
e^{-1/\alpha} & e^{-\sqrt{2}/\alpha} & 1 & e^{-1/\alpha} \\
e^{-\sqrt{2}/\alpha} & e^{-1/\alpha} & e^{-1/\alpha} & 1
\end{pmatrix}.
$$

This matrix is symmetric, and it is spatially structured in the same sense as $\mathbf{R}$ was in the previous example, but it is specified only up to the unknown values of σ^2 and α; in the previous example, α was assumed to equal one. The parameter space for $\boldsymbol{\theta} \equiv (\sigma^2, \alpha)^T$ within which $\boldsymbol{\Sigma}(\boldsymbol{\theta})$ is positive definite is $\{(\sigma^2, \alpha) : \sigma^2 > 0, \alpha > 0\}$ (this will be established in a later chapter). Small values of α correspond to weak spatial correlation among the observations, and as α increases the spatial correlation among observations becomes stronger.

Like $\boldsymbol{\beta}$, $\boldsymbol{\theta}$ must be estimated from the data. The advantage of estimating $\boldsymbol{\theta}$ is that it allows the data themselves to inform the modeler as to the nature and strength of the spatial covariance between observations, so that the modeler need not guess (possibly poorly). The disadvantage is that it considerably complicates inference for $\boldsymbol{\beta}$. The spatial general linear model is a specialization (to spatial data) of a linear model known as the general linear model, for which inference procedures are already well-developed (see Harville (2018) and Zimmerman (2020) for two recent treatments). But owing to its greater complexity, the general linear model and its associated estimation and other inferential procedures are not typically included in the content of a first applied regression course.

This book is focused on spatial general linear models and procedures for making statistical inferences for them. Because of the important role played by the covariance matrix of the observed data, some general types of covariance structures for spatial linear models are introduced in Chapter 2. Chapter 3 presents exploratory data analytic methods for formulating a spatial general linear model appropriate for any given spatial dataset. Chapter 4 considers the provisional estimation of the mean structure of a spatial linear model by ordinary least squares. Ordinary least squares generally does not yield optimal inference for spatial linear models because it ignores possible heteroscedasticity and residual spatial correlation. However, the residuals from the ordinary least squares fit allow for additional model formulation, using methods also described in Chapter 4. Generalized least squares for a spatial Aitken model is the topic of Chapter 5. An entire chapter is devoted to this topic in spite of the inadequacy of the spatial Aitken model in practice because it turns out that a reasonable procedure for estimating $\boldsymbol{\beta}$ for a spatial general linear model is to apply GLS to a spatial Aitken model obtained from the spatial general linear model by substituting estimates of the variance-covariance parameters for their true values. Chapters 6 and 7 describe parametric models for the covariance structure of spatial general linear models for geostatistical data and areal data, respectively. Inferential methods for spatial general linear models are the topics of the next two chapters, with likelihood-based methods for estimating model parameters and comparing models considered in Chapter 8, and spatial prediction of predictands featured in Chapter 9. Spatial sampling design and the design and analysis of spatial experiments based on a spatial general linear model are presented in Chapters 10 and 11, respectively. Finally, Chapter 12 briefly reviews extensions to non-Euclidean spatial domains (surfaces of spheres and river networks, in particular), space-time data, multivariate data, and discrete data.

2

An Introduction to Covariance Structures for Spatial Linear Models

As described in Chapter 1, a linear model is a spatial linear model if its covariance structure is a spatially structured function of the data sites. Covariance structures for geostatistical data typically are specified directly, either in terms of a function known as the covariance function, or in terms of slightly more general functions called the semivariogram and generalized covariance function. Covariance structures for areal data have been specified both directly and indirectly, the latter via a spatial weights matrix from which the covariance structure may be derived using linear algebra and/or mathematical statistics. This chapter presents an introduction to these specifications; later chapters (6 and 7) present parametric models that conform to these specifications. But we begin with a review of moments of random variables, of which covariances, of course, are a very important type.

2.1 First-order and second-order moments

2.1.1 Moments of one or two random variables

Suppose that Y is a random variable. If Y is discrete with probability mass function $f(y)$, where $f(y) > 0$ for $y \in \mathcal{Y}$, and the sum

$$\sum_{y \in \mathcal{Y}} u(y) f(y) \tag{2.1}$$

exists, then it (the sum) is called the **expectation** of $u(Y)$. If Y is (absolutely) continuous with probability density function $f(y)$, where $f(y) > 0$ for $y \in \mathcal{Y}$, and

$$\int_{\mathcal{Y}} u(y) f(y) \, dy \tag{2.2}$$

exists, then it (the integral) is called the **expectation** of $u(Y)$. A more general definition of expectation can be given for any type of random variable, but for our purposes the two definitions above will suffice. In either case, the expectation of $u(Y)$ is written as $\mathrm{E}(u(Y))$.

Two very important types of expectations are the **moments** and **central moments** of order k ($k = 1, 2, \ldots$), corresponding to setting $u(Y) = Y^k$ and $u(Y) = (Y - \mathrm{E}(Y))^k$, respectively, in either (2.1) or (2.2) (depending on whether Y is discrete or continuous). When they exist, the first-order moment $\mathrm{E}(Y)$ and the second-order central moment $\mathrm{E}[(Y - \mathrm{E}(Y))^2]$ are called the **mean** and the **variance**, respectively, of Y. The mean is a measure of the center of a distribution, while the variance is a measure of dispersion. Conventionally, the mean of Y is often represented by the Greek symbol μ, and the variance of Y is often represented by σ^2. We follow these conventions throughout, using subscripts or other devices

DOI: 10.1201/9780429060878-2

as necessary to distinguish means and variances for different random variables (such as those corresponding to different spatial locations) from one another. The variance of Y may also be written as $\text{Var}(Y)$.

It follows easily from the definition of expectation that

$$E(c_1 u_1(Y) + c_2 u_2(Y)) = c_1 E(u_1(Y)) + c_2 E(u_2(Y)), \tag{2.3}$$

where c_1 and c_2 are nonrandom real numbers and $u_1(\cdot)$ and $u_2(\cdot)$ are functions whose expectations exist. Using (2.3), many important facts about means and variances can be obtained. For example, a well-known alternative formula for the variance of Y may be obtained via the following calculation:

$$\text{Var}(Y) = E[(Y - \mu)^2] = E[Y^2 - 2\mu Y + \mu^2] = E(Y^2) - \mu^2. \tag{2.4}$$

Two other important facts derived as special cases of (2.3) are

$$E(aY + b) = a\mu + b$$

and

$$\text{Var}(aY + b) = a^2 \sigma^2 \tag{2.5}$$

for any real numbers a and b.

Now suppose that Y_1 and Y_2 are two random variables, either both discrete or both continuous, with joint probability mass function or density function $f(y_1, y_2)$ on support $\mathcal{Y}$. Then the expectation of $u(Y_1, Y_2)$, when it exists, is defined by a straightforward extension of definitions (2.1) and (2.2). In the discrete case,

$$E(u(Y_1, Y_2)) = \sum_{(y_1, y_2) \in \mathcal{Y}} u(y_1, y_2) f(y_1, y_2),$$

and in the continuous case,

$$E(u(Y_1, Y_2)) = \iint_{\mathcal{Y}} u(y_1, y_2) f(y_1, y_2) \, dy_1 \, dy_2.$$

From the definition, it follows that

$$E(c_1 u_1(Y_1, Y_2) + c_2 u_2(Y_1, Y_2)) = c_1 E(u_1(Y_1, Y_2)) + c_2 E(u_2(Y_1, Y_2)). \tag{2.6}$$

It turns out that the means μ_1 and μ_2, and variances σ_1^2 and σ_2^2, of Y_1 and Y_2 (when they exist) may be obtained using either the bivariate expectation just described [with $u(Y_1, Y_2) = Y_i$ for μ_i and $u(Y_1, Y_2) = (Y_i - E(Y_i))^2$ for σ_i^2, for $i = 1$ or 2] or the previous univariate expectation [with the marginal distributions of Y_1 and Y_2 playing the role previously played by $f(y)$]. However, there are other important expectations in spatial statistics that require the bivariate approach. One of these is the covariance. The **covariance** of (or between) Y_1 and Y_2, when it exists, is defined as

$$\text{Cov}(Y_1, Y_2) = E[(Y_1 - \mu_1)(Y_2 - \mu_2)] \equiv \sigma_{12}.$$

The covariance of Y_1 and Y_2 exists if σ_1^2 and σ_2^2 exist because (as a consequence of the Cauchy-Schwartz inequality) $\sigma_{12} \leq \sqrt{\sigma_1^2 \sigma_2^2}$. Observe that $\text{Cov}(Y_1, Y_1) = \text{Var}(Y_1)$ and $\text{Cov}(Y_2, Y_1) = \text{Cov}(Y_1, Y_2)$, so symbols σ_1^2 and σ_{11} can be used interchangeably, as can σ_{12} and σ_{21}. Using (2.6), the following additional facts can be obtained:

$$\text{Cov}(Y_1, Y_2) = E(Y_1 Y_2) - \mu_1 \mu_2, \tag{2.7}$$

$$\text{Cov}(a_1 Y_1 + b_1, a_2 Y_2 + b_2) = a_1 a_2 \sigma_{12}, \tag{2.8}$$

$$\text{Var}(a_1 Y_1 + a_2 Y_2) = a_1^2 \sigma_1^2 + a_2^2 \sigma_2^2 + 2a_1 a_2 \sigma_{12} \tag{2.9}$$

for any real numbers a_1, a_2, b_1, and b_2. In the special case $Y_1 = Y_2$, (2.7) reduces to (2.4); if, in addition $a_1 = a_2$ and $b_1 = b_2$, then (2.8) reduces to (2.5).

As its name suggests, the covariance of Y_1 and Y_2 measures, in a certain sense, the extent to which Y_1 and Y_2 co-vary or, in other words, how they "depend" on each other. However, there are two limitations to this interpretation. First, the magnitude of σ_{12} depends on the scale of measurement, so by itself it cannot be regarded as a measure of dependence. A unitless version of the covariance is the **correlation**, written as $\mathrm{Corr}(Y_1, Y_2)$ and defined as

$$\mathrm{Corr}(Y_1, Y_2) = \frac{\mathrm{Cov}(Y_1, Y_2)}{\sqrt{\mathrm{Var}(Y_1)\mathrm{Var}(Y_2)}} = \frac{\sigma_{12}}{\sqrt{\sigma_1^2 \sigma_2^2}} \equiv \rho_{12},$$

provided that σ_1^2 and σ_2^2 are positive. Observe that $\mathrm{Corr}(Y_1, Y_1) = 1$. It can be shown that $-1 \leq \rho_{12} \leq 1$ (as another consequence of the Cauchy-Schwartz inequality), so the correlation solves the scaling issue. The second limitation is that ρ_{12} measures the strength of only the *linear* dependence between Y_1 and Y_2. That is, if Y_1 and Y_2 are independent, then $\rho_{12} = 0$; but the converse is not generally true. For example, consider a discrete random variable Y_2 that is equal to 1, 0, or -1, each with probability $\frac{1}{3}$, and define $Y_1 = Y_2^2$. Then it can be easily verified that $\rho_{12} = 0$ despite the perfect but nonlinear dependence between Y_1 and Y_2. Thus the correlation is an imperfect measure of the dependence between two variables, but it is still quite useful. Crucially, in the special case in which Y_1 and Y_2 have a bivariate normal distribution, zero correlation *does* imply independence.

Another bivariate moment used frequently in spatial statistics, but seldom introduced in mathematical statistics courses, is the **semivariance**. The semivariance of Y_1 and Y_2 is defined as

$$\gamma_{12} = \frac{1}{2}\mathrm{Var}(Y_1 - Y_2),$$

provided that this variance exists. It is clear from this definition that $\gamma_{12} = \gamma_{21}$;

$$\gamma_{12} \geq 0 \tag{2.10}$$

with equality if and only if

$$Y_2 = Y_1 + b \quad \text{(with probability one) for some constant } b; \tag{2.11}$$

and

$$\gamma_{12} = \frac{1}{2}\mathrm{E}[(Y_1 - Y_2)^2] \quad \text{if and only if } \mathrm{E}(Y_1 - Y_2) = 0. \tag{2.12}$$

(To obtain (2.12), we have applied (2.4) to $Y \equiv Y_1 - Y_2$.) In contrast to the covariance, the semivariance measures *lack of dependence* in some sense, as its value is smallest (0) when Y_1 and Y_2 are related as in (2.11) and increases with the variance of the difference of the two variables. Using (2.9), it is easy to show that

$$\gamma_{12} = \frac{1}{2}(\sigma_1^2 + \sigma_2^2 - 2\sigma_{12}) \quad \text{if } \sigma_1^2 \text{ and } \sigma_2^2 \text{ exist,} \tag{2.13}$$

and

$$\gamma_{12} = \sigma^2 - \sigma_{12} = \sigma^2(1 - \rho_{12}) \quad \text{if } \sigma_1^2 = \sigma_2^2 \equiv \sigma^2 > 0. \tag{2.14}$$

By (2.13), the semivariance of two variables exists if their variances exist, and in this case it is bounded above by $\frac{1}{2}(\sigma_1^2 + \sigma_2^2) + |\sigma_1 \sigma_2|$. The converse is not generally true, however. To see this, let $Y_1 = X_1 + W$ and $Y_2 = X_2 + W$, where X_1 and X_2 are random variables with finite variances but W is a random variable, independent of X_1 and X_2, whose variance does not exist (e.g., a Cauchy random variable). Then neither $\mathrm{Var}(Y_1)$ nor $\mathrm{Var}(Y_2)$ exist, but $\frac{1}{2}\mathrm{Var}(Y_1 - Y_2) = \frac{1}{2}\mathrm{Var}(X_1 - X_2)$, which exists because X_1 and X_2 have finite variances.

2.1.2 Moments of a random vector

The definitions of moments may be extended from one- and two-variable settings to situations with any finite number n of random variables $Y_1, \ldots, Y_n$, i.e., to a random n-vector $\mathbf{y}$. We skip formal definitions of expectation and moments in this setting and instead immediately give extensions of some of the facts pertaining to moments of linear combinations of random variables that we provided earlier for the bivariate case. Let $Y_1, \ldots, Y_n$ be random variables whose variances $\sigma_1^2, \ldots, \sigma_n^2$ exist, whose means are $\mu_1, \ldots, \mu_n$, whose covariances are σ_{ij} $(i \neq j = 1, \ldots, n)$, and whose correlations (assuming the variances are positive) are ρ_{ij} $(i \neq j = 1, \ldots, n)$. Then, the following rules hold for the mean, variance, and covariance of arbitrary linear combinations of the variables:

$$\mathrm{E}\left(\sum_{i=1}^n a_i Y_i\right) = \sum_{i=1}^n a_i \mu_i, \tag{2.15}$$

$$\mathrm{Var}\left(\sum_{i=1}^n a_i Y_i\right) = \sum_{i=1}^n a_i^2 \sigma_i^2 + 2\sum_{i=1}^n \sum_{j=i+1}^n a_i a_j \sigma_{ij}$$

$$= \sum_{i=1}^n a_i^2 \sigma_i^2 + 2\sum_{i=1}^n \sum_{j=i+1}^n a_i a_j \sigma_i \sigma_j \rho_{ij}, \tag{2.16}$$

$$\mathrm{Cov}\left(\sum_{i=1}^n a_i Y_i, \sum_{j=1}^n b_j Y_j\right) = \sum_{i=1}^n \sum_{j=1}^n a_i b_j \sigma_{ij}$$

$$= \sum_{i=1}^n \sum_{j=1}^n a_i b_j \sigma_i \sigma_j \rho_{ij}. \tag{2.17}$$

For economy of expression, it can be helpful to organize the means of the elements of $\mathbf{y}$ in an n-dimensional vector $\boldsymbol{\mu}$, called the **mean vector** of $\mathbf{y}$:

$$\boldsymbol{\mu} = \begin{pmatrix} \mu_1 \\ \mu_2 \\ \vdots \\ \mu_n \end{pmatrix}.$$

Likewise, it is helpful to organize all of their variances and covariances in an $n \times n$ matrix $\boldsymbol{\Sigma}$:

$$\boldsymbol{\Sigma} = \begin{pmatrix} \sigma_1^2 & \sigma_{12} & \cdots & \sigma_{1n} \\ \sigma_{21} & \sigma_2^2 & \cdots & \sigma_{2n} \\ \vdots & \vdots & & \vdots \\ \sigma_{n1} & \sigma_{n2} & \cdots & \sigma_n^2 \end{pmatrix}.$$

Provided that $\sigma_i^2 > 0$ for all i, we can similarly construct an $n \times n$ matrix of correlations $\boldsymbol{\rho}$:

$$\boldsymbol{\rho} = \begin{pmatrix} 1 & \rho_{12} & \cdots & \rho_{1n} \\ \rho_{21} & 1 & \cdots & \rho_{2n} \\ \vdots & \vdots & & \vdots \\ \rho_{n1} & \rho_{n2} & \cdots & 1 \end{pmatrix}.$$

$\boldsymbol{\Sigma}$ is called the **variance-covariance matrix**, or simply the **covariance matrix**, of $\mathbf{y}$, and $\boldsymbol{\rho}$ is called the **correlation matrix** of $\mathbf{y}$. The two matrices are related via the equation

$$\boldsymbol{\rho} = \mathrm{diag}\left(\frac{1}{\sqrt{\sigma_{11}}}, \ldots, \frac{1}{\sqrt{\sigma_{nn}}}\right) \boldsymbol{\Sigma} \, \mathrm{diag}\left(\frac{1}{\sqrt{\sigma_{11}}}, \ldots, \frac{1}{\sqrt{\sigma_{nn}}}\right), \tag{2.18}$$

where for any n-dimensional row vector $\mathbf{c}$, diag$(\mathbf{c})$ is defined here and subsequently as the $n \times n$ matrix with main diagonal elements given by the elements of $\mathbf{c}$ (in the order specified) and off-diagonal elements equal to zero.

Using the matrices just defined, equations (2.15), (2.16), and (2.17) may be re-expressed as follows:

$$E(\mathbf{a}^T\mathbf{y}) = \mathbf{a}^T\boldsymbol{\mu}, \tag{2.19}$$
$$\text{Var}(\mathbf{a}^T\mathbf{y}) = \mathbf{a}^T\boldsymbol{\Sigma}\mathbf{a}, \tag{2.20}$$
$$\text{Cov}(\mathbf{a}^T\mathbf{y}, \mathbf{b}^T\mathbf{y}) = \mathbf{a}^T\boldsymbol{\Sigma}\mathbf{b}. \tag{2.21}$$

(Here and throughout, $\mathbf{a}^T$ represents the transpose of an arbitrary vector $\mathbf{a}$.)

Now, suppose that the means and variances of the random variables $Y_1, \ldots, Y_n$ do not necessarily exist, but their semivariances $\gamma_{ij} \equiv (1/2)\text{Var}(Y_i - Y_j)$ $(i \neq j = 1, \ldots, n)$ do. Then for any real numbers $a_1, \ldots, a_n$ and $b_1, \ldots, b_n$ such that $\sum_{i=1}^n a_i = 0$ and $\sum_{j=1}^n b_j = 0$,

$$\text{Var}\left(\sum_{i=1}^n a_i Y_i\right) = -\sum_{i=1}^n \sum_{j=1}^n a_i a_j \gamma_{ij} = -2\sum_{i=1}^{n-1}\sum_{j=i+1}^n a_i a_j \gamma_{ij}, \tag{2.22}$$

$$\text{Cov}\left(\sum_{i=1}^n a_i Y_i, \sum_{j=1}^n b_j Y_j\right) = -\sum_{i=1}^n \sum_{j=1}^n a_i b_j \gamma_{ij} = -\sum_{i=1}^n \sum_{j \neq i} a_i b_j \gamma_{ij}. \tag{2.23}$$

We emphasize that these last two results do not hold for arbitrary linear combinations of the variables, but only for those linear combinations whose coefficients sum to zero. In fact, the variances and covariances of linear combinations other than those whose coefficients sum to zero need not even exist.

For random variables whose semivariances exist, we may define, analogously to the covariance matrix, the $n \times n$ **semivariance matrix**

$$\boldsymbol{\Gamma} = \begin{pmatrix} 0 & \gamma_{12} & \cdots & \gamma_{1n} \\ \gamma_{21} & 0 & \cdots & \gamma_{2n} \\ \vdots & \vdots & & \vdots \\ \gamma_{n1} & \gamma_{n2} & \cdots & 0 \end{pmatrix}.$$

All of the main diagonal elements of $\boldsymbol{\Gamma}$ are equal to 0 by the extension of (2.10) to n variables. Equations (2.22) and (2.23) may be re-expressed in terms of the semivariance matrix as follows:

$$\text{Var}(\mathbf{a}^T\mathbf{y}) = -\mathbf{a}^T\boldsymbol{\Gamma}\mathbf{a} \text{ for all } \mathbf{a} \text{ such that } \mathbf{a}^T\mathbf{1} = 0, \tag{2.24}$$
$$\text{Cov}(\mathbf{a}^T\mathbf{y}, \mathbf{b}^T\mathbf{y}) = -\mathbf{a}^T\boldsymbol{\Gamma}\mathbf{b} \text{ for all } \mathbf{a} \text{ and } \mathbf{b} \text{ such that } \mathbf{a}^T\mathbf{1} = \mathbf{b}^T\mathbf{1} = 0.$$

(Here and throughout, $\mathbf{1}$ is a vector, of appropriate dimension, whose elements are all ones.)

The matrices of moments defined above satisfy several important properties. First, because $\sigma_{ij} = \sigma_{ji}$, $\rho_{ij} = \rho_{ji}$, and $\gamma_{ij} = \gamma_{ji}$ for all i and j (when they exist), the covariance, correlation, and semivariance matrices are **symmetric**. Second, because a variance is always nonnegative (when it exists), (2.20) implies that

$$\mathbf{a}^T\boldsymbol{\Sigma}\mathbf{a} \geq 0 \quad \text{for all } n\text{-vectors } \mathbf{a}, \tag{2.25}$$

or in words, that a covariance matrix is **nonnegative definite**. By (2.18), a correlation matrix is also nonnegative definite. Third, as a straightforward consequence of (2.24), a semivariance matrix is **conditionally nonpositive definite**, meaning that

$$\mathbf{a}^T\boldsymbol{\Gamma}\mathbf{a} \leq 0 \quad \text{for all } n\text{-vectors } \mathbf{a} \text{ satisfying } \mathbf{a}^T\mathbf{1} = 0. \tag{2.26}$$

For the spatial linear models we consider herein, the covariance and correlation matrices are assumed to be not merely nonnegative definite but **positive definite**, i.e., $\mathbf{a}^T\boldsymbol{\Sigma}\mathbf{a} > 0$ and $\mathbf{a}^T\boldsymbol{\rho}\mathbf{a} > 0$ for all $\mathbf{a} \neq \mathbf{0}$. Symmetric positive definite matrices have several useful properties, some of which are stated in Appendix A and used throughout this book. One of these properties pertains to the **eigenvalues** of the matrix. Eigenvalues of an $n \times n$ matrix $\mathbf{A}$ are defined as those numbers $\lambda_1, \ldots, \lambda_n$ (possibly complex and not necessarily all distinct) for which a vector $\mathbf{x} \neq \mathbf{0}$ exists such that $\mathbf{A}\mathbf{x} = \lambda\mathbf{x}$. A necessary and sufficient condition for an $n \times n$ symmetric matrix to be positive definite is that all of its eigenvalues are positive (Theorem A.1b in Appendix A). This suggests a convenient method to ascertain whether a symmetric matrix with given numerical entries is positive definite: merely compute the eigenvalues of the matrix, using widely available software for this purpose, and then check whether they are all positive. A positive definite matrix is nonsingular (Theorem A.2 in Appendix A), hence invertible, but the converse is false.

The following straightforward extensions of (2.19)–(2.21) will be useful throughout. Let $\mathbf{A}$ and $\mathbf{B}$ be $n \times m$ and $n \times \ell$ matrices of constants. Then

$$
\begin{aligned}
\mathrm{E}(\mathbf{A}^T\mathbf{y}) &= \mathbf{A}^T\boldsymbol{\mu}, & (2.27) \\
\mathrm{Var}(\mathbf{A}^T\mathbf{y}) &= \mathbf{A}^T\boldsymbol{\Sigma}\mathbf{A}, & (2.28) \\
\mathrm{Cov}(\mathbf{A}^T\mathbf{y}, \mathbf{B}^T\mathbf{y}) &= \mathbf{A}^T\boldsymbol{\Sigma}\mathbf{B}.
\end{aligned}
$$

A final set of relevant quantities related to the second moments of a random vector is their **partial variances** and **partial covariances**. To describe these, suppose that $n \geq 3$, and let A and B represent a partition of the set of indices $\{1, 2, \ldots, n\}$ of the elements of $\mathbf{y}$ such that A has at least 2 but not more than $n-1$ elements. Then let $\mathbf{y}_A$ and $\mathbf{y}_B$ denote the subvectors of $\mathbf{y}$ consisting of the elements of A and B, respectively, and denote the covariance matrix of $\begin{pmatrix} \mathbf{y}_A \\ \mathbf{y}_B \end{pmatrix}$ by $\begin{pmatrix} \boldsymbol{\Sigma}_{AA} & \boldsymbol{\Sigma}_{AB} \\ \boldsymbol{\Sigma}_{AB}^T & \boldsymbol{\Sigma}_{BB} \end{pmatrix}$. Furthermore, define the **partial covariance matrix** of $\mathbf{y}_A$, adjusted for (or partialling out) $\mathbf{y}_B$, as $\boldsymbol{\Sigma}_{AA \cdot B} \equiv \boldsymbol{\Sigma}_{AA} - \boldsymbol{\Sigma}_{AB}\boldsymbol{\Sigma}_{BB}^{-1}\boldsymbol{\Sigma}_{AB}^T$. The (i, j)th element of $\boldsymbol{\Sigma}_{AA \cdot B}$, denoted by $\sigma_{ij \cdot B}$, is the **partial covariance** between the ith and jth elements of $\mathbf{y}_A$ (or the **partial variance** of the ith element of $\mathbf{y}$ if $i = j$), adjusted for (or partialling out) the elements of $\mathbf{y}_B$. The **partial correlation** between the ith and jth elements of $\mathbf{y}_A$, adjusted for (or partialling out) the elements of $\mathbf{y}_B$, is given by

$$
\rho_{ij \cdot B} = \frac{\sigma_{ij \cdot B}}{(\sigma_{ii \cdot B}\sigma_{jj \cdot B})^{1/2}}.
$$

When the distribution of $\mathbf{y}$ is multivariate normal, the partial variances and covariances of $\mathbf{y}_A$, adjusted for $\mathbf{y}_B$, are not merely functions of the variances and covariances of the two subvectors, but are *bona fide* moments themselves. Specifically, they are the variances and covariances of the conditional distribution of $\mathbf{y}_A$, given $\mathbf{y}_B$.

As shown later in this chapter, the particular relevance of partial covariances for spatial linear models stems from the following result: the partial covariance (and partial correlation) between the ith and jth elements of $\mathbf{y}$, adjusted for the remaining elements of $\mathbf{y}$, is equal to zero if and only if the (i, j)th element of $\boldsymbol{\Sigma}^{-1}$ is equal to zero. [For a proof of this result, see, for example, pp. 249–250 of Zimmerman & Núñez-Antón (2010)]. Thus, in the case where $\mathbf{y}$ has a multivariate normal distribution, it can be said that the ith and jth elements of $\mathbf{y}$ are conditionally independent, given the remaining elements, if and only if the (i, j)th element of $\boldsymbol{\Sigma}^{-1}$ is equal to zero.

2.2 Covariance structures for geostatistical data

2.2.1 First-order and second-order moments of a geostatistical process

The vectors and matrices of moments described in the previous section are sufficient for models for a random vector, such as the observational vector $\mathbf{y}$ of spatial linear model, but they are insufficient for models of a geostatistical process such as $\{Y(\mathbf{s}) : \mathbf{s} \in \mathcal{A}\}$. For the latter, there is a mean and a variance corresponding to the variable located at each of uncountably many points, and a covariance, correlation, and semivariance corresponding to the variables at each of uncountably many pairs of points (assuming that these moments exist). It is natural, then, to represent the mean, variance, covariance (or correlation), and semivariance of a geostatistical process as functions of those points. In particular, the **mean function** and **covariance function** of $\{Y(\mathbf{s}) : \mathbf{s} \in \mathcal{A}\}$ may be defined as

$$\mu(\mathbf{s}) = \mathrm{E}(Y(\mathbf{s})), \quad \mathbf{s} \in \mathcal{A}$$

and

$$C(\mathbf{s}, \mathbf{s}') = \mathrm{Cov}(Y(\mathbf{s}), Y(\mathbf{s}')), \quad \mathbf{s}, \mathbf{s}' \in \mathcal{A},$$

respectively. The covariance function provides not only the covariance between the random variables at any distinct locations $\mathbf{s}$ and $\mathbf{s}'$, but also the variance at any location $\mathbf{s}$ (by substituting $\mathbf{s}$ for $\mathbf{s}'$ in the function). Along the same lines, the **correlation function** and **semivariogram** are defined as

$$\rho(\mathbf{s}, \mathbf{s}') = \mathrm{Corr}(Y(\mathbf{s}), Y(\mathbf{s}')), \quad \mathbf{s}, \mathbf{s}' \in \mathcal{A}$$

and

$$\gamma(\mathbf{s}, \mathbf{s}') = \frac{1}{2}\mathrm{Var}(Y(\mathbf{s}) - Y(\mathbf{s}')), \quad \mathbf{s}, \mathbf{s}' \in \mathcal{A},$$

respectively. Noting that the correlation function and semivariogram are represented by Greek symbols (ρ and γ), it would seem more consistent to represent the covariance function by another Greek symbol, say, σ. However, the use of C for a spatial covariance function is so widespread that we feel compelled to conform to the established norm.

It follows immediately from the definitions of the covariance function, correlation function, and semivariogram that they are **symmetric**, i.e., they satisfy $f(\mathbf{s}, \mathbf{s}') = f(\mathbf{s}', \mathbf{s})$ for all $\mathbf{s}, \mathbf{s}' \in \mathcal{A}$ (for each of $f = C, \rho, \gamma$). Furthermore, as a consequence of the nonnegative definiteness of the covariance matrix of a random vector noted in Section 2.1.2, the covariance function is **positive definite**, meaning that

$$\sum_{i=1}^{n}\sum_{j=1}^{n} a_i a_j C(\mathbf{s}_i, \mathbf{s}_j) \geq 0 \quad \text{for all } n, \text{ all real numbers } a_1, \ldots, a_n, \text{ and all } \mathbf{s}_1, \ldots, \mathbf{s}_n \in \mathcal{A}.$$

The correlation function likewise is positive definite. And, as a consequence of the conditional nonpositive definiteness of the semivariance matrix of a random vector noted in Section 2.1.2, the semivariogram is **conditionally negative definite**, meaning that

$$\sum_{i=1}^{n}\sum_{j=1}^{n} a_i a_j \gamma(\mathbf{s}_i, \mathbf{s}_j) \leq 0 \text{ for all } n, \text{ all real numbers } a_1, \ldots, a_n \text{ such that } \sum_{i=1}^{n} a_i = 0,$$
$$\text{and all } \mathbf{s}_1, \ldots, \mathbf{s}_n \in \mathcal{A}.$$

The reader will surely have noticed a difference in the terminology used to describe the properties of covariance and semivariance matrices in the previous section (nonnegative definite and conditionally nonpositive definite), and the analogous properties of the covariance function and semivariogram in this section (positive definite and conditionally negative definite). This difference reflects a well established, though regrettable, tradition in the spatial statistics literature, to which we again feel we must adhere. If a covariance function or semivariogram satisfies a strict version of the inequality in (2.25) or (2.26), i.e., if the inequality is strictly positive or negative unless $\mathbf{a} = \mathbf{0}$, then the spatial statistics term used to describe it is *strictly* positive definite or *strictly* conditionally negative definite, respectively.

2.2.2 Spatially structured moments

In the classical statistical sampling framework, the moments (when they exist, which is assumed henceforth) of a random variable (or random vector) can often be estimated effectively by the corresponding moments, or minor modifications thereof, of a sample drawn from the distribution of the random variable (or random vector). For example, if

$$\begin{pmatrix} Y_{11} \\ Y_{21} \end{pmatrix}, \begin{pmatrix} Y_{12} \\ Y_{22} \end{pmatrix}, \ldots, \begin{pmatrix} Y_{1n} \\ Y_{2n} \end{pmatrix}$$

are a sample (of size n) drawn from a bivariate distribution, then one may estimate μ_i by the ith sample mean $(1/n)\sum_{j=1}^{n} Y_{ij}$, σ_i^2 by the ith sample variance $[(1/(n-1))]\sum_{j=1}^{n}(Y_{ij}-\overline{Y}_i)^2$ ($i = 1, 2$), and σ_{12} by the sample covariance $[(1/(n-1))\sum_{j=1}^{n}(Y_{1j}-\overline{Y}_1)(Y_{2j}-\overline{Y}_2)]$. The success of this estimation approach rests on two features of the classical framework: low dimensionality of the random vector relative to the sample size n, and independence across the replicates in the sample. However, as noted previously, in the spatial context there is (usually) just one response vector, whose dimension is equal to the sample size and whose elements are not reasonably regarded as independent. How can we possibly estimate moment functions successfully from such limited data in this context? In general we cannot. However, if enough additional structure is imposed on those functions, so that replication or quasi-replication of their values occurs among the data, then it may become possible to do so.

For spatial data, (quasi) replication may be introduced via **spatially structured** moment functions. For example, a continuous mean function is often assumed, which makes the mean structure spatially smooth. For covariance functions and semivariograms, a variety of spatial structures have been contemplated. One type, called **full symmetry**, specifies (for the case of a covariance function) that

$$C(\mathbf{s}, \mathbf{s}') = C(J(\mathbf{s}, \mathbf{s}')) \quad \text{for all } \mathbf{s}, \mathbf{s}' \in \mathcal{A},$$

where $J(\mathbf{s}, \mathbf{s}')$ swaps any element(s) of $\mathbf{s}$ with the corresponding element(s) of $\mathbf{s}'$. In two dimensions, this condition simplifies to

$$C\left(\begin{pmatrix} s_1 \\ s_2 \end{pmatrix}, \begin{pmatrix} s_1' \\ s_2' \end{pmatrix}\right) = C\left(\begin{pmatrix} s_1 \\ s_2 \end{pmatrix}, \begin{pmatrix} s_1' \\ s_2 \end{pmatrix}\right) \quad \text{for all } \begin{pmatrix} s_1 \\ s_2 \end{pmatrix}, \begin{pmatrix} s_1' \\ s_2' \end{pmatrix} \in \mathcal{A}.$$

For a fully symmetric process, the value of the covariance between responses at any pair of sites with distinct coordinates in each dimension necessarily occurs at one (if $d = 2$) or three (if $d = 3$) additional pairs of locations. Figure 2.1A illustrates a two-dimensional case with six data sites on a rectangular lattice. In general, the number of distinct variances and covariances among six sites could be as large as 21 ($= 6 \times 7/2$), but in this case that number is reduced by three, to 18, because of the full symmetry and the regularity of the grid. Hence, depending on the spatial configuration of data sites, under full symmetry some replication of covariances may occur that may be exploited for estimation purposes. But the extent of replication under full symmetry, even in conjunction with a regular configuration,

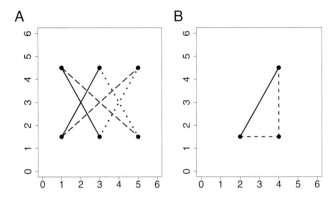

FIGURE 2.1
A) Depiction of full symmetry using six point sites (closed circles) on a rectangular grid in $\mathbb{R}^2$, and line segments connecting six pairs of those sites. Line segments of the same type (solid, dashed, or dotted) connect pairs of sites whose responses have equal covariances under full symmetry; covariances of all other pairs need not be equal. B) Depiction of separability (or additivity) using three sites in $\mathbb{R}^2$, with the covariance between responses at sites $\mathbf{s}$ and $\mathbf{s}'$ (connected by a solid line segment) equal to the product (or sum) of the covariances between responses at two pairs of sites (connected by dashed line segments), one pair aligned horizontally and the other pair aligned vertically.

is very limited. Two related types of spatial structure that confer somewhat more replication on the covariance are **separability** and **additivity**. These specify that

$$C(\mathbf{s}, \mathbf{s}') = \prod_{i=1}^{d} C_i(s_i, s_i')$$

or

$$C(\mathbf{s}, \mathbf{s}') = \sum_{i=1}^{d} C_i(s_i, s_i'),$$

respectively, for the covariance functions $\{C_i(\cdot, \cdot) : i = 1, \ldots, d\}$ of d one-dimensional processes, for all $\mathbf{s} = (s_1, \ldots, s_d)^T$ and $\mathbf{s}' = (s_1', \ldots, s_d')^T \in \mathcal{A}$. The two-dimensional version of these properties is depicted in Figure 2.1B. Separability implies full symmetry, as is easily verified, and so does additivity (Exercise 2.7). But they are stronger; besides the replication noted for full symmetry, additional quasi-replication can occur of a type similar to that present in main effects models for unreplicated two-way experimental designs. For example, for data sites on a 2×3 rectangular lattice, as in Figure 2.1A, the variances and covariances of a separable or additive process at those sites may be represented as functions of at most 9, rather than 21 (in general) or 18 (under full symmetry) quantities. Furthermore, separability in particular confers some computational advantages to estimation that will be described in Chapters 5 and 8. Observe that all three types of structure are defined in reference to a particular Euclidean coordinate system. Also note that full symmetry or additivity of the covariance function implies that the same property holds for the semivariogram, but separability of the covariance function does not imply that the semivariogram may be written as the product of semivariograms of one-dimensional processes.

Less complete multiplicative or additive decompositions of the covariance function are possible. In particular, a covariance function $C(\mathbf{s}, \mathbf{s}')$ in $\mathbb{R}^d$ is said to be **partially separable** if

$$C(\mathbf{s}, \mathbf{s}') = \prod_{i=1}^{k} C_i(\mathbf{s}_i, \mathbf{s}_i') \quad \text{for all } \mathbf{s}, \mathbf{s}' \in \mathcal{A}, \tag{2.29}$$

where $\mathbf{s} = (\mathbf{s}_1^T, \dots, \mathbf{s}_k^T)^T$ is a partitioning of $\mathbf{s}$ into k subvectors of specified dimensions $m_1, \dots, m_k$ which sum to d, $\mathbf{s}' = (\mathbf{s}_1'^T, \dots, \mathbf{s}_k'^T)^T$ is the same partitioning of $\mathbf{s}'$, and $C_i(\mathbf{s}_i, \mathbf{s}_i')$ is a covariance function on $\mathbb{R}^{m_i}$, $i = 1, \dots, k$. Partial additivity of a covariance function is defined similarly, with the product in (2.29) replaced by a sum. Clearly, separability implies partial separability and additivity implies partial additivity (for any partitioning of $\mathbf{s}$), but the converses are not true.

Perhaps the most important, and certainly the most frequently assumed, type of replication-inducing spatial moment structure is **second-order stationarity**, which specifies that the mean is constant and that the covariance function depends on the spatial locations only through the displacement of one location from the other. More formally, a geostatistical process on $\mathcal{A}$ with mean function $\mu(\mathbf{s})$ and covariance function $C(\mathbf{s}, \mathbf{s}')$ is said to be second-order stationary if $\mu(\mathbf{s})$ is identically equal to μ (a common mean) for all $\mathbf{s} \in \mathcal{A}$ and

$$C(\mathbf{s}, \mathbf{s}') = C(\mathbf{s} - \mathbf{s}') \quad \text{for all } \mathbf{s}, \mathbf{s}' \in \mathcal{A}.$$

The covariance function itself in this case is said to be **stationary**. Whereas a covariance function for a d-dimensional process generally is a function of $2d$ arguments (the elements of $\mathbf{s}$ and $\mathbf{s}'$), a stationary covariance function can be written in terms of only d arguments; that is, the function can be represented as $C(\mathbf{h})$ for $\mathbf{h} \in H(\mathcal{A})$, where $\mathbf{h} = (h_i)$, the **lag vector** or simply the **lag**, is a d-dimensional vector of displacements in each of the coordinate directions and $H(\mathcal{A})$ is the set of all lags generated by points in $\mathcal{A}$. It is an abuse of notation, of course, to write $C(\cdot)$ as a function of both $2d$ arguments (in general) and d arguments (in the second-order stationary case), but this is common practice and should not cause any confusion.

Second-order stationarity implies that $C(\mathbf{s}, \mathbf{s}) = C(\mathbf{0})$, i.e., that the variance is constant over the spatial domain $\mathcal{A}$; this property, by itself, is sometimes called **variance stationarity**, and the constant variance is often denoted by σ^2. Second-order stationarity also implies that the correlation function likewise depends on the spatial locations only through their **lag**, and that both the covariance and correlation functions are even functions of the lag [a function $f(\mathbf{x})$ of d variables is said to be **even** if $f(-\mathbf{x}) = f(\mathbf{x})$]; this latter property follows from the symmetry of the more general functions noted in the previous subsection. Second-order stationarity in concert with regular spacing between data locations results in substantial replication of covariances in a dataset. For the six-site case depicted in Figure 2.1A, the number of distinct variances and covariances is at most eight.

When combined with either full symmetry, separability, or additivity, second-order stationarity confers even stronger structure upon the covariance function. For example, a stationary, fully symmetric covariance function on $\mathbb{R}^d$ satisfies

$$C(h_1, \dots, h_j, \dots, h_d) = C(h_1, \dots, -h_j, \dots, h_d) \quad \text{for all } j = 1, \dots, d \text{ and all } \mathbf{h} \in H(\mathcal{A}),$$

which is known as **quadrant symmetry** (Van Marcke, 2010). When $d = 2$, quadrant symmetry reduces to

$$C(h_1, h_2) = C(-h_1, h_2) \quad \text{for all } \mathbf{h} \in H(\mathcal{A}),$$

a property known as **axial symmetry**. Another property of a two-dimensional covariance function, **diagonal symmetry**, is defined as

$$C(h_1, h_2) = C(h_2, h_1) \quad \text{for all } \mathbf{h} \in H(\mathcal{A}),$$

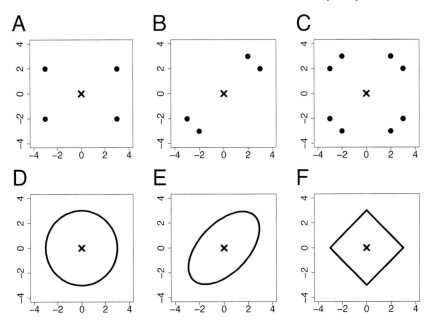

FIGURE 2.2
Loci of equal correlation (solid points or curves) with a variable at the origin ($\times$), for stationary covariance functions with various types of directional symmetry: A) axial symmetry; B) diagonal symmetry; C) complete symmetry; D) isotropy; E) geometric anisotropy; F) L_1-isotropy.

and the combination of axial and diagonal symmetry is called **complete symmetry**. Despite their similar connotations, complete symmetry and full symmetry are different. Stationarity for a two-dimensional covariance function, in tandem with either separability or additivity, implies axial symmetry (Exercise 2.8). The top three panels of Figure 2.2 display loci of equal correlation for such covariance functions, each of which is a set of points at which the corresponding variables have a specified level of correlation with the variable at an arbitrary point.

A stationary covariance function is said to be **compactly supported** if, for each direction defined by a unit vector $\mathbf{h}/\|\mathbf{h}\|$, a real number $r_0(\mathbf{h}/\|\mathbf{h}\|)$ exists such that

$$C(\mathbf{h}) = 0 \quad \text{whenever} \quad \|\mathbf{h}\| \geq r_0(\mathbf{h}/\|\mathbf{h}\|);$$

otherwise it is said to be **globally supported**. In the former case, the quantity $r_0(\mathbf{h}/\|\mathbf{h}\|)$ is called the **range** of the covariance function in direction $\mathbf{h}/\|\mathbf{h}\|$. Like separability, compact support may make estimation more computationally efficient, as will also be described in Chapter 5. If a stationary covariance function is globally supported, but $\lim_{u \to \infty} C(u\mathbf{h}) = 0$, then the value $u(\mathbf{h}/\|\mathbf{h}\|)$ at which $C(u\mathbf{h})/C(\mathbf{0}) < 0.05$ for all $u \geq u(\mathbf{h}/\|\mathbf{h}\|)$ is called the **practical range** or **effective range** (we will use the former term) in direction $\mathbf{h}/\|\mathbf{h}\|$.

A different type of spatial structure involves continuity rather than quasi-replication of the covariance function. This type of structure, known as **mean-square continuity**, is said to hold if the variances exist at all locations and

$$\lim_{\mathbf{s}' \to \mathbf{s}} C(\mathbf{s}, \mathbf{s}') = C(\mathbf{s}, \mathbf{s}) \quad \text{for all } \mathbf{s} \in \mathcal{A}.$$

For a stationary covariance function, this may be rewritten as

$$\lim_{\mathbf{h}\to\mathbf{0}} C(\mathbf{h}) = C(\mathbf{0}) \quad \text{for all } \mathbf{h} \in H(\mathcal{A}).$$

Stationary covariance functions can differ with respect to their continuity and differentiability behavior at the origin, with consequences for the smoothness of process realizations and for inference. We postpone discussions of the consequences to later chapters; for now we merely describe a feature related to the continuity/discontinuity of the covariance function at the origin. If a stationary covariance function $C(\mathbf{h})$ exists and is continuous at $\mathbf{h} = \mathbf{0}$ (implying that $Y(\cdot)$ is mean-square continuous), then it is continuous everywhere, and the covariance function is said to be **nuggetless**. If the covariance function is stationary but discontinuous at the origin, then the difference,

$$C(\mathbf{0}) - \lim_{u\to 0} C(u\mathbf{h}),$$

which is nonnegative, is called the **nugget effect** in direction $\mathbf{h}/\|\mathbf{h}\|$. It is possible for such a covariance function to be discontinuous at additional points (see Chiles & Delfiner (2009), p. 59, for an example), but we do not consider such functions for spatial linear models. In practice, if data suggest that a (possibly direction-dependent) nugget effect seems to exist, it may be attributed to either measurement error or to very small-scale irregularities in the underlying process. As noted by Ripley (1981), the distinction between the two possibilities is as follows: if a measurement was repeated at exactly the same data location, we would obtain the same observation only in the second case, but if we sample close to the original location we would likely obtain a substantially different observation in both cases. The attribution of the nugget effect to measurement error is explored in Exercise 2.12.

A form of spatial structure somewhat weaker than second-order stationarity is **intrinsic stationarity**. A geostatistical process on $\mathcal{A}$ with mean function $\mu(\mathbf{s})$ and semivariogram $\gamma(\mathbf{s}, \mathbf{s}')$ is said to be intrinsically stationary if $\mu(\mathbf{s}) = \mu$ for all $\mathbf{s} \in \mathcal{A}$ and

$$\gamma(\mathbf{s}, \mathbf{s}') = \gamma(\mathbf{s} - \mathbf{s}') \quad \text{for all } \mathbf{s}, \mathbf{s}' \in \mathcal{A}.$$

To see that intrinsic stationarity is weaker than second-order stationarity, consider an arbitrary second-order stationary process with semivariogram $\gamma(\mathbf{s}, \mathbf{s}')$. Because the variances of such a process exist and are all equal to a common value σ^2, we find, using (2.14), that

$$\gamma(\mathbf{s}, \mathbf{s}') = \sigma^2 - C(\mathbf{s} - \mathbf{s}') \quad \text{for all } \mathbf{s}, \mathbf{s}' \in \mathcal{A}, \tag{2.30}$$

which clearly depends on $\mathbf{s}$ and $\mathbf{s}'$ only through $\mathbf{h} = \mathbf{s} - \mathbf{s}'$. Thus, second-order stationarity implies intrinsic stationarity. However, the converse is not true; a counterexample is provided in Exercise 2.13. For future reference, we write (2.30) as

$$\gamma(\mathbf{h}) = \sigma^2 - C(\mathbf{h}) \quad \text{for all } \mathbf{h} \in H(\mathcal{A}). \tag{2.31}$$

By definition $\gamma(\mathbf{0}) = 0$, but the semivariogram need not be continuous at $\mathbf{0}$. The jump in the semivariogram in direction $\mathbf{h}/\|\mathbf{h}\|$ at the origin, i.e., $\lim_{u\to 0} \gamma(u\mathbf{h})$, is the semivariogram's nugget effect in direction $\mathbf{h}/\|\mathbf{h}\|$. Replacing $\mathbf{h}$ in (2.31) with $u\mathbf{h}$ and taking the limit of both sides as $u \to 0$ shows that the semivariogram's nugget effect in a given direction coincides with that of the covariance function in the same direction under second-order stationarity.

Axial, diagonal, and complete symmetry may be defined for intrinsically stationary processes also, by simply extending the same property previously defined in terms of the covariance function to the semivariogram. However, the notion of compact support only holds for intrinsically stationary processes that are also second-order stationary. If the covariance function of such a process is compactly supported, then by (2.31) the corresponding semivariogram satisfies

$$\gamma(\mathbf{h}) = \sigma^2 \quad \text{whenever } \|\mathbf{h}\| \geq r_0(\mathbf{h}/\|\mathbf{h}\|).$$

In this context, the process variance σ^2 is called the **sill**, and $r_0(\mathbf{h}/\|\mathbf{h}\|)$ is called the **range**, of the semivariogram (in direction $\mathbf{h}/\|\mathbf{h}\|$). The range of the semivariogram in any direction, when it exists, coincides with the range of the covariance function in that direction, and the same is true of the practical range. Unlike the range and practical range, however, the sill of a compactly supported intrinsically stationary process cannot be direction-dependent. In fact, if the limit, as distance increases, of the semivariogram of an intrinsically and second-order stationary process varies with direction, then in at least one direction the spatial correlation does not vanish with increasing distance (Zimmerman, 1993); see Exercise 2.16. While the limit of the semivariogram of such a process may not exist, by (2.31) the semivariogram necessarily is bounded between 0 and $2\sigma^2$.

If the semivariogram of an intrinsically stationary process is unbounded, then the process cannot be second-order stationary. However, the growth of the semivariogram of an intrinsically stationary process is subquadratic, meaning that $\gamma(\mathbf{h})/\|\mathbf{h}\|^2 \to 0$ as $\|\mathbf{h}\| \to \infty$ in any direction (Matheron, 1973). If the growth of the semivariogram of a process is not subquadratic, then it cannot be intrinsically stationary, and a more general class of processes and a more general function characterizing spatial dependence are needed. The natural extension of intrinsically stationary processes is the class of **intrinsic random functions of order k** (IRF-k), where k is any nonnegative integer (the case $k = 0$ corresponds to intrinsic stationarity). A geostatistical process $\{Y(\mathbf{s}) : \mathbf{s} \in \mathcal{A}\}$ is an IRF-k if certain linear combinations called **generalized increments of order** k (GI-k's) are second-order stationary. A GI-k is a finite linear combination $\sum_{i=1}^{m} a_i Y(\mathbf{s}_i)$ such that

$$\sum_{i=1}^{m} a_i s_{i1}^{p_1} s_{i2}^{p_2} \cdots s_{id}^{p_d} = 0 \tag{2.32}$$

for all nonnegative integers $p_1, \ldots, p_d$ such that $p_1 + p_2 + \cdots + p_d \leq k$, where $s_{i1}, \ldots, s_{id}$ are the elements (coordinates) of $\mathbf{s}_i$. For example, if $d = 2$ and $k = 0$, then it is necessary that $\sum_{i=1}^{m} a_i = 0$; if $d = 2$ and $k = 1$, then it is also necessary that $\sum_{i=1}^{m} a_i s_{i1} = 0$ and $\sum_{i=1}^{m} a_i s_{i2} = 0$; and if $d = 2$ and $k = 2$, then it is also necessary that $\sum_{i=1}^{m} a_i s_{i1}^2 = 0$, $\sum_{i=1}^{m} a_i s_{i2}^2 = 0$, and $\sum_{i=1}^{m} a_i s_{i1} s_{i2} = 0$. Every continuous d-dimensional kth-order intrinsically stationary process has associated with it a continuous and even function $G_k(\cdot)$ of d variables called a **kth-order generalized covariance function**, which satisfies

$$\text{Cov}\left(\sum_{i=1}^{n} a_i Y(\mathbf{s}_i), \sum_{j=1}^{n} b_j Y(\mathbf{s}_j')\right) = \sum_{i=1}^{n} \sum_{j=1}^{n} a_i b_j G_k(\mathbf{s}_i - \mathbf{s}_j')$$

for all GI-ks $\sum_{i=1}^{n} a_i Y(\mathbf{s}_i)$ and $\sum_{j=1}^{n} b_j Y(\mathbf{s}_j')$. Furthermore, a continuous and even function $G(\cdot)$ of d variables is a generalized covariance function of order k if and only if it is k-**conditionally positive definite**, i.e., if and only if

$$\sum_{i=1}^{n} \sum_{j=1}^{n} a_i a_j G(\mathbf{s}_i - \mathbf{s}_j) \geq 0 \quad \text{for all } n, \text{ all } \mathbf{s}_1, \ldots, \mathbf{s}_n \in \mathcal{A}, \text{ and all } a_1, \ldots, a_n \text{ satisfying (2.32)}.$$

Consideration of the case $k = 0$ reveals that a generalized covariance function of order 0 is the negative of a semivariogram, and vice versa.

Stronger versions of second-order, intrinsic, and kth-order intrinsic stationarity are obtained by requiring that the covariance function, semivariogram, or kth-order generalized covariance function be not merely translation-invariant, but also direction-invariant. This property is called **isotropy** (either second-order, intrinsic, or kth-order intrinsic). Thus, a second-order stationary process and its covariance function $C(\mathbf{h})$ are said to be isotropic if

$$C(\mathbf{h}) = C(\|\mathbf{h}\|) \quad \text{for all } \mathbf{h} \in H(\mathcal{A}),$$

with analogous definitions for a kth-order intrinsically stationary process and its generalized covariance function. Isotropic second-order stationarity implies complete symmetry and implies further that $\rho(\mathbf{h}) = \rho(\|\mathbf{h}\|)$ for all $\mathbf{h} \in H(\mathcal{A})$. Under isotropy, the covariance, semivariogram, generalized covariance, and correlation functions may be written as a function of a single, scalar argument $r = \|\mathbf{s} - \mathbf{s}'\|$, representing the Euclidean distance between two points, and we follow common practice by doing so throughout, despite the abuse of notation. Under isotropic second-order stationarity, the nugget effect and range (or practical range), if they exist, are not direction-dependent, and the loci of equal correlation with a variable at any given point are circles (if $d = 2$, see Figure 2.2D) or spheres (if $d = 3$). Similar statements apply to the loci of equal semivariances and equal generalized covariances in the intrinsically stationary and kth-order intrinsically stationary cases.

If a geostatistical process having one of the stationarity properties we have described is not isotropic, it is said to be **anisotropic**. There are a variety of structured forms of anisotropy, the most useful of which is called **geometric anisotropy**. A second-order stationary process and its covariance function are said to be geometrically anisotropic if a positive definite matrix $\mathbf{B}$ exists such that

$$C(\mathbf{h}) = C((\mathbf{h}^T \mathbf{B} \mathbf{h})^{1/2}) \quad \text{for all } \mathbf{h} \in H(\mathcal{A}).$$

Geometrically anisotropic versions of intrinsically stationary and kth-order intrinsically stationary processes, and the corresponding semivariogram or kth-order generalized covariance function, are defined analogously. Note that isotropy corresponds to the special case $\mathbf{B} = \mathbf{I}$. Loci of equal correlations (or semivariances/generalized covariance functions) of a geometrically anisotropic second-order (or intrinsically/kth-order intrinsically) stationary process are d-dimensional ellipsoids. If $d = 2$, the ellipsoid is an ellipse; the direction of the major axis of this ellipse is called the **anisotropy angle**, and the ratio of the length of the minor axis to that of the major axis is called the **anisotropy ratio**. Figure 2.2E displays a locus of equal correlation for a geometrically anisotropic covariance function.

Another useful structured form of anisotropy is $\mathbf{L}_1$**-isotropy**, for which the covariance function satisfies

$$C(\mathbf{h}) = C\left(\sum_{i=1}^{d} |h_i|\right) \quad \text{for all } \mathbf{h} \in H(\mathcal{A}).$$

L_1-isotropy can be viewed as isotropy with distance defined as "city block distance" (or the L_1 norm, in mathematical jargon) rather than Euclidean distance. Its loci of equal correlation are squares, rotated from alignment with the coordinate axes by 45 degrees. Figure 2.2F shows such a locus.

The types of stationarity introduced to this point impose restrictions only on the first- and second-order moments of a geostatistical process or generalized increments of a geostatistical process. As might be expected, stronger types of stationarity may impose restrictions on higher-order moments of the process. A much stronger form, called **strict stationarity**, requires that the entire joint probability distribution of $\{Y(\mathbf{s}) : \mathbf{s} \in \mathcal{A}\}$, not merely its moments up to some order, depends only on the relative positions of sites. That is, letting

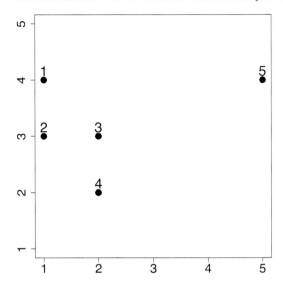

FIGURE 2.3
Spatial locations of sites for three examples of geostatistical covariance functions. Numbers next to sites are the site indices.

$F_{Y(\mathbf{s}_1),\ldots,Y(\mathbf{s}_n)}(x_1,\ldots,x_n)$ denote the cumulative distribution function of the n variables $Y(\mathbf{s}_1),\ldots,Y(\mathbf{s}_n)$, it is required that

$$F_{Y(\mathbf{s}_1+\mathbf{h}),\ldots,Y(\mathbf{s}_n+\mathbf{h})}(x_1,\ldots,x_n) = F_{Y(\mathbf{s}_1),\ldots,Y(\mathbf{s}_n)}(x_1,\ldots,x_n)$$

for all n, all locations and displacements such that $\mathbf{s}_1,\ldots,\mathbf{s}_n \in \mathcal{A}$ and $\mathbf{h} \in H(\mathcal{A})$, and all $x_1,\ldots,x_n$. This implies, for example, that $\Pr[Y(\mathbf{s}+\mathbf{h}) \leq x] = \Pr[Y(\mathbf{s}) \leq x]$ for all $\mathbf{s}$, all $\mathbf{h}$, and all x; that is, that the marginal distributions of all the variables are identical. It also implies that bivariate distributions are identical for any pairs of variables whose spatial locations are displaced by the same vector.

Thus, strict stationarity is stronger than second-order stationarity for a process that has second-order moments. However, in the case of a Gaussian geostatistical process, which is completely characterized by its mean and covariance functions, the two forms of stationarity coincide. In other words, a second-order stationary Gaussian process is strictly stationary. This makes such processes very useful for modeling spatial data. For economy of expression, henceforth we refer to such processes as **stationary Gaussian processes**.

We conclude this section with three numerical examples of covariance matrices for geo-statistical data. For all three examples, the observations are located at the five points depicted in Figure 2.3, but the covariance functions differ, as follows:

- $C_1(\mathbf{s},\mathbf{s}') = \begin{cases} 1 - \frac{3}{8}\|\mathbf{s}-\mathbf{s}'\| + \frac{1}{128}\|\mathbf{s}-\mathbf{s}'\|^3 & \text{if } \|\mathbf{s}-\mathbf{s}'\| \leq 4, \\ 0 & \text{otherwise,} \end{cases}$

- $C_2(\mathbf{s},\mathbf{s}') = \begin{cases} (1 - \frac{1}{5}|s_1 - s_1'|)(1 - \frac{1}{2}|s_2 - s_2'|) & \text{if } |s_1 - s_1'| \leq 5 \text{ and } |s_2 - s_2'| \leq 2, \\ 0 & \text{otherwise,} \end{cases}$

- $C_3(\mathbf{s},\mathbf{s}') = \begin{cases} 2 & \text{if } \mathbf{s} = \mathbf{s}', \\ \exp(-\|\mathbf{s}-\mathbf{s}'\|/2) & \text{otherwise.} \end{cases}$

All of these covariance functions are stationary. The first and third are isotropic also, while the second is anisotropic and separable. The first and second, but not the third, are compactly supported and nuggetless. The third has a nugget effect equal to 1.0. The corresponding covariance matrices, labeled using the same subscripts used for the covariance functions, are as follows:

$$\Sigma_1 = \begin{pmatrix} 1.0 & 0.633 & 0.492 & 0.249 & 0 \\ & 1.0 & 0.633 & 0.492 & 0 \\ & & 1.0 & 0.633 & 0.061 \\ & & & 1.0 & 0.014 \\ & & & & 1.0 \end{pmatrix},$$

$$\Sigma_2 = \begin{pmatrix} 1.0 & 0.500 & 0.400 & 0 & 0.200 \\ & 1.0 & 0.800 & 0.400 & 0.100 \\ & & 1.0 & 0.500 & 0.200 \\ & & & 1.0 & 0 \\ & & & & 1.0 \end{pmatrix},$$

$$\Sigma_3 = \begin{pmatrix} 2.0 & 0.607 & 0.493 & 0.327 & 0.135 \\ & 2.0 & 0.607 & 0.493 & 0.127 \\ & & 2.0 & 0.607 & 0.206 \\ & & & 2.0 & 0.165 \\ & & & & 2.0 \end{pmatrix}.$$

(The elements of these matrices have been rounded off to three decimal places, and all elements below the main diagonal have been suppressed for improved readability without any loss of information, due to the symmetry of a covariance matrix—a practice we will follow henceforth when displaying symmetric matrices.) Within each matrix, note the equality of entries corresponding to sites that are lagged the same distance apart in the same direction or, in the isotropic cases, to sites that are merely lagged the same distance apart. Also note the presence of some zeros within the matrices corresponding to the compactly supported covariance functions.

2.3   Covariance structures for areal data

2.3.1   Spatial weights matrices

Covariance structures for areal data may be constructed in at least three different ways. One is to act as though the data are actually geostatistical by replacing the areal data site A_i with the coordinates of a representative point within A_i, such as its centroid. Then it is supposed that the elements of Σ are given directly by one of the spatially structured covariance functions described in the previous section, taking as its arguments the representative points. A second approach is also based on a geostatistical covariance function, but instead of obtaining covariances by evaluating the function at points representing A_i and A_j, they are obtained by integrating the function over the sites themselves. That is, it is supposed that

$$\text{Cov}(Y(A_i), Y(A_j)) = \int_{A_i} \int_{A_j} C(\mathbf{s}, \mathbf{s}') \, d\mathbf{s} \, d\mathbf{s}'.$$

We call these two approaches the **pseudo-geostatistical** and **aggregate-geostatistical** approaches, respectively, to modeling an areal-data covariance structure. The former is computationally simpler, but the latter is possibly more realistic. The third and most common approach to specifying the covariance structure of areal data is to do so indirectly through a **spatial weights matrix**, denoted generically by $\mathbf{\Omega}$, whose diagonal elements are all zeros and whose (i, j)th element ω_{ij} $(i \neq j)$ is a function of A_i and A_j. We now describe typical forms of the spatial weights matrix.

The fundamental construct for a spatial weights matrix is the **spatial neighborhood**, defined for site i as the subset $\mathcal{N}_i$ of other sites that are in sufficiently close proximity to site i to have an effect on y_i. Perhaps the simplest operationalization of "sufficiently close proximity" is adjacency, so that two sites are said to be neighbors of each other if and only if they share any nonzero-length portion of their boundaries. Alternatively, if some quantitative measure of proximity (e.g., distance between areal centroids) is available, the neighbors of a given site could be defined as those sites that lie within a certain distance of it, or as the closest q sites to it (for fixed $q < n$). Sites are not allowed to be neighbors of themselves, however, which is why the diagonal elements of $\mathbf{\Omega}$ are all zeros. Neighbors are usually symmetric, i.e., if A_i is a neighbor of A_j then A_j is a neighbor of A_i, but this is not always so. A counterexample is easily constructed by taking the sites to be three adjacent squares arranged linearly, left to right, with the rightmost somewhat smaller than the first two, and defining the only neighbor of A_i to be the site with which it shares the greatest proportion of its boundary. For this scenario, the rightmost square's neighbor is the middle square, but the middle square's neighbor is not the rightmost square.

Once a spatial neighborhood structure is defined for the sites, the elements of the spatial weights matrix may be specified in several ways. Perhaps the simplest weighting scheme is a binary one, i.e., equal to one if sites are neighbors and equal to zero otherwise. An advantage of binary weights is the consequent sparsity of $\mathbf{\Omega}$, which can be exploited by computer algorithms specialized to store and manipulate sparse data. Alternatively, if some neighbors are thought to have stronger effects than other neighbors on the response, weights may be defined by a step function corresponding to an ordering of neighbors, in which those sites with the strongest effect are called **first-order neighbors**, those with the next strongest effect are called **second-order neighbors**, etc. For example, spatial weights corresponding to first-order neighbors may be set equal to one, those corresponding to second-order neighbors may be set equal to some number α strictly smaller than one but strictly larger than zero, those corresponding to third-order neighbors may be set equal to some number δ strictly smaller than α but strictly larger than zero, and so on, with weights corresponding to non-neighbors set equal to zero. Weights may even be defined on a continuous scale, for example as the proportion of a site's boundary shared with another site, or as values of some monotone decreasing function of the distance between representative points of the sites. In such non-binary schemes, it is possible for all other sites to be neighbors of a given site.

Some of the same notions regarding spatially structured dependence that arise in conjunction with covariance functions for geostatistical data have analogs for spatial weights matrices. For instance, any distance-based ordering of neighbors would appear to be consistent with Tobler's first law. A definition of ordered neighbors that is consistent over the entire study area is analogous to second-order stationarity of a geostatistical covariance function, and if the definition is also invariant to direction it is analogous to isotropy. Furthermore, a definition in which the only sites that are neighbors of a given site are those for which the measure of their proximity to that site is positive (for examples, sites that share a boundary with the given site) is somewhat analogous to a compactly supported geostatistical covariance function.

To be a bit more specific, suppose that the sites are contiguous, square, and form a square grid, like the squares of a chessboard. For such sites, the **rook's definition** of neighbors specifies that two squares are neighbors if they share an edge, while the **bishop's definition** specifies that sites are neighbors if they share only a corner; the **queen's definition** is the union of the two. Each of these definitions is in some sense a stationary, isotropic, and compact-support definition of neighbor. Anisotropy could be introduced, for example, by assigning more (or less) weight to adjacencies within rows of the chessboard than to adjacencies within columns. As noted previously, the actual weights assigned to neighbors can be binary (1 if a neighbor, 0 otherwise), but need not be; step-functions to account for anisotropies or order of neighbors can be considered, as can continuous monotone decreasing functions of the distance between representative points. Weights may be standardized so that the weights of all neighbors of any given site sum to one, a feature known as **row-standardization**, but this is also not necessary. Sites satisfying the rook's definition could be considered first-order neighbors, with those satisfying the bishop's definition considered second-order neighbors. Alternatively, a site's second-order neighbors could be defined as the first-order neighbors of its first-order neighbors (excluding itself). Neighbors of order greater than two may be defined similarly, e.g., third-order neighbors may be defined as first-order neighbors of second-order neighbors, etc.

For the irregularly spaced and shaped areal sites more common in applied work (such as counties or other administrative units), first-order neighbors are most commonly defined as sites that share a boundary, and higher-order neighbors are defined in terms of the minimum number of boundaries that must be crossed to traverse from one site to another. That is, second-order neighbors are those sites that are not first-order neighbors and for which a connecting path exists that crosses two boundaries, third-order neighbors are those sites that are not first- or second-order neighbors and for which a connecting path exists that crosses three boundaries, and so on. Such a scheme results in symmetric neighbors of each order, and if the neighbor weights are binary, the second- and higher-order neighbor weights matrices may be computed from the first-order neighbor weights matrix rather easily, as we now describe; see Exercise 2.19 for a justification of this procedure. Let $\mathbf{A}_1$ be an $n \times n$ symmetric matrix whose (i,j)th element is equal to one if $i = j$ or if sites A_i and A_j are (first-order) neighbors and is equal to 0 otherwise. The matrix $\mathbf{A}_1$ may not be used in a spatial-weights model directly because no site is allowed to be a neighbor of itself; however, $\mathbf{\Omega} = \mathbf{A}_1 - \mathbf{I}$ may be used instead. Next, define $\mathbf{A}_2 = \mathbb{I}(\mathbf{A}_1\mathbf{A}_1 > 0)$ where $\mathbb{I}(\cdot)$ is the matrix-valued indicator function (the matrix obtained by application of the usual indicator function to each of the elements of a matrix). The (i,j)th element of $\mathbf{A}_2$ is equal to one if $i = j$ or if A_i and A_j are first-order or second-order neighbors. Thus, $\mathbf{A}_2 - \mathbf{I}$ is the second-order neighbor weights matrix, inclusive of first-order neighbors. To obtain a matrix whose nonzero elements correspond to second-order neighbors only, we can form $\mathbf{A}_2 - \mathbf{A}_1$. To obtain the third-order neighbor weights matrix, inclusive of all lower-order neighbors, we may set $\mathbf{A}_3 = \mathbb{I}(\mathbf{A}_1\mathbf{A}_2 > 0)$ or $\mathbf{A}_3 = \mathbb{I}(\mathbf{A}_2\mathbf{A}_1 > 0)$ (the order of multiplication does not matter) and then form $\mathbf{A}_3 - \mathbf{I}$. The fourth-order neighbor weights matrix (inclusive of all lower-order neighbors) can be created by setting $\mathbf{A}_4$ equal to either $\mathbb{I}(\mathbf{A}_1\mathbf{A}_3 > 0)$, $\mathbb{I}(\mathbf{A}_2\mathbf{A}_2 > 0)$, or $\mathbb{I}(\mathbf{A}_3\mathbf{A}_1 > 0)$ and then forming $\mathbf{A}_4 - \mathbf{I}$, etc. The neighbor weights matrix corresponding to a set of neighbors of any specified order(s) can be found by subtraction and addition. For example, the neighbor weights matrix corresponding to only second-order and fourth-order neighbors may be constructed as $(\mathbf{A}_2 - \mathbf{A}_1) + (\mathbf{A}_4 - \mathbf{A}_3)$.

Note that a row-standardized spatial weights matrix for a given set of areal sites may change if that set is embedded within a larger set. As an example, consider a 3×3 subset of sites situated well within the interior of a chessboard, and then embed that set within a larger 5×5 set obtained by including the sites bordering the original 3×3 set. If binary adjacency weights are used, the spatial weights matrix for the nine sites in the 3×3 set is the same regardless of whether the weights are determined in the context of the original 3×3 universe or the 5×5 universe that embeds it. If the binary weights are subsequently row-standardized, however, then some of the weights determined in the first context are different from their counterparts determined in the second context.

2.3.2   Covariance matrices derived from a spatial weights matrix

Once a spatial weights matrix is determined, it may be used to construct the covariance matrix of a spatial linear model in several ways. Here we briefly describe three; much more will be said about these and some related models in Chapter 7.

In a (Gaussian) **simultaneous autoregressive (SAR) model** (Whittle, 1954), the spatial weights matrix $\mathbf{\Omega}$ is denoted conventionally by $\mathbf{B} = (b_{ij})$, where $b_{ii} = 0$ and $b_{ij} = 0$ if $j \notin \mathcal{N}_i$. The covariance matrix of $\mathbf{y}$ is that which results from assuming that

$$y_i = \mathbf{x}_i^T \boldsymbol{\beta} + \sum_{j \in \mathcal{N}_i} b_{ij}(y_j - \mathbf{x}_j^T \boldsymbol{\beta}) + d_i \quad (i = 1, \ldots, n), \tag{2.33}$$

or in vector form,

$$\mathbf{y} = \mathbf{X}\boldsymbol{\beta} + \mathbf{B}(\mathbf{y} - \mathbf{X}\boldsymbol{\beta}) + \mathbf{d},$$

where $\mathbf{d}$ is a multivariate normal random vector with mean vector $\mathbf{0}$ and positive definite diagonal covariance matrix $\mathbf{K}$. In words, equation (2.33) regresses the (mean-corrected) observation at each site on the observations at its (mean-corrected) neighbors, and does so in one model equation for all observations; hence the model's name. The equation may be rearranged to yield

$$(\mathbf{I} - \mathbf{B})(\mathbf{y} - \mathbf{X}\boldsymbol{\beta}) = \mathbf{d};$$

and provided that $\mathbf{I} - \mathbf{B}$ is nonsingular this last equation may be manipulated [by premultiplying both sides by $(\mathbf{I} - \mathbf{B})^{-1}$ and then algebraically transposing $\mathbf{X}\boldsymbol{\beta}$ from one side to the other] to obtain

$$\mathbf{y} = \mathbf{X}\boldsymbol{\beta} + (\mathbf{I} - \mathbf{B})^{-1}\mathbf{d}. \tag{2.34}$$

By comparison with (1.1), it becomes apparent that (2.34) is a spatial linear model with $\mathbf{e} = (\mathbf{I} - \mathbf{B})^{-1}\mathbf{d}$, and using (2.28) we find that the covariance matrix of $\mathbf{e}$ (and $\mathbf{y}$) is

$$\mathbf{\Sigma} = (\mathbf{I} - \mathbf{B})^{-1}\mathbf{K}[(\mathbf{I} - \mathbf{B})^{-1}]^T. \tag{2.35}$$

The SAR model just described is actually just one of several SAR models that have been proposed in the literature. This one is sometimes called the **SAR error model**. Others will be presented in Chapter 7.

In a (Gaussian) **conditional autoregressive (CAR) model** (Besag, 1974), the spatial weights matrix is traditionally denoted by $\mathbf{C} = (c_{ij})$, where $c_{ii} = 0$ for all i and $c_{ij} = 0$ if $j \notin \mathcal{N}_i$. The model for $\mathbf{y}$ results from assuming that for $i = 1, \ldots, n$, the conditional distribution of y_i, given all the other responses y_j $(j \neq i)$, is normal with mean $\mathbf{x}_i^T \boldsymbol{\beta} + \sum_{j \in \mathcal{N}_i} c_{ij}(y_j - \mathbf{x}_j^T \boldsymbol{\beta})$ and positive variance κ_i^2. The model's name derives from the autoregressive form of the conditional mean. A joint distribution of $\mathbf{y}$ that satisfies a given set of conditional specifications (i.e., a given $\mathbf{C}$) does not necessarily exist, and when it does, is not as straightforward to derive as the distribution for the SAR. However, it can be shown (see Exercises 2.17 and 2.18) that if $\mathbf{K}^{-1}(\mathbf{I} - \mathbf{C})$ is symmetric and $\mathbf{I} - \mathbf{C}$ is positive definite, then $\mathbf{y}$ follows a spatial linear model with covariance matrix

$$\mathbf{\Sigma} = (\mathbf{I} - \mathbf{C})^{-1}\mathbf{K}, \tag{2.36}$$

where $\mathbf{K} = \text{diag}(\kappa_1^2, \ldots, \kappa_n^2)$. Thus

$$\mathbf{\Sigma}^{-1} = \mathbf{K}^{-1}(\mathbf{I} - \mathbf{C}). \tag{2.37}$$

By the result noted at the end of Section 2.1.2, this implies that y_i and y_j are conditionally independent, given the remaining responses, if and only if $c_{ij} = 0$. This lends a very useful

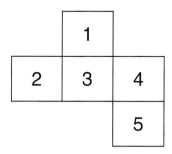

FIGURE 2.4
Layout of sites for the toy example in Section 2.3.2. Numbers within sites are the site indices.

interpretation to the zeros in the inverse of the covariance matrix that is not available for the SAR model.

In a (Gaussian) **spatial moving average (SMA) model**, the spatial weights matrix is denoted by $\mathbf{M} = (m_{ij})$, where $m_{ii} = 0$ for all i and $m_{ij} = 0$ if $j \notin \mathcal{N}_i$, and it is assumed that

$$y_i = \mathbf{x}_i^T \boldsymbol{\beta} + \sum_{j \in \mathcal{N}_i} m_{ij} d_j + d_i \quad (i = 1, \ldots, n)$$

or in vector form,

$$\mathbf{y} = \mathbf{X}\boldsymbol{\beta} + (\mathbf{M} + \mathbf{I})\mathbf{d},$$

where $\mathbf{d}$ is a multivariate normal random vector with mean vector $\mathbf{0}$ and positive definite diagonal covariance matrix $\mathbf{K}$. In this model, observations are related to one another somewhat less directly than in the SAR and CAR models, by sharing some of the same d_{ij}'s rather than by some form of autoregressive equation. Nevertheless, the SMA model is clearly a spatial linear model, and using (2.28) once again, we find that

$$\boldsymbol{\Sigma} = (\mathbf{M} + \mathbf{I})\mathbf{K}(\mathbf{M}^T + \mathbf{I}), \tag{2.38}$$

which is positive definite if and only if $\mathbf{M} + \mathbf{I}$ is nonsingular (Theorem A.3 in Appendix A).

Now we present a toy example that provides an illustration of a spatial weights matrix and the covariance matrices that result from the corresponding SAR, CAR, and SMA models. Consider the layout of square sites depicted in Figure 2.4, where the number inside the square is the site index. Suppose that neighbors are defined as adjacent sites (i.e., the rook's definition), and that weights for horizontal neighbors and vertical neighbors are equal to 1/4 and 1/2, respectively. Then

$$\mathbf{B} = \mathbf{C} = \mathbf{M} = \begin{pmatrix} 0 & 0 & 0.50 & 0 & 0 \\ 0 & 0 & 0.25 & 0 & 0 \\ 0.50 & 0.25 & 0 & 0.25 & 0 \\ 0 & 0 & 0.25 & 0 & 0.50 \\ 0 & 0 & 0 & 0.50 & 0 \end{pmatrix}.$$

It can be verified that the eigenvalues of $\mathbf{I} - \mathbf{B}$ (and thus also the eigenvalues of $\mathbf{I} - \mathbf{C}$) are positive. Thus, $\mathbf{I} - \mathbf{B}$ is nonsingular and $\mathbf{I} - \mathbf{C}$ is symmetric and positive definite, with $b_{ii} = c_{ii} = 0$ for all i. Furthermore, if we set $\mathbf{K} = \mathbf{I}$, then $(\mathbf{I} - \mathbf{C})^{-1}\mathbf{K}$ is symmetric. It can also be verified that $m_{ii} = 0$ for all i, and that $\mathbf{M} + \mathbf{I}$ is nonsingular. Then, the covariance matrices of the responses under the SAR, CAR, and SMA models are, using (2.35), (2.36), and (2.38) respectively,

$$
\boldsymbol{\Sigma}_{\mathrm{SAR}} = \begin{pmatrix} 2.82 & 0.91 & 2.82 & 1.40 & 0.84 \\ & 1.46 & 1.41 & 0.70 & 0.42 \\ & & 3.98 & 2.24 & 1.40 \\ & & & 3.28 & 2.40 \\ & & & & 2.58 \end{pmatrix}, \quad \boldsymbol{\Sigma}_{\mathrm{CAR}} = \begin{pmatrix} 1.41 & 0.21 & 0.83 & 0.28 & 0.14 \\ & 1.10 & 0.41 & 0.14 & 0.07 \\ & & 1.66 & 0.55 & 0.28 \\ & & & 1.52 & 0.76 \\ & & & & 1.38 \end{pmatrix},
$$

and

$$
\boldsymbol{\Sigma}_{\mathrm{SMA}} = \begin{pmatrix} 1.25 & 0.13 & 1.00 & 0.13 & 0.00 \\ & 1.06 & 0.50 & 0.06 & 0.00 \\ & & 1.38 & 0.50 & 0.13 \\ & & & 1.31 & 1.00 \\ & & & & 1.25 \end{pmatrix}.
$$

(Here we rounded off entries to two decimal places.) Observe that, despite the equal diagonal elements of $\mathbf{K}$, equal horizontal neighbor weights, and equal vertical neighbor weights, the variances of responses are heteroscedastic under all three models, and two responses may be correlated even when their corresponding sites are not neighbors. Furthermore, the correlations, while uniformly positive, are generally not identical for neighbors lagged the same distance apart in the same direction. For example, under the SAR model, the correlation between responses at sites 2 and 3 is equal to $1.41/\sqrt{(1.46)(3.98)} \doteq 0.58$, while that between responses at sites 3 and 4 is $2.24/\sqrt{(3.98)(3.28)} \doteq 0.62$.

We may also observe that, under the SMA model, the fifth observation is uncorrelated with the first and second observations, hence it is independent of them under normality. In contrast, no pair of observations are (marginally) independent under the SAR and CAR models. However, from the entries in $\mathbf{C}$ we see that the first and second observations are conditionally independent given the rest, and additional pairs of observations that are conditionally independent given the rest are the first and fourth, first and fifth, second and fourth, second and fifth, and third and fifth.

In the previous subsection, it was noted that a spatial weights matrix may be context-dependent in the sense that the weights matrix corresponding to a given set of sites may depend on whether that set is the entire universe of sites under consideration or is embedded within a larger universe of sites. This has consequences for the SAR/CAR/SMA covariance matrix of the variables corresponding to the embedded subset. That is, the covariance matrix of the variables corresponding to the embedded subset will generally be different if the SAR/CAR/SMA model is formulated for variables corresponding to the larger universe than if it is formulated for only those variables corresponding to the embedded subset. This is so, in fact, regardless of whether the weights themselves are context-dependent. For example, consider embedding the toy example of five sites displayed in Figure 2.4 within a larger universe of six sites obtained by adding a single site in the third row and second column. Retaining the rook's definition of neighbor and the anisotropy in the horizontal and vertical directions from the original example, the weights matrix in the context of the larger universe is

$$
\mathbf{B}^* = \mathbf{C}^* = \mathbf{M}^* = \begin{pmatrix} 0 & 0 & 0.50 & 0 & 0 & 0 \\ 0 & 0 & 0.25 & 0 & 0 & 0 \\ 0.50 & 0.25 & 0 & 0.25 & 0 & 0.5 \\ 0 & 0 & 0.25 & 0 & 0.50 & 0 \\ 0 & 0 & 0 & 0.50 & 0 & 0.25 \\ 0 & 0 & 0.5 & 0 & 0.25 & 0 \end{pmatrix}.
$$

Observe that the weights for the original five sites are the same, regardless of whether they are determined in the context of the five-site or six-site universe. However, the marginal covariance matrices for responses at the five sites within the context of the six-site universe, assuming that $\mathbf{K} = \mathbf{I}_6$, are respectively

$$\boldsymbol{\Sigma}^*_{\text{SAR}} = \begin{pmatrix} 8.89 & 3.94 & 14.03 & 7.75 & 6.96 \\ & 2.97 & 7.01 & 3.87 & 3.48 \\ & & 24.54 & 13.98 & 12.65 \\ & & & 9.92 & 8.81 \\ & & & & 8.73 \end{pmatrix}, \quad \boldsymbol{\Sigma}^*_{\text{CAR}} = \begin{pmatrix} 1.88 & 0.44 & 1.76 & 0.76 & 0.64 \\ & 1.22 & 0.88 & 0.38 & 0.32 \\ & & 3.52 & 1.52 & 1.28 \\ & & & 2.02 & 1.28 \\ & & & & 1.92 \end{pmatrix},$$

and

$$\boldsymbol{\Sigma}^*_{\text{SMA}} = \begin{pmatrix} 1.25 & 0.13 & 1.00 & 0.13 & 0.00 \\ & 1.06 & 0.50 & 0.06 & 0.00 \\ & & 1.63 & 0.50 & 0.25 \\ & & & 1.31 & 1.00 \\ & & & & 1.31 \end{pmatrix}.$$

Thus, the marginal distribution of the observations corresponding to the original five-site configuration depends on whether the model is formulated only for that original configuration or for the expanded six-site configuration. This feature is not shared by (Gaussian) geostatistical models, for which the marginal distribution of any set of variables is invariant to whatever larger set in which it is embedded. The lack of such invariance for the SAR, CAR, and SMA models has led some statisticians to recommend restricting the kinds of inference that should be performed for such models, or even to discourage their use altogether; see, for example, McCullagh (2002).

2.4   Spatial linear models, revisited

Having developed the notions of a spatially structured covariance function (or semivariogram or generalized covariance function) and a spatial weights matrix in this chapter, we may now supplement the spatial linear model, as it was originally introduced in Section 1.3, with a little more detail. Recall that a spatial linear model may be written as

$$\mathbf{y} = \mathbf{X}\boldsymbol{\beta} + \mathbf{e},$$

where $\mathbf{y}$ is the response vector, $\mathbf{X}$ is the model matrix whose columns correspond to the explanatory variables, $\boldsymbol{\beta}$ is a vector of unknown parameters, and $\mathbf{e}$ is the error vector. The error vector is assumed to have mean vector $\mathbf{0}$ and a spatially structured, positive definite covariance matrix $\boldsymbol{\Sigma}$. The spatial structure might include such features as stationarity, separability, monotonic correlation decay with distance, compact support, or spatial weights that depend on site adjacency, to name a few. Usually, the error vector is further assumed to have a multivariate normal distribution, although this is not necessary for some purposes.

In the spatial general linear model, the spatial structure in $\boldsymbol{\Sigma}$ is modeled parametrically, as $\boldsymbol{\Sigma}(\boldsymbol{\theta})$, where $\boldsymbol{\theta}$ is required to belong to a parameter space within which $\boldsymbol{\Sigma}(\boldsymbol{\theta})$ is positive definite. Although this model is flexible enough to encompass the covariance structures typically used for both geostatistical and areal data, there is a noteworthy difference in the two cases. For geostatistical data, the matrix-valued function $\boldsymbol{\Sigma}(\boldsymbol{\theta})$ has the relatively simple form of a matrix of values of a single real-valued, parametric function of pairs of data sites (i.e., the spatial covariance function). That is,

$$\boldsymbol{\Sigma}(\boldsymbol{\theta}) = (C(\mathbf{s}_i, \mathbf{s}_j; \boldsymbol{\theta})). \tag{2.39}$$

The covariance matrix in the example of a spatial general linear model given in Section 1.3 was of this type, with $\boldsymbol{\theta} = (\sigma^2, \alpha)^T$ and $C(\mathbf{s}_i, \mathbf{s}_j; \boldsymbol{\theta}) = \sigma^2 \exp(-\|\mathbf{s}_i - \mathbf{s}_j\|/\alpha)$. We call a spatial linear model with $\boldsymbol{\Sigma}$ modeled in this fashion a **geostatistical linear model**.

For the pseudo-geostatistical and aggregate-geostatistical approaches to modeling an areal-data covariance structure described in the previous section, the elements of $\boldsymbol{\Sigma}(\boldsymbol{\theta})$ are either also specified by (2.39) at representative points or given by integrals of the covariance function in (2.39) over data sites. For the SAR/CAR/SMA and related models for areal data, however, the form of $\boldsymbol{\Sigma}(\boldsymbol{\theta})$ is not quite so simple. For those models,

$$\boldsymbol{\Sigma}(\boldsymbol{\theta}) = f(\boldsymbol{\Omega}(\boldsymbol{\theta}), \mathbf{K}(\boldsymbol{\theta})) = f((\omega^*(A_i, A_j; \boldsymbol{\theta}))), \tag{2.40}$$

where f is a matrix-valued function of the spatial weights matrix $\boldsymbol{\Omega} = (\omega_{ij})$ and the error covariance matrix $\mathbf{K} = \mathrm{diag}(\kappa_1^2, \ldots, \kappa_n^2)$, and

$$\omega^*(A_i, A_j; \boldsymbol{\theta}) = \begin{cases} \kappa_i^2(\boldsymbol{\theta}) & \text{if } i = j, \\ \omega_{ij}(\boldsymbol{\theta}) & \text{if } i \neq j. \end{cases}$$

As an example, consider once again the spatial configuration of sites in Figure 2.4. Suppose that the observations at those sites follow a SAR model for which

$$\mathbf{B} = \mathbf{B}(\rho) = \rho \begin{pmatrix} 0 & 0 & 0.50 & 0 & 0 \\ 0 & 0 & 0.25 & 0 & 0 \\ 0.50 & 0.25 & 0 & 0.25 & 0 \\ 0 & 0 & 0.25 & 0 & 0.50 \\ 0 & 0 & 0 & 0.50 & 0 \end{pmatrix}$$

and $\mathbf{K} = \sigma^2 \mathbf{I}$ where ρ and σ^2 are unknown parameters confined to the parameter space for which $\mathbf{I} - \mathbf{B}$ is nonsingular. Then, the covariance matrix of the data is of the form specified by (2.40) with $\boldsymbol{\theta} = (\sigma^2, \rho)^T$, $\boldsymbol{\Omega} = \mathbf{B}$, and $f(\mathbf{B}, \mathbf{K}) = (\mathbf{I} - \mathbf{B})^{-1} \mathbf{K} [(\mathbf{I} - \mathbf{B})^{-1}]^T$ according to (2.35). Thus, it is possible to write the elements of $\boldsymbol{\Sigma}$ as a function (f) of two matrices whose (i, j)th elements are functions of sites i and j only, but it is not possible to obtain the (i, j)th element of $\boldsymbol{\Sigma}$ by evaluating a single common function of sites i and j. We call a spatial linear model of this type a **spatial-weights linear model**.

This is also an appropriate place to note that in some spatial linear models, the error vector $\mathbf{e}$ may be decomposed into two or more additive components, each with zero mean vector and finite covariance matrix. In particular, we may suppose that

$$e(S_i) = Z(S_i) + \delta(S_i),$$

where $Z(S_1), \ldots, Z(S_n)$ are zero-mean, finite-variance, and possibly spatially correlated random variables representing "noiseless" versions of $y_1 - \mathbf{x}_1^T \boldsymbol{\beta}, \ldots, y_n - \mathbf{x}_n^T \boldsymbol{\beta}$ (i.e., versions that are not contaminated with measurement error) and $\delta(S_1), \ldots, \delta(S_n)$ are zero-mean, finite-variance measurement errors that are independent of the $Z(S_i)$'s. Ordinarily, the measurement errors are assumed to be independent and identically distributed but this is not strictly necessary. Either $Z(S_i)$ or $\delta(S_i)$ or both may be further decomposed into sums of other zero-mean, finite-variance random variables. For example, $Z(S_i)$ may be envisioned as the sum of a zero-mean mean-square continuous process observed at S_i, say $W(S_i)$, whose range of correlation exists and exceeds the minimum distance between data sites, and an independent zero-mean process observed at S_i, say $\eta(S_i)$, whose range of correlation also exists but is smaller than the minimum distance between data sites. These two component processes are often called the **smooth small-scale variation** and the **microscale variation**, respectively (Cressie, 1988). As a second example, consider the dataset of heavy metal concentrations of mosses introduced in Example 1.4. It will be seen later that measurement errors for these data can be ascribed to two distinct sources, namely replicate-within-site sampling error and laboratory error; hence, it may make sense (and improve understanding) in this case to regard $\delta(S_i)$ as the sum of two independent measurement errors, one corresponding to each source. In any such example, the parameter vector $\boldsymbol{\theta}$ comprises the variance and spatial dependence parameters of $W(\cdot)$, the variance of the microscale variation, and the variances of measurement errors from all sources.

The spatial general linear model, as described so far, can serve as the basis for making inferences about spatial trend and dependence as embodied by $\boldsymbol{\beta}$ and $\boldsymbol{\theta}$, respectively. In subsequent chapters, however, we will generalize or expand upon this model as needed. For instance, to deal with certain types of nonstationarity, the model will be generalized in Chapter 6 to allow for the modeling of spatial dependence via the semivariogram or a generalized covariance function. Furthermore, to accommodate the spatial prediction objective, the model will be augmented in Chapter 9 to include spatial predictands.

2.5 Exercises

1. Verify (2.7)-(2.9).

2. Verify (2.13) and (2.14).

3. Consider 16 observations taken at points on a 4×4 square grid, indexed by rows and columns as $Y(i, j)$. Suppose that the observations are jointly normally distributed with common mean μ and common variance 1, and

$$\text{Corr}[Y(i, j), Y(k, l)] = 0.5^{|i-k|+|j-l|}.$$

   Show that $\text{Var}(\overline{Y}) \doteq 0.266$. If a standard confidence interval for the mean (i.e., an interval based on the false assumption that the observations are uncorrelated) with nominal coverage probability 95% is constructed, what is its true coverage probability?

4. Let $\{Y(\mathbf{s}) : \mathbf{s} \in \mathbb{R}^d\}$ be an intrinsically stationary process. Show that $\overline{Y} - Y(\mathbf{s}_0)$ (where $\mathbf{s}_0$ is an arbitrary point in $\mathbb{R}^d$) is a linear combination whose coefficients sum to zero, and give a simplified expression for its variance in terms of the semivariogram.

5. Verify (2.23). [Hint: Consider $\frac{1}{2}\sum_{i=1}^{n}\sum_{j=1}^{n} a_i b_j \text{Var}(Y_i - Y_j)$ and rewrite $Y_i - Y_j$ as $(Y_i - Y_k) + (Y_k - Y_j)$ for fixed $k \in \{1, \ldots, n\}$.]

6. Let $Y(\mathbf{s}) = \prod_{i=1}^{d} Y_i(s_i)$ where the $\{Y_i(s_i)\}$'s are independent random processes with zero means and covariance functions $C_i(s_i, s_i')$ on $\mathbb{R}$ ($i = 1, \ldots, d$). Show that the covariance function of $\{Y(\mathbf{s}) : \mathbf{s} \in \mathbb{R}^d\}$ is separable.

7. Show that separability implies full symmetry, and that additivity likewise implies full symmetry.

8. Show that if a two-dimensional stationary covariance function is either separable or additive, then it is also axially symmetric.

9. Show that a two-dimensional, stationary, separable covariance function $C(h_1, h_2) = C_1(h_1)C_2(h_2)$ is completely symmetric if $C_1(h) = C_2(h)$ for all h.

10. Let $C(\mathbf{h})$ be a stationary covariance function in $\mathbb{R}^2$.

    (a) If $C(\mathbf{h})$ is diagonally symmetric, is it necessarily geometrically anisotropic? Explain why or why not.

    (b) If $C(\mathbf{h})$ is axially symmetric and geometrically anisotropic, and its value at $\mathbf{h} = (2, 3)^T$ is smaller than its value at $\mathbf{h} = (3, 2)^T$, what is its anisotropy angle (i.e., the direction in which the spatial correlation decays most slowly)?

    (c) Show that if $C(\mathbf{h})$ is additive, then so is the semivariogram.

11. Consider data sites in $\mathbb{R}^2$ that lie on the nodes of an $r \times c$ rectangular (not necessarily square) lattice. Determine the maximum number of distinct quantities upon which the variances and covariances depend when the covariance function is:

    (a) fully symmetric;

    (b) additive or separable;

    (c) stationary; or

    (d) isotropic.

12. Suppose that the residual spatial variation of a geostatistical process can be decomposed as follows:
$$e(\mathbf{s}) = Z(\mathbf{s}) + \delta(\mathbf{s})$$
where $\delta(\mathbf{s})$ is the measurement error at $\mathbf{s}$ and $Z(\mathbf{s})$ accounts for all other sources of variation. Assume that the measurement errors are independent and identically distributed random variables, with variance σ_0^2 say, and that they are independent of $\{Z(\mathbf{s}): \mathbf{s} \in \mathcal{A}\}$. Finally, suppose that $\{Z(\mathbf{s}) : \mathbf{s} \in \mathcal{A}\}$ is mean-square continuous and second-order stationary with covariance function $C_Z(\mathbf{h})$.

    (a) Show that the covariance function of $\{e(\mathbf{s}) : \mathbf{s} \in \mathcal{A}\}$ is
$$C(\mathbf{h}) = \begin{cases} C_Z(\mathbf{0}) + \sigma_0^2 & \text{for } \mathbf{h} = \mathbf{0}, \\ C_Z(\mathbf{h}) & \text{for } \|\mathbf{h}\| > 0. \end{cases}$$

    (b) Show that σ_0^2 satisfies the definition of the nugget effect given in Section 2.2.2.

13. Show that intrinsic stationarity does not imply second-order stationarity by considering a one-dimensional geostatistical process $\{Y(s): s \geq 0\}$ that satisfies $\mathrm{E}(Y(s)) \equiv 0$ and $\mathrm{Cov}(Y(s), Y(s')) = \sigma^2 \min(s, s')$. What is the semivariogram of the process?

14. For a second-order stationary geostatistical process, show that positive definiteness of the covariance function implies conditional negative definiteness of the semivariogram.

15. For a two-dimensional, second-order stationary separable process for which both one-dimensional processes have unit variances, express the two-dimensional semivariogram in terms of its one-dimensional semivariograms.

16. Consider a second-order stationary process with semivariogram $\gamma(\mathbf{h})$, and suppose that $\lim_{u \to \infty} \gamma(u\mathbf{h})$ exists in each direction $\mathbf{h}/\|\mathbf{h}\|$. Show that if these limits are different in at least two directions, then in at least one direction the spatial correlation does not vanish with increasing distance.

17. The following result is known as **Brook's lemma** (Brook, 1964). Let $\mathbf{y}$ be an n-dimensional random vector with joint probability density (or mass) function $f_{1:n}(\mathbf{y})$ on support $\mathcal{Y}$, and let $f_{i,-i}(y_i|\mathbf{y}_{-i})$ represent the conditional pdf (or pmf) of y_i, given the remaining variables. (These conditional pdfs are known as the **full conditional distributions**.) For $\mathbf{y}, \mathbf{y}' \in \mathcal{Y}$,

$$\frac{f_{1:n}(\mathbf{y})}{f_{1:n}(\mathbf{y}')} = \prod_{i=1}^{n} \frac{f_{i,-i}(y_i|y_1,\ldots,y_{i-1},y'_{i+1},\ldots,y'_n)}{f_{i,-i}(y'_i|y_1,\ldots,y_{i-1},y'_{i+1},\ldots,y'_n)} \qquad (2.41)$$

$$= \prod_{i=1}^{n} \frac{f_{i,-i}(y_i|y'_1,\ldots,y'_{i-1},y_{i+1},\ldots,y_n)}{f_{i,-i}(y'_i|y'_1,\ldots,y'_{i-1},y_{i+1},\ldots,y_n)}. \qquad (2.42)$$

Prove Brook's lemma by proceeding through the following steps:

(a) Using standard rules relating a conditional pdf to the corresponding joint and marginal pdfs, show that

$$f_{1:n}(y_1,\ldots,y_n) = \frac{f_{n,-n}(y_n|y_1,\ldots,y_{n-1})}{f_{n,-n}(y'_n|y_1,\ldots,y_{n-1})} f_{1:n}(y_1,\ldots,y_{n-1},y'_n)$$
$$\text{for } \mathbf{y}, (y_1,\ldots,y_{n-1},y'_n) \in \mathcal{Y}.$$

(b) Express the last term on the right-hand side of the equality in part (a) similarly to obtain

$$f_{1:n}(y_1,\ldots,y_n) = \frac{f_{n,-n}(y_n|y_1,\ldots,y_{n-1})}{f_{n,-n}(y'_n|y_1,\ldots,y_{n-1})} \frac{f_{(n-1),-(n-1)}(y_{n-1}|y_1,\ldots,y_{n-2},y'_n)}{f_{(n-1),-(n-1)}(y'_{n-1}|y_1,\ldots,y_{n-2},y'_n)}$$
$$\times f_{1:n}(y_1,\ldots,y_{n-2},y'_{n-1},y'_n)$$
$$\text{for } \mathbf{y}, (y_1,\ldots,y_{n-1},y'_n), (y_1,\ldots,y_{n-2},y'_{n-1},y'_n) \in \mathcal{Y},$$

and then note how repeating the same process $n-2$ more times yields (2.41).

(c) Show that (2.42) holds by an argument similar to that used to obtain (2.41), but starting with

$$f_{1:n}(y_1,\ldots,y_n) = \frac{f_{1,-1}(y_1|y_2,\ldots,y_n)}{f_{1,-1}(y'_1|y_2,\ldots,y_n)} f_{1:n}(y'_1,y_2,\ldots,y_n)$$
$$\text{for all } \mathbf{y}, (y'_1,y_2,\ldots,y_n) \in \mathcal{Y}.$$

18. Consider the Gaussian CAR model

$$y_i|\mathbf{y}_{-i} \sim \mathrm{N}\left(\mathbf{x}_i^T\boldsymbol{\beta} + \sum_{j=1}^{n} c_{ij}(y_j - \mathbf{x}_j^T\boldsymbol{\beta}), \kappa_i^2\right) \qquad (i=1,\ldots,n),$$

where $c_{ii} = 0$ and $\kappa_i^2 > 0$ for all i. By carrying out the following steps, show that the joint distribution of $\mathbf{y}$ under this model is multivariate normal, with mean vector $\mathbf{X}\boldsymbol{\beta}$ and covariance matrix $(\mathbf{I} - \mathbf{C})^{-1}\mathbf{K}$, where

$$\mathbf{X} = \begin{pmatrix} \mathbf{x}_1^T \\ \vdots \\ \mathbf{x}_n^T \end{pmatrix},$$

$\mathbf{C} = (c_{ij})$ and $\mathbf{K} = \mathrm{diag}(\kappa_1^2,\ldots,\kappa_n^2)$, provided that $\mathbf{I} - \mathbf{C}$ is positive definite and $\mathbf{K}^{-1}(\mathbf{I} - \mathbf{C})$ is symmetric.

(a) Define $\mathbf{z} = \mathbf{y} - \mathbf{X}\boldsymbol{\beta}$ and let $f_{1:n}(\cdot)$ represent the pdf of $\mathbf{z}$, with support $\mathcal{Z}$. Use (2.41) in Brook's lemma to show that

$$\log \frac{f_{1:n}(\mathbf{z})}{f_{1:n}(\mathbf{0})} = -\frac{1}{2} \sum_{i=1}^{n} \frac{z_i^2}{\kappa_i^2} + \sum_{i=2}^{n} \sum_{j=1}^{i-1} \frac{c_{ij} z_i z_j}{\kappa_i^2} \quad \text{for } \mathbf{z} \in \mathcal{Z}.$$

(b) Using part (a), show that

$$f_{1:n}(\mathbf{z}) = f_{1:n}(\mathbf{0}) \exp\left(-\frac{1}{2} \sum_{i=1}^{n} \frac{z_i^2}{\kappa_i^2} + \frac{1}{2} \sum_{i=2}^{n} \sum_{j=1}^{i-1} \frac{c_{ij} z_i z_j}{\kappa_i^2} \right) \quad \text{for } \mathbf{z} \in \mathcal{Z}.$$

(c) Similarly, use (2.42) in Brook's lemma to show that

$$\log \frac{f_{1:n}(\mathbf{z})}{f_{1:n}(\mathbf{0})} = -\frac{1}{2} \sum_{i=1}^{n} \frac{z_i^2}{\kappa_i^2} + \sum_{i=1}^{n-1} \sum_{j=i+1}^{n} \frac{c_{ij} z_i z_j}{\kappa_i^2}$$

and hence that

$$f_{1:n}(\mathbf{z}) = f_{1:n}(\mathbf{0}) \exp\left(-\frac{1}{2} \sum_{i=1}^{n} \frac{z_i^2}{\kappa_i^2} + \frac{1}{2} \sum_{i=1}^{n-1} \sum_{j=i+1}^{n} \frac{c_{ij} z_i z_j}{\kappa_i^2} \right) \quad \text{for } \mathbf{z} \in \mathcal{Z}.$$

(d) Show that in order to reconcile the two expressions for $f_{1:n}(\mathbf{z})$ derived in parts (b) and (c), necessarily

$$c_{ij}/\kappa_i^2 = c_{ji}/\kappa_j^2 \quad \text{for all } i, j = 1, \ldots, n,$$

and that in that case

$$f_{1:n}(\mathbf{z}) = f_{1:n}(\mathbf{0}) \exp\left(-\frac{1}{2} \mathbf{z}^T \mathbf{K}^{-1} (\mathbf{I} - \mathbf{C}) \mathbf{z} \right) \quad \text{for } \mathbf{z} \in \mathcal{Z}$$

where $\mathbf{C} = (c_{ij})$ is such that $\mathbf{K}^{-1}(\mathbf{I} - \mathbf{C})$ is symmetric.

(e) Observe that the function on the right-hand side in part (d) has the form of the pdf of a multivariate normal vector with mean $\mathbf{0}$, provided that $\mathbf{I} - \mathbf{C}$ is positive definite, and then complete the proof using a well-known property of linear transformations of a multivariate normal vector.

19. Consider an areal data setting with data sites $A_1, \ldots, A_n$, in which first-order neighbors are defined as sites sharing a boundary and higher-order neighbors are defined in terms of the minimum number of boundaries that must be crossed to traverse from one site into another, and the neighbor weights are binary. For each positive integer $q < n$, define $\mathbf{A}_q$ as the $n \times n$ symmetric matrix whose (i, j)th element is equal to 1 if $i = j$ or if sites A_i and A_j are neighbors of any order less than or equal to q, and is equal to 0 otherwise. Let p be a positive integer greater than 1. Show that if A_i and A_j are pth-order neighbors, then $\mathbb{I}(\mathbf{A}_1 \mathbf{A}_{p-1} > 0) = \mathbb{I}(\mathbf{A}_2 \mathbf{A}_{p-2} > 0) = \cdots = \mathbb{I}(\mathbf{A}_{p-1} \mathbf{A}_1 > 0) = \mathbf{A}_p$ where $\mathbb{I}(\cdot)$ is the matrix-valued indicator function defined in Section 2.3.1. (Hint: consider the elements of these matrices for neighbors of order less than p, neighbors of order p, and neighbors of order greater than p.)

20. Obtain the correlation matrices corresponding to the five-site SAR, CAR, and SMA covariance matrices that are displayed in Section 2.3.2. Is the correlation discernibly stronger in the vertical direction than in the horizontal direction, as is true of the spatial weights?

21. Consider the five-site spatial layout with a rook's definition of neighbors as described in Section 2.3.2, but modify the weights so that they are equal to $\frac{1}{2}$ in both the vertical and horizontal directions.

    (a) Obtain the covariance matrix of $\mathbf{y}$ for each of the SAR, CAR, and SMA models using the modified weights.

    (b) Obtain the correlation matrices corresponding to the three covariance matrices obtained in part (a). Are the correlations in the vertical and horizontal directions of roughly equal strength?

22. Consider the model

$$y_i | \mathbf{y}_{-i} \sim \mathrm{N}\left(\sum_{j \in \mathcal{N}_i} y_j / m_i,\ \kappa^2 / m_i \right) \quad \text{for}\ i = 1, \ldots, n,$$

which is known as the **intrinsic conditional autoregressive (ICAR) model**. This model specifies (among other things) that the conditional mean of the ith response is the average of its neighbors.

    (a) Use the distributional result stated in Exercise 2.18d to show that the joint density function of $\mathbf{y}$ is proportional to $\exp[-(\kappa^2/2)\mathbf{y}^T\mathbf{Q}\mathbf{y}]$ for some $n \times n$ matrix $\mathbf{Q}$ whose elements you are to determine.

    (b) Show that $\mathbf{Q}$ is not positive definite. (Hint: consider the sum of the elements in each of its rows.)

3

Exploratory Spatial Data Analysis

Before formulating a spatial linear model that is likely to be appropriate for a given dataset, it is a good idea to first carry out some exploratory spatial data analyses (ESDA). ESDA methods may guide the user to make a reasonable choice of mean structure and covariance structure of a spatial linear model, and they may indicate whether the assumptions of a particular model or inferential method are plausible. They may also suggest the need for additional observations and indicate where those observations should be taken.

Throughout this chapter, we describe methods of ESDA as they apply to the raw data. Many of the methods can also be applied to fitted residuals from spatial linear models, to guide the process of model building and reformulation at later stages of the analysis. These re-applications of the methods will be described in Chapter 4.

3.1   Aspatial data summaries

Standard numerical descriptive statistics such as the sample mean, standard deviation, and range, or multidimensional summaries such as the **five-number summary** (minimum, first quartile, median, third quartile, maximum) are often computed for nonspatial data, and can and should be computed for the observed responses and explanatory variables (if any) of spatial data also. These summaries inform the data analyst of "typical" or representative values and other data attributes, including dispersion and skewness. Similarly, graphical summaries of the responses, such as boxplots, histograms, and stem-and-leaf displays, and scatterplots of the response versus each nonspatial explanatory variable, can facilitate understanding; in addition, they can identify potential outliers. For example, the parallel boxplots displayed in Figure 1.6 indicate that a higher level of nitrogen in willows is associated with shaded plots in the caribou forage experiment.

However, for spatial data, all aspatial summaries, be they numerical or graphical, are useful only up to a point. Because they ignore (marginalize over) the spatial locations of data sites, they give no indication of the data's spatial structure. Furthermore, they cannot, as they can in a classical statistical sampling framework, be interpreted as sample-based estimates of meaningful population quantities. For example, in this context the histogram of observed responses generally does not represent an estimate of a probability distribution from which the responses were drawn, and the temptation to give it this interpretation (because we are so accustomed to doing so within a classical framework) should be strongly resisted.

As an example, consider Table 3.1, which displays the five-number summary and stem-and-leaf plot of the wet sulfate deposition data. The five-number summary indicates that the response ranges from a minimum of about 0.2 kg/ha to a maximum of about 44 kg/ha, with about half of the observations lying between 3.3 and 19 kg/ha. The stem-and-leaf plot shows that the responses are skewed marginally, and perhaps not unimodal. Also, the

DOI: 10.1201/9780429060878-3

TABLE 3.1
Five-number summary (top), and stem-and-leaf plot (bottom) of the wet sulfate deposition data. The vertical bar (|) marks the decimal point of the response. The units of measurement are kg/ha.

Minimum	First quartile	Median	Third quartile	Maximum
0.186	3.319	10.824	19.247	44.023

```
 0 | 24466789999901234455667999
 2 | 1222222334455577889901133467778
 4 | 01556889991122377799
 6 | 112459369
 8 | 177814677
10 | 4448901257789
12 | 099924477
14 | 23468847
16 | 11244888889
18 | 0033457901223458
20 | 2260011358889
22 | 13569
24 | 4689112
26 | 24689
28 | 1149113
30 | 46
32 | 2935
34 |
36 | 5
38 |
40 |
42 |
44 | 0
```

sample maximum is somewhat "off by itself" in the upper tail of the marginal distribution, so it seems that it could be an outlier deserving additional scrutiny.

In spite of the skewness, multiple modes, and a possible outlier present in the marginal distribution of these data, it is possible that the residuals from a spatial linear model for these data could be normally distributed; we will revisit this possibility later.

3.2 Maps of data sites

Aspatial summaries of the data allow for exploration of the responses stripped of their spatial labels. The complementary activity is to explore the spatial data locations stripped of their responses, i.e., to make a map of the data sites. A map can reveal variation in the sampling density across the study area and the regularity or irregularity of the spacing between sites. In the areal case, it also indicates the sizes and shapes of the sites. Figure 1.8, for example, shows that the sampling density of data locations in the moss heavy metals study is greater near the road than elsewhere, and Figure 1.2 reveals the extremely irregularity of the shape and spacing of the areal sites in the harbor seal trends study. Such information in isolation is, like an aspatial summary, not particularly helpful for formulating spatial linear

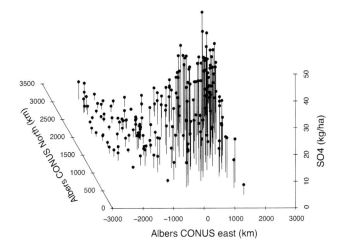

FIGURE 3.1
Three-dimensional scatterplot of the wet sulfate deposition data.

models. However, as will be seen in later chapters, this information can be quite useful for understanding the behavior of fitted spatial surfaces and the spatial variation in quality of spatial predictions. It may also be useful for guiding the placement of additional data sites (spatial design).

3.3 Exploring large-scale variation

The primary tools for exploring the large-scale, or global, variation of the response over a (d-dimensional) spatial domain are the ($d + 1$)-**dimensional scatterplot** of response versus spatial location and various plots derived from it. As an example, consider the three-dimensional scatterplot of the wet sulfate deposition data displayed in Figure 3.1. This plot indicates that wet sulfate deposition tends to be considerably larger in the eastern United States, especially in the Ohio Valley region, than elsewhere. However, the plot is somewhat cluttered, making it difficult to discern much detail in densely sampled, high-deposition areas. Plots derived from the ($d + 1$)-dimensional scatterplot by projecting the data locations to one less dimension, so that the scatterplot can be plotted in d dimensions, may be considerably easier to "read" and interpret. An example is provided by Figure 1.9, which plots (log) lead concentration in moss as a function of (log) distance from the road and reveals a strong association between these variables. Alternatively, a plot in d dimensions may be derived from the ($d + 1$)-dimensional scatterplot by using color or shading to code the response, yielding what is called a **choropleth map**. Figures 1.3 and 1.5 are examples of choropleth maps for the harbor seal trends and caribou forage data, respectively, and Figure 1.1 is a variation suitable for geostatistical data such as, in this instance, the sulfate wet deposition data. Yet another alternative is a **bubble plot**, in which a circle with center at S_i and radius proportional to y_i is plotted for each i; Figure 3.2 is such a plot for the sulfate deposition data. All of these derived plots convey the same information as a three-dimensional scatterplot, albeit with slightly less detail as to the exact magnitude of the response.

FIGURE 3.2
Bubble plot of the wet sulfate deposition data.

3.4 Detecting outliers

Outliers are atypical observations. In a spatial context, outliers may be of two types: **distributional outliers** and **spatial outliers**. Distributional outliers are observations that are unusual with respect to the data's marginal distribution; hence they are easily detected by viewing a histogram or stem-and-leaf plot. Spatial outliers are observations that are unusual in comparison to observations at nearby sites; they may or may not be distributional outliers. Thus, a histogram may not reveal a spatial outlier.

A graphical display that can reveal spatial outliers is the **nearest-neighbor scatterplot**, also called the spatial lag scatterplot (Anselin, 1994). This plot consists of all points $\{(y_i, y_j) : S_j$ is a neighbor of $S_i\}$, where a "neighbor" of S_i may be defined in the geostatistical case as any S_j for which the Euclidean distance between S_i and S_j is minimized or is less than a certain threshhold, and in the areal case as any S_j that shares a boundary with S_i (other definitions are possible, of course). Another version of the plot uses the average response over all neighbors, rather than the response of each individual neighbor, as the ordinate. A 45-degree line may be superimposed on these plots for reference, and responses corresponding to points that are sufficiently far from this line may be declared (spatial) outliers. Points below the 45-degree line correspond to responses that are larger than those of their neighbors, and points above the line correspond to responses that are smaller than those of their neighbors. Thus, in the non-averaged version of the plot, mutual neighbors of one another will generate two points (unless the responses at those sites are equal) located symmetrically about the line. Typically not all neighbors are mutual, however, so the plot may have some asymmetry.

The nearest-neighbor scatterplot is designed to discover individual spatial outliers, so it may not reveal clusters of spatial outliers. Consider, for example, a case in which two sites are nearest neighbors of each other and the responses at those sites are similar to each other but very different than responses at other sites in the same general vicinity. Then the points in the plot corresponding to the two outliers will lie close to the 45-degree line and escape notice. A pocket plot can detect "pockets" of multiple outliers; see subsection 3.6.1. Another method for identifying such pockets was proposed by Cerioli & Riani (1999), but that method is rather more complicated hence not considered herein.

A commonly used criterion for deciding whether an observation is a distributional outlier is the absolute value of its standardized value, with 3.0 often used as a threshold. Another common criterion is the absolute value of the difference between the observation and either the first or third quartile (depending on whether the observation lies to the left or right of the median), scaled by the interquartile range, with 1.5 used as a threshold. Similarly, criteria

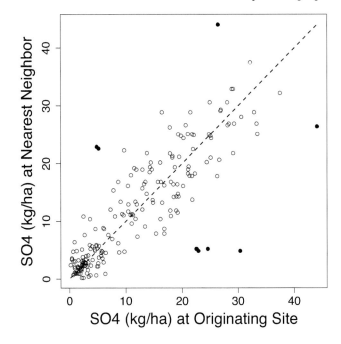

FIGURE 3.3
Nearest-neighbor scatterplot of the wet sulfate deposition data. Points corresponding to originating sites whose standardized absolute difference with its nearest neighbor exceeds 3.0 are represented by closed circles.

for determining whether an observation is a spatial outlier include the value of $|d_i - \overline{d}|/s_d$, where d_i is the absolute value of the difference between the abscissa and ordinate in the nearest-neighbor scatterplot and $\overline{d}$ and s_d are the mean and standard deviation of those absolute differences, or the analogous scaled distance from the first or third quartile. Again, values of 3.0 or 1.5, respectively, may be used as crude thresholds. A more formal method for detecting spatial outliers was developed by Chen *et al.* (2008).

The maximum wet sulfate deposition value of 44.023 kg/ha (from Table 3.1), when standardized in the usual fashion, is equal to $(44.023 - 12.183)/\sqrt{91.907} = 3.32$, and when scaled using the third quartile and interquartile range is equal to $(44.19.247 - 19.247)/(19.247 - 3.319) = 1.56$. The value thus exceeds the aforementioned crude thresholds and may be deemed a (right-tail) distributional outlier. So, necessarily, it is also a spatial outlier, but is it the only one? To address this question, consider Figure 3.3, which is a nearest-neighbor scatterplot of the wet sulfate deposition data. The plot reveals nine points attributable to potential spatial outliers, whose (originating response, nearest neighbor response) are as follows: (22.492, 5.225), (22.584, 5.225), (24.567, 5.225), (5.225, 22.584), (22.889, 4.875), (30.365, 4.875), (4.875, 22.889), (44.023, 26.408), (26.408, 44.023). These points, which all lie in the eastern United States, are plotted in Figure 3.4. Comparison of the responses at the nine sites in this figure to responses at nearby sites indicates that there are really three potential spatial outliers, which are circled in the figure. The site represented by the red dot in northern Pennsylvania corresponds to the sample maximum, which we have already determined to be a (right-tail) distributional and spatial outlier. The other two sites, one close to the Tennessee-North Carolina border and the other in northern Virginia, have low depositions (4.875 and 5.225 kg/ha) relative to their first-, second-, and third-nearest neighbors, hence they correspond to potential left-tail spatial outliers. It turns out that both

FIGURE 3.4
Site-pairs including potential spatial outliers for the wet sulfate deposition data. Arrows indicate the direction from the originating site to its nearest neighbor, and the circled sites are the members of the site-pairs that are determined to be spatial outliers.

criteria for both of these sites exceed the corresponding crude thresholds, so the responses at these sites may indeed be regarded as (left-tail) spatial outliers.

Figure 3.5 is another nearest-neighbor scatterplot, in this case for the harbor seal trends data. This time we plot the response (trend value) at the originating site against the average response of its first-order (adjacent) neighbors. Again there is evidence that some observations should receive scrutiny. Five different polygons have standardized absolute differences greater than 3.0, hence might be declared as spatial outliers.

If outliers are discovered in a dataset, what should be done about them? One option (Nirel *et al.*, 1998; Park *et al.*, 2012) is to replace each outlier with a pseudo-datum, such as the median of its neighbors. Another is to adopt methods of statistical analysis that are robust/resistant to outliers (Cressie & Hawkins, 1980; Hawkins & Cressie, 1984; Dowd, 1984; Cressie, 1986; Yildirim & Kantar, 2020). Still another recourse, for later stages of analysis, is to enrich the spatial linear model so that it will accommodate the outliers, perhaps by changing the mean structure or adding a separate error term that allows the outliers to have a larger variance than the rest of the observations (Baba *et al.*, 2022). A final possibility is to perform the desired analyses of the data both with and without the outliers, and compare the results. We employ this last option in subsequent modeling and prediction activities.

3.5   Exploring spatial variation in the variance

Just as the response's mean may vary across space, so may its variance (or equivalently, its standard deviation). To investigate this possibility, the sample standard deviation may be

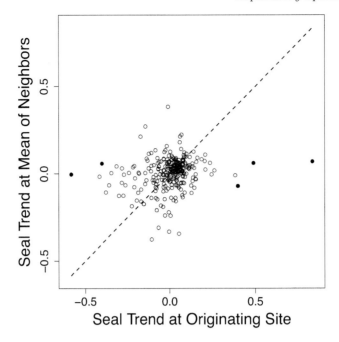

FIGURE 3.5
Nearest-neighbor scatterplot of the harbor seal trends data. Points corresponding to originating sites whose standardized absolute difference with the average of its neighbors exceeds
3.0 are represented by closed circles.

computed for all responses whose sites lie within a specified subregion of the study area,
and then these standard deviations may be compared across subregions. The subregions
may be overlapping, as when a (usually circular) window of fixed size is moved over the
d-dimensional study area, or disjoint. In either case, a $(d+1)$-dimensional scatterplot of the
sample standard deviation versus the spatial coordinates of the region's centroid may be
constructed. The plot may indicate not only the degree of variance nonstationarity, but also
the portion(s) of the study area where it occurs. Using disjoint subregions reduces the degree
of spatial dependence between sample standard deviations from proximate subregions, at
the cost of a sparser plot.

A test statistic corresponding to any one of several available tests for equality of group
variances may be used as an overall measure of the degree of heterogeneity in the subregion-
specific standard deviations. Among those, we recommend Levene's test statistic (Levene,
1960), as modified by Brown & Forsythe (1974), for its overall robustness properties, including robustness to nonnormality. This statistic is

$$H = \frac{n-k}{k-1} \frac{\sum_{i=1}^{k} n_i (\overline{d}_{i.} - \overline{d}_{..})^2}{\sum_{i=1}^{k} \sum_{j=1}^{n_i} (d_{ij} - \overline{d}_{i.})^2},$$

where k is the number of subregions, n_i is the number of observations in the ith subregion,
$n = \sum_{i=1}^{k} n_i$, d_{ij} is the absolute difference between the jth response and the median of
all responses in the ith subregion, and $\overline{d}_{i.}$ and $\overline{d}_{..}$ are the means of the d_{ij}'s over the ith
subregion and over the entire region, respectively. The extremeness of H may be evaluated
by comparing its value to a chosen percentile of an F distribution with $k-1$ and $n-k$
degrees of freedom (which is the approximate null distribution of H in a classical sampling

TABLE 3.2

Sample standard deviations of the wet sulfate deposition data (in kg/ha) within eight rectangular subregions formed by partitioning the Albers projections of latitude and longitude. Sample sizes are given in parentheses.

	Longitude class			
Latitude class	Far west	Near west	Near east	Far east
North	1.69	2.71	8.67	9.21
	(25)	(25)	(24)	(25)
South	1.69	5.03	5.03	6.66
	(25)	(24)	(24)	(25)

framework), although this is a very crude threshold in this context because the observations, both within and between regions, are likely to be correlated and thereby in violation of the assumptions of the classical sampling framework. Hence the result of this evaluation should not be accorded the same status as the result of a formal hypothesis test.

In addition to the scatterplot of sample standard deviation versus spatial location, a two-dimensional scatterplot of the subregion-specific sample standard deviation versus the subregion-specific sample mean may be constructed. This plot may also reveal nonconstant variance, but of a type that depends on the mean response rather than on spatial location *per se*. Depending on the nature of the dependence of the standard deviation on the mean, a transformation of the response may be warranted. For example, if the standard deviation is proportional to (i.e., a linear function of) the mean, then taking the logarithm of the response will render its standard deviation constant with respect to the mean. Appropriate transformations for other types of mean-variance relationships are well-known; see, for example, Weisberg (2005).

Table 3.2 is a tabular version of the scatterplot of standard deviation versus region centroid for the wet sulfate deposition data, and Figure 3.6 is a scatterplot of standard deviation versus mean for the same data. There are eight subregions, which are defined by partitioning the Albers projection of latitudes and longitudes into two and four intervals, respectively, with median latitude and the quartiles of longitude serving as interval

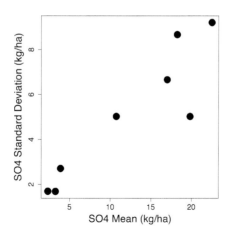

FIGURE 3.6

Scatterplot of standard deviation versus the mean for the wet sulfate deposition data.

endpoints. The table shows that the standard deviation tends to increase from west to east, to the extent that the ratio of largest to smallest standard deviation is greater than 5. The Brown-Forsythe test statistic for variance homogeneity, H, is equal to 49.9, while the 99th percentile of the reference F distribution with 7 and 189 degrees of freedom is 2.74. This indicates extreme variance nonstationarity in the data. Moreover, the strong signal in the plot of standard deviation versus mean suggests that transforming the response, possibly to the square root or log scale, would be wise if one wishes to use a model with constant variance.

3.6   Variography for geostatistical data

Local, small-scale spatial dependence among responses can exist irrespective of any large-scale variation and can have a significant impact on various types of inference, especially spatial prediction. Hence it is just as important to explore the nature and strength of the small-scale dependence as it is to explore the large-scale variation. The nearest-neighbor scatterplot, introduced previously for the purpose of identifying potential spatial outliers, is also relevant for exploring small-scale dependence. In particular, the degree of scatter about the 45-degree line in the plot is related (inversely) to the strength of the small-scale dependence, and the sample correlation coefficient computed from the points in the plot is a numerical summary of that strength. (For example, Pearson's and Spearman's correlation coefficients corresponding to the nearest-neighbor scatterplot of the wet sulfate deposition data displayed in Figure 3.3 are 0.83 and 0.84, respectively, indicating strong small-scale dependence.) To better understand the nature and extent of the dependence in geostatistical data, however, two interrelated statistics, the **sample autocovariance function** and the **sample semivariogram**, are useful. This section introduces these tools; analogous tools for areal data are described in the following section.

3.6.1   The sample autocovariance function and sample semivariogram

The mean-corrected product of two responses, i.e., $(y_i - \overline{y})(y_j - \overline{y})$, is a measure of their similarity: it is large (and positive) if y_i and y_j are both substantially larger, or both substantially smaller, than the sample mean. To explore whether the similarity of responses is associated with site proximity, this similarity measure may be plotted against the corresponding intersite distance $\|\mathbf{s}_i - \mathbf{s}_j\|$ for all pairs of observations, yielding the **autocovariance cloud**. Recall Tobler's first law of geography from Section 1.2, which specifies that values of an attribute variable at nearby sites tend to be more alike than values at sites further apart. For data that conform to this law, a smooth curve drawn through the points in the cloud will tend to decrease as distance increases. Furthermore, the distance at which the curve appears to reach the horizontal axis may be interpreted as the spatial extent, or **sample range**, of the dependence. A closely related plot is the **semivariogram cloud**, which consists of points $(\|\mathbf{s}_i - \mathbf{s}_j\|, (1/2)(y_i - y_j)^2)$ for all pairs of observations. Observe that $(1/2)(y_i - y_j)^2$ is a measure of dissimilarity rather than similarity, so a smooth curve drawn through the points in the semivariogram cloud will tend to *increase* with distance if the data conform to Tobler's law. If the curve appears to reach a plateau at some distance, that plateau is the **sample sill** and the distance at which it is reached is the **sample range** of the spatial dependence, as determined by the semivariogram cloud. The value of the curve as it nears zero (from the right, moving left) is the **sample nugget effect**, and the difference between the sample sill and the sample nugget effect is the **sample partial sill**.

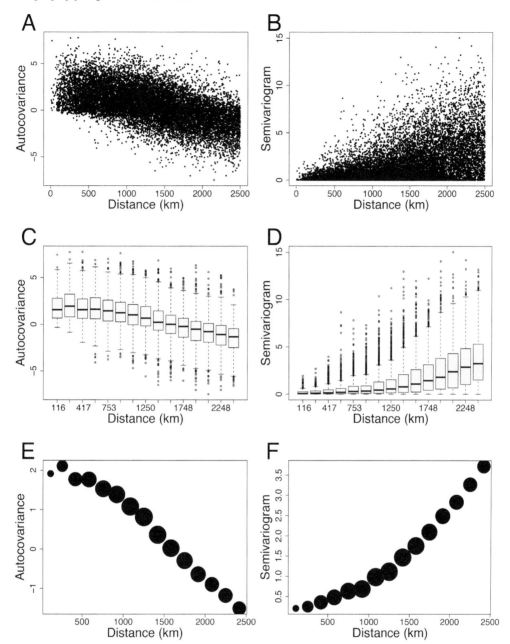

FIGURE 3.7
Measures of spatial dependence for the clean wet sulfate deposition data. A) autocovariance cloud; B) semivariogram cloud; C) binned autocovariance boxplots; D) binned semivariogram boxplots; E) sample acf; F) sample semivariogram.

Truth be told, however, the autocovariance and semivariogram clouds are often such a mass of points that few of the aforementioned features can be identified; all that can be discerned from them may be an overall downward or upward trend. Figure 3.7A,B displays both clouds for the wet sulfate deposition data, after (in reaction to findings noted earlier in this section) transforming the response to the square-root scale and setting aside the three spatial outliers. Henceforth we refer to these data as the "clean" wet sulfate deposition data. Euclidean distance between points in the Albers projection is the distance metric used. Although a decrease in the autocovariances with distance is discernible in the autocovariance cloud, and a concomitant increase in the semivariances with increasing distance can be seen in the semivariogram cloud, the skewed marginal distributions of the squared differences in the latter and the sheer number of points in both make more detailed conclusions impossible. For improved interpretation, the intersite distances may be "binned," i.e., grouped into distance classes $R_1, R_2, \ldots, R_K$, and the results displayed in the form of parallel boxplots, as in Figure 3.7C,D. These parallel boxplots show the trends more clearly and suggest that the sill, if one exists, equals or exceeds 3 $(\text{kg/ha})^2$, with a corresponding range that is no less than 2000 km.

How should the distance classes be chosen in general? A longstanding recommendation (Journel & Huijbregts, 1978), motivated by the tendency for the variability of squared response differences to increase with distance, is to form classes only from those distances that are smaller than half the maximum possible distance in the dataset, and ignore the remainder. Equally-spaced distance classes over this restricted range usually work well in practice, except when the data sites are highly clustered. As for the number of classes, a tradeoff is involved: the more classes used, the narrower the class widths, and hence, the better the intersite distances in the kth class are approximated by the class midpoint (reducing the blurring effect, see below), but the fewer the number of site-pairs in the class (resulting in a boxplot that is a less reliable representation of the population distribution of similarity or dissimilarity values for the class). As a consequence of this last point, Journel and Huijbregts also recommended that no class contain fewer than 30 site-pairs. Within these constraints, 10–20 classes, as is commonly recommended for constructing histograms of large datasets, usually strikes a good balance between summarization and detail. These recommendations were followed in obtaining Figure 3.7.

Summarization within bins can be taken a step further by replacing the boxplot of mean-corrected products or squared differences for each distance class with the average of those quantities. This yields the **omnidirectional sample autocovariance function (acf)**

$$\hat{C}(r_k) = \frac{1}{N_k} \sum_{\|\mathbf{s}_i - \mathbf{s}_j\| \in R_k} (y_i - \overline{y})(y_j - \overline{y}) \qquad (k = 1, \ldots, K)$$

and the **omnidirectional sample semivariogram**

$$\hat{\gamma}(r_k) = \frac{1}{2N_k} \sum_{\|\mathbf{s}_i - \mathbf{s}_j\| \in R_k} (y_i - y_j)^2 \qquad (k = 1, \ldots, K), \tag{3.1}$$

where r_k is a representative distance, and N_k is the number of site-pairs, for the kth distance class. The kth component of the latter function, i.e., $\hat{\gamma}(r_k)$, is referred to as the kth (omni-directional) sample semivariance. Some software packages take the representative distance to be the class midpoint but, following Zimmerman & Stein (2010), we recommend using the class mean instead. For irregularly spaced data sites, the differences between the two may often be small, but for regularly spaced sites the differences can be substantial.

If the underlying process is intrinsically stationary and isotropic, then the summand $(y_i - y_j)^2$ in (3.1) is unbiased for $2\gamma(\|\mathbf{s}_i - \mathbf{s}_j\|)$, hence $\hat{\gamma}(r_k)$ is unbiased for $(1/N_k) \sum_{\|\mathbf{s}_i - \mathbf{s}_j\| \in R_k} \gamma(\|\mathbf{s}_i - \mathbf{s}_j\|)$ but not necessarily for $\gamma(r_k)$. The bias, $E[\hat{\gamma}(r_k)] - \gamma(r_k)$, which exists solely as a result of the binning, is called the **blurring effect**. Under certain conditions, the blurring effect can vanish; for example, it vanishes for the kth class if $\gamma(r)$ is constant for all site-pairs in that class. One way this can happen is if the data sites lie on a regular grid. While the same vanishing of bias does not hold exactly for the sample acf, it holds asymptotically. That is, in the special case where the underlying process is second-order stationary and isotropic, and $C(r)$ is constant for all site-pairs in the kth distance class, the bias of $\hat{C}(r_k)$ tends to zero as the sample size increases. In any case, the blurring effect is usually quite small if the aforementioned recommendations for choosing classes and representative distances are followed.

However, if the underlying process is not intrinsically (or second-order) stationary, then "all bets are off" as far as using the sample semivariogram (or acf) to characterize the small-scale dependence is concerned. In particular, if the process mean is not constant, then these estimators can be badly biased. A strong linear global trend in the response yields a sample semivariogram that tends to increase quadratically without bound, a phenomenon that is showcased in Exercise 3.5. If the process mean is constant, but intrinsic stationarity is otherwise not satisfied, the sample semivariogram need not exhibit unbounded quadratic behavior, but must be interpreted not as an estimate of a single underlying process semivariogram but as an estimate of a (weighted) average of site-specific semivariograms. A method similar to that described previously for investigating spatial variability in the variance may be used to explore how the small-scale dependence varies spatially. The method consists of computing a local sample semivariogram corresponding to each of several (possibly overlapping) subregions of the study area and then comparing them across subregions, with the range being perhaps the most salient feature for comparison. The subregion-specific sample semivariograms may be computed in the standard way using only the data within the subregion (Haas, 1990) or in a geographically weighted fashion in which semivariances from the entire dataset are used but each is weighted inversely proportional to, or by applying a Gaussian kernel to, the average distance from the subregion's centroid to the two points involved (Harris *et al.*, 2010; Machuca-Mory & Deutsch, 2013). Unfortunately, however, this approach is not as useful in practice as it is for investigating the variance because (a) much larger sample sizes are required to estimate the range and other semivariogram features well enough to make effective comparisons across subregions than are required to reliably estimate the variance, and (b) a threshold for assessing the degree of heterogeneity is not readily available.

The omnidirectional sample acf and semivariogram of the clean wet sulfate deposition data are displayed in Figure 3.7E,F. Several comments should be made about these plots. First, note that their vertical scale is different than in the clouds and boxplots, in order to show more detail. Second, observe that the plotting symbols are of different sizes. Here and in all sample semivariograms displayed subsequently, we take the area of the plotting symbol to be proportional to the number of site-pairs in the corresponding distance class. Third and most importantly, the sample semivariogram reveals exactly the kind of unbounded behavior described in the previous paragraph. A sample semivariogram that appears to increase without bound is not necessarily a symptom of a problem; recall from Section 2.2.2 that a semivariogram is permitted to be unbounded, subject to the restriction that its rate of growth is subquadratic. However, in this case, it is likely, in light of our earlier exploratory investigation of large-scale variation of these data, that what we are seeing is actually the effects of the strong trend in wet sulfate deposition from the western to the eastern U.S., which is overwhelming the semivariogram's ability to reveal small-scale spatial dependence. Incidentally, spatially varying variability is not a viable explanation, since the data have been transformed to make the variance more nearly constant. The remedy for the unboundedness is just as obvious as the cause: remove the trend from the data and recompute the sample acf/semivariogram for the detrended data.

Many detrending methods could be considered. Here, with an eye toward Chapter 4, we detrend by fitting a second-order (quadratic) polynomial to the data by ordinary least squares and subtracting the fitted polynomial from the data. The resulting detrended data are subsequently referred to as the "clean wet sulfate deposition residuals." The

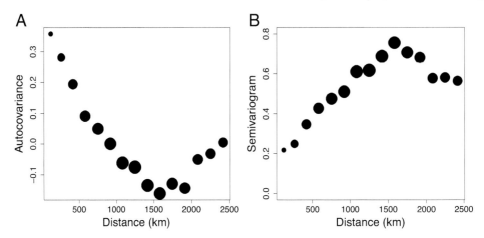

FIGURE 3.8
Measures of spatial dependence for the clean wet sulfate deposition residuals. A) Omnidirectional sample acf; B) Omnidirectional sample semivariogram.

omnidirectional sample acf and semivariogram of the clean wet sulfate deposition residuals are displayed in Figure 3.8. Several features of these plots are of interest. First and foremost, both plots suggest that dependence among the residuals from a fitted second-order polynomial surface does indeed exist at small scales; more formal tests to complement this graphical diagnostic will be presented in Chapter 4. However, the reduction in scale of the limits on the vertical axis, relative to what they were for the original data, indicates that the small-scale spatial variation is considerably smaller than the large-scale (global) spatial variation. Furthermore, a "flattening of the curve" occurs as distance increases, suggesting that the clean wet sulfate deposition residuals are second-order stationary, with a range of small-scale spatial dependence lying somewhere near 1500 km. The sample nugget effect and sill appear to be equal to approximately 0.2 and 0.6 (kg/ha)^2, respectively.

Recall from Chapter 2 that the semivariogram and covariance function of an isotropic second-order stationary random process are perfectly related via the equation

$$\gamma(r) = C(0) - C(r).$$

A similar relationship between the omnidirectional sample semivariogram and sample acf does not hold; that is, in general

$$\hat{\gamma}(r_k) \neq \hat{C}(0) - \hat{C}(r_k),$$

where $\hat{C}(0) = \frac{1}{n}\sum_{i=1}^{n}(y_i - \overline{y})^2$. Nor does the result hold if a divisor of $n-1$ is used in place of n in $\hat{C}(0)$. Although the difference between the two estimated functions tends to zero as the sample size increases under weak conditions, it is fair to ask if one is better than the other for finite samples. Compared to the sample acf, the sample semivariogram has the advantages of unbiasedness (under intrinsic stationarity and isotropy, and apart from the blurring effect) and greater applicability, as the class of intrinsically stationary processes for which it is meaningful is (as noted in Chapter 2) larger than the class of second-order stationary processes. Moreover, the sample acf exhibits some non-intuitive behaviors, such as guaranteed negative estimates at some lags even when the process covariance function is strictly nonnegative (see Exercise 3.4). For these reasons, the sample semivariogram is used much more frequently than the sample acf as an exploratory tool in spatial statistics, and

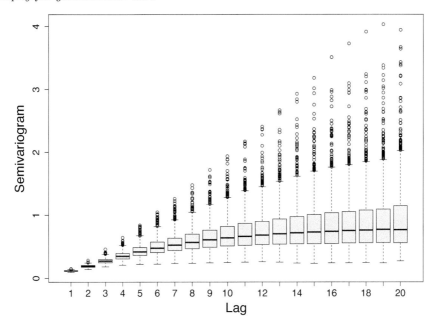

FIGURE 3.9
Parallel boxplots summarizing the marginal distributions of the omnidirectional sample semivariances as a function of lag for a scenario described in Section 3.6.1.

henceforth we focus relatively more attention on it. Nevertheless, many of the comments concerning the sample semivariogram made in the remainder of this section apply to the sample acf as well.

It is not uncommon in practice for a plot of the omnidirectional sample semivariogram to reveal semivariances that are generally well-behaved at small lags, in the sense that they generally increase with lag at short to moderate lags, but become much more erratic at large lags. In fact, sometimes the semivariances at large lags will suddenly careen upwards or plunge downwards after seeming to have settled at a sill at moderate lags. There are two factors that work together to cause this undesirable behavior. The first is heterogeneity of variance among the sample semivariances, with semivariances at larger lags being much more variable. Figure 3.9 provides a visualization of the sampling variabilities of the omnidirectional sample semivariances and how they depend on lag. The figure shows parallel boxplots summarizing the marginal distributions of $\hat{\gamma}(1), \ldots, \hat{\gamma}(20)$, which were obtained from 1000 simulated realizations of a mean-zero, second-order stationary Gaussian process on a 20×20 square grid with unit spacing. The covariance function of the process was a nuggetless isotropic exponential covariance function with a sill of 1.0 and a correlation decay parameter of 8.0, i.e., $C(r) = \exp(-r/8)$ for $r \geq 0$. The 20 distance classes are bins of equal width 1.0. The boxplots clearly reveal an increase in the variance of the estimated semivariances with increasing distance. The second factor that can cause undesirable behavior in the omnidirectional sample semivariogram is strong positive correlations among semivariances corresponding to successive distance classes. This feature, which is examined in Exercise 3.10, can cause several successive semivariances to all lie above (or all lie below) their long-run averages, giving the appearance of a new "signal" in the semivariogram after it had already seemed to reach a sill. Together, the influences of these factors suggest that when interpreting the omnidirectional sample semivariogram, typically more weight should

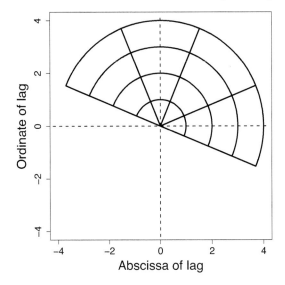

FIGURE 3.10
A polar coordinates partition of the lag space in two dimensions. This particular partition, which is common, yields directional sample semivariograms in the NW-SE, N-S, NE-SW, and E-W directions.

be given to semivariances at small lags than at large lags, and in particular that a sudden excursion in the sample semivariogram at large lags should be discounted.

The omnidirectional sample semivariogram averages squared response differences within a distance class with no regard for the orientations of the site-pairs in the class. This is fine if the underlying process is isotropic (and intrinsically stationary), but not otherwise. A broader definition of the sample semivariogram is needed if proper account is to be taken of anisotropy. The **sample semivariogram**, more broadly defined, is

$$\hat{\gamma}(\mathbf{h}_k) = \frac{1}{2N(\mathbf{h}_k)} \sum_{\mathbf{s}_i - \mathbf{s}_j \in H_k} (y_i - y_j)^2 \qquad (k = 1, \ldots, K).$$

Here, $\mathbf{h}_k$ is a representative lag for the kth lag class, such as the centroid of the observed lags falling into the class; $H_1, \ldots, H_k$ are lag classes formed by partitioning the lags with respect to not only distance but also direction; $N(\mathbf{h}_k)$ is the number of lags within the kth class; and K is the number of classes. For data sites lying on a regular rectangular grid (or nearly so), it seems most natural to partition the lag space into rectangles formed as Cartesian products of intervals in the directions aligned with the grid. For more general spatial configurations of data sites, however, a "polar coordinates partition" of lags into distance $\times$ angle classes (i.e., radial sectors), such as that depicted in Figure 3.10, is more common. Because the sample semivariogram is an even function, i.e., $\hat{\gamma}(\mathbf{h}_k) = \hat{\gamma}(-\mathbf{h}_k)$, only a half-plane within the lag space needs to be partitioned. The sample semivariogram constructed using a polar coordinates partition may be displayed graphically in the form of either a $(d+1)$ dimensional scatterplot or as a series of d-dimensional scatterplots in selected directions (e.g., N-S, E-W, NE-SW, NW-SE) called **directional semivariogram plots**. Directional semivariograms may be superimposed on the same graph to facilitate comparison. To illustrate, Figure 3.11 shows directional sample semivariograms for the clean wet sulfate deposition residuals. The noisiness and meandering nature of these directional semivariogram plots do not provide clear evidence of either nugget or range anisotropy, but this is investigated further in

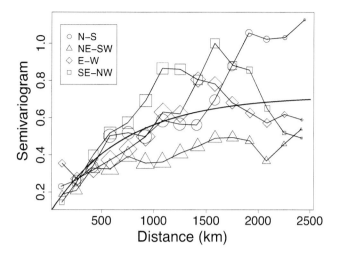

FIGURE 3.11
Directional sample semivariograms of the clean wet sulfate deposition residuals.

Section 3.6.3 and Chapter 4. Some additional guidance for interpreting directional semivariogram plots may be found in Zimmerman (1993).

Because squares of differences are involved in its calculation, the sample semivariogram is sensitive to outliers. An alternative estimator of the semivariogram, introduced by Cressie & Hawkins (1980), is

$$\bar{\gamma}(\mathbf{h}_k) = \frac{1}{0.914 + 0.988/N(\mathbf{h}_k)} \left(\frac{1}{N(\mathbf{h}_k)} \sum_{\mathbf{s}_i - \mathbf{s}_j \in H_k} (y_i - y_j)^{1/2} \right)^4 \qquad (k = 1, \ldots, K).$$

This estimator, being an average of square-root differences rather than squared differences, is more robust (less sensitive to outliers) than the sample semivariogram. However, another way to prevent outliers from affecting the exploration of small-scale dependence is, as noted previously, to simply identify them and set them aside when computing the sample acf or semivariogram.

Another graphical tool, the **pocket plot** (Cressie, 1991), can identify localized pockets of nonstationarity in the spatial dependence. Various versions of this plot are possible, depending on the number and spatial configuration of sites considered for inclusion in the pockets; here we describe a version for determining pockets of individual sites. Consider the Cressie-Hawkins robust sample semivariogram at a fixed lag $\mathbf{h}_k$ for the complete data, as described in the previous paragraph. Now consider setting aside the ith observation and recomputing the estimator, yielding

$$\bar{\gamma}_{-i}(\mathbf{h}_k) = \frac{1}{0.914 + 0.988/N_{-i}(\mathbf{h}_k)} \left(\frac{1}{N_{-i}(\mathbf{h}_k)} \sum_{\mathbf{s}_l - \mathbf{s}_j \in H_k} (y_l - y_j)^{1/2} \right)^4 \qquad (k = 1, \ldots, K),$$

where $N_{-i}(\mathbf{h}_k)$ is the number of lags in the kth lag class after removing the ith observation. Repeat this for each $i = 1, \ldots, n$ and $k = 1, \ldots, K$, and then construct, for each i, the boxplot of values $\{\bar{\gamma}_{-i}(\mathbf{h}_k) - \bar{\gamma}(\mathbf{h}_k) : k = 1, \ldots, K\}$. If no observations are "unusual," then the scatter within each boxplot will be roughly centered at zero, but if the ith observation is unusually high or low, then a large proportion of the corresponding boxplot will lie above zero. A threshold for declaring that a pocket of nonstationarity exists at site i is whether the box containing the middle half of the distribution lies completely above zero. The same procedure may be carried out for pairs, triples, etc. of nearest neighbors, or for entire rows and columns if the sites form a rectangular grid. The individual-site version of the pocket plot will tend to identify the same spatial outliers identified by the nearest-neighbor scatterplot, but when two (or more) unusual observations occur at neighboring sites, a multiple-site version may reveal such a pocket when the nearest-neighbor scatterplot does not.

3.6.2   Smoothing the sample semivariogram

For many datasets, the sample semivariogram, be it directional or omnidirectional, is quite "bumpy" and therefore not so easy to interpret. Factors that contribute to the difficulty in interpretation were noted in the previous subsection. Because of these, a smoothed version of the sample semivariogram(s) may improve understanding. Smoothing may be performed either nonparametrically or by fitting a parametric model. Classically, methods for smoothing the semivariogram were viewed by geostatisticians as estimation procedures, and results of the smoothing were used to make inferences and perform spatial prediction. But because they are not based on a probabilistic model for the underlying spatial process, smoothing methods are suboptimal and do not rest on as firm a theoretical foundation as the likelihood-based approaches to parameter estimation to be featured later in this book. Hence, we prefer to view such methods as mere smoothers, useful primarily for exploratory purposes. Nevertheless, for at least some geostatistical processes, their estimation performance may not be greatly inferior to those that are likelihood-based (Zimmerman & Zimmerman, 1991; Lark, 2000).

Methods for smoothing the sample semivariogram that treat it as an unconstrained regression function include kernel-based local mean smoothing and locally linear smoothing. For the omnidirectional case, a general class of semivariogram smoothers has form

$$\widetilde{\gamma}(r) = \frac{1}{2} \sum_{i=1}^{n} \sum_{j=i+1}^{n} w_{ij}(y_i - y_j)^2,$$

where the weights $\{w_{ij}\}$ sum to one and decrease with increasing distance of $\|\mathbf{s}_i - \mathbf{s}_j\|$ from the point of estimation r. In particular, a local-mean smoother of the semivariogram takes

$$w_{ij} \propto \frac{1}{b}\phi\left(\frac{r - \|\mathbf{s}_i - \mathbf{s}_j\|}{b}\right),$$

where ϕ represents the standard normal pdf and b is a user-supplied bandwidth parameter. This local-mean smoother is a solution, for α, to the problem of minimizing

$$\sum_{i<j}\left[\frac{1}{2}(y_i - y_j)^2 - \alpha\right]^2 \frac{1}{b}\phi\left(\frac{r - \|\mathbf{s}_i - \mathbf{s}_j\|}{b}\right).$$

A locally linear smoother, which has better properties at the extremes of the range of distances, is of the same form except that the weights are determined by minimizing

$$\sum_{i<j}\left[\frac{1}{2}(y_i - y_j)^2 - \alpha - \beta(\|\mathbf{s}_i - \mathbf{s}_j\|)\right]^2 \frac{1}{b}\phi\left(\frac{r - \|\mathbf{s}_i - \mathbf{s}_j\|}{b}\right)$$

with respect to α and β. The function `sm.variogram` in the `sm` package of R computes and plots the locally linear smoother of the omnidirectional Cressie-Hawkins robust sample

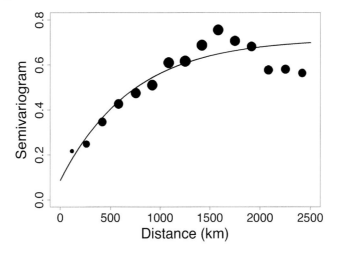

FIGURE 3.12
Weighted least squares smooth of the omnidirectional sample semivariogram for the clean wet sulfate deposition residuals. An exponential model with nugget effect was used for the smoothing.

semivariogram. Note that this method of smoothing does not account for the sampling variation of, and correlation among, the sample semivariances, nor does it account for the conditional negative definiteness constraint on the semivariogram.

Nonparametric smoothing methods exist which enforce the conditional negative definiteness constraint [see, for example, Shapiro & Botha (1991), Lele (1995), and Choi *et al.* (2013)], but they are more complicated than is typical of exploratory methods, so we do not consider them here. Instead, we consider parametric smoothing, which also enforces conditional negative definiteness provided that a valid semivariogram model is used, and which conforms more closely to the parametric estimation approach taken for spatial linear models throughout the rest of this book. (Valid semivariogram models are described in detail in Chapter 6.) Several parametric smoothing methods are available; of these, the weighted (nonlinear) least squares (WLS) method of Cressie (1985) is most popular among practitioners because it accounts (to some degree) for the sampling distribution of the semivariances while still being relatively simple. Let $\gamma(\mathbf{h}_k; \boldsymbol{\theta})$ denote the parametric semivariogram model to be used for the smoothing, evaluated at $\mathbf{h}_k$. The WLS smooth of the sample semivariogram is $\gamma(\mathbf{h}; \hat{\boldsymbol{\theta}})$, where $\hat{\boldsymbol{\theta}}$ is the value of $\boldsymbol{\theta}$ that minimizes a weighted residual sum of squares criterion,

$$\sum_{k=1}^{K} \frac{N(\mathbf{h}_k)}{[\gamma(\mathbf{h}_k; \boldsymbol{\theta})]^2} [\hat{\gamma}(\mathbf{h}_k) - \gamma(\mathbf{h}_k; \boldsymbol{\theta})]^2. \qquad (3.2)$$

Observe that the weights, $N(\mathbf{h}_k)/[\gamma(\mathbf{h}_k; \boldsymbol{\theta})]^2$, are small if either $N(\mathbf{h}_k)$ is small or $\gamma(\mathbf{h}_k; \boldsymbol{\theta})$ is large. Because most commonly used parametric semivariogram models are monotone increasing in each direction and (for most spatial configurations of data sites) $N(\mathbf{h}_k)$ is relatively small for large lags, the WLS smooth typically assigns less weight to ordinates of the sample semivariogram corresponding to large lags. Since the weights depend on $\boldsymbol{\theta}$, we recommend that $\hat{\boldsymbol{\theta}}$ be obtained iteratively, updating the weights on each iteration until the procedure converges. In fact, if (3.2) is minimized directly (without reweighting), the minimizer is not a consistent estimator of $\boldsymbol{\theta}$ (Müller, 1999).

For illustration, Figure 3.12 includes a superimposed WLS smooth of the omnidirectional semivariogram of the clean wet sulfate deposition residuals. An exponential model with nugget effect (see Chapter 6) was used for the smoothing. The WLS estimates of the nugget effect and partial sill are 0.086 $(\text{kg/ha})^2$ and 0.632 $(\text{kg/ha})^2$, respectively, and the WLS estimate of the practical range (the distance where the curve reaches 95% of the sill) is 2142 km.

3.6.3   Exploring additional structure in the dependence

As noted in Chapter 2, the second-order dependence of a geostatistical process can have additional properties beyond second-order or intrinsic stationarity, such as isotropy or separability. The sample acf and sample semivariogram may be used to assess whether the spatial dependence within a particular dataset has any of these properties. For example, examination of the directional sample semivariograms in multiple directions can be used to informally assess whether the data conform with isotropy. At least three directions are needed to distinguish isotropy from geometric anisotropy, and four or more are recommended to distinguish between geometric and nongeometric anisotropy. A **rose diagram** is another tool for graphically exploring how the dependence varies with direction. To construct a rose diagram for data in $\mathbb{R}^2$, lag vectors in various directions at which the smoothed directional sample semivariograms are roughly equal to an arbitrary constant g (within the range of observed sample semivariances) are determined, then plotted in $\mathbb{R}^2$. A smooth closed curve that passes through the tips of the plotted lag vectors constitutes an estimated isosemivariogram contour. Roughly circular (elliptical) concentric curves for several choices of g may be taken as evidence in support of isotropy (geometric anisotropy). Unfortunately, however, variability in the sample semivariogram often renders the rose diagram ineffective. For example, a rose diagram for the clean wet sulfate deposition residuals (not shown) adds little to what can be gleaned from Figure 3.11.

A graphical diagnostic for assessing separability of the covariance function may be constructed from the sample acf. Consider the two-dimensional case, and let $\hat{C}(h_u, h_v)$ denote the sample acf, preferably based upon a rectangular partition of the lag space. Under separability, we should expect the sample acf $\{\hat{C}(h_{uk}, h_{vk'}) : k, k' = 1, \ldots, K\}$ to be approximately proportional to $\{\hat{C}(h_{uk}, 0)\hat{C}(0, h_{vk'}) : k, k' = 1, \ldots, K\}$, and consequently that the matrix whose (k, k')th element is $\hat{C}(h_{uk}, h_{vk'})$ should be well-approximated by a rank-one matrix. A **scree plot** of the eigenvalues of that matrix, i.e., a plot of the eigenvalues versus their index (where eigenvalues are indexed according to their ranking from largest to smallest), provides the separability assessment. A large drop from the first to the second eigenvalue with much smaller drops thereafter, or equivalently a strong bend or "elbow" in the plot at the second index, supports a separability assumption.

More formal assessments of spatial structure in the small-scale dependence, based upon the sample acf or semivariogram of residuals from an ordinary least squares fit of a mean structure, are considered in Chapter 4.

3.7   Exploring spatial correlation in areal data

In the areal setting, spatial adjacencies rather than distances are commonly used to measure "proximity." Thus, the methods for exploring local spatial dependence described in the previous section are not directly applicable. However, analogous methods exist, which we now describe.

3.7.1   Neighbor graphs

Recall that, in the areal context, the neighbors of the ith site correspond to nonzero elements in the ith row of the spatial weights matrix. A **neighbor graph** is a graphical display of these neighbor relationships. Specifically, it consists of a map of the sites, upon which is superimposed a graph, i.e., a collection of nodes (represented by points) and edges. The nodes are located at the centroids of each site, and the edges connect any two nodes whose corresponding sites are neighbors. Ordinarily the edges are of uniform width, but if the weights are symmetric but not binary, the width of the edge may be used to convey the magnitude of the weight. Figure 1.4 is a neighbor graph for the harbor seal trends data. Such graphs can get rather "busy" and, in the case of neighbors defined by mere adjacencies, may not add much of anything to what is already discernible in a simple site map. Perhaps they are most useful when seeking a definition of neighbors in terms of distances between site centroids, and the distance threshold that yields a suitable number of neighbors (not too many and not too few) is unknown. In that case, examination of the neighbor graphs for various thresholds may help the user make an appropriate choice of threshold.

3.7.2   Moran's I and Geary's c

Moran (1950) introduced an autocorrelation statistic for areal data, which Cliff & Ord (1981) extended as follows:

$$I = \frac{n \sum_{i=1}^{n} \sum_{j=1}^{n} w_{ij}(y_i - \overline{y})(y_j - \overline{y})}{w_{..} \sum_{i=1}^{n}(y_i - \overline{y})^2},$$

where $w_{..} = \sum_{i=1}^{n} \sum_{j=1}^{n} w_{ij}$. (Here and throughout, the replacement of a subscript by a dot indicates that we have summed over that subscript.) The statistic is commonly called **Moran's** I. Because its numerator and denominator are proportional to a (weighted) sum of products and a sum of squares of mean-corrected responses, respectively, I bears a superficial resemblance to the ordinary sample correlation coefficient between two variables. Under the assumption that responses are independent and identically distributed, the expectation of I is $-1/(n-1)$, which is close, but not exactly equal, to 0. Values of I greater than $-1/(n-1)$ correspond to positive spatial correlation, and values less than $-1/(n-1)$ correspond to negative spatial correlation. Unlike the extremes of the ordinary sample correlation coefficient, which are -1 and 1, the extremes of I generally are different than -1 and 1 and depend heavily on the spatial weights matrix (de Jong *et al.*, 1984).

Another autocorrelation statistic is **Geary's** c (Geary, 1954), defined by

$$c = \frac{(n-1) \sum_{i=1}^{n} \sum_{j=1}^{n} w_{ij}(y_i - y_j)^2}{2w_{..} \sum_{i=1}^{n}(y_i - \overline{y})^2}.$$

Whereas I resembles the ordinary sample correlation coefficient, the sum of squared differences in the numerator of c, in concert with the factor of two in the denominator, is reminiscent of the semivariogram. In fact, c may be interpreted as a weighted average of the semivariances. The expectation of c is 1 under the assumption of independent and identically distributed responses. In contrast to I, values larger than the expectation are indicative of negative spatial correlation and values smaller than the expectation are indicative of positive spatial correlation. Again, the extremes of c vary with the spatial weights matrix, though it is obvious that $c \geq 0$. Both I and c are location- and scale-free, meaning that adding a constant to every observation or multiplying every observation by the same constant (or doing both) does not affect the statistic.

The literature describes four distinct ways to assess the extremeness of I and c. The first is via comparison to their randomization distributions. This entails computing the statistic,

TABLE 3.3
Spatial autocorrelation indices for the harbor seal trends data. The first P-value listed uses 10,000 Monte Carlo samples from the randomization distribution for reference, while the second uses the limiting normal distribution.

Neighbor scheme	Moran's I	P-values	Geary's c	P-values
Order 1	0.102	2.71e-02, 2.69e-02	0.678	2.00e-04, 4.89e-07
Order 1 and 2	0.076	1.97e-02, 1.82e-02	0.723	5.00e-04, 2.61e-07
Order 1, 2, and 4	0.118	2.00e-04, 3.19e-06	0.629	2.00e-04, 2.85e-12

I or c, for each possible permutation of the observed responses among the data sites, ranking these values, and determining the P-value as the proportion of values as extreme or more extreme than the value computed from the observed data. One-sided or two-sided tests can be carried out, using the appropriate tail(s) of the randomization distribution. The second method is motivated by the fact that the number of permutations required by the first method can be computationally prohibitive, even for moderately sized datasets. So, rather than completely enumerating all possible permutations, a random sample of size M is taken from the randomization distribution by computing the autocorrelation statistic for M random reassignments of the observed responses to data sites. The P-value obtained by this Monte Carlo approach is computed as

$$P = \frac{m+1}{M+1},$$

where m represents the number of random reassignments for which the autocorrelation statistic was at least as extreme as its value for the observed data. The third method for evaluating extremeness is based on a normal approximation, i.e., an assumption that I and c are approximately distributed as $N(-1/(n-1), \text{Var}(I))$ and $N(1, \text{Var}(c))$, respectively; expressions for $\text{Var}(I)$ and $\text{Var}(c)$, which are rather complicated and not particularly transparent or illuminating, may be found in Cliff & Ord (1981). In this method, the autocorrelation statistic, say I, is judged to be extreme if its standardized value,

$$\frac{I + \frac{1}{n+1}}{\sqrt{\text{Var}(I)}},$$

exceeds a chosen percentile of the standard normal distribution. The fourth and final method for assessing extremeness is something of a hybrid between the second and third methods. It is based on a normal approximation to the distribution of a standardized value of the autocorrelation statistic exactly like the one just given for the third method, except that the variance of the statistic is estimated by simulating from the randomization distribution rather than computed using the theoretical expression. For the latter two methods to be regarded as trustworthy, the sample size should be no smaller than 25.

Functions `moran.mc()`, `moran.test()`, `geary.mc()`, and `geary.test()` in the R package `spdep` may be used to compute I and c and to obtain corresponding P-values by all methods described in the previous paragraph except the first. Functions `moran.mc()` and `geary.mc()` implement the second method, while `moran.test()` and `geary.test()` implement the other two.

As an example, Table 3.3 gives Moran's I and Geary's c for the harbor seal trends data using a spatial weights matrix corresponding to each of the schemes depicted in the neighbor graphs of Figure 1.4. P-values for each statistic, obtained via the second method (with 10,000 permutations) and third method are also given. Both autocorrelation statistics

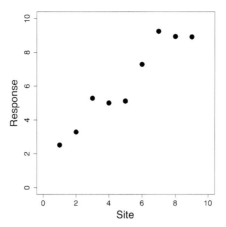

FIGURE 3.13
A plot of responses simulated according to model (3.3) on a regular transect of nine sites.

indicate significant evidence of spatial correlation for each neighbor scheme, with Geary's statistic and the scheme of order 1, 2, and 4 being the most highly significant.

Which statistic, I or c, is to be preferred? Simulation experiments conducted by Cliff & Ord (1973, 1975) indicated that I is slightly more powerful than c for detecting spatial correlation, but more recent studies argue for a more qualified assessment. Although Luo *et al.* (2017) showed analytically that the asymptotic variance of I is more reliable than that of c over a range of sample sizes, distributions, and neighborhood definitions, they also showed that when the joint distribution of the observations is normal, which test is more powerful depends on both the neighborhood definition and the sign of the spatial correlation. In particular, I was generally more powerful for detecting positive spatial correlation and c was sometimes more powerful for detecting negative spatial correlation.

Although an extreme Moran's I or Geary's c indicates that the observed responses are spatially correlated, it does not necessarily imply that there is spatial correlation in the underlying random mechanism that gave rise to the responses; we made a similar point earlier in regard to a nonconstant sample semivariogram. The proper interpretation of an extreme autocorrelation statistic is that some aspect of the assumption of independent and identically distributed responses is dubious. Thus, rather than being spatially correlated, the underlying random variables could be non-identically distributed; for example, they could have different means. To illustrate, consider the data displayed in Figure 3.13, which were generated by the following mechanism:

$$y_i = 1 + i + e_i \quad (i = 1, \dots, 9). \tag{3.3}$$

Here, the e_i's are independent and identically distributed standard normal variables. This is merely a simple linear regression model, with intercept and slope both equal to 1, except that here we take i to represent the "x-coordinate" of the centroid of a site along a one-dimensional spatial transect of nine contiguous, equally spaced sites. With neighbors defined as adjacent sites, $I = 0.75$ and $c = 0.13$ for these data, which are both highly statistically significant (in fact $P < 0.005$) regardless of which of the aforementioned methods for assessing extremeness is used. But in this case, the positive sample spatial correlation, i.e., the tendency for high values of the observed response to be proximate to other high values of the observed response and vice versa, is due entirely to a spatially structured mean (in this case a linear trend) rather than to spatial correlation among the errors of the model.

The upshot of this example for the spatial linear modeler is that the detection of spatial correlation in the data (or, later, in the fitted residuals) does not necessarily mean that a model should be adopted that has spatially correlated errors; an alternative may be to include the sites' spatial coordinates (or functions thereof) as explanatory variables in the model's mean structure.

3.7.3   Correlograms

Moran's and Geary's indices of autocorrelation reduce all the information about the spatial correlation to a single number. An alternative, more refined approach is to divide the range of spatial weights into q classes, and then compute the index using only those pairs of observations whose weights fall into a given class. The class-specific indices, say I_w or c_w, where w indexes the class, are then plotted against the class index. Such a plot is called a **correlogram** and may be regarded as an areal-data analog of the sample acf (in the case of I_w) or the sample semivariogram (in the case of c_w). This is especially so if proximity between areal sites is measured by the actual physical distance between representative points (e.g. centroids) within the sites. If proximity is measured in the classical way, by adjacencies, then the situation is a bit more complicated, so in order to understand correlograms in this case let us take a deeper look at spatial neighborhoods.

First consider time series models, where the neighborhood definitions depend on the notion of a "lag" in time. In time series where data are indexed by the integers, $i = 1, \ldots, n$, "lag-1" refers to only the immediate neighbor, so we compute autocorrelation only between data indexed by pairs $(i, i+1)$, for $i = 1, \ldots, n-1$. The "lag-2" autocorrelation is computed between observations indexed by pairs $(i, i+2)$ for $i = 1, \ldots, n-2$, etc., continuing up to some maximum lag of interest.

This idea generalizes to spatial data based on neighborhoods defined by adjacency. Lag-1 is the basic neighborhood definition, i.e., adjacent areal sites. Lag-2 is a neighbor of a neighbor, exclusive of the lag-1 neighbors and higher order lags. Lag-3 is a neighbor of a neighbor of a neighbor, exclusive of lag-1, lag-2, and higher order lags, etc. For the harbor seal trends data, Figure 3.14 shows the same close-up of the "polygons" (data sites) that was shown in Figure 1.4B. Here, the white polygons are missing estimated trend values, while the grey polygons have estimates. We need values of the response to compute autocorrelation, so we compute neighborhoods among only those polygons with estimated trends. The centroid of each polygon is shown by the solid black circle, and their neighbors are shown by the light green lines. These are the same neighbors that were shown in Figure 1.4A, but only for polygons with estimated trend values.

If we visualize Figure 3.14 without the polygons, then we have what is known as a "graph" in mathematics, where the black circles are termed *nodes* and the light green lines are *edges*, which connect the nodes. For our purposes, a lag-2 neighbor is one that can be connected most directly by two edges, which are the yellow lines in Figure 3.14. A lag-3 neighbor is one that can be connected most directly by three edges, which are the lavender lines in Figure 3.14. Higher order lags are similarly defined. Each node has a number of lag-1 neighbors, lag-2 neighbors, etc, and they can vary from node to node. For lag-1, all nodes have at least one neighbor, but it is possible for a node to have no higher-order neighbors.

A correlogram computes Moran's I (or Geary's c, or simple correlation) by lag. Note that polygons with more connections of a certain lag will be used more in the computation of Moran's I at that lag. For the harbor seal data, the correlogram using Moran's I is given in Figure 3.15. The expected values under the null hypothesis of no autocorrelation are near zero. Approximate confidence intervals around the null hypothesis are shown, and there is evidence of positive spatial correlation, especially where the estimated Moran's I is outside of the confidence intervals for lags 3 and 4. This correlogram was computed using the `spdep`

FIGURE 3.14
Neighbor lags for a portion of the harbor seal study area. Polygons with missing data are shown as white, while those with estimated trends are shown as grey. Connections are shown only between polygons with observed data. Neighbors as originally defined are connected by light green lines. Second-order, or lag-2 neighbors, are connected by yellow lines. Third-order, or lag-3 neighbors, are connected by lavender lines.

package in R with simple binary weights (a 1 indicating a neighbor, and 0 otherwise). Other weighting schemes are discussed when we introduce statistical models for these data.

3.7.4   Local indicators of spatial association

The correlogram can be regarded as a decomposition of an index of autocorrelation into components corresponding to neighbor-lag classes. An alternative decomposition of an index of autocorrelation is into components corresponding to data sites, yielding what Anselin (1995) called a **local indicator of spatial association (LISA)**. The ith Moran-I LISA is

$$I_i = \frac{n}{\sum_{j=1}^{n}(y_j - \overline{y})^2}(y_i - \overline{y})\sum_{j=1}^{n} w_{ij}(y_j - \overline{y}) \qquad (i = 1, \ldots, n),$$

for which the sum over i is proportional to I; specifically, $\sum_{i=1}^{n} I_i = w_{..}I$, as is easily verified. The expected value of I_i is $-\sum_{j=1}^{n} w_{ij}/(n-1)$, and as was the case for I, the sign of the difference, $I_i - \mathrm{E}(I_i)$, is indicative of the sign of the spatial correlation. That is, an I_i larger (smaller) than $\mathrm{E}(I_i)$ indicates positive (negative) spatial correlation at A_i. The ith Geary-c LISA is

$$c_i = \frac{1}{2}\sum_{j=1}^{n} w_{ij}(y_i - y_j)^2 \qquad (i = 1, \ldots, n),$$

the sum of which is proportional to c.

A version of the nearest-neighbor scatterplot introduced earlier in this chapter that is particularly relevant to the Moran-I LISAs is the **Moran scatterplot** (Anselin, 1996). In

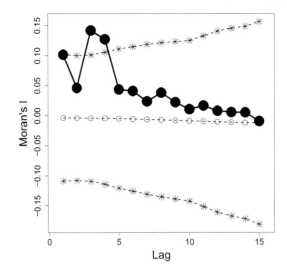

FIGURE 3.15
Correlogram using Moran's I for the harbor seal trends data. The expected value, under the null hypothesis of no autocorrelation, is given by the open circles connected by the dashed line. The 95% confidence interval for the null hypothesis at each lag, using a normal distribution approximation, is given by the star symbols connected by the dotted line. The computed Moran's I is given by solid black circles connected by a solid line.

this plot, the responses are centered (by subtracting their average) and the spatial weights matrix is used in the calculation of the ordinates; specifically, the points plotted are $\{(y_i - \overline{y}, \sum_{j=1}^{n} w_{ij}(y_j - \overline{y})) : i = 1, \ldots, n\}$. It turns out (see Exercise 3.13) that the slope of the least squares regression line fitted to the points in this plot is a known multiple of I; in fact, the multiple is 1 if the spatial weights matrix is row-standardized. Thus, the Moran scatterplot is a visualization of the Moran-I LISAs and, when augmented with the least squares line, may be used like any bivariate scatterplot to assess goodness of fit and to identify possible outliers and points with high leverage.

LISAs may also be plotted *in situ*, using a choropleth map, to visualize the extent to which the spatial association of areal data varies over the study area. If a LISA shows little spatial variation, then the underlying random process is relatively stable (stationary); if not, then those LISA values that are extreme may serve to identify which sites are outliers or which localized groups of sites are pockets of nonstationarity. In this sense the plot is an areal-data analogue of the pocket plot for geostatistical data.

To assess the extremeness of a LISA, one possibility is a conditional randomization approach. This approach differs from the aforementioned (unconditional) randomization approach for assessing the extremeness of the global spatial autocorrelation statistic only by "freezing" y_i at A_i; the remaining $n - 1$ responses are randomly permuted to the remaining sites a large number of times. This process is repeated for each site. In this case, however, exact determination of a P-value is not feasible, due to the multiplicity of tests (one for each site) and the dependence among them (due to the overlap of neighborhoods). An alternative approach, which allows for correction for multiplicity, is based on a normal approximation; see Bivand & Wong (2018).

Figure 3.16 is the Moran scatterplot, and Figure 3.17 is a choropleth map of the I_i's, for the harbor seal trends data. Both figures were created using the `spdep` package with row-standardized weights, where Figure 3.16 was created using the `moran.plot` function

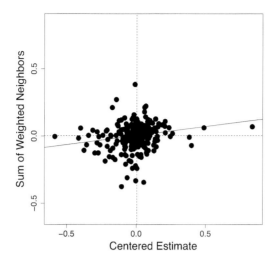

FIGURE 3.16
Moran scatterplot for the harbor seal trends data. The solid line is the least squares regression line through the points in the plot.

and LISAs for Figure 3.17 were computed using the `localmoran` function. The Moran scatterplot, with its corresponding least squares regression line, has most of its points in the first and third quadrants, and none appear to be individually responsible for the significant positive spatial autocorrelation reported previously in Table 3.3. In fact, the most extreme

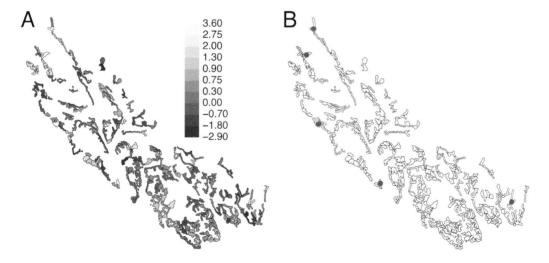

FIGURE 3.17
LISAs for exploratory data analysis of harbor seal trends. A) Choropleth map of standardized Moran LISAs. B) The green circle shows the potential outlier from Figure 3.16 with a high centered estimate. The other circles show standardized LISAs that have a P-value that is smaller than 0.01. Additionally, the blue circle shows the site with the only P-value smaller than 0.20 when using a Bonferroni correction, but none are significant at the 0.05 level.

centered values along the horizontal axis (at about -0.6 and 0.85) actually serve to reduce *I* from what it would be otherwise. In Figure 3.17, the displayed LISAs were standardized (by subtracting their expectation under the null hypothesis of no spatial correlation and dividing the result by the standard error), and colors (Figure 3.17A) show sites with high and low values relative to their neighbors. Tests of significance were carried out using the aforementioned normal approximation. Using a Bonferroni correction and a two-sided test of size 0.05, no LISAs were found to be significantly different from zero. Using a two-sided test of size 0.10 or 0.20, the datum at the single site colored blue near the top of Figure 3.17B might be considered an outlier. Going site by site, and testing at level 0.01, the data at sites colored red, in addition to the datum at the blue site, might be considered outliers. The single site with a high positive value in Figure 3.16, where the centered estimate was far to the right, was colored green in Figure 3.17. Considered together, the high value in Figure 3.16 was adjacent to one of the sites identified from the standardized LISA values using a level 0.01 test, which makes sense because it caused the average of the neighbors to be quite different from the observed value. Given the information in Figures 3.16 and 3.17, henceforth we will regard the datum at the single site identified with the large positive value in Figure 3.16, near 0.85 on the abscissa, as a potential outlier.

3.8   Exercises

1. Obtain bubble plots for lead (Pb), zinc (Zn), and cadmium (Cd) using the moss heavy metals data, with one overall plot for each element that includes the Cape Krusenstern National Preserve boundary, and one plot zoomed into the area around the road. State your findings. Do lead, zinc, and cadmium have the same pattern?

2. Suppose that five lags are used to obtain the omnidirectional sample semivariogram for a certain dataset. Suppose further that the five omnidirectional sample semivariances are as follows: $\hat{\gamma}(1) = 1$, $\hat{\gamma}(2) = 2$, $\hat{\gamma}(3) = \hat{\gamma}(4) = \hat{\gamma}(5) = 3$. Suppose that an additional observation is taken which is a spatial outlier, but not a distributional outlier. How would the nugget effect, range, and sill of the sample semivariogram of the dataset with the additional observation compare to those quantities for the sample semivariogram of the original dataset.

3. Show, using the following "data" on a 2×3 grid with unit spacing between grid points, that $\hat{\gamma}(r_k)$ need not equal $\hat{C}(0) - \hat{C}(r_k)$.

$$1 \quad 2 \quad 4$$

$$2 \quad 4 \quad 5$$

4. Consider a geostatical process $\{Y(s) : s \in \mathbb{R}\}$ that is observed at n regularly spaced locations $s_i = i$ $(i = 1, \dots, n)$. For such a process, consider an $n \times n$ matrix $\mathbf{H}$ whose (i, j)th element is $(y_i - \overline{y})(y_j - \overline{y})$ for $1 \leq i, j \leq n$, i.e.,

$$\mathbf{H} = \begin{pmatrix} (y_1 - \overline{y})(y_1 - \overline{y}) & (y_1 - \overline{y})(y_2 - \overline{y}) & \cdots & (y_1 - \overline{y})(y_n - \overline{y}) \\ (y_2 - \overline{y})(y_1 - \overline{y}) & (y_2 - \overline{y})(y_2 - \overline{y}) & \cdots & (y_2 - \overline{y})(y_n - \overline{y}) \\ \vdots & \vdots & \ddots & \vdots \\ (y_n - \overline{y})(y_1 - \overline{y}) & (y_n - \overline{y})(y_2 - \overline{y}) & \cdots & (y_n - \overline{y})(y_n - \overline{y}) \end{pmatrix}.$$

By carrying out the following steps, show that $\hat{C}(r) < 0$ for at least one h, even if $C(r) > 0$ for all $r = 1, \ldots, n-1$.

(a) Show that the sum of the elements on the main diagonal of $\mathbf{H}$ is equal to $n\hat{C}(0)$.

(b) Show that the sum of the elements on the hth superdiagonal (the rth diagonal of elements above and to the right of the main diagonal) of $\mathbf{H}$ is equal to $(n-r)\hat{C}(r)$.

(c) Show that the sum of the elements in each row of $\mathbf{H}$ is equal to 0, and therefore the sum of all the elements in $\mathbf{H}$ is equal to zero.

(d) Use parts (a), (b), and (c) to argue that $\hat{C}(r) < 0$ for at least one $r \in \{1, \ldots, n-1\}$.

5. Let $y_i = i$ for $i = 1, \ldots, 50$.

(a) Plot the sample autocovariance function $\hat{C}(r)$, $r = 0, 1, \ldots, 49$.

(b) For the smallest value of r for which $\hat{C}(r)$ is negative, plot y_{i+r} versus y_i $(i = 1, \ldots, 50 - r)$. In light of this value being negative, does anything about this plot surprise you?

(c) Plot the sample semivariogram $\hat{\gamma}(r)$ for $r = 0, 1, \ldots, 49$. How, if at all, does the sample semivariogram for these artificial data differ from an "ideal" sample semivariogram?

6. Consider a one-dimensional geostatistical process $\{Y(s): s \in \mathbb{R}\}$ observed at ten equally-spaced points $s = 1, 2, \ldots, 10$. Suppose that $Y(s) = \beta_0 + \beta_1 s + e(s)$, where $\beta_1 \neq 0$ and $\{e(s): s \in \mathbb{R}\}$ is a zero-mean, intrinsically stationary process with semivariogram $\gamma(r)$. [Note that $\{Y(s): s \in \mathbb{R}\}$ is not second-order or intrinsically stationary.] Determine the bias of the sample semivariogram (which is computed as if $\{Y(s): s \in \mathbb{R}\}$ was intrinsically stationary).

7. Data from a 4×4 square grid of sites was as follows:

4	5	6	7
3	4	5.5	6
2	2.5	4	5
1	2	3	4

(a) Compute the directional sample semivariograms in the N-S and E-W directions, and superimpose them on one plot. Does this plot provide any evidence for anisotropy?

(b) Now add the directional sample semivariograms in the NW-SE and NE-SW directions to the plot. Does the plot provide evidence of anisotropy now? If so, is there evidence that the anisotropy is non-geometric?

8. Consider a geostatistical process in $\mathbb{R}^2$ that is intrinsically stationary and has a semivariogram with range 0.5 units. Suppose further that the process is observed at data locations which form a square grid with grid spacing 1.0 units.

(a) Apart from chance fluctuations (noise), what would the sample semivariogram tend to look like? Sketch a picture.

(b) What does your answer to part (a) tell you about an appropriate sampling design for estimating a semivariogram?

9. Repeat the simulation study that generated the marginal distributions of sample semivariances displayed in Figure 3.9, but this time compute and plot the omni-directional sample autocovariance function. How do the variances of the sample autocovariances depend on lag, and are there any notable differences between that dependence and the dependence of the variances of the sample semivariances on lag?

10. Compute the sample correlation matrix of the sample semivariances whose marginal distributions are plotted in Figure 3.9. Similarly, compute the sample correlation matrix of the sample autocovariances obtained in the previous problem. Which correlations, if any, are stronger than others, and is the behavior of the correlations between the sample semivariances different than that of the correlations between the sample autocovariances?

11. We often act as though the spatial locations of geostatistical data are measured without error. In reality, the location we are recording as (u, v) is usually $(u + \delta_u, v + \delta_v)$; that is, there is positional error in our recorded data locations. By carrying out the following steps, investigate, via simulation, the effect of positional error on the quality of the sample semivariogram as an estimate of the true semivariogram.

Step 1: Simulate a second-order stationary Gaussian process, with mean 0 and a nuggetless isotropic exponential covariance function on a 20×20 square grid with unit spacing. Take the sill to be 1.0 and the range argument to be 8.0.

Step 2: Obtain (and save to a vector) the isotropic sample semivariogram at 14 equally-spaced lags with unit spacing.

Step 3: Perturb or "jitter" the data locations by adding iid Uniform$(-1, 1)$ random variables to each of their Euclidean coordinates, and repeat Step 2.

Step 4: Repeat Steps 1-3 1000 times, saving the sample semivariograms based on the original data locations and the jittered data locations from each replication.

Step 5: Repeat Steps 1-4 except that this time take the random variables in Step 3 to be iid Uniform(-2,2).

Step 6: Construct side-by-side boxplots of the sample semivariogram at the default lags, based on the original data locations. Do the same for the sample semivariogram based on the jittered data locations.

Compare the three side-by-side boxplots, and describe the effect of positional error on the quality of the sample semivariogram as an estimate of the true semivariogram. Be as specific as possible.

12. Proceed through the following steps to show that $E(I) = -1/(n-1)$ and $E(c) = 1$ under the assumption that the areal observations are independent and identically distributed with finite second moments.

(a) Let $Z_i = Y_i - \overline{Y}$. Observe that conditional on the observed values $y_1, y_2, \ldots, y_n$,

$$P(z_i \text{ is assigned to } A_k) = 1/n \quad \text{for all } i \text{ and } k,$$

where $z_i = y_i - \overline{y}$. Use this to show that $E(Z_i | y_1, \ldots, y_n) = 0$, where the expectation is taken over the $n!$ random permutations of $y_1, \ldots, y_n$ to $A_1, \ldots, A_n$.

(b) In a similar fashion, show that $E(Z_i^2 | y_1, \ldots, y_n) = (1/n) \sum_{i=1}^n z_i^2 \equiv m_2$.

(c) Similarly once again, show that $E(Z_i Z_j | y_1, \ldots, y_n) = -m_2/(n-1)$ for $i \neq j$.

(d) Complete the task for Moran's I using part (c), the representation of I as $\frac{n \sum_{i=1}^{n} \sum_{j=1}^{n} w_{ij} z_i z_j}{w_{..} m_2}$, and a justification for the equality of the conditional and unconditional expectations.

(e) Apply the same general approach to Geary's c.

13. Show that when the spatial weights matrix is row-standardized, the slope of the least squares regression line in a Moran scatterplot is equal to I.

14. Using the `nb2listw` function in `spdep`, verify computationally that the slope of the regression line obtained from the Moran scatterplot for the harbor seal trends data displayed in Figure 3.16 coincides with Moran's I (when spatial weights are row-standardized).

15. The objective of this exercise is to explore, via simulation, how the weighted least squares estimator of the semivariogram is affected by using different upper distance bounds and different partitions of the lag space to obtain the sample semivariogram. For this purpose, carry out the following sequence of steps.

Step 1: Simulate a second-order stationary Gaussian process, with mean 0 and a nuggetless isotropic exponential covariance function, on a 20×20 square grid with unit spacing. Take the sill to be 1.0 and the range to be either 3.0 or 6.0.

Step 2: Obtain the omnidirectional sample semivariogram (assuming a constant mean), taking the maximum distance to be either 10 or 25, and using ℓ equally-spaced lag classes, where ℓ is either 5 or 15.

Step 3: From the omnidirectional sample semivariogram obtained in Step 2, obtain the weighted least squares estimators of the nugget effect, sill, and range parameters, and save them to a file.

Step 4: Repeat Steps 1, 2, and 3 10000 times, each time appending the new weighted least squares estimates to the file.

Step 5: Compute the empirical bias, variance, and mean squared error of the weighted least squares estimates and display them in a table, for each of the 8 combinations of range parameter, upper distance bound, and number of bins.

From the results in the table, describe how the upper distance bound, number of bins, and strength of spatial correlation (as characterized by the range) affect the sampling distribution of the WLS estimates.

16. The following are soil surface pH data taken on an 11×11 square grid with 10-meter spacing, reported by Laslett *et al.* (1987).

4.800	4.380	4.330	4.310	4.490	4.380	4.440	4.455	4.540	4.500	4.235
4.420	4.290	4.190	4.280	4.585	4.890	4.740	4.675	4.545	4.865	4.330
4.300	4.295	4.870	4.700	4.675	5.045	5.025	4.860	4.430	4.135	4.325
4.260	4.635	4.540	4.540	4.640	4.755	4.420	4.605	4.300	4.535	4.295
4.205	4.420	4.495	4.795	4.760	4.525	4.230	4.255	4.580	4.120
4.185	4.400	4.320	4.485	4.590	4.670	4.500	4.800	4.285	4.500	4.435
4.345	4.535	4.520	4.730	4.320	4.900	4.340	4.360	4.235	4.310	4.300
4.535	4.205	4.445	4.600	4.840	4.455	4.395	4.355	4.270	4.025	4.370
4.560	4.645	4.645	4.645	4.690	4.360	4.365	4.625	4.305	4.340	4.475
4.440	4.830	4.800	4.840	4.505	4.305	4.290	4.490	4.210	4.155	4.645
4.525	4.395	4.740	4.700	4.355	4.515	4.345	4.440	4.305	4.295	4.155

Perform a thorough exploratory spatial data analysis of these data.

17. Heavy metal concentration data in soil at 155 sites in a flood plain of the Meuse River near Stein, Netherlands are saved in the `meuse` dataset in the `gstat` package of `R`. Perform a thorough exploratory spatial data analysis of the data on the following metals:

(a) cadmium
(b) copper
(c) lead
(d) zinc

4

Provisional Estimation of the Mean Structure by Ordinary Least Squares

Often, when a scatterplot of spatial data or a derived plot such as a choropleth map or bubble plot is examined, the responses appear as if they could have arisen from adding random noise to a smooth trend. In a spatial linear model, the trend is modeled by a parametric mean function that is linear in its parameters. This chapter considers the estimation of parametric mean functions for spatial linear models by the classical method of ordinary least squares (OLS). Ordinary least squares estimators are linear combinations of the observed responses and are unbiased for the unknown parameters, but generally are sub-optimal for spatial linear models in the sense that they do not minimize the variance among all linear unbiased estimators. Improving upon the OLS estimators requires, however, that the covariance structure be known, which it is not, in practice. Ordinary least squares estimation, which does not require a covariance structure to be specified, is therefore a practical alternative. Although OLS estimators generally are sub-optimal, their unbiasedness makes it possible, using procedures described in this chapter, to formulate a suitable parametric covariance structure for the model errors from the residuals from the OLS fit. The covariance structure chosen on the basis of this preliminary step may be used in subsequent re-estimation of the trend by a better method, such as empirical generalized least squares (to be described in Chapter 5) or (what is essentially the same) one of the likelihood-based methods to be presented in Chapter 8.

We begin this chapter by presenting some theory and methodology associated with OLS. Because it is assumed that the reader has already had some training in regression analysis including ordinary least squares estimation, this presentation is rather brief and focuses on those aspects that either are useful for preliminary model formulation or have important extensions to generalized least squares estimation. Then, we examine some properties of OLS estimators and OLS-based inferential methods under a spatial linear model. Next, we describe some tests for the existence and properties of spatial correlation among spatial linear model errors using the residuals from the OLS fit. Following that, we describe the two most commonly used mean structures in spatial linear models: polynomials and additive coordinate effects. The final section consists of examples.

4.1 Some classical OLS theory and methodology

Least squares estimators of the coefficients of a linear model are so named because they minimize a (generalized) residual sum of squares criterion. In the case of **ordinary least squares (OLS)**, the criterion is merely the sum of squares of the residual vector $\mathbf{y} - \mathbf{X}\boldsymbol{\beta}$, regarded as a function of $\boldsymbol{\beta}$, i.e.,

$$RSS(\boldsymbol{\beta}; \mathbf{y}) = (\mathbf{y} - \mathbf{X}\boldsymbol{\beta})^T (\mathbf{y} - \mathbf{X}\boldsymbol{\beta}).$$

DOI: 10.1201/9780429060878-4

It turns out [see, e.g., Harville (2018)] that a p-vector $\hat{\boldsymbol{\beta}}_{OLS}$ is a minimizer of $RSS(\boldsymbol{\beta}; \mathbf{y})$ if and only if it is a solution to the **normal equations**

$$\mathbf{X}^T\mathbf{X}\boldsymbol{\beta} - \mathbf{X}^T\mathbf{y}.$$

If $\mathbf{X}$ has full column rank p (equivalently, if the p columns of $\mathbf{X}$ are linearly independent), as we assume henceforth, then $\mathbf{X}^T\mathbf{X}$ is nonsingular. Thus, the unique solution to the normal equations, and hence the unique OLS estimator of $\boldsymbol{\beta}$, is

$$\hat{\boldsymbol{\beta}}_{OLS} = (\mathbf{X}^T\mathbf{X})^{-1}\mathbf{X}^T\mathbf{y}.$$

Furthermore, the OLS estimator of an arbitrary linear function of $\boldsymbol{\beta}$, $\boldsymbol{\ell}^T\boldsymbol{\beta}$, is $\boldsymbol{\ell}^T\hat{\boldsymbol{\beta}}_{OLS}$. Consideration of the special case in which $\boldsymbol{\ell}$ is the ith row $\mathbf{x}_i^T$ of $\mathbf{X}$ (i.e., the vector of explanatory variables observed at site S_i) yields the **OLS fitted values** $\hat{y}_{i,OLS} = \mathbf{x}_i^T\hat{\boldsymbol{\beta}}_{OLS}$ $(i = 1, \ldots, n)$. Consideration of another special case in which $\boldsymbol{\ell} = \mathbf{x}$ is an arbitrary combination of the explanatory variables yields the OLS predicted value for that combination, $\hat{y}(\mathbf{x}) = \mathbf{x}^T\hat{\boldsymbol{\beta}}_{OLS}$. And consideration of the further special case where all of the explanatory variables are spatial coordinates or functions thereof, i.e., where $\mathbf{x}(\mathbf{s}) = [f_1(\mathbf{s}), f_2(\mathbf{s}), \ldots, f_p(\mathbf{s})]^T$ for some functions $f_1, \ldots, f_p$, and writing $\hat{y}(\mathbf{x}(\mathbf{s}))$ in this case as $\hat{y}(\mathbf{s})$, we may obtain the (infinite) collection of predicted values $\{\hat{y}_{OLS}(\mathbf{s}) : \mathbf{s} \in \mathcal{A}\}$, which is called the **fitted OLS surface** over the study area. Creating this surface using least squares estimation is one of several types of **spatial smoothing** to be described in this chapter and subsequently. This type uses the assumed functional form of the mean to perform the smoothing.

For future reference, let $\mathbf{P_X} = \mathbf{X}(\mathbf{X}^T\mathbf{X})^{-1}\mathbf{X}^T$. $\mathbf{P_X}$ is sometimes called the **hat matrix** because it "puts a hat" on $\mathbf{y}$, in the sense that pre-multiplication of $\mathbf{y}$ by $\mathbf{P_X}$ yields the vector of OLS fitted values, $\hat{\mathbf{y}}_{OLS} = \mathbf{X}\hat{\boldsymbol{\beta}}_{OLS}$. $\mathbf{P_X}$ is also referred to as the **orthogonal projection matrix onto the column space of X**, as pre-multiplication of $\mathbf{y}$ by $\mathbf{P_X}$ also "projects" (moves) $\mathbf{y}$ (which lies in $\mathbb{R}^n$) to the column space of $\mathbf{X}$ (which is a p-dimensional subspace of $\mathbb{R}^n$) in a direction orthogonal to that column space.

Because OLS estimators are defined merely in terms of minimizing a numerical criterion, they are well-defined under a linear model with any covariance structure. For the remainder of this section, however, let us consider their properties under a Gauss-Markov model, i.e., a model with covariance matrix $\mathrm{Var}(\mathbf{y}) = \sigma^2\mathbf{I}$ for some $\sigma^2 > 0$. Using (2.27), we find that

$$\mathrm{E}(\hat{\boldsymbol{\beta}}_{OLS}) = \mathrm{E}[(\mathbf{X}^T\mathbf{X})^{-1}\mathbf{X}^T\mathbf{y}] = (\mathbf{X}^T\mathbf{X})^{-1}\mathbf{X}^T\mathrm{E}(\mathbf{y}) = (\mathbf{X}^T\mathbf{X})^{-1}\mathbf{X}^T\mathbf{X}\boldsymbol{\beta} = \boldsymbol{\beta}, \qquad (4.1)$$

i.e., $\hat{\boldsymbol{\beta}}_{OLS}$ is unbiased. Furthermore, by (2.28),

$$\begin{aligned} \mathrm{Var}(\hat{\boldsymbol{\beta}}_{OLS}) &= \mathrm{Var}[(\mathbf{X}^T\mathbf{X})^{-1}\mathbf{X}^T\mathbf{y}] = (\mathbf{X}^T\mathbf{X})^{-1}\mathbf{X}^T\mathrm{Var}(\mathbf{y})[(\mathbf{X}^T\mathbf{X})^{-1}\mathbf{X}^T]^T \\ &= (\mathbf{X}^T\mathbf{X})^{-1}\mathbf{X}^T(\sigma^2\mathbf{I})\mathbf{X}(\mathbf{X}^T\mathbf{X})^{-1} = \sigma^2(\mathbf{X}^T\mathbf{X})^{-1}. \end{aligned}$$

From these results, it follows easily [again by (2.27) and (2.28)] that the OLS estimator, $\boldsymbol{\ell}^T\hat{\boldsymbol{\beta}}_{OLS}$, of any linear function $\boldsymbol{\ell}^T\boldsymbol{\beta}$ is unbiased and has variance $\sigma^2\boldsymbol{\ell}^T(\mathbf{X}^T\mathbf{X})^{-1}\boldsymbol{\ell}$.

The OLS estimator of $\boldsymbol{\ell}^T\boldsymbol{\beta}$ is the best (minimum variance) linear unbiased estimator (BLUE) under the assumed Gauss-Markov model. Furthermore, if $n > p$, as we assume henceforth, then σ^2 may be estimated by the **residual mean square**,

$$\hat{\sigma}^2_{OLS} = (\mathbf{y} - \mathbf{X}\hat{\boldsymbol{\beta}}_{OLS})^T(\mathbf{y} - \mathbf{X}\hat{\boldsymbol{\beta}}_{OLS})/(n - p),$$

which is the best quadratic (i.e., a quadratic function of the observations) unbiased estimator. An alternative representation of $\hat{\sigma}^2_{OLS}$ is $\mathbf{y}^T(\mathbf{I} - \mathbf{P_X})\mathbf{y}/(n - p)$. The variance of $\hat{\boldsymbol{\beta}}_{OLS}$ may then be estimated unbiasedly by $\hat{\sigma}^2_{OLS}(\mathbf{X}^T\mathbf{X})^{-1}$. If the model errors are normally distributed, these optimality results may be strengthened by removing the restrictions to the class of linear (for $\boldsymbol{\ell}^T\boldsymbol{\beta}$) or quadratic (for σ^2) estimators. Moreover, under normality, $\boldsymbol{\ell}^T\hat{\boldsymbol{\beta}}_{OLS}$ is normally distributed with mean $\boldsymbol{\ell}^T\boldsymbol{\beta}$ and variance $\sigma^2\boldsymbol{\ell}^T(\mathbf{X}^T\mathbf{X})^{-1}\boldsymbol{\ell}$, $(n-p)\hat{\sigma}^2_{OLS}/\sigma^2$ has a chi-square distribution with $n-p$ degrees of freedom, and $\boldsymbol{\ell}^T\hat{\boldsymbol{\beta}}_{OLS}$ and $\hat{\sigma}^2_{OLS}$ are independent.

Of primary interest in most scientific work using linear models is the characterization of how the response depends on the explanatory variables. **Partial regression plots** address this question graphically by revealing how the response depends on each explanatory variable *after adjusting for all the others*. The partial regression plot for the jth explanatory variable, x_j, is constructed by plotting the residuals from the OLS regression of the response on all explanatory variables except x_j against the residuals from the OLS regression of x_j on all the other explanatory variables. The OLS estimate of the slope in the plot is equal to the OLS estimate of the regression coefficient corresponding to x_j in the full model (i.e., the model that includes all the observed explanatory variables). Furthermore, the plot will show whether the dependence of the response on x_j, adjusted for the remaining variables, is linear or nonlinear, positive or negative, and weak or strong; it can also reveal outliers or influential sites with respect to that dependence. The formal inferential equivalent to the partial regression plot for x_j is the **adjusted t-test** of the null hypothesis $H_0 : \beta_j = 0$ versus the alternative hypothesis $H_a : \beta_j \neq 0$. The t test statistic is

$$t_{\beta_j} = \frac{\hat{\beta}_{j,OLS}}{\hat{\sigma}_{OLS}\sqrt{[(\mathbf{X}^T\mathbf{X})^{-1}]_{jj}}}. \tag{4.2}$$

If the model errors are normally distributed, then by the aforementioned normality of $\boldsymbol{\ell}^T\hat{\boldsymbol{\beta}}_{OLS}$ (and hence $\hat{\beta}_{j,OLS}$), the chi-squaredness of $(n-p)\hat{\sigma}^2_{OLS}/\sigma^2$, and the independence of these two quantities, t_{β_j} has Student's t distribution with $n-p$ degrees of freedom. Therefore, the absolute value of the computed t_{β_j} may be compared to $t_{\epsilon/2,n-p}$ (the $100[1-(\epsilon/2)]$th percentile of Student's t distribution with $n-p$ degrees of freedom, where $0 < \epsilon < 1$) to assess the strength of evidence against H_0. Furthermore, the interval whose endpoints are

$$\hat{\beta}_{j,OLS} \pm t_{\epsilon/2,n-p}\hat{\sigma}_{OLS}\sqrt{[(\mathbf{X}^T\mathbf{X})^{-1}]_{jj}}$$

is a $100(1-\epsilon)\%$ confidence interval for β_j. More generally, letting ℓ_0 be any constant, the null hypothesis $H_0 : \boldsymbol{\ell}^T\boldsymbol{\beta} = \ell_0$ may be tested against the alternative hypothesis $H_a : \boldsymbol{\ell}^T\boldsymbol{\beta} \neq \ell_0$ by comparing the test statistic

$$t_{\boldsymbol{\ell}^T\boldsymbol{\beta}} = \frac{\boldsymbol{\ell}^T\hat{\boldsymbol{\beta}}_{OLS} - \ell_0}{\hat{\sigma}_{OLS}\sqrt{\boldsymbol{\ell}^T(\mathbf{X}^T\mathbf{X})^{-1}\boldsymbol{\ell}}}$$

to $t_{\epsilon/2,n-p}$. And, the interval whose endpoints are

$$\boldsymbol{\ell}^T\hat{\boldsymbol{\beta}}_{OLS} \pm t_{\epsilon/2,n-p}\hat{\sigma}_{OLS}\sqrt{\boldsymbol{\ell}^T(\mathbf{X}^T\mathbf{X})^{-1}\boldsymbol{\ell}}$$

is a $100(1-\epsilon)\%$ confidence interval for $\boldsymbol{\ell}^T\boldsymbol{\beta}$.

For some types of inference on the regression parameters, simultaneously testing whether two or more of the parameters are equal to zero is desired rather than individually testing whether each is zero. For example, one might wish to test simultaneously whether all three second-order terms in a quadratic surface over a planar region are zero, or equivalently whether a planar surface is adequate to describe the observations. To test the joint linear hypothesis $H_0 : \mathbf{L}^T\boldsymbol{\beta} = \boldsymbol{\ell}_0$ against $H_a : \mathbf{L}^T\boldsymbol{\beta} \neq \boldsymbol{\ell}_0$, where $\mathbf{L}^T$ is a specified $q \times p$ matrix with linearly independent rows, the test statistic

$$F = \frac{(\mathbf{L}^T\hat{\boldsymbol{\beta}}_{OLS} - \boldsymbol{\ell}_0)^T[\mathbf{L}^T(\mathbf{X}^T\mathbf{X})^{-1}\mathbf{L}]^{-1}(\mathbf{L}^T\hat{\boldsymbol{\beta}}_{OLS} - \boldsymbol{\ell}_0)}{q\hat{\sigma}^2_{OLS}} \tag{4.3}$$

may be compared to the upper ϵ percentage point of an F distribution with q and $n - p$ degrees of freedom. In the special case in which the null hypothesis specifies that q of the elements of $\boldsymbol{\beta}$ are equal to zero, this test statistic may be written as

$$F = \frac{[RSS_{\mathrm{reduced}}(\hat{\boldsymbol{\beta}}_{OLS}; \mathbf{y}) - RSS_{\mathrm{full}}(\hat{\boldsymbol{\beta}}_{OLS}; \mathbf{y})]/[df_{\mathrm{reduced}} - df_{\mathrm{full}}]}{RSS_{\mathrm{full}}(\hat{\boldsymbol{\beta}}_{OLS}; \mathbf{y})/df_{\mathrm{full}}}, \qquad (4.4)$$

where "df" is short for degrees of freedom and the subscripts "full" and "reduced" on RSS and df refer to the full model (with p explanatory variables, including an intercept if there is one) and the reduced model obtained by imposing the equality specified in the null hypothesis.

Determining whether an explanatory variable is needed to adequately represent the mean surface is part of a broader enterprise known as **model selection**, i.e., selecting the best set of explanatory variables from those available. As in nonspatial contexts, for a spatial linear model there is a trade-off between parsimony (which favors the choice of a smaller model) and goodness-of-fit (which favors the opposite). The classical test for lack of fit based on partitioning the error sum of squares into a "lack-of-fit" portion affected by inadequacy of the explanatory variables in the model and a "pure error" portion not so affected typically is not possible in a spatial context because there is no replication at sites. Instead, adjusted t-tests may be used, with partial regression plots as graphical accompaniments. These tests may be implemented using either a forward selection/backward elimination strategy. In forward selection, all one-variable models (with an intercept) are fitted first. If none of the t-ratios are deemed sufficiently large, the procedure stops with the intercept-only model; otherwise, the model with the largest t-ratio becomes the first provisional model. In the latter case, all two-variable models obtained by adding one of the remaining variables to the provisional model are then fitted. If none of the t-ratios are deemed sufficiently large, the procedure stops with the first provisional model; otherwise, the two-variable model with the largest added t-ratio becomes the new provisional model. The procedure continues in this fashion, adding variables one at a time, until none of the variables added have sufficiently large t-ratios. Backward elimination begins by fitting the model with all p explanatory variables (plus an intercept). If none of the t-ratios are deemed sufficiently small to warrant the removal of the corresponding variables from the model, the procedure stops with the complete model; otherwise, the variable with the smallest t-ratio is removed and the model with the remaining $p - 1$ variables becomes the first provisional model. This process continues, at each stage removing the variable with the smallest t-ratio if it is sufficiently small, and stopping when there are no such remaining t-ratios.

A tabular summary known as the **analysis of variance**, or **ANOVA**, is often used to apportion the total variability in the responses to that which can be explained by the model, and that which cannot. We consider here two ANOVAs, where in both cases the models have an intercept and are in "corrected-for-the-mean form." The first, called the **overall ANOVA**, is a decomposition of the total mean-corrected sum of squares of the responses into two parts: one part measuring the variability explained by the explanatory variables, and the other part measuring the remaining (unexplained) variability. Table 4.1 shows the overall ANOVA table, with columns for the sources of variability, degrees of freedom (df), and sums of squares. Three equivalent expressions are given for each sum of squares. The first is in terms of $\mathbf{y}$, $\mathbf{P_X}$, and $\mathbf{P_0}$, where the latter two are the orthogonal projection matrices onto the column spaces of $\mathbf{X}$ and $\mathbf{1}$, respectively. The second is in terms of $\mathbf{y}$ $\mathbf{X}$, $\mathbf{1}$, and the OLS estimates of $\boldsymbol{\beta}$ and its first element β_0 (representing the intercept). The third, which is free of matrix or vector notation, is in terms of the scalar observations y_i, their fitted values $\hat{y}_{i,OLS}$, and their overall mean $\bar{y}$ (which is identical to $\hat{\beta}_0$ under the mean-only model). An F test of the hypothesis that the coefficients on all explanatory variables except

TABLE 4.1

Overall ANOVA table, corrected for the mean, and assuming that $\mathbf{X}$ has full column rank. Three equivalent expressions are given for each source of variation in the model.

Source	df	Sum of Squares
Model	$p-1$	$\mathbf{y}^T(\mathbf{P_X} - \mathbf{P}_0)\mathbf{y} = (\mathbf{X}\hat{\boldsymbol{\beta}}_{OLS} - \mathbf{1}\hat{\beta}_{0,OLS})^T(\mathbf{X}\hat{\boldsymbol{\beta}}_{OLS} - \mathbf{1}\hat{\beta}_{0,OLS})$ $= \sum_{i=1}^n (\hat{y}_{i,OLS} - \overline{y})^2$
Residual	$n-p$	$\mathbf{y}^T(\mathbf{I} - \mathbf{P_X})\mathbf{y} = (\mathbf{y} - \mathbf{X}\hat{\boldsymbol{\beta}})^T(\mathbf{y} - \mathbf{X}\hat{\boldsymbol{\beta}})$ $= \sum_{i=1}^n (y_i - \hat{y}_{i,OLS})^2$
Total	$n-1$	$\mathbf{y}^T(\mathbf{I} - \mathbf{P}_0)\mathbf{y} = (\mathbf{y} - \mathbf{1}\hat{\beta}_{0,OLS})^T(\mathbf{y} - \mathbf{1}\hat{\beta}_{0,OLS})$ $= \sum_{i=1}^n (y_i - \overline{y})^2$

the intercept are equal to zero may be constructed by computing the ratio of the model sum of squares divided by its df, $p-1$, to the residual sum of squares divided by its df, $n-p$, and comparing this ratio to percentiles of the F distribution with $p-1$ and $n-p$ degrees of freedom. This test is equivalent to the special case of (4.4) in which the full model includes the intercept and all of the explanatory variables and the reduced model consists of only the intercept.

Furthermore, from the information in the overall ANOVA table a statistic known as the **coefficient of determination**, also commonly called "R-squared," can be defined. Conceptually, R-squared is the proportion of the total variability in the response that is explained by the explanatory variables included in the model, and is given by

$$R^2 = \frac{\text{Model sum of squares}}{\text{Total sum of squares}}.$$

Equivalently,

$$R^2 = \frac{(\mathbf{X}\hat{\boldsymbol{\beta}}_{OLS} - \mathbf{1}\hat{\beta}_{0,OLS})^T(\mathbf{X}\hat{\boldsymbol{\beta}}_{OLS} - \mathbf{1}\hat{\beta}_{0,OLS})}{(\mathbf{y} - \mathbf{1}\hat{\beta}_{0,OLS})^T(\mathbf{y} - \mathbf{1}\hat{\beta}_{0,OLS})} = \frac{\sum_{i=1}^n (\hat{y}_{i,OLS} - \overline{y})^2}{\sum_{i=1}^n (y_i - \overline{y})^2}. \tag{4.5}$$

The second ANOVA we present here is the **sequential ANOVA**. This is essentially a more detailed version of the overall ANOVA in which the model sum of squares is partitioned into component sums of squares, each of which may be associated with one or more of the explanatory variables in the model. Let the model matrix $\mathbf{X}$ be partitioned as $\mathbf{X} = (\mathbf{1}|\mathbf{X}_1|\mathbf{X}_2|\ldots|\mathbf{X}_q)$, where each $\mathbf{X}_i$ represents a column or a subset of columns of $\mathbf{X}$, and let $\boldsymbol{\beta} = (\beta_0|\boldsymbol{\beta}_1^T|\boldsymbol{\beta}_2^T|\ldots|\boldsymbol{\beta}_q^T)^T$ be the corresponding partition of $\boldsymbol{\beta}$. Table 4.2 displays the sequential ANOVA table corresponding to this partition. The rows labeled as "Residual" and "Total" are identical to the rows with the same names in the overall ANOVA table, and the other rows sum to the row labeled "Model" in that table. The component sums of squares in Table 4.2 correspond to submatrices of the partitioned model matrix $(\mathbf{1}|\mathbf{X}_1|\mathbf{X}_2|\ldots|\mathbf{X}_q)$ and for the sake of simplicity are given only in terms of the orthogonal projection matrices onto the column spaces of $\mathbf{1}, (\mathbf{1}|\mathbf{X}_1), (\mathbf{1}|\mathbf{X}_1|\mathbf{X}_2), \ldots, (\mathbf{1}|\mathbf{X}_1|\mathbf{X}_2|\cdots|\mathbf{X}_q)$.

A **mean square** for each source (row) in the sequential ANOVA table is defined as that row's sum of squares divided by its df. The particular mean square known as the mean squared error is the residual sum of squares divided by its df. An F statistic for testing the null hypothesis that the coefficients corresponding to the columns in $\mathbf{X}_i$ are equal to zero, in the context of a model whose model matrix contains those columns and all columns listed above it in the table, may be formed as the ratio of the ith row's mean square to the mean squared error. To assess statistical significance, this statistic is compared to quantiles of an F distribution with the ith row's df and the mean squared error's df. Such F-tests,

TABLE 4.2

Sequential ANOVA table, with sums of squares expressed in terms of orthogonal projection matrices. Here, $r_{01\cdots q} = \mathrm{rank}(\mathbf{1}|\mathbf{X}_1|\cdots|\mathbf{X}_q)$ and $\mathbf{P}_{01\cdots q} = \mathbf{P}_{(\mathbf{1}|\mathbf{X}_1|\cdots|\mathbf{X}_q)}$.

Source	df	Sum of Squares
$\mathbf{X}_1$ after $\mathbf{1}$	$r_{01} - 1$	$\mathbf{y}^T(\mathbf{P}_{01} - \mathbf{P}_0)\mathbf{y}$
$\mathbf{X}_2$ after $(\mathbf{1}, \mathbf{X}_1)$	$r_{012} - r_{01}$	$\mathbf{y}^T(\mathbf{P}_{012} - \mathbf{P}_{01})\mathbf{y}$
$\vdots$	$\vdots$	$\vdots$
$\mathbf{X}_q$ after $(\mathbf{1}, \mathbf{X}_1, \ldots, \mathbf{X}_{q-1})$	$r_{01\cdots q} - r_{01\cdots(q-1)}$	$\mathbf{y}^T(\mathbf{P}_{01\cdots q} - \mathbf{P}_{01\cdots(q-1)})\mathbf{y}$
Residual	$n - r_{01\cdots q}$	$\mathbf{y}^T(\mathbf{I} - \mathbf{P}_{01\cdots q})\mathbf{y}$
Total	$n - 1$	$\mathbf{y}^T(\mathbf{I} - \mathbf{P}_0)\mathbf{y}$

performed sequentially by moving down the columns of Table 4.2, are called Type I tests of hypotheses. By way of comparison, F tests obtained by placing each $\mathbf{X}_i$, in turn, in the last position within $\mathbf{X}$, after all others, are called Type III tests. This terminology was made popular by the SAS software system. Note that Type I sums of squares always sum to the total sum of squares, but Type III sums of squares generally do not, unless all parts of the partitioned model matrix are orthogonal to one another (Zimmerman, 2020, Section 9.3.4).

Various graphics and statistics known as regression diagnostics are often computed as part of an OLS regression analysis for the purpose of checking the Gauss-Markov model assumptions. Many of these are based on the residuals from the fitted model, i.e., the vector $\hat{\mathbf{e}}_{OLS} = (\hat{e}_{1,OLS}, \ldots, \hat{e}_{n,OLS})^T = \mathbf{y} - \mathbf{X}\hat{\boldsymbol{\beta}}_{OLS}$. One such graphical diagnostic is the **OLS residual plot**, which is a plot of $\hat{e}_{i,OLS}$ versus either $\hat{y}_{i,OLS}$, or the ith spatial location, or the ith value of any one of the explanatory variables. Nonlinearity in the plot can be interpreted as an indication of a need to add explanatory variables to the model or to transform the response and/or explanatory variables. A systematic pattern in the vertical spread of the residuals in the plot, e.g., a rightward-opening or leftward-opening funnel shape, indicates that the residuals are heteroscedastic, and an accompanying formal hypothesis test for heteroscedasticity is available; see Myers (1990). Many, though not all, types of nonlinearity or heteroscedasticity may be remedied by making a power transformation of the response suggested by the well-known Box-Cox procedure. Another residual-based graphical diagnostic is the normal probability plot, in which the residuals, ranked from smallest to largest, are plotted against their expected values under normality. A substantial departure from linearity in the plot indicates that the model errors are unlikely to be normally distributed.

Several numerical regression diagnostics are of a type known as case diagnostics because they have values for each case (observation). Included among these are the residuals, of course, but also some related quantities computed after excluding each case, one by one, from the data. To describe these latter quantities, sometimes called **case-deletion diagnostics**, let $\mathbf{y}_{-i}$ and $\mathbf{X}_{-i}$ represent the response vector and model matrix without the ith observation. Furthermore, let $\hat{\boldsymbol{\beta}}_{-i,OLS} = (\hat{\beta}_{1,-i,OLS}, \ldots, \hat{\beta}_{p,-i,OLS})^T$, $\hat{\mathbf{y}}_{-i,OLS} = (\hat{y}_{1,-i,OLS}, \ldots, \hat{y}_{n,-i,OLS})^T$, $\hat{\mathbf{e}}_{-i,OLS} = (\hat{e}_{1,-i,OLS}, \ldots, \hat{e}_{n,-i,OLS})^T$, and $\hat{\sigma}^2_{-i,OLS}$ represent the OLS estimator of $\boldsymbol{\beta}$, fitted values vector, residual vector, and residual mean square from an OLS fit without the ith observation. The ith case-deleted residual is then

$$\hat{e}_{i,-i,OLS} = y_i - \mathbf{x}_i^T \hat{\boldsymbol{\beta}}_{-i,OLS},$$

or equivalently,

$$\hat{e}_{i,-i,OLS} = \frac{\hat{e}_{i,OLS}}{1 - h_{ii}},$$

where $h_{ii} = \mathbf{x}_i^T(\mathbf{X}^T\mathbf{X})^{-1}\mathbf{x}_i$, which is the ith main diagonal element of the hat matrix. This second expression for $\hat{e}_{i,-i,OLS}$ shows that it is possible to compute all of the case-deleted residuals from one OLS fit using all the observations rather than from n separate OLS fits, each using all but one observation. The quantity h_{ii} appearing in that expression is called the **leverage** of the ith observation, and measures the (non-Euclidean) distance from $\mathbf{x}_i$ to the "center" of the $\mathbf{x}_i$'s, $\overline{\mathbf{x}} = (1/n)\sum_{i=1}^n \mathbf{x}_i$. Leverages serve the useful purpose of identifying which, if any, of the data sites have the potential to give rise to responses that exert an undue influence (much more than their fair share) on the OLS estimate of at least one regression coefficient. A widely used yardstick for this assessment is the exceedance of $2p/n$. For models in which the explanatory variables include or consist solely of the spatial coordinates and functions thereof, sites at or near the boundary of the study area tend to have high leverage.

Case-deleted residuals, when properly standardized, may be used to test for outliers. Specifically, the null hypothesis that the ith case is not an outlier may be tested against the alternative hypothesis that it is an outlier by comparing the externally studentized residual,

$$t_i = \frac{\hat{e}_{i,OLS}}{\hat{\sigma}_{-i,OLS}\sqrt{1-h_{ii}}},$$

to percentiles of a t distribution with $n - p - 1$ degrees of freedom. A two-sided test of size ϵ results from rejecting the null hypothesis if the absolute value of the externally studentized residual exceeds the upper $\epsilon/2$ percentile of that t distribution. To test for the existence of any outliers among the n cases at level ϵ, the maximum absolute value of the externally studentized residuals among the n cases may be compared to the upper $\epsilon/2n$ percentile of that same t distribution, which adjusts conservatively (using a Bonferroni adjustment) for the multiplicity of hypotheses being tested.

Some case-deletion diagnostics are called **influence diagnostics** because they measure the amount of actual influence a case has on the analysis. Influence diagnostics combine leverage (which measures potential influence, as noted previously) with some function of the corresponding residual's magnitude. Four influence diagnostics are listed below by their "definition formulas"; computational formulas, which allow them to be computed from one OLS fit using all the observations, may be found in various regression analysis textbooks including Myers (1990):

$$\text{DFBETAS}_{j,i} = \frac{\hat{\beta}_{j,OLS} - \hat{\beta}_{j,-i,OLS}}{\hat{\sigma}_{-i,OLS}\sqrt{[(\mathbf{X}^T\mathbf{X})^{-1}]_{jj}}} \quad (j = 1,\ldots,p;\ i = 1,\ldots,n),$$

$$\text{DFFITS}_i = \frac{\hat{y}_{i,OLS} - \hat{y}_{i,-i,OLS}}{\hat{\sigma}_{-i,OLS}\sqrt{h_{ii}}} \quad (i = 1,\ldots,n),$$

$$\text{Cook's } D_i = \frac{(\hat{\boldsymbol{\beta}}_{OLS} - \hat{\boldsymbol{\beta}}_{-i,OLS})^T\mathbf{X}^T\mathbf{X}(\hat{\boldsymbol{\beta}}_{OLS} - \hat{\boldsymbol{\beta}}_{-i,OLS})}{p\hat{\sigma}_{OLS}^2} \quad (i = 1,\ldots,n),$$

$$\text{COVRATIO}_i = \frac{|(\mathbf{X}_{-i}^T\mathbf{X}_{-i})^{-1}\hat{\sigma}_{-i,OLS}^2|}{|(\mathbf{X}^T\mathbf{X})^{-1}\hat{\sigma}_{OLS}^2|} \quad (i = 1,\ldots,n).$$

$\text{DFBETAS}_{j,i}$ measures the influence of the ith case on the jth estimated regression coefficient; DFFITS_i measures the influence of the ith case on the ith fitted value; Cook's D_i has a dual interpretation as a measure of influence of the ith case on either the entire vector of estimated regression coefficients or the entire vector of fitted values; and COVRATIO_i measures the influence of the ith case on the usual unbiased estimate of the covariance matrix of estimated regression coefficients. The use of strict "cutoffs" to categorize observations as influential on the basis of these diagnostics is generally discouraged; a combination of rough yardsticks, disparities of the measure among the cases, and practical experience is suggested instead. For reasonably large samples, Belsley *et al.* (1980) suggested yardsticks of $2/\sqrt{n}$ for DFBETAS$_{j,i}$ and $2\sqrt{p/n}$ for DFFITS$_i$; that is, a case for which the statistic exceeds these yardsticks would be regarded as influential. They also suggested yardsticks of less than $1 - (3p/n)$ or greater than $1 + (3p/n)$ for COVRATIO$_i$. Various yardsticks for Cook's D_i have been suggested, including $4/n$, 0.5, and any value that "sticks out" from the others.

Not all regression diagnostics involve the residuals, nor are all diagnostics case diagnostics. One such exception is the **variance inflation factor**

$$VIF_j = \frac{1}{1 - R_j^2},$$

where R_j^2 is the coefficient of determination from the OLS regression of x_j on all the other explanatory variables. Thus, there is a variance inflation factor associated with each explanatory variable. VIFs are measures of multicollinearity; large values correspond to explanatory variables that are nearly a linear combination of the others. Overly large VIFs can render the OLS estimators highly unreliable (although they are still unbiased), so all other things being equal it is desirable for the VIFs to be as small as possible. This can often be achieved by eliminating or modifying some of the explanatory variables. Further details will be given later in this chapter for the specific case of polynomial regression models.

Throughout this section, it was assumed that the regressors were nonrandom. As we noted in Section 1.2, in many applications it may be more reasonable to regard one or more of the regressors as random; this is particularly so when the investigator has little or no control over the values of the regressors, as in an observational study. Fortunately, it turns out that many of the products of OLS-based estimation obtained under a classical fixed-regressors model retain the same properties under a random-regressors model, provided that the regressors and model errors are independent. For example, the OLS estimator of $\boldsymbol{\beta}$ is still unbiased, as are the residual variance $\hat{\sigma}^2_{OLS}$ and the classical estimator of the covariance matrix of $\hat{\boldsymbol{\beta}}_{OLS}$, viz. $\hat{\sigma}^2_{OLS}(\mathbf{X}^T\mathbf{X})^{-1}$. Furthermore, the distribution of $(n-p)\hat{\sigma}^2_{OLS}/\sigma^2$ is still $\chi^2(n-p)$, and the null distributions of the t- and F-statistics, (4.2) and (4.3), are still $t(n-p)$ and $F(q, n - p)$, respectively. Thus, under the assumption that the regressors and model errors are independent, inferences for model parameters in the random-regressors model may proceed just as if the regressors were nonrandom. The validity of this assumption, and the consequences when it is violated, are considered in Section 5.5.

4.2   Properties of OLS under a spatial linear model

Several properties of OLS estimators and associated inference procedures described in the previous section no longer hold, in general, under a spatial linear model. However, some do. In this section we describe which do and which do not. Unless noted otherwise, suppose throughout this subsection that $\mathrm{Var}(\mathbf{y}) = \boldsymbol{\Sigma} = \sigma^2\mathbf{R}$, an arbitrary positive definite covariance matrix.

First, note that $\hat{\boldsymbol{\beta}}_{OLS}$ and $\boldsymbol{\ell}^T\hat{\boldsymbol{\beta}}_{OLS}$ remain unbiased under a spatial linear model because the derivation of $\mathrm{E}(\hat{\boldsymbol{\beta}}_{OLS})$ in (4.1) does not involve the covariance matrix of $\mathbf{y}$. However, the variances of these quantities change, as

$$
\begin{aligned}
\mathrm{Var}_{SLM}(\hat{\boldsymbol{\beta}}_{OLS}) &= \mathrm{Var}_{SLM}[(\mathbf{X}^T\mathbf{X})^{-1}\mathbf{X}^T\mathbf{y}] \\
&= (\mathbf{X}^T\mathbf{X})^{-1}\mathbf{X}^T\mathrm{Var}_{SLM}(\mathbf{y})[(\mathbf{X}^T\mathbf{X})^{-1}\mathbf{X}^T]^T \\
&= \sigma^2(\mathbf{X}^T\mathbf{X})^{-1}\mathbf{X}^T\mathbf{R}\mathbf{X}(\mathbf{X}^T\mathbf{X})^{-1},
\end{aligned}
\tag{4.6}
$$

and thus $\mathrm{Var}_{SLM}(\boldsymbol{\ell}^T\hat{\boldsymbol{\beta}}_{OLS}) = \sigma^2\boldsymbol{\ell}^T(\mathbf{X}^T\mathbf{X})^{-1}\mathbf{X}^T\mathbf{R}\mathbf{X}(\mathbf{X}^T\mathbf{X})^{-1}\boldsymbol{\ell}$. (Here we have subscripted "Var" with SLM to indicate that the variance is being taken with respect to the distribution specified by the spatial linear model. This practice will continue throughout and will also be applied to other models such as the Gauss-Markov model, for which the subscript will be GMM). This last variance generally is not smallest among all unbiased estimators of $\boldsymbol{\ell}^T\boldsymbol{\beta}$ under the spatial linear model, so the OLS estimator is no longer the best linear unbiased estimator. But it is still normally distributed if the joint distribution of model errors is multivariate normal.

Next, consider the estimation of σ^2. It turns out [see, e.g., Theorem 12.2.1d in Zimmerman (2020)] that the residual variance generally is not unbiased for σ^2; in fact,

$$
\mathrm{E}_{SLM}(\hat{\sigma}^2_{OLS}) = \frac{\sigma^2}{n-p}\mathrm{tr}\{[\mathbf{I} - \mathbf{X}(\mathbf{X}^T\mathbf{X})^{-1}\mathbf{X}^T]\mathbf{R}\},
\tag{4.7}
$$

where $\mathrm{tr}(\mathbf{A})$ denotes the trace (the sum of the elements on the main diagonal) of a square matrix $\mathbf{A}$. Furthermore, $(n-p)\hat{\sigma}^2_{OLS}/\sigma^2$ generally does not have a chi-square distribution, and $(n-p)\hat{\sigma}^2_{OLS}/\sigma^2$ and $\hat{\boldsymbol{\beta}}_{OLS}$ generally are not independent. Thus, the distributional results that are foundational to the hypothesis tests and confidence intervals described in the previous subsection generally do not hold. Consequently, those inference procedures should be used cautiously for spatial linear models and the results regarded as merely provisional, to be improved upon after one has properly estimated and accounted for spatial correlation.

Although formal inference for a spatial linear model based on OLS estimation is not advisable, the fact that the OLS estimators retain their unbiasedness under the spatial linear model means that some by-products of OLS estimation are still of some value for diagnostic purposes. For instance, residual plots continue to be very useful graphical diagnostics because the residuals are unbiased estimators of zero even in the presence of heteroscedasticity/spatial correlation. Any systematic trend or pattern (e.g., nonlinearity or variance monotonically increasing or decreasing with the mean) in the plot might be interpreted, just as it would for a model with uncorrelated errors, as suggesting the need for additional explanatory variables or for transforming the response and/or explanatory variables. However, another possible interpretation of a systematic trend could be strong spatial correlation itself, as will be seen in a different context in Section 4.4.1. In any case, formal methods for testing for heteroscedasticity (which are based on an assumption of independent errors) are not advisable in this setting. Instead, the exploratory methods described in Chapter 3 for diagnosing spatial variability in the variance should be used. Either problem, nonlinearity or monotonic heteroscedasticity, may be addressed using the aforementioned Box-Cox procedure to suggest a power transformation of the response, albeit without relying too heavily on the confidence interval it produces for the power because that confidence interval is derived under OLS assumptions. Once those issues are addressed, a histogram and a normal probability plot of the residuals may be used to check for nonnormality, as the marginal distributions of the OLS residuals are normal if the joint distribution of the spatial linear model errors is multivariate normal, regardless of the degree of spatial correlation. However, once again, standard hypothesis tests for normality should not be used because they are affected by spatial correlation.

OLS-based case diagnostics lose quite a bit of their usefulness when the data are spatially correlated. Diagnostics that are functions of the explanatory variables only, such as leverages

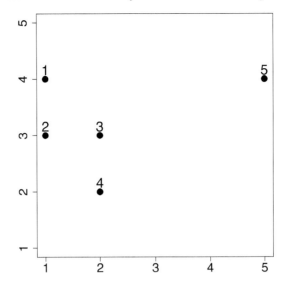

FIGURE 4.1
Spatial locations of sites for the toy example of Section 4.2. Numbers next to sites are the
site indices.

and variance inflation factors, retain the most value because they are not affected by the
presence of spatial correlation in model errors, but even they can be improved. Testing for
outliers and computing influence diagnostics for a spatial linear model is best saved for
a later stage of the analysis, i.e., after a model with spatially correlated errors is fitted.
Methods for performing these tasks will be described in Chapter 5.

   To gain a better understanding of how the sampling variation of OLS estimators under a
Gauss-Markov model compares to that under a spatial linear model, consider the following
toy example, for which the data sites are the five points displayed in Figure 4.1. Take the
spatial linear model for responses at those sites to have constant mean μ and covariance
matrix derived from the isotropic covariance function

$$C(r) = \begin{cases} \sigma^2 \left(1 - \frac{3r}{8} + \frac{r^3}{128}\right) & \text{for } 0 \leq r \leq 4, \\ 0 & \text{for } r > 4. \end{cases}$$

(This is the same spatial configuration of sites and one of the same covariance functions that
were considered previously in Section 2.2. Moreover, this covariance function is a case of
the so-called spherical covariance function, which is positive definite and will be described
in more detail in Chapter 6.) Then $\mathbf{X} = \mathbf{1}_5$ and $\mathbf{\Sigma} = \sigma^2 \mathbf{R}$, where

$$\mathbf{R} = \begin{pmatrix} 1.000 & 0.633 & 0.492 & 0.249 & 0 \\ & 1.000 & 0.633 & 0.492 & 0 \\ & & 1.000 & 0.633 & 0.061 \\ & & & 1.000 & 0.014 \\ & & & & 1.000 \end{pmatrix}.$$

Here we have ordered the five observations by moving down within the vertical columns of
sites in Figure 4.1, beginning with the leftmost column. The positive definiteness of $C(r)$
(as a function) passes on to $\mathbf{R}$ and $\mathbf{\Sigma}$ (as matrices) also. Furthermore, take $\sigma^2 = 1$.

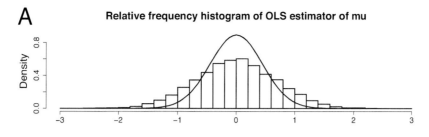

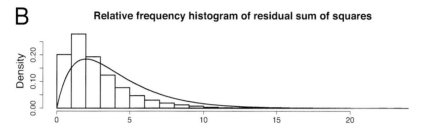

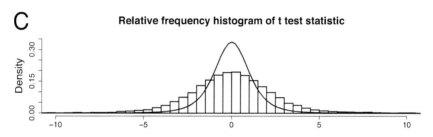

FIGURE 4.2

Histograms of A) OLS estimates of μ, B) residual sum of squares, and C) test statistics given by (4.2), based on 10,000 simulated realizations of observations from the toy example of Section 4.2. Solid curves superimposed on each histogram are the corresponding distributions of the statistic under a Gauss-Markov model (and under $H_0 : \mu = 0$).

Because $(\mathbf{X}^T\mathbf{X})^{-1}\mathbf{X}^T = (1/5)\mathbf{1}_5^T$, the OLS estimator of μ is $(1/5)\mathbf{1}^T\mathbf{y} = \overline{y}$, and using (4.6) its variance under the spatial linear model is given by

$$(1/5)\mathbf{1}_5^T\mathbf{R}\mathbf{1}_5(1/5) = 0.456.$$

This is more than twice as large as the variance of $\overline{y}$ under a Gauss-Markov model, which is merely $(\mathbf{1}_5^T\mathbf{1}_5)^{-1} = 0.200$. The residual mean square is $\hat{\sigma}_{OLS}^2 = \sum_{i=1}^{5}(y_i - \overline{y})^2/4$, which by (4.7) has expectation

$$(1/4)\text{tr}\{[\mathbf{I} - (1/5)\mathbf{1}_5\mathbf{1}_5^T]\mathbf{R}\} = (1/4)[\text{tr}(\mathbf{R}) - (1/5)\mathbf{1}_5^T\mathbf{R}\mathbf{1}_5] = 0.679.$$

Since $\sigma^2 = 1.0$, this calculation shows that the residual mean square is biased downwards, i.e., it tends to underestimate σ^2. Figure 4.2 displays histograms of the OLS estimator $\overline{y}$, the residual sum of squares $4\hat{\sigma}_{OLS}^2$, and the t test statistic given by (4.2), based on 10,000 simulated realizations of a multivariate normal vector following this spatial linear model with $\mu = 0$. Superimposed upon the histograms are pdfs of the corresponding statistics under a Gauss-Markov model. Comparison of the histograms to the pdfs clearly shows the effect of spatial correlation on standard statistics used to perform inference under the Gauss-Markov model. Note especially that the t test statistic is poorly approximated by the corresponding t distribution. Although it is centered properly, it tends to have much greater variability than the corresponding t distribution, owing to the dual effects of underestimating $\text{Var}(\bar{y})$ and σ^2.

4.3　Formal tests for residual spatial correlation and its properties

As noted previously, the optimality of ordinary least squares estimation rests on an assumption of uncorrelated model errors. For data ordered linearly, such as in time, a commonly used procedure for formally testing this assumption is the Durbin-Watson test (Durbin & Watson, 1950, 1951, 1971). The Durbin-Watson test is based on the statistic

$$DW = \frac{\sum_{i=2}^{n}(\hat{e}_{i,OLS} - \hat{e}_{i-1,OLS})^2}{\sum_{i=1}^{n} \hat{e}_{i,OLS}^2},$$

where $\hat{e}_{i,OLS}$, as in Section 4.1, is the OLS residual corresponding to the ith observation. Observe that the numerator of DW is proportional to the lag-one sample semivariance of residuals (corresponding to equally spaced time points) and, provided that there is an intercept in the model (so that the residuals sum to 0), the denominator is proportional to the sample variance of the residuals. Various analogs of the Durbin-Watson test have been developed for testing for residual spatial correlation in spatial linear models. In fact, the Moran and Geary statistics presented in Chapter 3, when applied to residuals rather than the original observations, are rather direct analogs of DW for areal data. (Recall that the Geary statistic is a weighted average of semivariances among neighbors.) Applying the Moran and Geary tests to residuals from a fitted regression surface adds another level of approximation to the evaluation of statistical significance using percentiles from the standard normal distribution, so we recommend relying solely upon the permutation-based testing approach in this case.

For geostatistical data, we recommend a semivariogram-based method for testing for spatial correlation due to Diblasi & Bowman (2001). The rationale for this method is that the semivariogram of a second-order stationary process is flat (pure nugget effect) if and only if the correlation of the process is zero at all nonzero distances. The approach is based on nonparametrically smoothing the square-root difference cloud and measuring how well that smooth curve can be fit by a flat (zero-slope) line. Square-root, rather than squared, differences are used because they are less skewed and behave more like normal variates, improving the performance of the test. Nonparametric smoothing of the square-root differences cloud may proceed in a manner exactly like that described for the squared differences cloud in Section 3.6.2; in particular, Diblasi and Bowman recommended the use of local linear smoothing. The method for testing for spatial correlation is then a nonparametric-regression version of the "reduction in residual sum of squares" testing approach used so often to compare nested regression models [as, for example, in (4.4)], with test statistic

$$T = \frac{\sum_{i<j}(d_{ij} - \bar{d}_{..})^2 - \sum_{i<j}(d_{ij} - \tilde{d}_{ij})^2}{\sum_{i<j}(d_{ij} - \tilde{d}_{ij})^2}.$$

Here, $d_{ij} = |y_i - y_j|^{1/2}$, $\bar{d}_{..}$ is the average of the d_{ij}'s, and $\tilde{d}_{ij}$ is the smoothed square-root-difference at distance $\|\mathbf{s}_i - \mathbf{s}_j\|$. Diblasi and Bowman derived a method for approximating the P-value of the test under the assumption that the errors in the spatial linear model are normally distributed. The test is implemented in the `sm.variogram` function of the `sm` package in R and runs sufficiently fast that a permutation-based approach to obtain P-values is feasible for data sets of several hundred observations.

Hypothesis tests based on the sample acf and semivariogram have also been developed for formal assessment of other properties of the second-order dependence. Behind all such tests is the following idea, originating with Lu (1994) and Lu & Zimmerman (2001): because a given second-order dependence property imposes some equality constraints on the true co-variances or semivariances (or functions thereof), a sensible test statistic for the dependence property is one that measures the extent to which the sample autocovariance function or sample semivariogram violates those constraints. For example, if data are on a square grid aligned with the compass directions, and we let $\gamma(0,1)$ and $\gamma(1,0)$ denote the semivariances at one unit of distance in the N-S and E-W directions, respectively, then the statistic

$$T_I = \frac{[\hat{\gamma}(0,1) - \hat{\gamma}(1,0)]^2}{\widehat{\text{Var}}[\hat{\gamma}(0,1) - \hat{\gamma}(1,0)]}$$

is a scale-invariant measure of the extent of anisotropy between those two directions at one unit of distance. A formula that can be used to evaluate $\widehat{\text{Var}}[\hat{\gamma}(0,1) - \hat{\gamma}(1,0)]$ can be found in Lu & Zimmerman (2001). Provided that the observations satisfy isotropy and have finite fourth-order moments (as they do when they are normally distributed), T_I is asymptotically distributed as a chi-square variable with one degree of freedom, implying that the isotropy assumption may be assessed by comparing the statistic to percentiles of the $\chi^2(1)$ distribution. This approach to testing for isotropy and other symmetry properties of the dependence structure was extended for use with any number of sample autocovari-ances or semivariances, for spatial data on a regular grid, by Lu & Zimmerman (2001). It was developed further for testing specifically for isotropy using irregularly spaced data by Guan *et al.* (2004), and subsequently for testing for separability and several other spatial correlation properties by many authors (Lu & Zimmerman, 2005; Scaccia & Martin, 2005; Fuentes, 2005; Maity & Sherman, 2012). These and a few other tests for isotropy and other directional symmetry properties are reviewed by Weller & Hoeting (2016), and some have been implemented in the `spTest` package in R.

Another approach to testing for isotropy, due to Bowman & Crujeiras (2013), is based on an idea similar to that which motivated the Diblasi-Bowman test for independence. In this approach, the square-root-scale semivariogram cloud, with square-root differences plotted against lags in their polar coordinates, is smoothed nonparametrically, and that smoothed surface is compared to the smoothed omnidirectional square-root-scale semivariogram cloud obtained for the Diblasi-Bowman test. The comparison is made using another reduction-in-residual-sum-of-squares test statistic. The test is implemented in the `sm` package. Yet another testing option available in that package, also based on comparisons of nonparametric smooths, is a test for stationarity of the covariance structure. Unlike the Diblasi-Bowman test, however, these tests are not permutation-based and are valid only to the extent that the errors in the spatial model are normally distributed. Furthermore, they are much more computationally intensive than the Diblasi-Bowman test and the tests described in the previous paragraph.

4.4 Trend surface analysis

In this section, we specialize ordinary least squares estimation methodology to cases of spatial linear models in which the only explanatory variables are the spatial coordinates and functions thereof. The selection and fitting of such a mean structure is known as **trend surface analysis**.

4.4.1   Polynomial regression models

Without a doubt, the most useful all-purpose parametric family of models for trend over a Euclidean space is the family of **polynomial regression models**. Polynomials are continuous and dense, the latter term meaning that any continuous function over a bounded study region can be approximated arbitrarily well by a polynomial of sufficiently high order. Thus, a polynomial model may be regarded more as an approximation to an underlying, not-well-understood mean structure than as a mechanistic model of that structure.

A polynomial model in one spatial dimension u is a linear model in which each explanatory variable is a monomial, i.e., a nonnegative integer power, of u. A **complete polynomial** model of order k in one dimension is given by

$$y_i = \beta_0 + \beta_1 u_i + \beta_2 u_i^2 + \cdots + \beta_k u_i^k + e_i,$$

while an **incomplete polynomial** of order k lacks at least one of the terms of order less than k. More generally, a complete polynomial of order k in d dimensions is a parametric linear combination of all d-fold products of monomials for which the sum of exponents across terms is less than or equal to k. For example, in two dimensions complete polynomials of order one, two, and three are given by

$$
\begin{aligned}
y_i &= \beta_{00} + \beta_{10} u_i + \beta_{01} v_i + e_i, \\
y_i &= \beta_{00} + \beta_{10} u_i + \beta_{01} v_i + \beta_{20} u_i^2 + \beta_{11} u_i v_i + \beta_{02} v_i^2 + e_i, \\
y_i &= \beta_{00} + \beta_{10} u_i + \beta_{01} v_i + \beta_{20} u_i^2 + \beta_{11} u_i v_i + \beta_{02} v_i^2 + \beta_{30} u_i^3 + \beta_{21} u_i^2 v_i + \beta_{12} u_i v_i^2 \\
&\quad + \beta_{03} v_i^3 + e_i,
\end{aligned}
$$

respectively. These are commonly referred to as planar, quadratic, and cubic surfaces, respectively, in $\mathbb{R}^2$. Complete polynomials of the same orders in three dimensions have additional terms that include monomials in the variable (say z) representing the third dimension; for example, the complete third-order polynomial includes such terms as $\beta_{021} v_i^2 z_i$ and $\beta_{111} u_i v_i z_i$. The number of terms in a complete polynomial of order 1, 2, or 3 in three dimensions is 4, 10, or 20, respectively.

In a polynomial regression model, some of the regression variables, particularly those that are powers of the same coordinate (e.g., u^2 and u^3), tend to be highly collinear, making the $\mathbf{X}^T\mathbf{X}$ matrix nearly singular. Modern computing packages solve the least squares problem in a way that is not adversely affected numerically by multicollinearity, hence they yield reliable fitted values at data sites. However, the estimators of individual polynomial coefficients are affected statistically; they become unstable, with variances inflated to an extreme. Consequently, when fitting polynomials, little stock should be placed in the estimated coefficients unless steps have been taken to reduce the multicollinearity. A partial step in this direction is to center each coordinate by subtracting its overall mean, and then replace each term in the polynomial with the corresponding d-fold product of monomials in the centered coordinates. For example, the quadratic version of this modification in two dimensions is

$$y_i = \beta_{00} + \beta_{10}(u_i - \overline{u}) + \beta_{01}(v_i - \overline{v}) + \beta_{20}(u_i - \overline{u})^2 + \beta_{11}(u_i - \overline{u})(v_i - \overline{v}) + \beta_{02}(v_i - \overline{v})^2 + e_i.$$

Centering eliminates the collinearities between u and u^2, and between v and v^2, and reduces the others. To illustrate, for the clean wet sulfate deposition data, the variance inflation factors range from 1.17–34.6 and from 7.91–2411 for the uncentered quadratic and cubic models, and from 1.17–1.26 and 1.67–8.24 for the centered quadratic and cubic models. To completely eliminate multicollinearity, the terms can be orthogonalized, but if the order of the centered model is three or less the variance inflation factors are usually small enough that this is unnecessary.

In practice, the modeler has to choose an order for the polynomial. Forward selection or backward elimination procedures may be used to make a tentative determination of order. In this application of those procedures, only complete polynomials should be considered, so joint F-test statistics replace t-ratios (when $d \geq 2$) as the quantities on which the strategies are based. With one exception, there is no point in attempting to add to or reduce the number of terms within a given order. This is because the corresponding individual coefficient estimates and their statistical significance are not invariant to a change in the coordinate system, which, after all, is arbitrary for most spatial data. To see this, consider a one-dimensional case in which the quadratic model

$$y_i = \beta_0 + \beta_1 u_i + \beta_2 u_i^2 + e_i$$

is fitted, and suppose that the OLS estimates of β_0 and β_2, but not β_1, are found to be significantly different from zero. For the sake of model parsimony, should the linear term u_i be removed from the model? Suppose it is removed so that the model becomes $\mathrm{E}(y_i) = \beta_0 + \beta_2 u_i^2$. Now consider what happens when the coordinate system is shifted to the right one unit, or equivalently, 1 is subtracted from each u_i. In that case,

$$\begin{aligned} \mathrm{E}(y_i) &= \beta_0 + \beta_2(u_i - 1)^2 \\ &= (\beta_0 + \beta_2) - 2\beta_2 u_i + \beta_2 u_i^2. \end{aligned}$$

Thus, the linear term must be present in this representation of the model; without it, the OLS fit would be different. Fitted surfaces of complete polynomials are invariant to such changes in the coordinate system (Cliff & Kelly, 1977). The exception to the rule for adding or eliminating individual terms in polynomials occurs after the order of the complete polynomial has been tentatively determined to be greater than or equal to two; at that stage, adjusted t-tests may be used to remove individual terms of the highest order or to add individual terms of the next highest order, as those changes do not have the effect just described. Thus the final order of the polynomial will either coincide with the order as determined by forward selection/backward elimination, or be one degree higher, and the final polynomial model need not be complete.

The sampling distributions of the F-statistics used to determine the polynomial's order may be affected by the presence of spatial correlation; in particular, positive correlation will tend to make the statistical evidence in favor of a higher-order model appear stronger than it really is. Furthermore, just as a strong trend may be misconstrued as spatial correlation (the example in Section 3.7.2 being a case in point), strong positive spatial correlation may be mistaken for trend. Figure 4.3 is a plot of spatially correlated data at 20 equally spaced locations along a one-dimensional transect. The response was generated by the equation

$$y_1 = 0, \quad y_i = .95y_{i-1} + e_i \quad (i = 2, \ldots, 20),$$

where the e_i's are independent normal random variables with mean 0 and standard deviation 0.1. Thus the generating mechanism has no trend, merely strong positive correlation among nearby observations that has "systematized the noise" to make it appear similar to a relatively low-order polynomial. A modeler who did not know how the data were generated might be tempted to fit a cubic polynomial to them. Because of these issues and the spectre of multicollinearity raised in the previous paragraph, when carrying out a forward selection/backward elimination strategy on polynomials for spatial data we recommend using

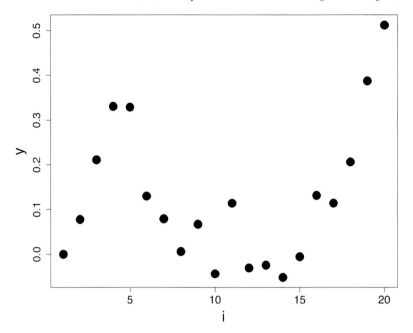

FIGURE 4.3
Data simulated from a model with strong positive correlation and no trend.

a threshold for the F statistic's P-value that is much smaller than usual, perhaps even as small as 1×10^{-6}. Moreover, in the absence of either a plausible scientific rationale or an extremely compelling F statistic, we recommend that the final choice of polynomial model have order no higher than three.

A final issue concerning polynomial regression models pertains to the sampling configuration of data sites. If there are no sites within a sizable subregion of the study area, the fitted surface over that subregion is unreliable, for there is no information there to "tie down" the surface; it can "flap in the wind," as it were, if that serves to improve the fit of the model at data sites. The problem is exacerbated if the unsampled area is near the boundary and/or there is substantial multicollinearity in the model. It also becomes more acute with increasing d because the proportion of the study area within a given distance of the boundary increases with d (this is one manifestation of the curse of dimensionality). Thus, the usual admonitions about extrapolation using fitted regression models apply, perhaps even more strongly in a spatial setting than in others. Furthermore, if there is just one data site (or a relative few) in a subregion near the boundary, that site will have large leverage and the response observed there is potentially highly influential. Clearly, the implication of these issues for sampling design is that data sites should be distributed more-or-less uniformly over the study area, with a substantial proportion close to its boundary. We revisit this issue in Chapter 10.

4.4.2   Additive coordinate effects models

The other family of models that has been used for trend surface analysis frequently, though not as often as polynomials, are **additive coordinate effects models**. As the name implies, these models suppose that the mean response at a site is the sum of effects attributable to the sites' marginal coordinates. We describe the model in detail for the two-dimensional

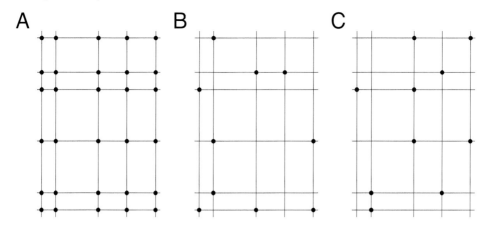

FIGURE 4.4
Examples of spatial data sites on a 6-row by 5-column unequally spaced rectangular grid: A) a complete row-column design; B) an incomplete but connected design; C) a disconnected design. Grid nodes with a closed circle represent data sites.

case; the three-dimensional case is a straightforward, but notationally more cumbersome, extension. Let $u_{(1)} < u_{(2)} < \cdots < u_{(n_c)}$ and $v_{(1)} < v_{(2)} < \cdots < v_{(n_r)}$ represent the distinct ordered values of the sites' u-coordinates and v-coordinates, respectively, so that the sites lie on a subset of the nodes of a rectangular, but possibly incomplete and unequally spaced, grid of n_c columns and n_r rows; see Figure 4.4 for three examples. In general, it is possible that there is little to no replication in some or all rows, and in some or all columns. Now, re-index the responses using double subscripts, where the first subscript indicates the row, and the second the column, of the corresponding site. Then the additive coordinate effects model is

$$y_{ij} = \mu + \alpha_i + \gamma_j + e_{ij},$$

where μ is an overall mean effect, α_i is the effect of the ith row, and γ_j is the effect of the jth column on the response.

Two major issues with this model must be addressed. First, in its current form, the model is overparameterized. To see this, suppose that the data sites form a complete grid, i.e., every grid node is a data site, as in Figure 4.4A. Then there are exactly two linear dependencies among the columns of $\mathbf{X}$: the sum of the columns corresponding to the row effects and the sum of columns corresponding to the column effects are both equal to the column corresponding to μ (a vector of all ones). Thus, the rank of the model matrix $\mathbf{X}$ is $n_c + n_r - 1$, while the model has $n_c + n_r + 1$ parameters. The model may be made to have full rank by imposing constraints $\sum_{i=1}^{n_r} \alpha_i = 0$ and $\sum_{j=1}^{n_c} \gamma_j = 0$, whence the parameters may be estimated uniquely by OLS. In the case of a complete grid, the OLS estimators under these constraints are as follows:

$$\hat{\mu}_{OLS} = \overline{y}_{..}, \qquad \hat{\alpha}_{i,OLS} = \overline{y}_{i.} - \overline{y}_{..} \ (i = 1, \ldots, n_r), \qquad \hat{\gamma}_{j,OLS} = \overline{y}_{.j} - \overline{y}_{..} \ (j = 1, \ldots, n_c).$$

The second issue, which applies only when the grid is incomplete, is that there may be insufficient replication within grid rows and columns to sensibly estimate all the parameters. A necessary and sufficient condition for estimability of all parameters of the model with the two constraints is that the two-way layout is **connected**. A two-way layout is disconnected if the rows can be separated into two or more groups, with the data sites in a row from one group never appearing in the same column with data sites in a row from another group. A connected two-way layout is one that is not disconnected. Of the three layouts in Figure 4.4, the first is complete (and thus connected), the second is incomplete yet connected, and the third is disconnected, as data sites in the first (from the top), third, and fourth rows never appear in the same column with data sites in the second, fifth, and sixth rows. If a two-way layout is not connected, one can try to find a grid with fewer rows and columns that, when overlaid upon the study area and sites are assigned to the nearest grid node, does yield a connected layout. If distances from sites to their nearest gridpoints are too large for this approach to be acceptable, then an additive coordinate effects model is not an option.

The additive coordinate effects model as described so far is unstructured, in the sense that no relationships among either the row effects or the column effects are assumed. This makes the model flexible but also, as noted above, nonestimable unless the layout is connected. The nonestimability issue disappears for more structured versions of the model, at the cost of reduced flexibility. One highly structured version results from assuming that $\alpha_i = \beta_1 v_{(i)}$ and $\gamma_j = \beta_2 u_{(j)}$, i.e., that the dependence on each coordinate is linear, which is merely another representation of a planar surface. Polynomials other than linear $\times$ linear may also be considered, as may other types of structure (e.g., step functions). The various possibilities may be compared using full-versus-reduced model F tests, subject to the same warning that was given for interpreting P-values of F tests for comparing polynomial models. A less structured version (but still more structured than the additive coordinate effects models presented here) is the family of generalized additive models (Hastie & Tibshirani, 1990), for which the dependence of the response on the coordinates is assumed to be additive but smooth, with the smoothness modeled nonparametrically. Such models are not linear models so we do not consider them herein.

The feature of additive coordinate effects models that limits their flexibility is, of course, their additivity (lack of interaction between row and column coordinate effects). Terms may be added to the model to give it some degree of nonadditivity. Most terms that have been proposed are multiplicative in nature. For example, the model corresponding to Tukey's classical one-degree-of-freedom test for nonadditivity (Tukey, 1949) is

$$y_{ij} = \mu + \alpha_i + \gamma_j + \lambda \alpha_i \gamma_j + e_{ij}.$$

Johnson & Graybill (1972) proposed making the interaction term functionally independent of the coordinate main effects via the model

$$y_{ij} = \mu + \alpha_i + \gamma_j + \lambda \psi_i \xi_j + e_{ij}.$$

Although relatively simple tests for nonadditivity may be based on these models, the models themselves are not linear. For this reason, we prefer an approach proposed by Cressie (1986), which is based upon the full-rank linear model

$$y_{ij} = \mu + \alpha_i + \gamma_j + \lambda (u_{(j)} - \overline{u})(v_{(i)} - \overline{v}) + e_{ij} \qquad (4.8)$$

(with the same two constraints as before). Here $\overline{u} = (1/n_c)\sum_{j=1}^{n_c} u_{(j)}$ and $\overline{v} = (1/n_r)\sum_{i=1}^{n_r} v_{(i)}$. This model may be fitted by OLS, with the adjusted t-test and partial regression plot for $(u_{(j)} - \overline{u})(v_{(i)} - \overline{v})$ used to determine whether the additional term should be included in the model.

A fitted additive coordinate effects model directly produces estimates of a spatial surface only at gridpoints. In a geostatistical setting, the fitted surface between gridpoints may be "filled out" by planar interpolation. For (u, v) such that $u_{(j)} < u < u_{(j+1)}$ and $v_{(i)} < v <$

$v_{(i+1)}$, the planar interpolant, assuming additivity, is

$$
\begin{aligned}
\hat{y}(u,v) \;=\;& \hat{\alpha}_{i,OLS} + \left(\frac{v - v_{(i)}}{v_{(i+1)} - v_{(i)}} \right) (\hat{\alpha}_{i+1,OLS} - \hat{\alpha}_{i,OLS}) + \hat{\gamma}_{j,OLS} \\
& + \left(\frac{u - u_{(j)}}{u_{(j+1)} - u_{(j)}} \right) (\hat{\gamma}_{j+1,OLS} - \hat{\gamma}_{j,OLS}) \\
& (i = 1, \ldots, n_r; \; j = 1, \ldots, n_c).
\end{aligned}
$$

For the portion of the study area lying outside the grid, the fitted surface may be defined by the planar extrapolant

$$
\begin{aligned}
\hat{y}(u,v) \;=\;& \hat{\alpha}_{i,OLS} + \left(\frac{v - v_{(i)}}{v_{(i+1)} - v_{(i)}} \right) (\hat{\alpha}_{i+1,OLS} - \hat{\alpha}_{i,OLS}) + \hat{\gamma}_{j,OLS} \\
& + \left(\frac{u - u_{(1)}}{u_{(2)} - u_{(1)}} \right) (\hat{\gamma}_{2,OLS} - \hat{\gamma}_{1,OLS}) \\
& \text{if } u < u_{(1)} \text{ and } v_{(j)} < v < v_{(j+1)} \; (i = 1, \ldots, n_r; \; j = 1, \ldots, n_c),
\end{aligned}
$$

with similar expressions for the other cases of (u,v) lying outside the grid. If the interaction term is deemed necessary, the fitted surface at (u,v) may be obtained by adding $\hat{\lambda}_{OLS}(u - \bar{u})(v - \bar{v})$, from the OLS fit of model (4.8), to these expressions. The usual admonition to not extrapolate too far outside the grid applies.

4.4.3 Other trend surface models

Although polynomial and coordinate effects models are the predominant models used for trend surface analysis, there are a few other possibilities. One possibility is to mix polynomial effects in one or more dimensions with coordinate effects in the other(s). For example, if $d = 2$ and the data sites lie along a small number of rows but are located rather haphazardly within rows, a model with unstructured row effects and a polynomial in the horizontal coordinate may be useful. Denoting the observations by y_{ij}, where i labels the row and $j = 1, \ldots, n_i$ labels the observations within the ith row, and denoting the u-coordinate of y_{ij}'s site by u_{ij}, two polynomial/coordinate effects models that could be contemplated for such a situation are

$$
y_{ij} = \mu + \alpha_i + \gamma_1 u_{ij} + \gamma_2 u_{ij}^2 + e_{ij} \quad (i = 1, \ldots, n_r; j = 1, \ldots, n_i)
$$

or

$$
y_{ij} = \mu + \alpha_i + \gamma_{1i} u_{ij} + \gamma_{2i} u_{ij}^2 + e_{ij} \quad (i = 1, \ldots, n_r; j = 1, \ldots, n_i).
$$

The first of these models assumes that the second-order polynomial describing the dependence of the response on u is the same in all rows, while the latter allows it to vary across rows. Alternatively, polynomials of order one (i.e., straight-line models) or of order three or higher within rows could be considered. Models like this can also be used when the data lie at the nodes of a regular rectangular grid; the assumption for this example that data sites are haphazardly located within rows (or columns) is not necessary.

Models that include trigonometric terms, specifically sines and cosines, may be useful when the data exhibit some periodicity (Kupper, 1972). Models with trigonometric terms only are unlikely to be useful for spatial data, but mixing them with low-order polynomial terms may produce a model that fits better than a pure higher-order polynomial (Eubank & Speckman, 1990). An example of such a polynomial-trigonometric model in two dimensions is

$$
\begin{aligned}
y_i \;=\; & \beta_{00} + \beta_{10}u_i + \beta_{01}v_i + \beta_{20}u_i^2 + \beta_{11}u_iv_i + \beta_{02}v_i^2 \\
& + \sum_{j=1}^{2}[\gamma_{1j}\sin(ju_i) + \gamma_{2j}\cos(ju_i) + \tau_{1j}\sin(jv_i) + \tau_{2j}\cos(jv_i)] + e_i \quad (i = 1, \ldots, n).
\end{aligned}
$$

This model includes a quadratic surface plus sine and cosine terms, up to order 2, in each of the two coordinates. It is worth noting that trigonometric terms do not create multicollinearity issues to the degree that higher-order polynomial terms do. Indeed, when the data locations are on a regular rectangular grid, $\sin(ju)$ and $\cos(ju)$, and $\sin(jv)$ and $\cos(jv)$, $j = 1, 2, \ldots$, are orthogonal, and when the data are not regularly spaced, the collinearities between these terms are usually small.

4.4.4   Models with spatial and nonspatial explanatory variables

Polynomial and additive coordinate effects models are appropriate not merely for trend surface analysis; they may also be useful when the explanatory variables are nonspatial or when some are spatial and the others nonspatial. When there is one nonspatial explanatory variable and it is categorical (e.g. a treatment effect), the fitted trend surface corresponding to each level of the treatment may be plotted for between-level comparisons of surfaces. When the nonspatial explanatory variable is measured instead on a discrete (with many levels) or continuous scale, the partial regression plot and partial t test for that variable may be examined to visualize and measure its effect on the response, adjusted for the effects of the spatial explanatory variables.

4.5   Examples

4.5.1   The wet sulfate deposition data

Fitted planar, quadratic, and cubic surfaces for the clean wet sulfate deposition data are displayed in Figure 4.5. (These surfaces were actually fitted to an Albers projection of this portion of the United States onto the plane.) While each surface appears considerably different than the others, all of them indicate a strong, mainly increasing trend from west to east, nuanced by a slight eastward decrease in the extreme west and extreme east for the cubic model. The fitted quadratic and cubic surfaces exhibit a north-south trend that is

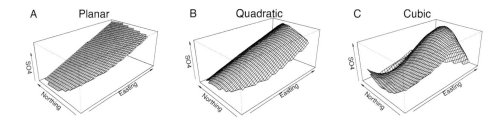

FIGURE 4.5
OLS-fitted surfaces for the clean wet sulfate deposition data: A) planar surface; B) quadratic surface; C) cubic surface.

TABLE 4.3

Summary statistics and results of F tests for polynomial fits to the clean wet sulfate deposition data.

Surface	R^2	$\hat{\sigma}^2_{OLS}$	Residual df	F	P-value
Planar	0.68	0.823	191	—	—
Quadratic	0.73	0.760	188	11.95	3.4×10^{-7}
Cubic	0.87	0.530	184	50.76	9.6×10^{-29}
Quartic	0.90	0.488	179	7.55	1.9×10^{-6}
Quintic	0.91	0.470	173	3.29	4.3×10^{-3}

distinctly curvilinear and nonmonotonic, increasing rapidly from southern to central latitudes and then declining somewhat to the north. The variance inflation factors for the fitted centered cubic surface are all less than 10, and VIFs for the lower-order surfaces are much less than that. Due to its semi-isolation near the boundary of the study area, the site in extreme southern Florida (see Figure 1.1) has about twice the leverage of any other site, but neither it nor any other site is sufficiently influential to be of concern. Residual plots for the quadratic and cubic fits reveal no identifiable patterns, apart from possible small-scale spatial correlation (especially among the quadratic residuals). Thus there are no major concerns that should prevent further analysis from proceeding.

Table 4.3 summarizes the fits of centered polynomial models from first- to fifth-order and the results of tests of each order against the next lowest order. The quadratic surface is reasonable based on scientific considerations (more population and more coal-burning power plants in the central latitudes of the eastern U.S. than at either more southern or more northern latitudes), the cubic surface perhaps less so. However, the result of the F test comparing the quadratic and cubic models strongly favors the cubic model. Evidence for the necessity of a surface of order greater than three is not compelling in light of previous discussion regarding the effect of positive spatial correlation on tests on regression coefficients. Hence we continue with an examination of the residual spatial correlation, entertaining both the quadratic and cubic models as possible mean structures.

Recall that the omnidirectional sample acf and sample semivariogram displayed in Figure 3.8 suggested that small-scale spatial dependence exists among the OLS residuals from a quadratic surface. The Diblasi-Bowman test confirms this suggestion (permutation-based $P = 0.006$). Figure 4.6 shows the omnidirectional sample semivariogram of the OLS residuals from a fitted cubic surface. These residuals exhibit considerably less spatial dependence than those from the fitted quadratic surface, both visually and on the basis of the Diblasi-Bowman test ($P = 0.22$). The Bowman-Crujeiras test for second-order stationarity finds it to be a plausible assumption for either the quadratic or cubic residuals ($P = 0.24$ and 0.46, respectively). As a consequence of these findings, in Chapter 8, models with either quadratic mean structure and stationary spatially correlated errors, or cubic mean structure and uncorrelated errors, among other models, will be estimated and compared for these data using likelihood-based methods.

Directional semivariograms (in the standard four directions) of residuals from the OLS fit of a cubic surface are shown in Figure 4.7. As was the case for the directional semivariograms of OLS residuals from the fitted quadratic surface shown in Figure 3.11, there is little evidence of anisotropy. In fact, since the cubic residuals were judged to be independent, it makes little sense to test them further for isotropy. It makes much more sense to do so for the residuals from the fitted quadratic surface. Using either the Guan-Sherman-Calvin test, the Bowman-Crujeiras test, or the Maity-Sherman test as they apply to the quadratic residuals in the standard four directions, the null hypothesis of isotropy is not rejected (the

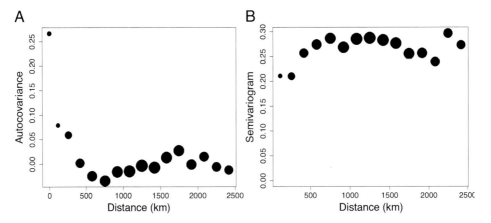

FIGURE 4.6
A) Omnidirectional sample acf and B) semivariogram of residuals, from the OLS fitted cubic surface for the clean wet sulfate deposition data.

P-values range from 0.31 to 0.44). Thus, for further analysis of the data that uses either a quadratic or cubic polynomial model for the mean structure, an isotropic model for the covariance structure seems justified.

4.5.2   The harbor seal trends data

One of the important questions about the harbor seals is whether there are differences in annual trends among the five stocks. A provisional answer to this question may be obtained by performing an ordinary least squares analysis of the data, taking the mean structure to

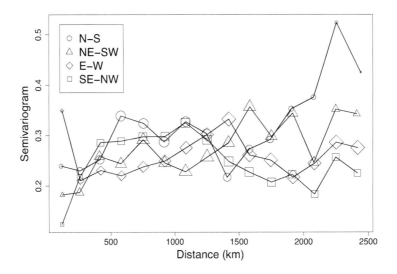

FIGURE 4.7
Directional sample semivariograms of residuals from the OLS fitted cubic surface for the clean wet sulfate deposition data.

TABLE 4.4

OLS estimates of trends in harbor seal populations (on the log scale), and estimated standard errors, for the five stocks.

Parameter	OLS estimate	Standard error
α_8	-0.079	0.023
α_9	-0.032	0.020
α_{10}	0.014	0.017
α_{11}	0.053	0.017
α_{12}	0.005	0.013

be a simple one-way classificatory model with stocks as the single factor of classification. Although the spatial coordinates of the haulouts are known, researchers believe they are unlikely to be helpful for explaining stock trends, so they are not included in the mean structure. Thus, after relabeling the observations using double subscripts, with the first subscript labeling the stocks (which are numbered from 8–12, cf. Section 1.1) and the second labeling the haulouts used by a given stock, the model is taken to be

$$y_{ij} = \alpha_i + e_{ij} \quad (i = 8, 9, 10, 11, 12; \ j = 1, 2, \ldots, n_i),$$

where α_i is the mean annual trend for the ith stock, n_i is the number of haulouts used by stock i and the e_{ij}'s are independent normal random variables with common mean zero and common variance σ^2.

Ordinary least squares estimates of annual trends for the five stocks are given in Table 4.4. Estimated standard errors of the ordinary least squares estimates are also provided, but these must be "taken with a grain of salt" because they are calculated under a model that ignores the possible presence of spatial correlation in the errors. Because the trends were measured on the log scale, the OLS estimates must be exponentiated to yield trend estimates in the original scale. Thus, for stock 8, we obtain $\exp(-0.079) = 0.92$, meaning that stock 8 is estimated to have decreased an average of 8% per year over the 14-year period. It appears that stocks 8 and 9 have decreased and stocks 10–12 have increased over the period, although the only trends that are statistically significant (at the 0.05 level) are those of stocks 8 and 11. Pairwise comparisons (also at the 0.05 level) suggest that the trends in stocks 10–12 are different from the trend in stock 8. Of course, these are only tentative assessments, based as they are on a model with independent errors. More reliable comparisons await improved estimates of stock trends and standard errors from models that account for spatial correlation, and will be performed in Chapter 8.

4.5.3 The caribou forage data

A linear model for the caribou forage data that is based purely on the experimental design is the two-factor crossed classificatory model with interaction, given by

$$y_{ijk} = \mu + \alpha_i + \gamma_j + \tau_{ij} + e_{ijk} \quad (i = 1, 2; \ j = 1, 2, 3; \ k = 1, 2, 3, 4, 5).$$

Here y_{ijk} is the percent nitrogen in the third clipping of prostrate willows in 1995 in the kth replicate of the ith level of the "Water" factor and the jth level of the "Tarp" factor, α_i represents the effect of adding water ($i = 1$) or not ($i = 2$), γ_j represents the effect of a clear tarp ($j = 1$), shade tarp ($j = 2$), or none ($j = 3$), and τ_{ij} represents the interaction effect of levels i and j of the two factors. This model is overparameterized, rendering the parameters nonestimable and resulting in a model matrix $\mathbf{X}$ of less than full column rank.

TABLE 4.5

Analysis of variance of the two-factor crossed model with interaction for the caribou forage data. Entries for sums of squares and mean squares are 10^3 times their actual values.

Source	df	Sum of squares	Mean square	F	*P*-value
Water	1	9.29	9.29	0.25	0.622
Tarp	2	353.01	176.50	4.74	0.018
Water × Tarp	2	52.15	26.08	0.70	0.506
Error	24	893.04	37.21		
Corrected Total	29	1307.49			

By imposing constraints on the parameters, those parameters may be "identified," making them estimable and yielding a model matrix of full column rank. One set of constraints that identifies the parameters is as follows:

$$\sum_{i=1}^{2} \alpha_i = \sum_{j=1}^{3} \gamma_j = \sum_{i=1}^{2} \tau_{ij} = \sum_{j=1}^{3} \tau_{ij} = 0, \tag{4.9}$$

where the latter two sums must hold for all j and all i, respectively. Another set of identifying constraints is

$$\alpha_1 = \gamma_1 = 0, \ \tau_{1j} = 0 \text{ for all } j, \ \tau_{i1} = 0 \text{ for all } i. \tag{4.10}$$

This second constraint set is the default in the lm function in R.

Table 4.5 gives a standard OLS-based analysis of variance of these data. This is a sequential ANOVA, but because the experimental design is balanced (i.e., all treatment combinations are equally replicated) the order of main effect terms, "Water" and "Tarp," in the ANOVA makes no difference provided that they both appear before their interaction, and the interpretation of their importance is unequivocal. The ANOVA indicates that the model explains about 32% of the overall variation in the data ($R^2 = 0.317$). There is not a significant interaction between the two factors, and neither is there a significant difference in the response to the Water factor. However, the Tarp factor is statistically significant. Level means for this factor are 1.942 for a clear tarp, 1.992 for no tarp, and 2.193 for a shaded tarp, and pairwise comparisons among these indicate that the only significant differences are between the shaded tarp and the other two. Of course, these results are tentative because they may be based on incorrect assumptions (homoscedasticity and independence) about the model errors.

Boxplots of residuals at each of the six factor combinations (Figure 1.6) suggest that there is no relationship between the variance and the mean, and that there are no outliers, so homoscedasticity of errors seems to be a reasonable assumption. Next we investigate whether there is spatial correlation among the residuals. Using the rook's definition of neighbors and row-standardized weights in the spatial weights matrix, Moran's and Geary's tests for positive spatial correlation yield permutation-based *P*-values of 0.021 and 0.016, respectively. Thus, there is considerable evidence that the residuals are spatially correlated. However, neither the test of Lu & Zimmerman (2001) nor that of Guan *et al.* (2004) indicate any evidence of anisotropy.

The two-factor model with interaction fitted above has no explicitly spatial terms in its mean structure. Since the data sites lie on a regular rectangular grid, and since there are ample degrees of freedom for error as a result of sufficient replication of the factor levels, additive row and column effects that sum to zero may be added to the model, yielding

$$y_{ijk} = \mu + \alpha_i + \gamma_j + \tau_{ij} + r_{(ijk)} + c_{(ijk)} + e_{ijk} \quad (i = 1, 2; \ j = 1, 2, 3; \ k = 1, 2, 3, 4, 5),$$

TABLE 4.6
Analysis of variance of the two-factor crossed model with interaction, plus row and column effects, for the caribou forage data. Sums of squares are adjusted for all other model effects. Entries for sums of squares and mean squares are 10^3 times their actual values.

Source	df	Sum of squares	Mean square	F	*P*-value
Row	5	237.95	47.59	1.70	0.195
Column	4	246.07	61.52	2.20	0.119
Water	1	38.22	38.22	1.36	0.261
Tarp	2	248.42	124.21	4.44	0.031
Water $\times$ Tarp	2	62.62	31.31	1.12	0.353
Error	15	420.08	28.01		
Corrected Total	29	1307.49			

where $r_{(ijk)}$ and $c_{(ijk)}$ are effects of the row and column, respectively, in which plot ijk lies. The analysis of variance of the resulting model is listed in Table 4.6; note that the sum of squares associated with each model term is adjusted for the presence of all other terms, and these sums of squares do not sum to the total sum of squares because of the nonorthogonality of the Water and Tarp factors with respect to rows and columns. Adding the coordinate effects increases the amount of variation explained by the model substantially, to 68%, though neither rows nor columns are statistically significant. As before, only the Tarp factor is statistically significant, and the least squares means for its three levels (which adjust for all other factors) are 1.909 for a clear tarp, 2.042 for no tarp, and 2.177 for the shade tarp. Hence there are some small differences among the estimated Tarp level means after adjusting for the coordinate effects, but the difference between the shade and clear tarps is still statistically significant.

Moran's and Geary's tests for autocorrelation were performed for the residuals from this model also. *P*-values were 0.78 for Moran's test and 0.75 for Geary's test. Thus, it appears that adding row and column effects to the model eliminates residual spatial correlation. In Chapter 8, it will be determined whether this model or a model without coordinate effects but with spatially correlated errors fits better.

4.5.4 The moss heavy metals data

For the moss heavy metals data, we would like to test whether heavy metals are more concentrated along the road, whether there is a difference in years (recall that better dust covers for trucks were implemented between 2001 and 2006, and dust from trucks is the likely contributor to lead contamination), and whether there is a difference between the two sides of the road (due to prevailing winds). Hence, we consider a linear model with two categorical variables and a continuous explanatory variable,

$$y_{ijk} = \mu + \alpha_i + \gamma_j + \beta x_k + \tau_j x_k + e_{ijk} \ (i = 1, 2; \ j = 1, 2; \ k = 1, 2, \dots, n_{ij}). \quad (4.11)$$

Here, y_{ijk} is the natural logarithm of lead (Pb) concentration in tissue samples of mosses for the ith year on the jth side of the road at log(distance-from-road) x_k for the kth replicate observation for each (i, j) combination, μ is an overall mean, α_i is a categorical effect for year, either 2001 or 2006, γ_j is a categorical effect indicating whether the sample was taken north or south of the haul road, x_k is log(distance-from-road) with regression parameter β, and τ_j is an interaction regression parameter that adjusts β for side-of-road. While we fit this model using OLS here, the data are a little more complicated than indicated by the model, even beyond spatial considerations. At some locations, a second sample was

TABLE 4.7

Analysis of variance of model (4.11) for the moss heavy metals data.

Source	df	Sum of squares	Mean square	F	P-value
Year	1	6.34	6.34	26.37	< 0.0001
Side-of-road	1	130.93	130.93	544.60	< 0.0001
log(Distance-from-road)	1	545.24	545.24	2267.96	< 0.0001
Side-of-road					
× log(Distance-from-road)	1	3.65	3.65	15.18	0.0001
Error	360	86.55	0.24		
Corrected Total	364	772.71			

collected, called a field duplicate, and some samples were analyzed twice at the laboratory, providing estimates of measurement error due to variation within the laboratory. A better approach would be based on a linear mixed model, where we treat location as a random effect and estimate parameters using generalized least squares (GLS). We will, in fact, fit such a model to these data later. For now, we simply ignore field duplicates and laboratory replicates, combining all errors and assuming them to be independent.

Table 4.7 gives an OLS-based sequential ANOVA for the model described above. It is clear that all factors are important, based on very high F-values and small P-values. We included the Side-of-road × log(Distance-from-road) interaction. This indicates that the slope of the line for log(Distance-from-road) is different on each side of the road. We also fitted a model with a Year × log(Distance-from-road) interaction (which would indicate a different slope for each year), but it was not statistically significant. The OLS estimates are given in Table 4.8. Recall that the model is overparameterized, so that μ is the intercept corresponding to 2001 and the north side of the road. However, for 2006, the intercept is 0.682 less, which suggests that the dust coverings on trucks may have helped to lower lead concentration in mosses between 2001 and 2006 (although other factors could be involved, and it should be emphasized that this was an observational study — not a designed experiment). The intercept is also lower, by 0.184, on the south side of the road, which makes sense because the prevailing wind is from the south. The slope on the south side of the road is more steeply negative by 0.087, indicating more rapid decay in lead concentration on that side. Hence, the fitted model makes sense based on our understanding of the data and the ecological setting from which they were collected.

As with all data analyses, it is important to evaluate the model with diagnostics. Because the model includes several factors, a plot of the data and the fitted model, as given by Figure 4.8, is helpful for interpretation and for providing assurance that the model is reasonable and there are no errors in computer code, etc. It is sometimes difficult to understand

TABLE 4.8

OLS estimates of the parameters of model (4.11) for the moss heavy metals data, and corresponding estimated standard errors.

Parameter	OLS estimate	Standard error
μ	7.426	0.131
$\alpha_{i=2006}$	−0.682	0.058
$\gamma_{j=\text{South}}$	−0.184	0.173
β	−0.515	0.017
$\tau_{j=\text{South}}$	−0.087	0.022

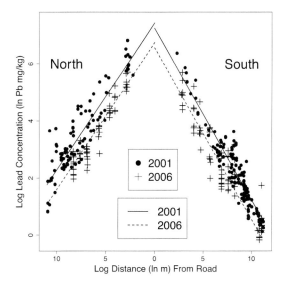

FIGURE 4.8
Log concentrations of lead in moss tissues as a function of year, side of the road (numbers to the left of 0 on the horizontal axis are north while numbers to the right are south), and log of the distance from the road, with the OLS fit of model (4.11) superimposed.

overparameterized models with multiple factors, some of which are categorical, and some continuous, so here we make the fits explicit. We have, essentially, four intercepts (two each for year and side-of-road), and hence, four lines, but the slopes only differ for the side of the road. The four intercepts are: 1) 7.426 for 2001 and north, 2) $7.426 - 0.682 = 6.744$ for 2006 and north, 3) $7.426 - 0.184 = 7.242$ for 2001 and south, and 4) $7.426 - 0.682 - 0.184 = 6.560$ for 2006 and south. The two slopes are: 1) -0.515 for north, and 2) $-0.515 - 0.087 = -0.602$ for south. These four lines are plotted over the raw data in the figure. The lines fit the data well, and show, at a glance, that concentration decreases with distance from the road, that concentrations went down between 2001 and 2006, and concentrations are lower to the south than the north.

Of course, this book is primarily concerned with spatial models, so, the next obvious question is, "Do we still need to fit a spatial model to these data?" Overall, the model explains about 89% of the variation ($R^2 = (772.71 - 686.16)/772.71 = 0.89$ using Table 4.7), and perhaps the explanatory variables have explained all of the spatial variation, leaving residuals that can be effectively modeled as independent. Just based on what we already know, the raw data *should* be strongly spatially correlated, because sites near the road will be more similar to each other than to sites far from the road, and vice versa. This is true when we look at a semivariogram of the raw data in Figure 4.9A. When we look at the semivariogram for residuals (Figure 4.9B), we see that the overall variance decreases about 10-fold (the sill is nearly 2.0 for the raw data, but only about 0.2 for the residuals), but that there is still some evidence of spatial correlation in the residuals. When breaking down the semivariogram by direction, notice that, in Figure 4.9C, the semivariogram of the raw data starts lower, and increases more slowly, in the 45-degree direction than in any of the other directions. Looking at the map of the data (Figure 1.8), this is not surprising because the road is oriented southwest to northeast, and we would expect the most similar values to be parallel to the road. However, for the residuals, which have accounted for the effect of

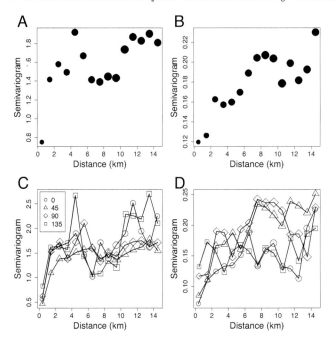

FIGURE 4.9
Sample semivariograms of the raw moss heavy metals data and residuals from the linear model: A) omidirectional semivariogram of raw data; B) omidirectional semivariogram of residuals; C) directional semivariograms of raw data at clockwise rotations 0, 45, 90, and 135 degrees from true north; D) directional semivariograms of residuals at same rotations as raw data.

distance-to-road, the 45-degree directional semivariogram is not necessarily the lowest, nor is it increasing more slowly than the semivariograms in other directions (Figure 4.9D).

4.6   Exercises

1. Consider a scenario in which four observations taken at sites $\mathbf{s}_1 = (u_1, v_1) = (1, 1)$, $\mathbf{s}_2 = (u_2, v_2) = (1, 2)$, $\mathbf{s}_3 = (u_3, v_3) = (3, 1)$, and $\mathbf{s}_4 = (u_4, v_4) = (3, 2)$. Suppose that the model for these observations is the Gauss-Markov model

$$y_i = \beta_0 + \beta_1 u_i + \beta_2 v_i + e_i \quad (i = 1, 2, 3, 4),$$

with $\mathbf{\Sigma} = \sigma^2 \mathbf{I}$.

   (a) Display the elements of the model matrix $\mathbf{X}$ for this scenario, and give expressions for the OLS estimators of β_0, β_1, and β_2 (as functions of the elements of $\mathbf{y}$) and for their covariance matrix.

   (b) If the third and fourth sites were moved to the left one unit (so that the four sites lie on the corners of a unit square rather than a rectangle), would you expect the variances of the ordinary least squares estimators of β_0, β_1, and

β_2 to increase or decrease from what they are for the original configuration? Explain.

(c) Verify your expectations stated in part (b), and show how, if at all, the OLS estimators change for the new configuration.

2. For the soil pH data introduced in Exercise 3.16:

   (a) Fit and plot polynomial surfaces, up to order four, by ordinary least squares.
   (b) Fit and plot an additive coordinate effects model. Is an additional term for nonadditivity warranted?
   (c) Select the best model.
   (d) For the selected model, compute the leverages, externally studentized residuals, and Cook's distance for each observation, and comment.
   (e) For the selected model, perform the Diblasi-Bowman test for spatial correlation, and if it is determined to exist, then test further for isotropy using the Guan-Sherman-Calvin test, the Lu-Zimmerman test, and the Bowman-Crujeiras test.

3. For the Meuse River zinc data introduced in Exercise 3.17:

   (a) Fit and plot polynomial surfaces, up to order four, by ordinary least squares.
   (b) Select the best polynomial model.
   (c) For the selected model, compute the leverages, externally studentized residuals, and Cook's distance for each observation, and comment.
   (d) For the selected model, perform the Diblasi-Bowman test for spatial correlation, and if it is determined to exist, then test further for isotropy using the Guan-Sherman-Calvin test and the Bowman-Crujeiras test.

4. For the wet sulfate deposition data:

   (a) Verify the results of Table 4.3.
   (b) Verify the results of the Diblasi-Bowman test for spatial correlation, and the Guan-Sherman-Calvin test, the Bowman-Crujeiras test, and the Maity-Sherman tests for isotropy presented in Section 4.5.1.

5. For the harbor seal trends data:

   (a) Verify the OLS estimates and standard errors listed in Table 4.4.
   (b) Perform Moran's test for spatial correlation on the residuals from the fitted one-factor classificatory model.

6. For the caribou forage data:

   (a) Verify the results (ANOVA table and means) of the OLS fit of the two-factor model with interaction, and verify the results of the analysis of the residuals from that fit reported in Section 4.5.3.
   (b) Verify the results (ANOVA table and least squares means) of the OLS fit of the two-factor model with interaction plus row and column effects, and verify the results of the analysis of the residuals from that fit reported in Section 4.5.3.

7. For the moss heavy metals data:

   (a) Verify the results (ANOVA table, OLS estimates, and standard errors) of the OLS-based analysis presented in Tables 4.7 and 4.8.
   (b) Verify the omnidirectional and directional sample semivariograms for the raw data and the OLS residuals, as displayed in Figure 4.9.

5

Generalized Least Squares Estimation of the Mean Structure

The previous chapter considered the estimation of the mean structure of a spatial linear model by the method of ordinary least squares (OLS). It was shown there that OLS can produce some useful results for data following a spatial linear model, but it is not fully satisfactory because it fails to account for spatial correlation. The extension of OLS known as generalized least squares (GLS) can properly account for spatial correlation and is the topic of this chapter. But GLS, like OLS, requires the specification, up to a multiplicative constant, of the covariance structure of model errors. Since the correlation structure generally is unknown, not even GLS is quite suited for the task. However, the likelihood-based estimation methods that are suitable for the task, which we will eventually consider in Chapter 8, are essentially equivalent to GLS insofar as the estimation of the mean structure is concerned, so it is important to gain an understanding of GLS estimation and inference procedures associated with it.

Our presentation begins with some theory and methodology associated with GLS estimation, which is followed by two illustrative toy examples that contrast GLS-based and OLS-based results. Then, the important class of spatial mixed linear models, to which GLS is applied, is introduced. Next is presented a careful study of spatial confounding, a phenomenon in which the covariance structure is purported to interfere, in some sense, with the estimation of the mean structure. After that, structures of the covariance matrix that can be exploited to carry out the necessary computations for GLS estimation most efficiently are considered. The chapter concludes with a brief introduction to empirical GLS, which is the form of GLS applicable to the spatial general linear model.

5.1 Some theory and methodology

Consider a spatial Aitken linear model with arbitrary positive definite covariance matrix $\boldsymbol{\Sigma} = \sigma^2 \mathbf{R}$, where $\mathbf{R}$ is known and σ^2 is an unknown positive parameter. If the underlying process is variance stationary with variance σ^2, then $\mathbf{R}$ is a correlation matrix, but generally not otherwise; therefore, we refer to $\mathbf{R}$ in general as the scale-free covariance matrix of $\mathbf{y}$. Because $\mathbf{R}$ is symmetric and positive definite, another symmetric and positive definite matrix, referred to as the **positive square root** of $\mathbf{R}$ and denoted by $\mathbf{R}^{\frac{1}{2}}$, exists such that $\mathbf{R}^{\frac{1}{2}}\mathbf{R}^{\frac{1}{2}} = \mathbf{R}$ (Theorem A.4 in Appendix A). Multiplying both sides of the spatial linear model equation by $\mathbf{R}^{-\frac{1}{2}} \equiv (\mathbf{R}^{\frac{1}{2}})^{-1}$ yields the "transformed" model equation

$$\mathbf{R}^{-\frac{1}{2}}\mathbf{y} = \mathbf{R}^{-\frac{1}{2}}\mathbf{X}\boldsymbol{\beta} + \mathbf{R}^{-\frac{1}{2}}\mathbf{e}.$$

Observe, by (2.27) and (2.28), that $\mathrm{E}(\mathbf{R}^{-\frac{1}{2}}\mathbf{e}) = \mathbf{0}$ and $\mathrm{Var}(\mathbf{R}^{-\frac{1}{2}}\mathbf{e}) = \mathbf{R}^{-\frac{1}{2}}(\sigma^2 \mathbf{R}^{\frac{1}{2}}\mathbf{R}^{\frac{1}{2}})\mathbf{R}^{-\frac{1}{2}} = \sigma^2 \mathbf{I}$. Thus, the model errors in the transformed model satisfy the Gauss-Markov model

DOI: 10.1201/9780429060878-5

assumptions. The sum of squares of the transformed model's errors is

$$[\mathbf{R}^{-\frac{1}{2}}\mathbf{y} - \mathbf{R}^{-\frac{1}{2}}\mathbf{X}\boldsymbol{\beta}]^T[\mathbf{R}^{-\frac{1}{2}}\mathbf{y} - \mathbf{R}^{-\frac{1}{2}}\mathbf{X}\boldsymbol{\beta}] = (\mathbf{y} - \mathbf{X}\boldsymbol{\beta})^T\mathbf{R}^{-1}(\mathbf{y} - \mathbf{X}\boldsymbol{\beta}) \equiv GRSS(\boldsymbol{\beta}; \mathbf{y}, \mathbf{R}),$$

which, as a function of $\boldsymbol{\beta}$, is called the **generalized residual sum of squares** function. As a generalization of the situation under Gauss-Markov assumptions, it turns out that a p-vector is a minimizer of $GRSS(\boldsymbol{\beta}; \mathbf{y}, \mathbf{R})$ if and only if it is a solution to the **Aitken equations**

$$\mathbf{X}^T\boldsymbol{\Sigma}^{-1}\mathbf{X}\boldsymbol{\beta} = \mathbf{X}^T\boldsymbol{\Sigma}^{-1}\mathbf{y},$$

or equivalently the equations

$$\mathbf{X}^T\mathbf{R}^{-1}\mathbf{X}\boldsymbol{\beta} = \mathbf{X}^T\mathbf{R}^{-1}\mathbf{y}.$$

The unique solution to the Aitken equations,

$$\hat{\boldsymbol{\beta}}_{GLS} \equiv (\mathbf{X}^T\boldsymbol{\Sigma}^{-1}\mathbf{X})^{-1}\mathbf{X}^T\boldsymbol{\Sigma}^{-1}\mathbf{y} = (\mathbf{X}^T\mathbf{R}^{-1}\mathbf{X})^{-1}\mathbf{X}^T\mathbf{R}^{-1}\mathbf{y}, \quad (5.1)$$

is called the **generalized least squares (GLS) estimator** of $\boldsymbol{\beta}$. Observe that we have replaced the subscript "*OLS*" with "*GLS*" to distinguish the GLS estimator from the OLS estimator. The GLS estimator of an arbitrary linear function $\boldsymbol{\ell}^T\boldsymbol{\beta}$ is $\boldsymbol{\ell}^T\hat{\boldsymbol{\beta}}_{GLS}$ and the **GLS fitted values** are given by $\hat{y}_{i,GLS} = \mathbf{x}_i^T\hat{\boldsymbol{\beta}}_{GLS}$ $(i = 1, \ldots, n)$. In the special case in which the vector of explanatory variables $\mathbf{x}$ comprises only the spatial coordinates or functions thereof, the **fitted GLS surface** is defined as $\{\hat{y}_{GLS}(\mathbf{s}) : \mathbf{s} \in \mathcal{A}\}$ where $\hat{y}_{GLS}(\mathbf{s}) = [\mathbf{x}(\mathbf{s})]^T\hat{\boldsymbol{\beta}}_{GLS}$. In contrast to spatial smoothing by OLS, this type of spatial smoothing is performed using the assumed mean structure *and* covariance structure.

The GLS estimator of an arbitrary linear function $\boldsymbol{\ell}^T\boldsymbol{\beta}$ is the best (minimum variance) linear unbiased estimator (BLUE) of that function under the Aitken model, with covariance matrix given by

$$\text{Var}(\hat{\boldsymbol{\beta}}_{GLS}) = (\mathbf{X}^T\boldsymbol{\Sigma}^{-1}\mathbf{X})^{-1} = \sigma^2(\mathbf{X}^T\mathbf{R}^{-1}\mathbf{X})^{-1}. \quad (5.2)$$

Furthermore, the **generalized residual mean square**

$$\hat{\sigma}^2_{GLS} = (\mathbf{y} - \mathbf{X}\hat{\boldsymbol{\beta}}_{GLS})^T\mathbf{R}^{-1}(\mathbf{y} - \mathbf{X}\hat{\boldsymbol{\beta}}_{GLS})/(n - p) \quad (5.3)$$

is the best quadratic unbiased estimator of σ^2 under the assumed model, and

$$\hat{\sigma}^2_{GLS}(\mathbf{X}^T\mathbf{R}^{-1}\mathbf{X})^{-1} \quad (5.4)$$

is an unbiased estimator of the variance of $\hat{\boldsymbol{\beta}}_{GLS}$. Under an assumption of normal errors, $\hat{\boldsymbol{\beta}}_{GLS}$ is normally distributed and $(n-p)\hat{\sigma}^2_{GLS}/\sigma^2$ is distributed independently as chi-square with $n - p$ degrees of freedom. Also under normality, a size-ϵ test of $H_0 : \boldsymbol{\ell}^T\boldsymbol{\beta} = \ell_0$ versus $H_a : \boldsymbol{\ell}^T\boldsymbol{\beta} \neq \ell_0$ may be performed by rejecting H_0 if and only if

$$\frac{|\boldsymbol{\ell}^T\hat{\boldsymbol{\beta}}_{GLS} - \ell_0|}{\hat{\sigma}_{GLS}\sqrt{\boldsymbol{\ell}^T(\mathbf{X}^T\mathbf{R}^{-1}\mathbf{X})^{-1}\boldsymbol{\ell}}} > t_{\epsilon/2, n-p}.$$

Furthermore, a $100(1 - \epsilon)\%$ confidence interval for $\boldsymbol{\ell}^T\boldsymbol{\beta}$ is

$$\boldsymbol{\ell}^T\hat{\boldsymbol{\beta}}_{GLS} \pm t_{\epsilon/2, n-p}\hat{\sigma}_{GLS}\sqrt{\boldsymbol{\ell}^T(\mathbf{X}^T\mathbf{R}^{-1}\mathbf{X})^{-1}\boldsymbol{\ell}}. \quad (5.5)$$

The graphical GLS equivalent of the accompanying partial regression plots are obtained by plotting $\mathbf{R}^{-1/2}$ times the residuals from the GLS regression of the response on all of the explanatory variables except x_j, against $\mathbf{R}^{-1/2}$ times the residuals from the GLS regression of x_j on all the other explanatory variables. To test the joint linear hypothesis $H_0 : \mathbf{L}^T\boldsymbol{\beta} = \boldsymbol{\ell}_0$ against $H_a : \mathbf{L}^T\boldsymbol{\beta} \neq \boldsymbol{\ell}_0$, where $\mathbf{L}^T$ is a specified $q \times p$ matrix with linearly independent rows, the test statistic

$$F = \frac{(\mathbf{L}^T\hat{\boldsymbol{\beta}}_{GLS} - \boldsymbol{\ell}_0)^T[\mathbf{L}^T(\mathbf{X}^T\mathbf{R}^{-1}\mathbf{X})^{-1}\mathbf{L}]^{-1}(\mathbf{L}^T\hat{\boldsymbol{\beta}}_{GLS} - \boldsymbol{\ell}_0)}{q\hat{\sigma}^2_{GLS}} \tag{5.6}$$

may be compared to the upper ϵ percentage point of an F distribution with q and $n - p$ degrees of freedom. Like its counterpart for a Gauss-Markov model, this statistic has an alternative representation in terms of (generalized) residual sums of squares under the full and reduced models. In this case, the statistic may be written as

$$F = \frac{[GRSS_{\text{reduced}}(\hat{\boldsymbol{\beta}}_{GLS}; \mathbf{y}, \mathbf{R}) - GRSS_{\text{full}}(\hat{\boldsymbol{\beta}}_{GLS}; \mathbf{y}, \mathbf{R})]/[df_{\text{reduced}} - df_{\text{full}}]}{GRSS_{\text{full}}(\hat{\boldsymbol{\beta}}_{GLS}; \mathbf{y}, \mathbf{R})/df_{\text{full}}}.$$

Such statistics may be used to select models using the same strategies described for Gauss-Markov models in Chapter 4.

The **generalized residual plot** plays a similar role for GLS that the (ordinary) residual plot plays for OLS. This is a plot of the elements of the vector $\mathbf{R}^{-1/2}(\mathbf{y} - \mathbf{X}\hat{\boldsymbol{\beta}}_{GLS})$ versus the corresponding elements of $\mathbf{R}^{-1/2}\mathbf{X}\hat{\boldsymbol{\beta}}_{GLS}$ (or versus $\mathbf{R}^{-1/2}$ times the vector of any one of the explanatory variables). Lack of trend or any other type of systematic pattern in the plot indicates that the fitted model is plausible with respect to both its mean structure and covariance structure. If the points do exhibit a trend or pattern, then the same actions described previously for improving a Gauss-Markov model, e.g., transforming the response or adding/transforming the explanatory variables, should be taken.

Once we have fitted the spatial linear model, it is possible to create overall and sequential ANOVA tables similar to Tables 4.1 and 4.2 that allow and account for spatial correlation. This is accomplished by replacing $\mathbf{y}$ and $\mathbf{X}$ in those tables with $\mathbf{R}^{-\frac{1}{2}}\mathbf{y}$ and $\mathbf{R}^{-\frac{1}{2}}\mathbf{X}$, respectively, and similarly replacing $\mathbf{P_X}$ with $\mathbf{P}_{\mathbf{R}^{-\frac{1}{2}}\mathbf{X}} = \mathbf{R}^{-\frac{1}{2}}\mathbf{X}[(\mathbf{R}^{-\frac{1}{2}}\mathbf{X})^T\mathbf{R}^{-\frac{1}{2}}\mathbf{X}]^{-1}(\mathbf{R}^{-\frac{1}{2}}\mathbf{X})^T = \mathbf{R}^{-\frac{1}{2}}\mathbf{X}(\mathbf{X}^T\mathbf{R}^{-1}\mathbf{X})^{-1}\mathbf{X}^T\mathbf{R}^{-\frac{1}{2}}$. Note that the rank of $\mathbf{R}^{-\frac{1}{2}}\mathbf{X}$ is equal to the rank of $\mathbf{X}$, so these ANOVA tables, as algebraic partitionings of degrees of freedom and sums of squares, are valid for the Aitken model.

The Aitken-extended overall ANOVA table also provides a way to generalize the coefficient of determination, R^2, to the Aitken model. In this case, the ratio of the model sum of squares to the total sum of squares may be written as

$$R^2(\boldsymbol{\Sigma}) = \frac{(\mathbf{X}\hat{\boldsymbol{\beta}}_{GLS} - \mathbf{1}\hat{\beta}_{0,GLS})^T\boldsymbol{\Sigma}^{-1}(\mathbf{X}\hat{\boldsymbol{\beta}}_{GLS} - \mathbf{1}\hat{\beta}_{0,GLS})}{(\mathbf{y} - \mathbf{1}\hat{\beta}_{0,GLS})^T\boldsymbol{\Sigma}^{-1}(\mathbf{y} - \mathbf{1}\hat{\beta}_{0,GLS})}, \tag{5.7}$$

where $\hat{\beta}_{0,GLS}$ is the generalized least squares estimator of the single intercept parameter β_0. Expression (5.7) may be compared to its OLS counterpart given by (4.5).

The case diagnostics introduced in Section 4.1 for use with OLS estimation can be extended for use with GLS estimation also. In fact, several distinct extensions have been proposed. For example, as generalized leverages, Pregibon (1981) and Langford & Lewis (1998) proposed the diagonal elements of $\mathbf{R}^{-1/2}\mathbf{X}(\mathbf{X}^T\mathbf{R}^{-1}\mathbf{X})^{-1}\mathbf{X}^T\mathbf{R}^{-1/2}$, denoted henceforth by $h_{ii}^{\#}$; Martin (1992) suggested the diagonal elements of $\mathbf{P} = \mathbf{R}^{-1}\mathbf{X}(\mathbf{X}^T\mathbf{R}^{-1}\mathbf{X})^{-1}\mathbf{X}^T\mathbf{R}^{-1}$ (which is different from $\mathbf{P_X}$); and Christensen *et al.* (1992) proposed

$$h_{ii}^* = \frac{(\mathbf{x}_i^T - \mathbf{r}_i^T\mathbf{R}_{-i}^{-1}\mathbf{X}_{-i})(\mathbf{X}^T\mathbf{R}^{-1}\mathbf{X})^{-1}(\mathbf{x}_i^T - \mathbf{r}_i^T\mathbf{R}_{-i}^{-1}\mathbf{X}_{-i})^T}{r_{ii} - \mathbf{r}_i^T\mathbf{R}_{-i}^{-1}\mathbf{r}_i} \quad (i = 1, \ldots, n),$$

where $\mathbf{x}_i^T$ is the ith row of $\mathbf{X}$, $\mathbf{X}_{-i}$ is $\mathbf{X}$ with the ith row deleted, r_{ii} is the ith diagonal element of $\mathbf{R}$, $\mathbf{R}_{-i}$ is $\mathbf{R}$ with the ith row and ith column deleted, and $\mathbf{r}_i$ is the ith row of $\mathbf{R}$ without r_{ii}. An alternative, more computationally convenient expression for h_{ii}^* is p_{ii}/r^{ii} where p_{ii} and r^{ii} are the ith diagonal elements of $\mathbf{P}$ and $\mathbf{R}^{-1}$, respectively. Of these generalizations of leverage, we consider $h_{ii}^{\#}$ and h_{ii}^* to be the most suitable because they are invariant to the scaling used to define $\mathbf{R}$ and σ^2. Also, because $\sum_{i=1}^n h_{ii}^{\#} = p$, a yardstick for evaluating $h_{ii}^{\#}$ is $2p/n$, the same as the yardstick for h_{ii} in a Gauss-Markov model. Although it is known that $0 \leq h_{ii}^* \leq 1$, their sum is not fixed, which may explain why no yardstick for evaluating h_{ii}^* seems to have been proposed. For h_{ii}^*, therefore, we recommend using the disparity among cases as the criterion for identifying those that are potentially influential. As a general rule, relatively isolated boundary sites and sites where the values of the explanatory values differ appreciably from those of their neighbors (i.e., values of the explanatory variables that are spatial outliers), tend to have high or disparate $h_{ii}^{\#}$ and h_{ii}^*.

Shi & Chen (2009) showed that a suitable extension to GLS of the previously defined notion of an externally studentized residual is

$$t_i^* = z_i \left(\frac{n-p-1}{n-p-z_i^2} \right)^{\frac{1}{2}},$$

where

$$z_i = \frac{q_i/\sqrt{r^{ii}}}{\hat{\sigma}_{GLS}\sqrt{1-h_{ii}^*}}$$

and q_i is the ith element of $\mathbf{R}^{-1}(\mathbf{y} - \mathbf{X}\hat{\boldsymbol{\beta}}_{GLS})$. The null hypothesis of no outliers among the n observations may be tested conservatively at the ϵ level of significance by rejecting this hypothesis if and only if $\max_i |t_i^*| > t_{\epsilon/(2n),n-p-1}$.

An appropriate extension of Cook's D_i, when viewed as a measure of influence on the regression coefficients, is

$$D_i^* = \frac{z_i^2}{p} \left(\frac{h_{ii}^*}{1-h_{ii}^*} \right),$$

and its magnitude may be evaluated by the same yardsticks used to evaluate the original Cook's D_i (Shi & Chen, 2009). The corresponding extension of COVRATIO$_i$ is

$$\text{COVRATIO}_i^* = \left(\frac{n-p-z_i^2}{n-p-1} \right)^p \left(\frac{1}{1-h_{ii}^*} \right),$$

and the OLS yardsticks for COVRATIO$_i$ apply here also [less than $1 - (3p/n)$ or greater than $1 + (3p/n)$].

5.2 Two toy examples

This section presents two toy examples for the purpose of gaining some understanding of how GLS estimators and related quantities compare to their OLS counterparts. Consider first the toy example considered previously in Section 4.2. Recall that the spatial linear

TABLE 5.1
Observation weights for OLS, GLS, and EGLS
estimation for the toy example of Section 4.2.

i	OLS weights	GLS weights	EGLS weights
1	0.200	0.261	0.246
2	0.200	0.079	0.084
3	0.200	0.014	0.084
4	0.200	0.266	0.246
5	0.200	0.380	0.340

model in that example had model matrix $\mathbf{X} = \mathbf{1}_5$ and covariance matrix

$$\sigma^2 \mathbf{R} = \begin{pmatrix} 1.000 & 0.633 & 0.492 & 0.249 & 0 \\ & 1.000 & 0.633 & 0.492 & 0 \\ & & 1.000 & 0.633 & 0.061 \\ & & & 1.000 & 0.014 \\ & & & & 1.000 \end{pmatrix},$$

where the latter was built from the isotropic spherical covariance function

$$C(r) = \begin{cases} \sigma^2 \left(1 - \frac{3r}{8} + \frac{r^3}{128}\right) & \text{for } 0 \le r \le 4, \\ 0 & \text{for } r > 4, \end{cases}$$

with $\sigma^2 = 1$. Suppose that the observed response vector was $\mathbf{y} = (1, 0, 3, 1, 5)^T$, with elements ordered as in Figure 4.1.

Tables 5.1, 5.2, and 5.3 list some results from applying OLS and GLS to these data to estimate μ and σ^2. The ith entries ($i = 1, 2, 3, 4, 5$) in the columns labeled "OLS weights" and "GLS weights" in Table 5.1 give the multipliers of y_i in the OLS and GLS estimators of μ, and sum to one within each column because the unbiasedness of a linear estimator $\mathbf{a}^T \mathbf{y}$ requires for this particular mean structure that

$$\mu = \mathrm{E}(\mathbf{a}^T \mathbf{y}) = \mathbf{a}^T \mathrm{E}(\mathbf{y}) = \mathbf{a}^T \mathbf{1}_5 \mu = \left(\sum_{i=1}^{5} a_i\right) \mu.$$

The OLS estimator, $\hat{\mu}_{OLS}$, weights each observation equally, and therefore is equal to their sample mean, as noted previously. In contrast, the GLS estimator, $\hat{\mu}_{GLS}$, tends to give more weight to observations at those sites that are more isolated spatially from others. This is intuitively sensible, as more isolated observations contain "less redundant" information than more highly correlated observations in close spatial proximity to one another. The fifth observation, which is the most spatially isolated and is nearly uncorrelated with all the others, receives the most weight. As a consequence of the differential weighting and

TABLE 5.2
OLS, GLS, and EGLS estimates of μ and σ^2
for the toy example of Section 4.2.

Estimation method	$\hat{\mu}$	$\hat{\sigma}^2$
OLS	2.000	4.000
GLS	2.467	5.555
EGLS	2.445	5.123

TABLE 5.3

Standardized variances of the OLS and GLS estimators of μ and σ^2 under the Gauss-Markov and spatial linear models, for the toy example of Section 4.2.

Variance	$\hat{\mu}_{OLS}$	$\hat{\mu}_{GLS}$
$(1/\sigma^2)\mathrm{Var}_{GMM}$	0.200	0.289
$(1/\sigma^2)\mathrm{Var}_{SLM}$	0.456	0.384

the relatively large positive magnitude of the fifth observation, the GLS estimate in this case is considerably larger than the OLS estimate (Table 5.2). The estimate of σ^2 obtained by GLS is also larger than that obtained by OLS, again because the fifth observation, for which the deviation from the GLS estimate of μ is relatively large, receives more weight. The standardized variances (Table 5.3) show that the OLS estimator of μ has smaller variance than the GLS estimator (0.200 versus 0.289) when OLS assumptions hold, and vice versa (0.457 versus 0.384) when GLS assumptions hold, in keeping with optimality properties noted previously. Moreover, these results also reveal how badly the variance of the estimator of μ would be underestimated if GLS assumptions held but OLS rather than GLS was used; specifically, the GLS-based estimate is $5.555 \times 0.384 = 2.133$, while the OLS-based estimate is $4.000 \times 0.200 = 0.800$. Consequently, confidence intervals for μ with a given nominal probability of coverage are appropriately much wider when computed in the context of GLS inference rather than OLS inference; for example, the 95% GLS-based and OLS-based confidence intervals are 2.467 ± 4.055 and 2.000 ± 2.483, respectively.

Table 5.4 lists GLS-based case diagnostics and, for comparison, their OLS-based counterparts (which were introduced in Section 4.1) for the five observations. In contrast to the OLS-based ordinary leverages, which are all equal to 0.200, the generalized leverages are quite disparate and point to the fifth case as potentially more influential than the others. The disparity in generalized leverages among cases is due to the positive spatial correlation among the responses, with cases that are most highly correlated with (or equivalently most proximate to) other cases tending to have less potential influence. Apart from the fifth case, the generalized leverages are about equal to or smaller than the ordinary leverages. The ordinary and generalized Cook's distances for the fifth case are considerably larger than their counterparts for the other cases, establishing the fifth case as actually influential. The generalized externally studentized residual for the fifth case is only about half as large as its OLS counterpart, but neither maximum absolute externally studentized residual approaches statistical significance. All values of COVRATIO$_i^*$ lie within the unremarkable range, though it is noteworthy that some cases whose removal would decrease the variance of the OLS estimator of μ under Gauss-Markov assumptions would increase the variance of the GLS estimator under the spatial model, and vice versa.

TABLE 5.4

OLS-based and GLS-based case diagnostics for the toy example of Section 4.2. Here, COVRATIO$_i$ and COVRATIO$_i^*$ are abbreviated to COV$_i$ and COV$_i^*$.

i	h_{ii}	t_i	D_i	COV$_i$	$h_{ii}^{\#}$	h_{ii}^*	t_i^*	D_i^*	COV$_i^*$
1	0.200	-0.504	0.078	1.536	0.212	0.100	-0.238	0.008	1.454
2	0.200	-1.168	0.313	1.146	0.114	0.007	-1.070	0.008	0.972
3	0.200	0.504	0.078	1.536	0.090	< 0.001	1.511	< 0.001	0.757
4	0.200	-0.504	0.078	1.536	0.211	0.104	-0.777	0.078	1.238
5	0.200	2.666	0.703	0.495	0.375	0.372	1.328	0.878	1.338

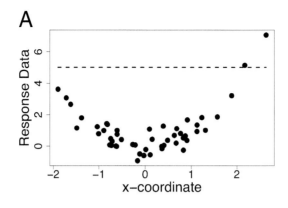

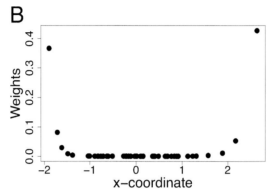

FIGURE 5.1
Simulated data with a bowl-shaped pattern. A) Simulated responses along a single x-coordinate, where the dashed line is the generalized least squares estimate of the mean; B) Weights used to compute the mean as a function of the x-coordinate of the datum.

As a second example, consider a "bowl-shaped" set of data along a transect, as displayed in Figure 5.1. These data were simulated from the model

$$y_i = x_i^2 + e_i \quad (i = 1, \ldots, 50),$$

where the x_i's and e_i's are independent and identically distributed normal random variates with zero means and standard deviations 1.0 and 0.5, respectively. For this example, we focus on the GLS weights and residuals. Suppose that a model with constant mean is fit to the data by generalized least squares, using a covariance structure obtained by first fitting an exponential covariance function to these data using methods to be described in later chapters. (Note that this fitted model does not coincide with the model used to generate the data.) The generalized least squares estimator of the mean is

$$\hat{\mu}_{GLS} = \frac{\mathbf{1}^T \mathbf{\Sigma}^{-1} \mathbf{y}}{\mathbf{1}^T \mathbf{\Sigma}^{-1} \mathbf{1}},$$

where $\mathbf{\Sigma}$ is the covariance matrix for the data corresponding to the fitted exponential covariance function. For these data, $\hat{\mu}_{GLS} = 5.01$, which is shown as a dashed line across Figure 5.1A. The weights applied to each datum are shown directly below, in Figure 5.1B, where the weights are computed as $\mathbf{w}^T = (\mathbf{1}^T \mathbf{\Sigma}^{-1})/(\mathbf{1}^T \mathbf{\Sigma}^{-1} \mathbf{1})$. There are two ways to understand Figure 5.1. First, from an intuitive standpoint, correlated data in time or space often "run away" from the mean for an extended span of time or distance, producing an oscillating pattern. However, if the covariance is proper (positive definite), then the data always return toward the mean. If we assume that the responses in Figure 5.1A are spatially correlated, then it makes sense, based on the shape of the data, that they are turning toward the mean at the endpoints. In that sense, the estimate of the mean is reasonable. Second, turning to Figure 5.1B, note that the weights at the edges are the largest. If we think about the "relative information content" of the observations, then the responses at sites in the middle are highly redundant, being correlated with their neighbors, while responses at the edges are most independent from all others, being the most distant from all others. Weighting of observations for estimating the mean, then, puts a premium on locations that are most independent, and gives them higher weights. Note that, after estimating the spatial covariance matrix (or assuming that it is known or fixed), the weights do not depend on the responses. The weights would be the same for any set of responses, such as those with a linear trend or an inverted bowl shape. Thus, from either an intuitive notion of returning toward the mean, or an understanding of spatial correlation and weighting, the GLS estimator of the mean makes sense.

This simple example (Figure 5.1) also illustrates that, for spatial linear models, generalized residuals generally *do not* sum to zero. This is important to keep in mind when performing model diagnostics, as many are computed from residuals. Model diagnostics developed from GLS residuals may behave quite differently than diagnostics developed from OLS residuals.

5.3 Why do OLS and GLS estimates and their variances differ?

The toy examples of the previous subsection revealed that the OLS and GLS estimates of a regression coefficient can be quite different. Numerous environmental and ecological studies have also demonstrated considerable differences (including flips in sign) in estimated regression coefficients, as well as changes in the relative importance of the corresponding explanatory variables, obtained by the two methods (Tognelli & Kelt, 2004; Dormann, 2007; Kühn, 2007). The first toy example and other studies have also shown that the variances of the OLS and GLS estimators often differ substantially. This has led many researchers to try to better understand mechanistically how and why the incorporation of spatial correlation in a GLS analysis changes the estimates and their variances (true and estimated).

Before considering this issue, it should be noted that it is not always the case that the OLS and GLS estimates differ. In fact, under certain conditions they are guaranteed to coincide, regardless of the observed responses; that is, the *estimators*, not merely the *estimates*, coincide. A necessary and sufficient condition for this is that a matrix $\mathbf{Q}$ exists such that $\mathbf{RX} = \mathbf{XQ}$, and an equivalent but often easier-to-verify condition is that $\mathbf{RX}(\mathbf{X}^T\mathbf{X})^{-1}\mathbf{X}^T$ is symmetric [see, e.g., Corollaries 11.3.1.1 and 11.3.1.2 of Zimmerman (2020)]. As an example of a scenario in which these conditions are satisfied, consider a spatial linear model for which the underlying process is second-order stationary (in which case $\mathbf{X} = \mathbf{1}_n$) and isotropic, the covariance function is compactly supported with range α, and the spatial configuration consists of any number of three-site clusters, where sites within a cluster are located at the vertices of an equilateral triangle, each triangle has the same size, and each pair of sites from different triangles lies more than α units apart. If observations located within the same cluster are placed consecutively into $\mathbf{y}$, then

$$\mathbf{R} = \begin{pmatrix} \mathbf{R}_\Delta & \mathbf{0} & \cdots & \mathbf{0} \\ \mathbf{0} & \mathbf{R}_\Delta & \cdots & \mathbf{0} \\ \vdots & \vdots & \ddots & \vdots \\ \mathbf{0} & \mathbf{0} & \cdots & \mathbf{R}_\Delta \end{pmatrix}, \quad \text{where } \mathbf{R}_\Delta = \begin{pmatrix} 1 & \rho & \rho \\ \rho & 1 & \rho \\ \rho & \rho & 1 \end{pmatrix}$$

and ρ is the common correlation among observations within the same cluster. Thus, $\mathbf{RX} = \mathbf{XQ}$ where $\mathbf{Q} = (1+2\rho)\mathbf{I}$, so the condition for the equality of the GLS and OLS estimators is satisfied. Furthermore, because the GLS and OLS estimators coincide, their variances under the spatial linear model also coincide, and are equal to $(1+2\rho)/n$. This is most easily seen using (4.6):

$$\begin{aligned} \text{Var}_{SLM}(\hat{\boldsymbol{\beta}}_{OLS}) &= \sigma^2(\mathbf{X}^T\mathbf{X})^{-1}\mathbf{X}^T\mathbf{RX}(\mathbf{X}^T\mathbf{X})^{-1} \\ &= \sigma^2(\mathbf{1}_n^T\mathbf{1}_n)^{-1}\mathbf{1}_n^T[(1+2\rho)\mathbf{I}]\mathbf{1}_n(\mathbf{1}_n^T\mathbf{1}_n)^{-1} \\ &= \sigma^2(1+2\rho)/n. \end{aligned}$$

The variance of the OLS estimator under the Gauss-Markov model is, of course, $\sigma^2(\mathbf{1}_n^T\mathbf{1}_n)^{-1} = \sigma^2/n$. This reveals that compared to the variance of the OLS estimator under the corresponding Gauss-Markov model, the variance of the GLS (and OLS) estimator under the spatial linear model is larger if ρ is positive, and smaller if ρ is negative. The vast majority of spatial linear models used in practice have only positive correlations, and the variance comparison is similar to what it is for this example: the variance of the GLS estimator under the spatial linear model is larger than the variance of the OLS estimator under a Gauss-Markov version of the model, but appropriately so.

Besides noting that the OLS and GLS estimators can coincide, it is worth reiterating that even when they do not coincide, they are both unbiased. Thus, neither estimator is systematically shifted to one side of the other, regardless of the nature of $\mathbf{X}$ and $\mathbf{R}$; their distributions are both centered on the true values of the regression coefficients. Where their distributions do generally differ is in their dispersions, with the OLS estimator being no more (and usually less) dispersed than the GLS estimator when the covariance matrix is $\sigma^2\mathbf{I}$, and vice versa when the covariance matrix is $\sigma^2\mathbf{R}$ (for $\mathbf{R} \neq \mathbf{I}$).

Because the OLS and GLS estimators are both linear combinations of the elements of $\mathbf{y}$, understanding why they are (usually) different requires an understanding of why the weights that the estimators assign to each observed response differ. In this regard, it is helpful to compare the ordinary and generalized leverages (h_{ii} and $h_{ii}^\#$) for each observation. If $h_{ii}^\#$ is larger than h_{ii}, then the ith case is potentially more influential for GLS than for OLS and thus receives relatively more weight in the GLS estimator than in the OLS estimator, and the opposite is true if $h_{ii}^\#$ is smaller than h_{ii}. For which cases is the discrepancy between $h_{ii}^\#$ and h_{ii} the greatest? Generally speaking, if the mean structure does not include spatial coordinates (or functions thereof), then $|h_{ii}^\# - h_{ii}|$ tends to be largest for cases that are either the least highly correlated, or the most highly correlated, with the others. When the process is second-order stationary and the spatial correlation decays with distance, cases that are the most spatially distant from the others, which are often among those lying on the convex hull of the set of data locations, tend to be the least highly autocorrelated and to have a positive value of $h_{ii}^\# - h_{ii}$, while cases lying within the interior of the convex hull of data locations tend to be the most highly autocorrelated and to have a negative value of $h_{ii}^\# - h_{ii}$. The first toy example of Section 5.2 provides an illustration. There, the case corresponding to the most isolated site (case 5) has the largest positive value of $h_{ii}^\# - h_{ii}$, and the only case lying inside the convex hull (case 3) has the largest (in magnitude) negative value. If, however, the mean structure includes spatial coordinates, then the situation is

TABLE 5.5
Ordinary and generalized leverages for an
extension of the first toy example in Section 5.2 to
a model with a planar mean structure.

i	h_{ii}	$h_{ii}^{\#}$
1	0.733	0.809
2	0.333	0.344
3	0.215	0.090
4	0.748	0.816
5	0.970	0.940

more complex because sites with high ordinary leverage also tend to be the most spatially isolated (and thus the least autocorrelated). Table 5.5 lists the ordinary and generalized leverages of the five cases of an example exactly like the first example of Section 5.2 except that it includes the coordinates of the sites as explanatory variables in addition to the overall constant, i.e., the mean structure is a planar surface. For this example, the ordinary and generalized leverages of the fifth case are much more similar (in fact, the ordinary leverage is the larger of the two), but there is still a substantial negative discrepancy for case 3, the "interior" site.

Precisely how the OLS and GLS estimates of a given regression coefficient compare to each other is determined not only by the weights that each estimator assigns to the observations, but also by the relative magnitudes of the observed elements of $\mathbf{y}$. If the elements of $\mathbf{y}$ that receive more weight in the GLS estimator tend to be larger than elements that receive more weight in the OLS estimator, then the GLS estimate will be larger than the OLS estimate (and vice versa). This is precisely the situation in the first toy example of Section 5.2: the case with the largest (by far) difference in generalized and ordinary leverages (case 5) coincides with the case with the largest response, rendering the GLS estimate of μ much larger than the OLS estimate.

As noted earlier in this section, the variances of the OLS estimator under the Gauss-Markov model and the GLS estimator under the spatial linear model are usually different. In fact, $\text{Var}_{SLM}(\hat{\beta}_{j,GLS})$ usually is larger than $\text{Var}_{GMM}(\hat{\beta}_{j,OLS})$ for $j = 1, 2, \ldots, p$ when the responses are positively spatially correlated. This phenomenon, referred to henceforth as the **variance inflation of GLS relative to OLS**, is often attributed to a reduction in the "effective number of independent observations," or equivalently a reduction in the **effective sample size** of the spatial linear model relative to the Gauss-Markov model, when the observations are positively correlated. To see this, consider once again the example with triangular clusters described earlier in this section. The variance of the GLS estimator of μ under the spatial linear model was shown to be $\sigma^2(1 + 2\rho)/n$, while the variance of the OLS estimator under the Gauss-Markov model was σ^2/n, revealing that the variance of the GLS estimator is inflated (when $\rho > 0$) by a factor of $1 + 2\rho$ relative to the OLS estimator. Or, equivalently, the effective sample size, defined for this example as the number of observations required for the variance of the OLS estimator under the Gauss-Markov model to equal the variance of the GLS estimator under the spatial linear model with n observations, is equal to $n/(1 + 2\rho)$, which is less than n if $\rho > 0$. For more general use, the effective sample size has been defined in various ways, depending on the inference objective [see, e.g., Griffith (2005); Vallejos & Osorio (2014); Acosta & Vallejos (2018)]. For inference on the regression coefficients in spatial multiple linear regression models, Acosta & Vallejos (2018) defined the effective sample size as

$$n^* = \frac{\text{tr}(\mathbf{X}^T \mathbf{R}^{-1} \mathbf{X})}{p}$$

for regressors that are normalized so that $\mathbf{x}_j^T \mathbf{x}_j = n$ for all $j = 2, \ldots, p$. When $\mathbf{X} = \mathbf{1}$, this definition reduces to the definition used for the triangular clusters example. Furthermore, using this definition, $n^* = (0.200/0.384)n \doteq 0.52n$ for the first toy example of Section 5.2, and if that example is extended to have a planar mean structure, then $n^* \doteq 0.66n$. Thus, the effective sample size depends not only on the covariance structure of the spatial linear model, but also on its mean structure.

Another interpretation of the variance inflation of GLS relative to OLS is in terms of multicollinearity between the fixed effects and the implicit random effects that generate the covariance structure of $\mathbf{y}$. We defer this interpretation to Section 5.5 because it requires some familiarity with spatial random and mixed effects models, which we introduce next.

5.4   Spatial random and mixed effects models

An important subclass of spatial linear models consists of those models in which the errors have a known linear structure. Such models are called **spatial random effects models** if no fixed-effects mean structure accompanies the error structure, i.e., if $\mathbf{X}$ is merely a vector whose elements are all equal. More generally, i.e., if $\mathbf{X}$ is arbitrary, they are called **spatial mixed effects models**. Because most models used in practice have a nonconstant mean structure, we present this subclass in terms of the general case. A spatial mixed effects model has the form

$$\mathbf{y} = \mathbf{X}\boldsymbol{\beta} + \mathbf{Zb} + \mathbf{d},$$

where $\mathbf{y}$, $\mathbf{X}$, and $\boldsymbol{\beta}$ are defined as for any spatial linear model, $\mathbf{Z}$ is a specified $n \times q$ matrix, $\mathbf{b}$ is a q-vector of zero-mean random variables (called random effects in this context), and $\mathbf{d}$ is an n-vector of zero-mean random variables. Furthermore,

$$\mathrm{Var}(\mathbf{b}) = \sigma^2 \mathbf{G}, \quad \mathrm{Var}(\mathbf{d}) = \sigma^2 \mathbf{D}, \quad \mathrm{Cov}(\mathbf{b}, \mathbf{d}) = \mathbf{0},$$

where $\mathbf{G}$ is a $q \times q$ symmetric positive definite matrix possessing spatial structure and $\mathbf{D}$ is a diagonal positive definite matrix. Thus, this model supposes that the error vector $\mathbf{e}$ of the spatial linear model may be decomposed additively as $\mathbf{Zb} + \mathbf{d}$ with the stated specifications of $\mathbf{Z}$, $\mathbf{b}$, and $\mathbf{d}$. Under this model,

$$\mathrm{Var}(\mathbf{e}) = \mathrm{Var}(\mathbf{Zb} + \mathbf{d}) = \sigma^2 (\mathbf{ZGZ}^T + \mathbf{D}),$$

which reveals that the scale-free covariance matrix is symmetric positive definite and has form

$$\mathbf{R} = \mathbf{ZGZ}^T + \mathbf{D}.$$

Now suppose that $\mathbf{G}$ and $\mathbf{D}$ are fully specified. (In reality, $\mathbf{G}$ and $\mathbf{D}$ may be unknown, in which case their elements, and thus those of $\mathbf{R}$ also, are taken to be known functions of an unknown parameter vector. We will deal with this later.) Then the spatial mixed effects model is merely an Aitken model with covariance matrix $\sigma^2(\mathbf{ZGZ}^T + \mathbf{D})$. By (5.1) and (5.3), the GLS estimator of $\boldsymbol{\beta}$ and the generalized residual variance are given by

$$\hat{\boldsymbol{\beta}}_{GLS} = [\mathbf{X}^T(\mathbf{ZGZ}^T + \mathbf{D})^{-1}\mathbf{X}]^{-1}\mathbf{X}^T(\mathbf{ZGZ}^T + \mathbf{D})^{-1}\mathbf{y}$$

and

$$\hat{\sigma}^2_{GLS} = \frac{(\mathbf{y} - \mathbf{X}\hat{\boldsymbol{\beta}}_{GLS})^T(\mathbf{ZGZ}^T + \mathbf{D})^{-1}(\mathbf{y} - \mathbf{X}\hat{\boldsymbol{\beta}}_{GLS})}{n - p}.$$

Prediction of the random effects, or of linear functions of the fixed effects and random effects, will be considered in Chapter 9.

One important special case of spatial mixed effects model is the classical spatial mixed linear model with measurement errors, for which $\mathbf{Z} = \mathbf{D} = \mathbf{I}_n$ so that $\mathbf{R} = \mathbf{G} + \mathbf{I}$ and

$$\mathbf{y} = \mathbf{X}\boldsymbol{\beta} + \mathbf{b} + \mathbf{d},$$

where $\mathbf{b}$ represents an n-vector of site-specific random effects and $\mathbf{d}$ represents a measurement error vector. By *site-specific* random effects, it is meant that there is a distinct random effect for each site. Geostatistical models and SAR/CAR/SMA models are important subclasses of such models. In other special cases, however, random effects may be fewer in number and shared among sites. For example, for a two-way experimental layout such as that of the caribou forage experiment, the row and column effects may be taken as random, with sites in the same row sharing the same random row effect, sites in the same column sharing the same random column effect, and all random effects independent. This introduces a degree of spatial dependence into the covariance structure, though not one that strictly decreases with increasing distance: if σ_r^2 and σ_c^2 represent the variances of the row and column effects, respectively, and $\mathbf{D} = \mathbf{I}$, then each observation is correlated by an amount $\sigma_r^2/(\sigma_r^2 + \sigma_c^2 + \sigma^2)$ with other observations in the same row and by $\sigma_c^2/(\sigma_r^2 + \sigma_c^2 + \sigma^2)$ with other observations in the same column, and is uncorrelated with all other observations (see Exercise 5.3).

Additional important special cases of spatial mixed effects models in which the random effects are shared among sites include the "fixed-rank kriging (FRK)" model of Cressie & Johannesson (2008), the Gaussian predictive process model of Banerjee *et al.* (2008), and the multiresolution Gaussian process model of Nychka *et al.* (2015). In these models, $\mathbf{b}$ represents a vector of random spatial location effects at a small number of locations (relative to n). More details about these models will be given in Chapter 8.

5.5 Spatial confounding and the red shift

Complete confounding in a classical Gauss-Markov model describes a situation in which two or more regression coefficients do not have unique least squares estimators because the regression variables to which they correspond are exactly collinear, i.e., at least one variable can be written as a linear combination of the other(s). Partial confounding in the same model refers to a less extreme situation in which the least squares estimators of the regression coefficients are unique but their variances are inflated as a result of approximate, rather than exact, collinearity among the regression variables. Confounding (complete or partial) due to collinearity (exact or approximate) among the regression variables can occur in a spatial linear model also, where it leads to variance inflation in the generalized least squares estimators. However, in a large class of spatial linear models another kind of confounding can occur, which is due to the approximate collinearity of the regression variables corresponding to the model's fixed effects with variables corresponding to implicit random effects that yield the model's covariance structure. In this section, we describe confounding of this other type, which is known as **spatial confounding**. Previous works that have described and studied spatial confounding include Clayton *et al.* (1993), Reich *et al.* (2006), Hodges & Reich (2010), Paciorek (2010), Hughes & Haran (2013), and Hanks *et al.* (2015), but the context for most of them was Bayesian inference for a class of spatial generalized linear mixed models. We present spatial confounding here in the context of frequentist OLS- and GLS-based inference for the corresponding class of spatial linear models, which is considerably more transparent. This inferential context may also be viewed as a special case of the Bayesian inferential context, with a flat prior and squared error loss function.

Consider the class of spatial linear models

$$\mathbf{y} = \mathbf{X}\boldsymbol{\beta} + \mathbf{b} + \mathbf{d}, \quad \begin{pmatrix} \mathbf{b} \\ \mathbf{d} \end{pmatrix} \sim \mathrm{N}_{2n}\left(\begin{pmatrix} \mathbf{0} \\ \mathbf{0} \end{pmatrix}, \sigma^2 \begin{pmatrix} \mathbf{G} & \mathbf{0} \\ \mathbf{0} & \mathbf{I} \end{pmatrix} \right), \tag{5.8}$$

where $\mathbf{G}$ is positive definite. This is the class of spatial mixed effects models introduced in Section 5.4, for which the random effects are site-specific and $\mathbf{R} = \mathbf{G} + \mathbf{I}$, supplemented by a normality assumption. Let $\mathbf{Q}\boldsymbol{\Lambda}\mathbf{Q}^T$ be the eigen (spectral) decomposition of $\mathbf{G}$, meaning that $\mathbf{G} = \mathbf{Q}\boldsymbol{\Lambda}\mathbf{Q}^T$ where $\boldsymbol{\Lambda}$ is an $n \times n$ diagonal matrix whose main diagonal elements are the eigenvalues of $\mathbf{G}$, and $\mathbf{Q}$ is the $n \times n$ matrix whose columns are the orthonormal eigenvectors of $\mathbf{G}$ (the ith column corresponding to the ith main diagonal element of $\boldsymbol{\Lambda}$). The existence of this eigen decomposition is guaranteed by Theorem A.5 in Appendix A. Define $\mathbf{Q}^* = \mathbf{Q}\boldsymbol{\Lambda}^{\frac{1}{2}}$. Then consider the ostensibly different class of spatial mixed effects models

$$\mathbf{y} = \mathbf{X}\boldsymbol{\beta} + \mathbf{Q}^*\mathbf{b}^* + \mathbf{d}, \quad \begin{pmatrix} \mathbf{b}^* \\ \mathbf{d} \end{pmatrix} \sim \mathrm{N}_{2n}\left(\begin{pmatrix} \mathbf{0} \\ \mathbf{0} \end{pmatrix}, \sigma^2 \begin{pmatrix} \mathbf{I} & \mathbf{0} \\ \mathbf{0} & \mathbf{I} \end{pmatrix} \right). \tag{5.9}$$

It can be shown (see Exercise 5.4) that $\mathrm{Var}(\mathbf{y}) = \sigma^2(\mathbf{G} + \mathbf{I})$ under model (5.9), so that this class of models is actually equivalent to the original class (5.8). Furthermore, if the random effects $\mathbf{b}^*$ in model (5.9) were actually fixed effects, then the model would be a Gauss-Markov model with model matrix $(\mathbf{X}|\mathbf{Q}^*)$, and the regression variables in the model would be exactly collinear because $\mathbf{Q}^*$ is nonsingular. Although $\mathbf{b}^*$ is actually random rather than fixed in (5.9), is it possible that the exact collinearity that would exist under a model in which it was fixed somehow acts to inflate the variances of the (generalized) least squares estimator of the elements of $\boldsymbol{\beta}$ under model (5.9), and hence under the equivalent model (5.8)? The answer to this question is yes, especially when the regression variables corresponding to the fixed effects are spatially patterned, as for example, when those variables are the spatial coordinates or smooth functions thereof.

To see this, let us visualize the eigenvectors by plotting their values at the spatial locations (like the response vector $\mathbf{y}$, each n-dimensional column $\mathbf{q}_i^*$ of $\mathbf{Q}^*$ has spatial coordinates associated with it). We computed the random effects covariance matrix, $\sigma^2\mathbf{G}$, corresponding to the isotropic covariance function $C(r) = \sigma^2 \exp(-r)$ on a 40×40 grid of points spaced evenly in the unit square. Plots of 25 different eigenvectors are shown in Figure 5.2. The 1600 eigenvalues are shown in Figure 5.3. Observe that the first eigenvector, which corresponds to the largest eigenvalue, is very smooth in space. As one proceeds through eigenvectors with ever smaller eigenvalues, their spatial patterns become oscillatory, with ever greater frequency. Because the eigenvectors in $\mathbf{Q}^*$ are spatially patterned, spatial patterning among the columns of $\mathbf{X}$ is likely to result in approximate collinearity in $(\mathbf{X}|\mathbf{q}_i^*)$ for some i.

We can create exact collinearity in $(\mathbf{X}|\mathbf{q}_i^*)$ by choosing one column of $\mathbf{X}$ to be $\mathbf{q}_i^*$ for any i. To demonstrate how spatial estimation of fixed effects behaves in this case, we generated 120 data sites by sampling from a uniform distribution on the unit square. We then simulated observations at those sites from a Gaussian process with covariance function $C(r) = \exp(-r)$, from which we created the covariance matrix, $\sigma^2\mathbf{R}$ and its eigen decomposition $\mathbf{R} = \mathbf{Q}\boldsymbol{\Lambda}\mathbf{Q}^T$. Next we set $\mathbf{Q}^* = \mathbf{Q}\boldsymbol{\Lambda}^{1/2}$. We then simulated the elements of $\mathbf{b}^*$ and $\mathbf{d}$ as independent zero-mean normal random variables as prescribed by (5.9), with variance $\sigma^2 = 1$. Furthermore, we created a three-column fixed effects model matrix with its first column $\mathbf{x}_1$ equal to all ones, its second column $\mathbf{x}_2$ equal to one of the columns of $\mathbf{Q}^*$, and its third column $\mathbf{x}_3$ consisting of elements simulated independently from a standard normal distribution. Thus, $\mathbf{X} = (\mathbf{1}|\mathbf{q}_i^*|\mathbf{x}_3)$. We fixed $\boldsymbol{\beta} = (0, 1, 1)^T$, and simulated the response

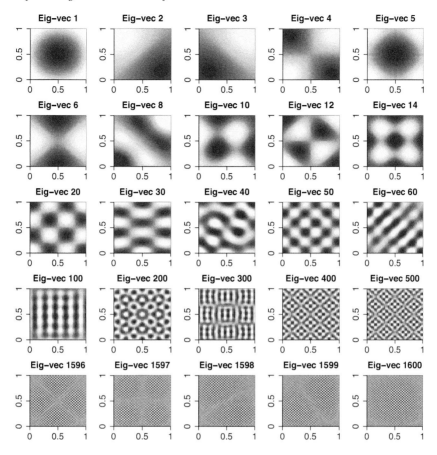

FIGURE 5.2
A selection of spatially patterned eigenvectors of the covariance matrix of a spatial process observed on a 40×40 grid.

variable $\mathbf{y}$ as
$$\mathbf{y} = \mathbf{X}\boldsymbol{\beta} + \mathbf{Q}^*\mathbf{b}^* + \mathbf{d}.$$

After simulating these data, we obtained OLS and GLS estimates of $\boldsymbol{\beta}$. We simulated and analyzed the data 200 times. Several different choices of $\mathbf{q}_i^*$ were considered, and we computed the root mean squared error (RMSE) of the estimates of β_2 and β_3 based on the 200 simulations for each choice.

Figure 5.4 shows that when $\mathbf{X}$ contains any eigenvector of $\mathbf{Q}^*$, there is substantial variance inflation in estimating β_2 compared to estimating β_3. The variance inflation is greatest when $\mathbf{x}_2 = \mathbf{q}_1^*$. Variance inflation is relatively stable for the middle eigenvectors, and then increases gradually for the eigenvectors with the smallest eigenvalues (most oscillations). Also, GLS is better than OLS for the first eigenvector, but thereafter there appears to be very little difference in the variance inflation until we get to eigenvectors with the smallest 60 eigenvectors (about half of them), when once again GLS is slightly better than OLS. Which eigenvector is included in $\mathbf{X}$ has little effect on the estimation of β_3, for which GLS had consistently lower RMSE than OLS estimation. Figure 5.5 shows that GLS-based confidence intervals for both β_2 and β_3 have good coverage, as do OLS-based confidence intervals for

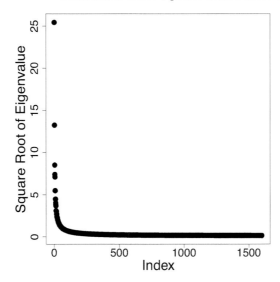

FIGURE 5.3
Ordered eigenvalues of the covariance matrix of a spatial process observed on a 40×40 grid.

β_3. However, the OLS-based intervals for β_2 badly undercover for eigenvectors with large eigenvalues and overcover for eigenvectors with small eigenvalues.

Hodges & Reich (2010) suggested that the variance inflation induced by spatial confounding is sufficiently large that something should be done about it. As a remedy, they proposed that fitting the spatial linear model (5.8) should be replaced with fitting the

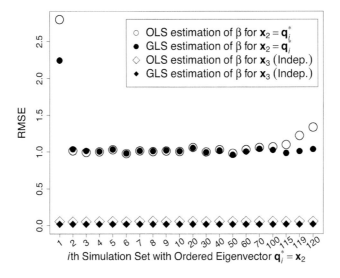

FIGURE 5.4
Root mean-squared error for OLS and GLS estimation of β_2, where $\mathbf{x}_2 = \mathbf{q}_i^*$ is exactly collinear with $\mathbf{Q}^*$, and OLS and GLS estimation of β_3, where $\mathbf{x}_3$ was generated independently of $\mathbf{Q}^*$. The horizontal axis shows which eigenvector was used for $\mathbf{q}_i^*$.

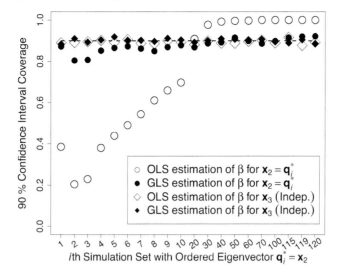

FIGURE 5.5
Coverage of nominal 90% confidence intervals for β_2 and β_3 based on OLS and GLS estimation, where $\mathbf{x}_2 = \mathbf{q}_i^*$ is exactly collinear with $\mathbf{Q}^*$ and $\mathbf{x}_3$ was generated independently of $\mathbf{Q}^*$. The horizontal axis shows which eigenvector was used for $\mathbf{q}_i^*$.

so-called **restricted spatial regression (RSR) model**

$$\mathbf{y} = \mathbf{X}\boldsymbol{\beta} + (\mathbf{I} - \mathbf{P_X})\mathbf{b} + \mathbf{d}, \quad \begin{pmatrix} \mathbf{b} \\ \mathbf{d} \end{pmatrix} \sim \mathrm{N}\left(\mathbf{0}, \sigma^2 \begin{pmatrix} \mathbf{G} & \mathbf{0} \\ \mathbf{0} & \mathbf{I} \end{pmatrix} \right),$$

where, as in Chapter 4, $\mathbf{P_X}$ is the orthogonal projection matrix onto the column space of $\mathbf{X}$ (and thus $\mathbf{I} - \mathbf{P_X}$ is the orthogonal projection matrix onto the orthogonal complement of the column space of $\mathbf{X}$). Under this model, the random effects, $(\mathbf{I} - \mathbf{P_X})\mathbf{b}$, are orthogonal to the regression variables associated with the fixed effects. Note also that $\mathrm{Var}(\mathbf{y}) = \sigma^2[(\mathbf{I} - \mathbf{P_X})\mathbf{G}(\mathbf{I} - \mathbf{P_X}) + \mathbf{I}]$ under the RSR model. Hodges and Reich made two claims about the posterior mean of $\boldsymbol{\beta}$ in a Bayesian analysis of this model that translate into the following claims about the GLS estimator of $\boldsymbol{\beta}$ based on this model: (1) it will be identical to the OLS estimator (and thus what remains of the spatial correlation after removing the portion that lies with the column space of $\mathbf{X}$ will not affect the point estimation of $\boldsymbol{\beta}$), and (2) its variance will be reduced somewhat from that of the GLS estimator under the spatial linear model, but will still be larger than the variance of the OLS estimator under the Gauss-Markov model, thereby appropriately "discounting the sample size to account for spatial correlation." The first of these claims is true, for the same reason that the OLS and GLS estimators of μ coincide in the triangular-clusters example of Section 5.3, i.e., that $\mathbf{RP_X}$, which in this case is equal to $[(\mathbf{I} - \mathbf{P_X})\mathbf{G}(\mathbf{I} - \mathbf{P_X}) + \mathbf{I}]\mathbf{P_X} = \mathbf{P_X}$, is symmetric. The second claim, however, is false, as proved by Zimmerman & Ver Hoef (2022). In fact, the variance of the GLS estimator based on the RSR model is identical to the variance of the OLS estimator based on a Gauss-Markov model. Furthermore, the *estimated* variance of the GLS estimator based on the RSR model tends to be smaller than the estimated variance of the OLS estimator based on a Gauss-Markov model. The estimated variance of the RSR-based GLS estimator is also smaller than the estimated variance of the GLS estimator based on the original spatial model, with the consequence that confidence intervals for the elements of $\boldsymbol{\beta}$ based on GLS estimation of the RSR model tend to undercover.

Hughes & Haran (2013) proposed a modification of restricted spatial regression that involves the adjacency matrix used in the computation of Moran's I. For their approach, the model is

$$\mathbf{y} = \mathbf{X}\boldsymbol{\beta} + \mathbf{M}\mathbf{b} + \mathbf{d}, \quad \begin{pmatrix} \mathbf{b} \\ \mathbf{d} \end{pmatrix} \sim \mathrm{N}_{q+n} \left(\mathbf{0}, \sigma^2 \begin{pmatrix} \mathbf{G} & \mathbf{0} \\ \mathbf{0} & \mathbf{I} \end{pmatrix} \right), \qquad (5.10)$$

where $\mathbf{M}$ is an $n \times q$ matrix whose columns are the first $q << n$ eigenvectors of the so-called "Moran operator" $(\mathbf{I} - \mathbf{P_X})\mathbf{A}(\mathbf{I} - \mathbf{P_X})$ and $\mathbf{A}$ is the $n \times n$ matrix whose (i,j)th element equals 1 if sites i and j are neighbors, and equals 0 otherwise. Hughes and Haran recommend choosing q to be no larger than the number of positive eigenvalues of the Moran operator because only positive spatial dependence is relevant. Similar to the situation with RSR, the GLS estimator of $\boldsymbol{\beta}$ and its variance under the Hughes–Haran model (5.10) coincide with the OLS estimator and its variance under a Gauss-Markov model, and confidence intervals for the elements of $\boldsymbol{\beta}$ again tend to undercover, though they are marginally better than those based on RSR (Zimmerman & Ver Hoef, 2022). Thus, although restricted spatial regression and the Hughes-Haran variant were put forward as methods for reducing the deleterious effects of spatial confounding, actually they yield inferior inferences and we strongly discourage using them. Instead, we recommend sticking with the GLS-based analysis of the original spatial linear model. Although the variance of the corresponding GLS estimator is larger than that of the GLS estimator associated with an RSR approach, the former, unlike the latter, is properly estimated. For those data analysts who are extremely concerned that spatial confounding has adversely affected their GLS-based analysis, we suggest that they either revert to an OLS-based analysis, or use the OLS estimator but properly account for spatial correlation in the estimation of its variance. Although inferences based on OLS in this setting are not valid (the confidence intervals tend to undercover, for example), they are better than those based on restricted spatial regression methods.

In addition to variance inflation, bias of the OLS and GLS estimators has been mentioned as a consequence of spatial confounding (Paciorek, 2010). However, bias due to spatial confounding only arises if the regression variables are regarded as random variables and are correlated with the spatial random effects or residual effects in a spatial linear mixed model. (Recall from Section 4.1 that the OLS estimator of $\boldsymbol{\beta}$ is unbiased, and so is the associated estimator of its covariance matrix, under a random Gauss-Markov model if the regressors and model errors are uncorrelated, and the same is true for the GLS estimator and the estimator of its covariance matrix under a spatial linear model with random regressors.) In a scenario in which there is one random regressor and one vector of random effects, Paciorek showed that if the dependence among the random effects is of the same magnitude as the dependence among the values of the random regressor across sites, then spatial confounding results in no additional bias beyond that incurred by the OLS estimator in a random regression model in which the regressors are correlated with the random effects. Additional bias occurs only when the regressor can be decomposed into two or more independent components, one of which is confounded with the random effects and the other which is not, and the dependence among the values of the unconfounded term is stronger than that among the values of the confounded term. Paciorek noted, however, that this case may be of limited practical interest.

Within a random-regressors framework, Lennon (2000) and Hanks *et al.* (2015) described a related phenomenon in which the significance levels of t-values associated with OLS-estimated regression coefficients tend to be overstated for more strongly spatially autocorrelated explanatory variables than other explanatory variables when the model residuals are also spatially correlated. Lennon referred to this phenomenon, using terminology from spectral analysis of time series, as a **red shift**, i.e., a shift in overstating the importance of

TABLE 5.6
Empirical Type I error rates of partial t tests for the importance of three explanatory variables in the simulation study of the red shift described in Section 5.5.

			Type I error rate		
y autocorrelated?	*X* autocorrelated?	OLS or GLS?	x_1	x_2	x_3
no	no	OLS	0.0503	0.0511	0.0496
yes	no	OLS	0.0465	0.0465	0.0450
yes	no	GLS	0.0489	0.0507	0.0502
no	yes	OLS	0.0502	0.0496	0.0492
yes	yes	OLS	0.1215	0.0856	0.0490
yes	yes	GLS	0.0472	0.0497	0.0491

those variables that are spatially smooth (toward the red end of the spectrum) compared to those that are more spatially discontinuous (bluer).

Table 5.6 summarizes a simulation study that demonstrates the red shift. This study considers a situation with $n = 25$ observations located at the nodes of a 5×5 square grid with unit spacing. Responses were simulated according to the model

$$\mathbf{y} = \mathbf{1} + \mathbf{e},$$

where $\mathbf{e}$ is distributed either as $N(\mathbf{0}, \sigma^2 \mathbf{I})$ or $N(\mathbf{0}, \sigma^2 \mathbf{R})$ and the covariance function corresponding to $\mathbf{R}$ is exponential with correlation 0.5 at unit distance. The model fitted to the data, however, has an intercept and three random regressors x_1, x_2, and x_3 (whose corresponding true regression coefficients are all equal to 0). Two cases of regressors were considered. In the first case, the elements of $\mathbf{x}_1 = (x_{i1})$, $\mathbf{x}_2 = (x_{i2})$, and $\mathbf{x}_3 = (x_{i3})$ were drawn independently from a standard normal distribution. In the second case, $\mathbf{x}_1$ was drawn from $N(\mathbf{0}, \mathbf{V}_1)$ where $\mathbf{V}_1$ is the covariance matrix corresponding to an exponential covariance function with variance 1.0 and correlation 0.5; $\mathbf{x}_2$ was drawn from $N(\mathbf{0}, \mathbf{V}_2)$ where $\mathbf{V}_2$ is the covariance matrix corresponding to an exponential function with variance 1.0 and correlation 0.25; and the elements of $\mathbf{x}_3$ were drawn from a standard normal distribution. One hundred thousand realizations of $\mathbf{y}$ and $\mathbf{X}$ were obtained for each case. Each realization of $(\mathbf{x}_1, \mathbf{x}_2, \mathbf{x}_3)$ was rescaled by post-multiplying it by the inverse square root of its sample covariance matrix, so that the regressors always had sample variances equal to 1.0 and sample correlations equal to 0. This scaling makes any differences between results for spatially autocorrelated regressors and uncorrelated regressors completely attributable to the spatial autocorrelation.

Size-0.05 partial t-tests for the statistical significance of each regressor were carried out for the data from each simulation. Entries in Table 5.6 are empirical Type I error rates, i.e., the proportions of realizations for which the null hypothesis that the regression coefficient equals zero was rejected. The results show that when either the response or the explanatory variables, or neither, is spatially autocorrelated, the empirical Type I error rate of the OLS-based t test is within two standard errors (0.0043) of its nominal 0.05 rate. However, when both the response and the explanatory variables are spatially autocorrelated, the OLS-based Type I error rate is inflated; furthermore, the more highly autocorrelated the explanatory variable, the greater the inflation. The last line of the table shows that the GLS-based partial t test (using the correct covariance matrix) incurs no Type I error rate inflation.

In practice, the red shift may be either diminished or exacerbated by other factors that were ignored in the simulation study just presented and that also were not considered by Lennon (2000). These factors include nonstationarity and other misspecification of the true covariance matrix, collinearity among the explanatory variables, and scale dependence; for

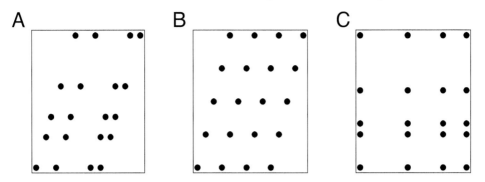

FIGURE 5.6
Spatial configurations: A) nonorthogonal nonregular lattice; B) nonorthogonal regular lattice; C) orthogonal nonregular lattice.

further discussion see Diniz-Filho *et al.* (2003), Diniz-Filho *et al.* (2007), and Hawkins *et al.* (2007).

5.6   Computationally exploitable covariance structures

From (5.1) and (5.3), it is clear that the inverse of the covariance matrix $\boldsymbol{\Sigma}$ of $\mathbf{y}$ or its scale-free version $\mathbf{R}$ must be available before one can evaluate the GLS estimator, $\hat{\boldsymbol{\beta}}_{GLS}$, and the generalized residual variance, $\hat{\sigma}^2_{GLS}$. From (5.2) it is also clear that the matrix $\mathbf{X}^T\mathbf{R}^{-1}\mathbf{X}$ must be inverted to obtain the covariance matrix of $\hat{\boldsymbol{\beta}}_{GLS}$ and its unbiased estimator (5.4). In general, the computations required for inverting an $m \times m$ positive definite matrix are of order m^3, which can be formidable if m is large. The dimensions of $\mathbf{X}^T\mathbf{R}^{-1}\mathbf{X}$, which are equal to the number of explanatory variables, are sufficiently small in the vast majority of practical applications of spatial linear models for its inversion to be accomplished easily (given $\mathbf{R}^{-1}$). The dimensions of $\boldsymbol{\Sigma}$ and $\mathbf{R}$, however, are equal to the sample size, so those matrices can be so large that they are computationally difficult or impossible to invert.

In the case of geostatistical models, some efficiencies in the inversion of $\boldsymbol{\Sigma}$ may be possible if the covariance function is spatially structured and the spatial configuration of data sites is well-suited to that structure. For example, consider a geostatistical process observed at point sites that lie at the nodes of an r-row by c-column lattice in two dimensions (see Figure 5.6). The rows and columns need not be orthogonal, and the lattice need not be regular (i.e., it need not have equal spacing between rows and between columns), as in Figure 5.6A. If, however, the lattice *is* regular (as in Figure 5.6B) and the process is second-order stationary, then $\boldsymbol{\Sigma}$ is a symmetric block **Toeplitz** matrix with $c \times c$ Toeplitz blocks, and the computations required to obtain its inverse are of order r^2c^3, compared to r^3c^3 in general, using a specialized inversion algorithm (Zimmerman, 1989b). (A Toeplitz matrix is a square matrix whose scalar elements are constant along diagonals, and a block Toeplitz matrix is a square matrix of square submatrices which are equal along diagonals of blocks.) If, in addition, the lattice is rectangular (i.e., the rows and columns are orthogonal) and the process is isotropic, then the blocks are symmetric in addition to being Toeplitz, which can further reduce the computational burden.

A covariance matrix with a structure particularly beneficial for computations arises when the lattice is orthogonal (but not necessarily regular), as in Figure 5.6C, and the process is separable; it need not be second-order stationary. In this case, $\boldsymbol{\Sigma}$ is equal to the Kronecker product of $r \times r$ and $c \times c$ symmetric matrices (Zimmerman, 1989b), for which the inverse is the Kronecker product of the inverses. Thus, the computational order of the inversion problem is reduced to $r^3 + c^3 + r^2c^2$ in this case, where the last summand is associated with the multiplications needed to form the Kronecker product of the two inverses. When the lattice is not only orthogonal (rectangular) but regular as well, and the process is second-order stationary, the two matrices being inverted are Toeplitz and the inversion problem is of order r^2c^2, with significant additional reduction in the computational burden. And, if the lattice is square (i.e., the spacing between rows is equal to the spacing between columns) and the underlying process is isotropic, still greater reduction is possible.

A covariance matrix corresponding to a compactly supported covariance function may, depending on the magnitude of the range of dependence and the spacing between sites, be **sparse**, i.e., it may contain many zeros. Efficient computational methods for solving sparse systems of equations may be exploited in this case and may be particularly useful if the responses can be arranged in $\mathbf{y}$ in such a way that the covariance matrix is banded, with a relatively small bandwidth (see, e.g., Golub & Loan (1996). (A square matrix is banded with bandwidth b if the elements of all diagonals more than b diagonals from the main diagonal are equal to zero.) Even sparse matrix algorithms that do not account for banding may be computationally efficient. For example, consider a situation in which there are 10000 observations whose data sites are generated by randomly displacing the nodes of a 100×100 unit-spaced square grid by amounts equal to independent realizations of a uniform distribution on the square $(-1/2, 1/2) \times (-1/2, 1/2)$, and suppose that the underlying process has a spherical covariance function with range equal to 2.0. The `SparseM.solve` function in the R package `SparseM` was used to invert the covariance matrix of this example, and it produced the inverse in about 5% of the time it took the `Matrix` package function `chol2inv`, which inverts a symmetric positive definite matrix, to perform the same task (12.455 seconds versus 242.627 seconds on an 11th Gen Intel (R) Core (TM) i7-11800H @ 2.30 GHz computer with 16 cores and 32 Gb RAM.). Incidentally, the all-purpose matrix inversion function `solve` required more than twice as much time as `chol2inv`, and more than 44 times as much time as `SparseM.solve`, to accomplish the same task.

An important subclass of spatial linear mixed models has a covariance matrix with computationally exploitable structure. The useful subclass consists of those models for which the random effects are shared among sites rather than site-specific. For any spatial mixed effects model, the inverse of the scale-free covariance matrix $\mathbf{Z}\mathbf{G}\mathbf{Z}^T + \mathbf{D}$ satisfies the identity (e.g., Harville (1997), Theorem 18.2.8),

$$(\mathbf{Z}\mathbf{G}\mathbf{Z}^T + \mathbf{D})^{-1} = \mathbf{D}^{-1} - \mathbf{D}^{-1}\mathbf{Z}(\mathbf{G}^{-1} + \mathbf{Z}^T\mathbf{D}^{-1}\mathbf{Z})^{-1}\mathbf{Z}^T\mathbf{D}^{-1}. \tag{5.11}$$

Thus, the inversion of the covariance matrix may be accomplished indirectly by the inversion of $\mathbf{D}$, $\mathbf{G}$, and $\mathbf{G}^{-1} + \mathbf{Z}^T\mathbf{D}^{-1}\mathbf{Z}$ followed by some matrix additions and multiplications. In general, this method for obtaining the inverse does not provide any computational advantages, but when random effects are shared among sites the dimensions of $\mathbf{G}$ and $\mathbf{G}^{-1} + \mathbf{Z}^T\mathbf{D}^{-1}\mathbf{Z}$ can be much smaller than those of $\mathbf{Z}\mathbf{G}\mathbf{Z}^T + \mathbf{D}$ (and $\mathbf{D}$ is diagonal hence easily inverted), so that $\mathbf{Z}\mathbf{G}\mathbf{Z}^T + \mathbf{D}$ may be inverted much more efficiently in this way.

The focus of the foregoing has been on computationally exploitable structure of the covariance matrix for geostatistical data. However, the semivariance matrix and generalized covariance matrix can also have computationally exploitable structure. In Chapter 8 we will note how the structure of these matrices can be exploited for greater computational efficiency in conjunction with a likelihood-based approach to parameter estimation.

Now let us consider the inversion of the covariance matrices corresponding to spatial-weights linear models for areal data. For SAR and CAR models specified for the observations only, the only computing needed to obtain the inverse of the covariance matrix is some scalar reciprocation and matrix multiplication. Recall from Section 2.3.2 that $\boldsymbol{\Sigma}_{\text{SAR}} = (\mathbf{I} - \mathbf{B})^{-1}\mathbf{K}[(\mathbf{I} - \mathbf{B})^{-1}]^T$ and $\boldsymbol{\Sigma}_{\text{CAR}} = (\mathbf{I} - \mathbf{C})^{-1}\mathbf{K}$. It follows easily that the inverses of these matrices are $\boldsymbol{\Sigma}_{\text{SAR}}^{-1} = (\mathbf{I} - \mathbf{B})^T\mathbf{K}^{-1}(\mathbf{I} - \mathbf{B})$ and $\boldsymbol{\Sigma}_{\text{CAR}}^{-1} = \mathbf{K}^{-1}(\mathbf{I} - \mathbf{C})$. Since $\mathbf{K}$ is diagonal, computing its inverse involves mere scalar reciprocation, and then simple matrix multiplications yield $\boldsymbol{\Sigma}_{\text{SAR}}^{-1}$ or $\boldsymbol{\Sigma}_{\text{CAR}}^{-1}$. This simple approach is not available for the covariance matrix of a SMA model, however, because the functional dependence of $\boldsymbol{\Sigma}_{\text{SMA}}$ on its spatial weights matrix involves no matrix inversion.

For SAR and CAR models specified for additional variables beyond those actually observed, the preceding paragraph is not the complete story. Recall from Section 2.3.2 that either intentionally or unintentionally, the spatial weights matrix used in a SAR or CAR model may be defined for an areal lattice of sites that embeds the lattice of sites on which observations are actually taken, and that the covariance matrix of the observations based upon a model for all variables on the embedding lattice is generally different than the covariance matrix of the observations based upon a model for only those observations. When the SAR/CAR model being fitted is one that was originally formulated for all variables on the embedding lattice, additional computations beyond those described in the preceding paragraph are required to obtain the covariance matrix of the observations efficiently. To see this, let $\boldsymbol{\Sigma}$ denote the covariance matrix for the variables on the embedding lattice, which we may write in block-partitioned form as

$$\boldsymbol{\Sigma} = \left(\begin{array}{cc} \boldsymbol{\Sigma}_{11} & \boldsymbol{\Sigma}_{12} \\ \boldsymbol{\Sigma}_{21} & \boldsymbol{\Sigma}_{22} \end{array} \right),$$

where $\boldsymbol{\Sigma}_{11}$ is the covariance matrix of the observations. Then, to obtain the GLS estimator of $\boldsymbol{\beta}$ and its covariance matrix we need to obtain $\boldsymbol{\Sigma}_{11}^{-1}$ from $\boldsymbol{\Sigma}^{-1}$. Of course, one way to obtain $\boldsymbol{\Sigma}_{11}^{-1}$ is to invert $\boldsymbol{\Sigma}^{-1}$ to obtain $\boldsymbol{\Sigma}$, and then invert the latter's upper left block $\boldsymbol{\Sigma}_{11}$. But this approach requires two inversions, both of which are for matrices whose dimensions are at least as large as the number of observations. An alternative approach is based on writing the specified SAR/CAR inverse covariance matrix for the variables on the embedding lattice in block-partitioned form as

$$\boldsymbol{\Sigma}^{-1} = \left(\begin{array}{cc} \boldsymbol{\Sigma}^{11} & \boldsymbol{\Sigma}^{12} \\ \boldsymbol{\Sigma}^{21} & \boldsymbol{\Sigma}^{22} \end{array} \right).$$

Then by Theorem A.6e in Appendix A,

$$\boldsymbol{\Sigma}_{11}^{-1} = \boldsymbol{\Sigma}^{11} - \boldsymbol{\Sigma}^{12}(\boldsymbol{\Sigma}^{22})^{-1}\boldsymbol{\Sigma}^{21}.$$

Thus, using this second approach, $\boldsymbol{\Sigma}_{11}^{-1}$ may be obtained by inverting only one matrix, namely $\boldsymbol{\Sigma}^{22}$, followed by some matrix multiplication and addition. If the dimensions of $\boldsymbol{\Sigma}^{22}$ are smaller than the number of observations, then this second approach is computationally more efficient than the first.

5.7   Empirical generalized least squares

Throughout this chapter, we have taken the model to be a spatial Aitken model, for which the only unknown parameter in the covariance matrix $\boldsymbol{\Sigma} = \sigma^2\mathbf{R}$ is σ^2; $\mathbf{R}$ is assumed to be known. In reality, however, $\mathbf{R}$ is unknown, so the GLS estimator of $\boldsymbol{\beta}$ corresponding to any particular $\mathbf{R}$ that might be assumed by a modeler is not necessarily optimal or even close to being optimal. The spatial general linear model offers the possibility of better estimation by asserting, more realistically, that $\boldsymbol{\Sigma}$ (and hence $\mathbf{R}$) is unknown but may be modeled as $\boldsymbol{\Sigma} = \boldsymbol{\Sigma}(\boldsymbol{\theta})$, where $\boldsymbol{\Sigma}(\cdot)$ is a known matrix-valued function of the data locations and $\boldsymbol{\theta}$ is an unknown parameter vector. This provides much more flexibility and transmutes the estimation of $\boldsymbol{\Sigma}$ to the estimation of the parameter vector $\boldsymbol{\theta}$. Once an estimate $\tilde{\boldsymbol{\theta}}$ of $\boldsymbol{\theta}$ is obtained, the conventional approach to estimating $\boldsymbol{\beta}$ is to use an estimator of the same form as the GLS estimator, but with $\boldsymbol{\Sigma} = \boldsymbol{\Sigma}(\boldsymbol{\theta})$ replaced by its estimate $\tilde{\boldsymbol{\Sigma}} = \boldsymbol{\Sigma}(\tilde{\boldsymbol{\theta}})$. This method of estimation of $\boldsymbol{\beta}$ is referred to as **empirical generalized least squares**, or EGLS. Thus, the EGLS estimator of $\boldsymbol{\beta}$ is

$$\hat{\boldsymbol{\beta}}_{EGLS} = \{\mathbf{X}^T[\boldsymbol{\Sigma}(\tilde{\boldsymbol{\theta}})]^{-1}\mathbf{X}\}^{-1}\mathbf{X}^T[\boldsymbol{\Sigma}(\tilde{\boldsymbol{\theta}})]^{-1}\mathbf{y}.$$

Similarly, the conventional approach to estimating σ^2 is to substitute $\mathbf{R}(\tilde{\boldsymbol{\theta}})$ for $\mathbf{R}$ and $\hat{\boldsymbol{\beta}}_{EGLS}$ for $\hat{\boldsymbol{\beta}}_{GLS}$ in expression (5.3) for the generalized residual mean square, yielding the estimator

$$\hat{\sigma}^2_{EGLS} = \frac{(\mathbf{y} - \mathbf{X}\hat{\boldsymbol{\beta}}_{EGLS})^T[\mathbf{R}(\tilde{\boldsymbol{\theta}})]^{-1}(\mathbf{y} - \mathbf{X}\hat{\boldsymbol{\beta}}_{EGLS})}{n - p}. \tag{5.12}$$

To illustrate, let us revisit the first toy example of Section 5.2. In that example, the covariance matrix can be represented as $\sigma^2\mathbf{R}$, where $\mathbf{R}$ is a correlation matrix. Suppose now that the correlation function in that example was not known exactly, but was estimated by some procedure (the details of which are unimportant at present) to be

$$\tilde{\rho}(r) = \begin{cases} 1 - \frac{3r}{6} + \frac{r^3}{54} & \text{for } 0 \leq r \leq 3, \\ 0 & \text{for } r > 3. \end{cases}$$

Then, the corresponding estimate of $\mathbf{R}$, say $\tilde{\mathbf{R}}$, is

$$\tilde{\mathbf{R}} = \begin{pmatrix} 1.000 & 0.519 & 0.345 & 0.089 & 0 \\ & 1.000 & 0.519 & 0.345 & 0 \\ & & 1.000 & 0.519 & 0 \\ & & & 1.000 & 0 \\ & & & & 1.000 \end{pmatrix}.$$

Observe that range of the correlation here was estimated to be 3 (rather than 4 as previously assumed in Section 5.2), so the nonzero off-diagonal elements of $\tilde{\mathbf{R}}$ are slightly smaller than the corresponding elements of $\mathbf{R}$, and some have become equal to 0. Thus, if the data truly have covariance matrix $\sigma^2\mathbf{R}$ as specified in Example 5.2, then they have been estimated to be not quite as strongly correlated as they truly are. Tables 5.1 and 5.2 give the observation weights associated with EGLS, the EGLS estimate of μ, and the corresponding estimate of σ^2, which are counterparts to those given previously for OLS and GLS. We may observe that all of the EGLS-based quantities have values that are intermediate to those of OLS and GLS, but are relatively more similar to those for GLS because the estimated $\mathbf{R}$ more closely resembles the true $\mathbf{R}$ than it resembles $\mathbf{I}$ in this example.

It should be noted that the "GLS estimator" of the mean described in the second toy example of Section 5.2 was actually an empirical GLS estimator, as it was based on a covariance matrix whose elements were determined by fitting a spatial linear model with exponential covariance function to the data, using methods yet to be presented.

Empirical GLS versions of other quantities associated with GLS may also be defined, such as the F-statistic (5.6), the Aitken version of the coefficient of determination (5.7), etc., by replacing $\hat{\boldsymbol{\beta}}_{GLS}$, $\hat{\sigma}^2_{GLS}$, $\mathbf{R}$, and $\boldsymbol{\Sigma}$ with $\hat{\boldsymbol{\beta}}_{EGLS}$, $\hat{\sigma}^2_{EGLS}$, $\tilde{\mathbf{R}}$, and $\tilde{\boldsymbol{\Sigma}}$, as required.

Although EGLS is an intuitively reasonable method for estimating $\boldsymbol{\beta}$ for a spatial general linear model, it lacks most of the nice properties of GLS estimation. For example, the EGLS estimator of $\boldsymbol{\beta}$ is generally not a linear estimator. Nor is it normally distributed or "best" in any known sense. Remarkably, however, it is unbiased under very weak conditions (Kackar & Harville, 1981). Owing to the additional uncertainty introduced by estimating $\boldsymbol{\theta}$, the variance of the EGLS estimator tends to be larger than the variance of the GLS estimator using the true value of $\boldsymbol{\theta}$. The conventional approach to estimating the variance of the EGLS estimator is to substitute $\mathbf{R}(\tilde{\boldsymbol{\theta}})$ for $\mathbf{R}$ and $\hat{\sigma}^2_{EGLS}$ for $\hat{\sigma}^2_{GLS}$ in (5.4). This estimator tends to be negatively biased, sometimes considerably so (Harville & Jeske, 1992). Some alternative, less biased estimators of $\text{Var}(\hat{\boldsymbol{\beta}}_{EGLS})$ for use in confidence intervals and hypothesis tests for $\boldsymbol{\beta}$ have been proposed by Harville & Jeske (1992), Zimmerman & Cressie (1992), and Kenward & Roger (1997), but they are somewhat complicated so most practitioners continue to use the conventional estimator. It is incumbent upon those practitioners to inform their end users that their confidence intervals for the elements of $\boldsymbol{\beta}$ are probably too narrow and their hypothesis tests probably overstate statistical significance.

5.8   Exercises

1. Verify the entries in Tables 5.1–5.4.

2. Consider a scenario in which four observations are taken at sites $\mathbf{s}_1 = (u_1, v_1) = (1, 1)$, $\mathbf{s}_2 = (u_2, v_2) = (1, 2)$, $\mathbf{s}_3 = (u_3, v_3) = (3, 1)$, and $\mathbf{s}_4 = (u_4, v_4) = (3, 2)$. Suppose that the spatial linear model for these observations is

$$y_i = \beta_0 + \beta_1 u_i + \beta_2 v_i + e_i \quad (i = 1, 2, 3, 4),$$

with isotropic exponential covariance function with nugget effect,

$$C(r; \sigma_1^2, \sigma_0^2, \alpha) = \begin{cases} \sigma_1^2 + \sigma_0^2 & \text{if } r = 0, \\ \sigma_1^2 \exp(-r/\alpha) & \text{if } r > 0. \end{cases}$$

   (a) Display the elements of $\mathbf{X}$ and $\boldsymbol{\Sigma}$ for this scenario.

   (b) If the third and fourth sites were moved to the left one unit (so that the four sites lie on the corners of a unit square rather than a rectangle), would you expect the variances of the generalized least squares estimators of β_0, β_1, and β_2 to increase or decrease from what they are for the original configuration under the spatial linear model? Explain.

   (c) For a case with $\alpha = 2$, verify the expectations you stated in your answer to part (b), and compare these results to the results of Exercise 4.1, noting any differences.

3. Verify that the correlations among observations in a two-way layout following a spatial mixed effects model with random row and column effects and uncorrelated and homoscedastic measurement errors have the structure described in Section 5.4.

4. Show that under (5.8) and (5.9), $\text{Var}(\mathbf{y}) = \sigma^2(\mathbf{G} + \mathbf{I})$ as claimed in Section 5.5.

5. Verify all the claims made about the structures of the covariance matrix for the spatial configurations in Figure 5.6B,C.

6

Parametric Covariance Structures for Geostatistical Models

The previous two chapters featured estimation (by ordinary and generalized least squares, respectively) of a spatial linear model and presented some possibilities for modeling its mean structure. Now, our attention returns to the model's covariance structure. Covariance structures for geostatistical data are sufficiently different from covariance structures for areal data, and both are considerably more complicated than mean structures, so we devote a chapter to each. In this chapter, we consider parametric covariance structures for geostatistical data, which are specified by a parametric model for either the covariance function, semivariogram, or generalized covariance function of the underlying geostatistical process.

6.1 General properties of covariance functions

There are two important differences between the modeling of the mean structure and the modeling of the covariance structure for geostatistical data. The first is that a mean function does not have to satisfy any particular mathematical requirements, whereas a covariance function, as noted in Section 1.3, must be symmetric and positive definite . That is, a **valid** parametric covariance function $C(\mathbf{s}, \mathbf{s}'; \boldsymbol{\theta})$ must satisfy

$$C(\mathbf{s}, \mathbf{s}'; \boldsymbol{\theta}) = C(\mathbf{s}', \mathbf{s}; \boldsymbol{\theta}) \quad \text{for all } \mathbf{s}, \mathbf{s}' \in \mathcal{A}$$

and

$$\sum_{i=1}^{n} \sum_{j=1}^{n} a_i a_j C(\mathbf{s}_i, \mathbf{s}_j; \boldsymbol{\theta}) \geq 0$$

for all n, all real numbers $a_1, \ldots, a_n$, and all $\mathbf{s}_1, \ldots, \mathbf{s}_n \in \mathcal{A}$. (6.1)

In fact, these two conditions are not merely necessary for a function on $\mathbb{R}^d \times \mathbb{R}^d$ to be a covariance function; they are sufficient as well. The second major difference between modeling the mean structure and modeling the covariance structure is related to the first, but pertains specifically to constraints on the parameter space. Whereas the parameter space for a linear mean function is unconstrained in general, the positive definiteness requirement imposes constraints on the parameter space for a covariance function. That is, a parametric function on $\mathbb{R}^d \times \mathbb{R}^d$ may be positive definite for some values of its parameters but not others. For example, putting $n = 1$ and $a_1 = 1$ into (6.1) yields

$$C(\mathbf{s}, \mathbf{s}; \boldsymbol{\theta}) \geq 0 \quad \text{for all } \mathbf{s}, \tag{6.2}$$

DOI: 10.1201/9780429060878-6

which implies that if $C(\mathbf{s}, \mathbf{s}'; \theta) = \theta f(\mathbf{s}, \mathbf{s}')$ where $f(\mathbf{s}, \mathbf{s}')$ is a positive definite function on $\mathbb{R}^d \times \mathbb{R}^d$, then $C(\mathbf{s}, \mathbf{s}'; \theta)$ is positive definite if and only if $\theta \geq 0$. We define the parameter space Θ for a covariance function as the set of all $\boldsymbol{\theta}$ for which the function is symmetric and positive definite.

Where are we to find symmetric and positive definite functions? It is easy to construct real-valued symmetric functions on $\mathbb{R}^d \times \mathbb{R}^d$; one example is $f(\mathbf{s}, \mathbf{s}') = (\mathbf{s} - \mathbf{s}')^T (\mathbf{s} - \mathbf{s}')$, and another is $\max_{i=1,\dots,d}(s_i + s_i')$. But it is considerably more difficult to ascertain whether a given function satisfies (6.1), let alone to see how to "build it in" to a function. Fortunately, mathematicians and geostatisticians have studied extensively the problem of characterizing positive definite functions, so many pertinent results are known and many valid covariance functions have been derived. We will describe some of those functions in the next section, but here we provide just a little of the mathematical background, drawing heavily from Yaglom (1987) and Stein (1999).

It turns out that a symmetric real-valued function on $\mathbb{R}^d \times \mathbb{R}^d$ is positive definite if and only if it has the form

$$C(\mathbf{s}, \mathbf{s}'; \boldsymbol{\theta}) = \int_{\mathbb{R}^d} \int_{\mathbb{R}^d} \cos(\mathbf{x}_1^T \mathbf{s} - \mathbf{x}_2^T \mathbf{s}') dF_{\boldsymbol{\theta}}(\mathbf{x}_1, \mathbf{x}_2),$$

where $F_{\boldsymbol{\theta}}$ is the joint cumulative distribution function (cdf) of two exchangeable d-dimensional random vectors. For the important subclass of stationary covariance functions, this result specializes to Bochner's theorem (Bochner, 1959), which says essentially that an even continuous real-valued function on $\mathbb{R}^d$ is positive definite if and only if it has the form

$$C(\mathbf{h}; \boldsymbol{\theta}) = \int_{\mathbb{R}^d} \exp(i\mathbf{x}^T \mathbf{h}) dF_{\boldsymbol{\theta}}(\mathbf{x}),$$

where $i = \sqrt{-1}$ and here $F_{\boldsymbol{\theta}}$ is a cdf of a single d-dimensional random vector. Equivalently, a function defined on $\mathbb{R}^d$ is a stationary covariance function if and only if it is a real-valued characteristic function of a d-dimensional random vector. Thus, for example, $\cos(h/\theta)$ (which is the characteristic function of a two-point discrete uniform distribution on $\{-1/\theta, 1/\theta\}$) and $\exp(-|h|/\theta)$ (which is the characteristic function of a Cauchy distribution with scale parameter $1/\theta$) are covariance functions on $\mathbb{R}$ for $\theta > 0$, and $\exp(-\mathbf{h}^T \mathbf{B} \mathbf{h}/\theta)$ (which is the characteristic function of a d-variate normal random vector with mean vector $\mathbf{0}$ and positive definite covariance matrix $(2/\theta)\mathbf{B}$ is a covariance function on $\mathbb{R}^d$ for any d, provided that $\theta > 0$.

Still more specialization occurs for the further subclass of isotropic stationary covariance functions. A real-valued continuous function of $r = \|\mathbf{h}\| = (\mathbf{h}^T \mathbf{h})^{1/2}$, where $\mathbf{h} \in \mathbb{R}^d$, is an isotropic covariance function on $\mathbb{R}^d$ if and only if it has the form

$$C(r; \boldsymbol{\theta}) = \int_0^\infty \Omega_d(xr) dG_{\boldsymbol{\theta}}(x),$$

where $G_{\boldsymbol{\theta}}$ is any nondecreasing bounded function on $(0, \infty)$,

$$\Omega_d(x) = \left(\frac{2}{x}\right)^{(d/2)-1} \Gamma\left(\frac{d}{2}\right) J_{(d/2)-1}(x),$$

Γ is the gamma function, and J_v is the Bessel function of the first kind of order v, i.e.,

$$J_v(x) = \frac{(x/2)^v}{\sqrt{\pi}\Gamma(v + 1/2)} \int_0^\pi \cos(x \cos \tau) \sin(\tau)^{2v} \, d\tau$$

(Schoenberg, 1938). Although Bessel functions may not be as familiar to applied statisticians and environmental scientists as sines and cosines, they occur frequently in engineering and physics applications and they are available in many mathematical software packages. For $d = 1, 2,$ or 3, $\Omega_d(x) = \cos x$, $J_0(x)$, and $\sin(x)/x$, respectively. Isotropic covariance functions have the property that if they are valid in $\mathbb{R}^{d_1}$, then they are valid in $\mathbb{R}^{d_2}$ where $d_2 < d_1$; the converse is not generally true.

From one or more covariance functions obtained in the manner described in the preceding paragraph, others can be built via certain algebraic operations. Perhaps the two simplest and most useful of these operations are linear combinations and products. If $C_1(\mathbf{s}, \mathbf{s}'; \boldsymbol{\theta}_1)$ and $C_2(\mathbf{s}, \mathbf{s}'; \boldsymbol{\theta}_2)$ are covariance functions on $\mathbb{R}^{2d}$ for $\boldsymbol{\theta}_1 \in \Theta_1$ and $\boldsymbol{\theta}_2 \in \Theta_2$, then: (1) any linear combination $a_1 C_1(\mathbf{s}, \mathbf{s}'; \boldsymbol{\theta}_1) + a_2 C_2(\mathbf{s}, \mathbf{s}'; \boldsymbol{\theta}_2)$ with nonnegative coefficients a_1 and a_2 is a covariance function on $\mathbb{R}^{2d}$ for $\boldsymbol{\theta} = (\boldsymbol{\theta}_1^T, \boldsymbol{\theta}_2^T)^T \in \Theta \equiv \{(\boldsymbol{\theta}_1, \boldsymbol{\theta}_2) : \boldsymbol{\theta}_1 \in \Theta_1, \boldsymbol{\theta}_2 \in \Theta_2\}$ (Exercise 6.1); and (2) their product,

$$C(\mathbf{s}, \mathbf{s}'; \boldsymbol{\theta}) = C_1(\mathbf{s}, \mathbf{s}'; \boldsymbol{\theta}_1) C_2(\mathbf{s}, \mathbf{s}'; \boldsymbol{\theta}_2), \tag{6.3}$$

is also a covariance function on $\mathbb{R}^{2d}$ for $\boldsymbol{\theta} \in \Theta$. Moreover, if $C_1((\mathbf{s}, \mathbf{s}'); \boldsymbol{\theta}_1)$ and $C_2((\mathbf{u}, \mathbf{u}'); \boldsymbol{\theta}_2)$ are covariance functions on $\mathbb{R}^{2d_1}$ and $\mathbb{R}^{2d_2}$ respectively, then

$$C((\mathbf{s}, \mathbf{u}), (\mathbf{s}', \mathbf{u}'); \boldsymbol{\theta}_1, \boldsymbol{\theta}_2) = C_1((\mathbf{s}, \mathbf{s}'); \boldsymbol{\theta}_1) + C_2((\mathbf{u}, \mathbf{u}'); \boldsymbol{\theta}_2)$$

and

$$C^*((\mathbf{s}, \mathbf{u}), (\mathbf{s}', \mathbf{u}'); \boldsymbol{\theta}_1, \boldsymbol{\theta}_2) = C_1((\mathbf{s}, \mathbf{s}'); \boldsymbol{\theta}_1) C_2((\mathbf{u}, \mathbf{u}'); \boldsymbol{\theta}_2) \tag{6.4}$$

are covariance functions on $\mathbb{R}^{2d_1 + 2d_2}$.

Another technique for the construction of valid stationary covariance functions is **auto-convolution**. If $g(\mathbf{x}; \boldsymbol{\theta})$ is a square-integrable real-valued function on $\mathbb{R}^d$, sometimes called a "kernel" or "moving-average" function, then the auto-convolution of $g(\cdot)$, defined by

$$C(\mathbf{h}; \boldsymbol{\theta}) = \int_{\mathbb{R}^d} g(\mathbf{x}; \boldsymbol{\theta}) g(\mathbf{x} - \mathbf{h}; \boldsymbol{\theta}) \, d\mathbf{x}, \tag{6.5}$$

is a valid covariance function on $\mathbb{R}^d$. In fact, this function is the covariance function of the process $\{\epsilon(\mathbf{s})\}$ on $\mathbb{R}^d$ defined as the convolution of a d-dimensional Gaussian white-noise process $\{W(\mathbf{x})\}$ with $g(\cdot)$, i.e., as

$$\epsilon(\mathbf{s}) = \int_{\mathbb{R}^d} g(\mathbf{x} - \mathbf{s}) \, W(\mathbf{x}) \, d\mathbf{x}. \tag{6.6}$$

Figure 6.1, which is similar to Figure 1 of Higdon (2002), serves to illustrate the idea in one dimension. Figure 6.1A shows a double exponential moving average function, and Figure 6.1B displays realizations of independent and identically distributed mean-zero normal random variables on a portion of a regular lattice of points in $\mathbb{R}$. The latter also displays (as the solid curve) weighted sums of those realizations, where the weights are given by the double exponential moving average function displayed in the former. The integral in (6.6) is the limit (in law) of the weighted sum at $\mathbf{s}$ as the lattice becomes ever more dense.

The evenness and positive definiteness of a stationary covariance function imply several other properties. One of them, which follows immediately from (6.2), is

$$C(\mathbf{0}; \boldsymbol{\theta}) \geq 0 \quad \text{for all } \boldsymbol{\theta} \in \Theta.$$

Another, which follows from the Cauchy-Schwartz inequality, is

$$|C(\mathbf{h}; \boldsymbol{\theta})| \leq C(\mathbf{0}; \boldsymbol{\theta}) \quad \text{for all } \boldsymbol{\theta} \in \Theta.$$

Thus a stationary covariance function may take on negative as well as positive values, but the function is bounded both above and below (above by the variance and below by its

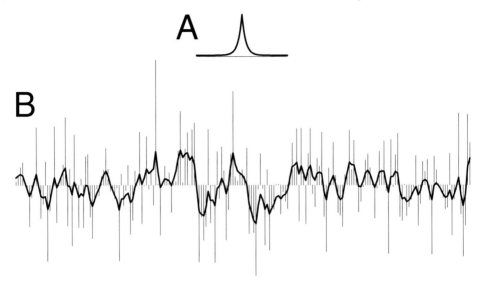

FIGURE 6.1
A) Double exponential moving average function, and B) a realization of a Gaussian process whose limit (in law) is obtained via (6.6) using that moving average function.

negative). The lower bound can be sharpened for isotropic covariance functions in $\mathbb{R}^d$ for $d \geq 2$ (Exercise 6.6). In fact, if an isotropic covariance function is valid in every dimension then it must satisfy

$$C(r; \boldsymbol{\theta}) \geq 0 \quad \text{for all } r.$$

Furthermore, it must be strictly decreasing, though it need not vanish in the limit, i.e., it need not be the case that $\lim_{r \to \infty} C(r; \boldsymbol{\theta}) = 0$; see Exercise 6.7.

The continuity and differentiability behavior at the origin of a stationary covariance function has consequences for the mean square continuity and differentiability of the underlying process and, in the isotropic case at least, for the smoothness of realizations of that process (Stein, 1999). These consequences will be illustrated in the next section; here we merely define the relevant continuity and differentiability properties for an isotropic second-order stationary process $\{Y(\mathbf{s}) : \mathbf{s} \in \mathcal{A}\}$. If $E[(Y(\mathbf{s}) - Y(\mathbf{s} + \mathbf{h})]^2 \to 0$ as $\|\mathbf{h}\| \to 0$, the process is said to be **mean-square continuous**, and if the kth-order derivative of $E[(Y(\mathbf{s}) - Y(\mathbf{s} + \mathbf{h})]^2$ exists, the process is said to be k-**times mean-square differentiable**. Because $E[(Y(\mathbf{s}) - Y(\mathbf{s} + \mathbf{h})]^2 = 2[C(\mathbf{0}) - C(\mathbf{h})]$ for a second-order stationary process, if that process is also isotropic then it is mean-square continuous if and only if its covariance function is continuous at $r = 0$. Furthermore, it turns out that such a process is k-times mean-square differentiable if and only if the $2k$th-order derivative of the covariance function exists and is finite at $r = 0$.

6.2 Isotropic covariance models

Tables 6.1 and 6.2 list several commonly used isotropic correlation functions in parametric form; the corresponding covariance function may be obtained merely by multiplying the

TABLE 6.1

Examples of compactly supported isotropic correlation models. The corresponding covariance models are obtained by multiplying the correlation model by a positive parameter, σ^2. The domain and parameter space for which the model is positive definite are listed for each model. For the generalized Wendland (Gen. Wendland) model, $_2F_1$ is the Gaussian hypergeometric function (Abramowitz & Stegun, 1972).

Model and dimension	Functional form and parameter space
No-correlation $\mathbb{R}^d$	$\rho(r) = \begin{cases} 1 & \text{for } r = 0 \\ 0 & \text{for } r > 0 \end{cases}$
Triangular $\mathbb{R}$	$\rho(r; \alpha) = \begin{cases} 1 - \frac{r}{\alpha} & \text{for } 0 \leq r \leq \alpha \\ 0 & \text{for } r > \alpha \end{cases}$ $\alpha > 0$
Circular $\mathbb{R}^2$	$\rho(r; \alpha) = \begin{cases} \left(\frac{2}{\pi}\right)\left(\arccos\left(\frac{r}{\alpha}\right) - \frac{r}{\alpha}\sqrt{1 - \frac{r^2}{\alpha^2}}\right) & \text{for } 0 \leq r \leq \alpha \\ 0 & \text{for } r > \alpha \end{cases}$ $\alpha > 0$
Spherical $\mathbb{R}^3$	$\rho(r; \alpha) = \begin{cases} 1 - \frac{3r}{2\alpha} + \frac{r^3}{2\alpha^3} & \text{for } 0 \leq r \leq \alpha \\ 0 & \text{for } r > \alpha \end{cases}$ $\alpha > 0$
Tetraspherical $\mathbb{R}^4$	$\rho(r; \alpha) = \begin{cases} \left(\frac{2}{\pi}\right)\left[\arccos\left(\frac{r}{\alpha}\right) - \frac{r}{\alpha}\sqrt{1 - \left(\frac{r}{\alpha}\right)^2} - \frac{2r}{3\alpha}\left(1 - \left(\frac{r}{\alpha}\right)^2\right)^{3/2}\right] \\ \quad \text{for } 0 \leq r \leq \alpha \\ 0 \text{ for } r > \alpha \end{cases}$ $\alpha > 0$
Pentaspherical $\mathbb{R}^5$	$\rho(r; \alpha) = \begin{cases} 1 - \frac{15r}{8\alpha} + \frac{5r^3}{4\alpha^3} - \frac{3r^5}{8\alpha^5} & \text{for } 0 \leq r \leq \alpha \\ 0 & \text{for } r > \alpha \end{cases}$ $\alpha > 0$
Cubic $\mathbb{R}^3$	$\rho(r; \alpha) = \begin{cases} 1 - \frac{7r^2}{\alpha^2} + \frac{35r^3}{4\alpha^3} - \frac{7r^5}{2\alpha^5} + \frac{3r^7}{4\alpha^7} & \text{for } 0 \leq r \leq \alpha \\ 0 & \text{for } r > \alpha \end{cases}$ $\alpha > 0$
Penta $\mathbb{R}^3$	$\rho(r; \alpha) = \begin{cases} 1 - \frac{22r^2}{3\alpha^2} + \frac{33r^4}{\alpha^4} - \frac{77r^5}{2\alpha^5} + \frac{33r^7}{2\alpha^7} - \frac{11r^9}{2\alpha^9} + \frac{5r^{11}}{6\alpha^{11}} \\ \quad \text{for } 0 \leq r \leq \alpha \\ 0 \text{ for } r > \alpha \end{cases}$ $\alpha > 0$
Askey $\mathbb{R}^d$	$\rho(r; \alpha, \mu) = \begin{cases} \left(1 - \frac{r}{\alpha}\right)^\mu & \text{for } 0 \leq r \leq \alpha \\ 0 & \text{for } r > \alpha \end{cases}$ $\alpha > 0, \; \mu \geq \frac{d+1}{2}$
C^2-Wendland $\mathbb{R}^d$	$\rho(r; \alpha, \mu) = \begin{cases} \left[1 + (\mu+1)\left(\frac{r}{\alpha}\right)\right]\left(1 - \frac{r}{\alpha}\right)^{\mu+1} & \text{for } 0 \leq r \leq \alpha \\ 0 & \text{for } r > \alpha \end{cases}$ $\alpha > 0, \; \mu \geq \frac{d+3}{2}$
C^4-Wendland $\mathbb{R}^d$	$\rho(r; \alpha, \mu) = \begin{cases} \left[1 + (\mu+2)\left(\frac{r}{\alpha}\right) + \frac{(\mu+2)^2-1}{3}\left(\frac{r}{\alpha}\right)^2\right]\left(1 - \frac{r}{\alpha}\right)^{\mu+2} \\ \quad \text{for } 0 \leq r \leq \alpha \\ 0 \text{ for } r > \alpha \end{cases}$ $\alpha > 0, \; \mu \geq \frac{d+5}{2}$
Gen. Wendland $\mathbb{R}^d$	$\rho(r; \alpha, \mu, \nu) = \begin{cases} \frac{\Gamma(\nu)\Gamma(2\nu+\mu+1)}{\Gamma(2\nu)\Gamma(\nu+\mu+1)2^{\mu+1}}\left(1 - \left(\frac{r}{\alpha}\right)^2\right)^{\nu+\mu} \\ \quad \times \; _2F_1\left(\frac{\mu}{2}, \frac{\mu+1}{2}; \nu+\mu+1; 1 - \left(\frac{r}{\alpha}\right)^2\right) \\ \quad \text{for } 0 \leq r \leq \alpha \\ 0 \text{ for } r > \alpha \end{cases}$ $\alpha > 0, \; \mu \geq \frac{d+2\nu+1}{2}, \; \nu \geq 0$

TABLE 6.2

Examples of globally supported isotropic correlation models. The corresponding covariance models are obtained by multiplying the correlation model by a positive parameter, σ^2. The domain and parameter space for which the model is positive definite are listed for each model, with two such domains and parameter spaces listed for the Dagum model. The conditions for positive definiteness of the Dagum model in the two cases are merely known to be sufficient; conditions that are both sufficient and necessary are not yet known.

Model and dimension	Functional form and parameter space
Cosine $\mathbb{R}$	$\rho(r;\alpha) = \cos(r/\alpha)$ $\alpha > 0$
Wave $\mathbb{R}^3$	$\rho(r;\alpha) = \frac{\sin(r/\alpha)}{r/\alpha}$ $\alpha > 0$
J-Bessel $\mathbb{R}^d$	$\rho(r;\alpha) = 2^{d/2-1}\Gamma(d/2)(r/\alpha)^{1-d/2}J_{d/2-1}(r/\alpha)$ $\alpha > 0$
Exponential $\mathbb{R}^d$	$\rho(r;\alpha) = \exp(-r/\alpha)$ $\alpha > 0$
Gaussian $\mathbb{R}^d$	$\rho(r;\alpha) = \exp[-(r/\alpha)^2]$ $\alpha > 0$
Powered exponential $\mathbb{R}^d$	$\rho(r;\alpha,\theta) = \exp[-(r/\alpha)^\theta]$ $\alpha > 0, 0 < \theta \le 2$
Radon transform of exponential, order 2 $\mathbb{R}^d$	$\rho(r;\alpha) = \left(1 + \frac{r}{\alpha}\right)\exp(-r/\alpha)$ $\alpha > 0$
Radon transform of exponential, order 4 $\mathbb{R}^d$	$\rho(r;\alpha) = \left(1 + \frac{r}{\alpha} + \frac{1}{3}\left(\frac{r}{\alpha}\right)^2\right)\exp(-r/\alpha)$ $\alpha > 0$
Matérn $\mathbb{R}^d$	$\rho(r;\alpha,\nu) = \frac{1}{2^{\nu-1}\Gamma(\nu)}\left(\frac{r}{\alpha}\right)^\nu K_\nu\left(\frac{r}{\alpha}\right)$ $\alpha > 0, \nu > 0$
Gravity $\mathbb{R}^d$	$\rho(r;\alpha) = [1 + (r/\alpha)^2]^{-1/2}$ $\alpha > 0$
Rational quadratic $\mathbb{R}^d$	$\rho(r;\alpha) = [1 + (r/\alpha)^2]^{-1}$ $\alpha > 0$
Magnetic $\mathbb{R}^d$	$\rho(r;\alpha) = [1 + (r/\alpha)^2]^{-3/2}$ $\alpha > 0$
Cauchy $\mathbb{R}^d$	$\rho(r;\alpha,\phi) = [1 + (r/\alpha)^2]^{-\phi}$ $\alpha > 0, \phi > 0$
Generalized Cauchy $\mathbb{R}^d$	$\rho(r;\alpha,\theta,\phi) = [1 + (r/\alpha)^\theta]^{-\phi}$ $\alpha > 0, 0 < \theta \le 2, \phi > 0$
Dagum $\mathbb{R}^3$ $\mathbb{R}^d$	$\rho(r;\alpha,\theta,\phi) = 1 - \left(\frac{(r/\alpha)^\theta}{1+(r/\alpha)^\theta}\right)^\phi$ $\alpha > 0, 0 < \theta \le (7-\phi)/(1+5\phi), 0 < \phi \le 7$ $\alpha > 0, 0 < \theta \le 1, 0 < \theta\phi \le 1$

correlation function by a constant variance parameter, σ^2. For each function, the tables give the parameter space and the (maximum) dimension of the Euclidean space within which it is positive definite, with $\mathbb{R}^d$ written for the latter if the dimension is arbitrary. Figures 6.2 and 6.3 display selected functions on a portion of $[0, \infty)$ for one or more choices of parameter vectors $\boldsymbol{\theta}$. Furthermore, Figures 6.4 and 6.5 display realizations, on a 100×100 square grid, of stationary Gaussian processes having some of these covariance functions (with $\sigma^2 = 1$).

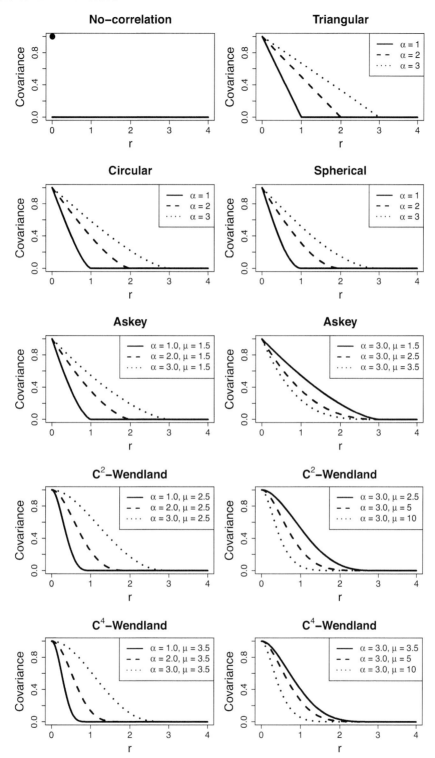

FIGURE 6.2
The no-correlation function and some other compactly supported isotropic correlation functions.

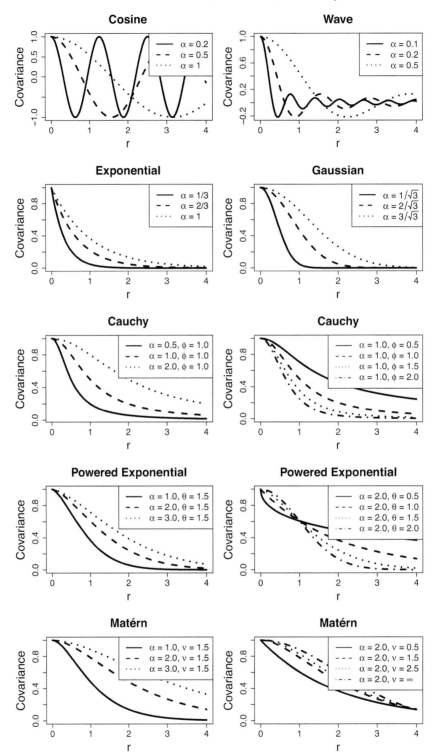

FIGURE 6.3
Some globally supported isotropic correlation functions.

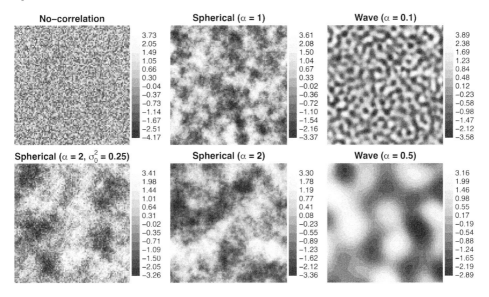

FIGURE 6.4
Simulations of some Gaussian processes with unit variance and various isotropic correlation functions.

The models listed in Table 6.1 are models for which the spatial correlation is positive at short distances, but vanishes completely beyond some finite distance. As noted in Chapter 2, such models are said to have **compact support**. The first model listed is the zero-correlation model, so named because it prescribes the complete absence of spatial

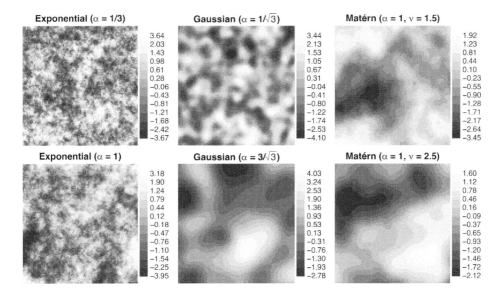

FIGURE 6.5
More simulations of Gaussian processes with unit variance and various isotropic correlation functions.

correlation: the function is equal to one at the origin and equal to 0 elsewhere. Thus, the function is discontinuous, there being a jump at $r = 0$, and the corresponding process is not mean-square continuous. The corresponding covariance matrix is merely a scalar multiple of the $n \times n$ identity matrix. Realizations of a Gaussian process with this covariance model, such as the one displayed in the upper left plot of Figure 6.4, appear as pure static.

The next five models listed in Table 6.1 are closely related to each other, being derived by taking $g(\mathbf{x})$ in the auto-convolution (6.5) to be the indicator function on unit-diameter (hyper)spheres of dimensions one through five. Each model is valid in Euclidean spaces of dimension up to that of the (hyper)sphere. For example, when $d = 1$ the domain of the indicator function is the unit interval and the auto-convolution is the length of overlap of two unit-length intervals, one translated from the other by distance r/α. This yields the triangular covariance function, which gets its name from the shape formed by the graph of the nonzero portion of the function and its reflection about the vertical axis. It is sometimes also called the piecewise linear, or tent, covariance function. When $d = 2$, the auto-convolution is the area of intersection of two unit-diameter circles, one whose center is offset from the other's by r/α; hence, the correlation function is called the circular correlation function. The three-, four-, and five-dimensional cases yield the spherical, tetraspherical, and pentaspherical correlation functions, respectively. Realizations of Gaussian processes corresponding to the spherical model with two different values of α are displayed in the upper and lower middle two plots of Figure 6.4. In contrast to the realization corresponding to the no-correlation model, these realizations exhibit spatial structure, which is more persistent over distance for the process with the larger value of α.

The next two models, the cubic and penta, are similar to the spherical model, except that they are smoother at the origin. Whereas all members of the spherical family are non-differentiable at the origin, the cubic and penta models are once and twice, respectively, differentiable there. They arise as Radon transforms of members of the spherical family. (A Radon transform integrates a function of n arguments over one of the arguments to produce a function of $n-1$ arguments, and this process can be repeated, say m times, to yield a Radon transform of order m.) The cubic and penta models are valid on Euclidean spaces up to dimension three.

The last four models in Table 6.1 are the generalized Wendland model and three special cases thereof. The general case combines the computational benefits of compact support with the modeling benefits afforded by continuously parameterizing the smoothness of the process via a parameter ν, called the **smoothness parameter**. In general, the model is 2ν times differentiable at the origin (Gneiting, 1999, 2002a); in particular the Askey, C^2-Wendland, and C^4-Wendland models are 0, 2, and 4 times differentiable there. Members of the class are valid in Euclidean spaces whose dimensions depend on the parameter μ (and on ν in the most general case). Note, interestingly, that the Askey model with $\mu = 1$ coincides with the triangular model.

The models listed in Table 6.2 are globally, rather than compactly, supported. The first three are sometimes called **hole-effect** models because they allow the correlation to be negative (a "hole") at some distances. These are the cosine, wave (also known as "cardinal-sine"), and J-Bessel functions. The cosine and wave models are the special cases of the J-Bessel function corresponding to $d = 1$ and $d = 3$, respectively. The upper right and lower right plots of Figure 6.4 display Gaussian process realizations corresponding to the wave model with two values of α. Periodicity over space is discernible in these realizations.

The next group of models listed in Table 6.2, six in number, are related to the exponential function. In fact, the first one is simply a parameterized version of the exponential function, so it is given that name. The second, the Gaussian model, is so named because it is a scaled version of the probability density function of a mean-zero Gaussian random variable. The powered exponential model raises the (scaled) distance to a power $\theta \in (0, 2]$ before exponentiating. The exponential and Gaussian models are special cases of the powered exponential model, corresponding to $\theta = 1$ and $\theta = 2$, respectively. The Matérn covariance function was named by Stein (Stein, 1999) after Bertil Matérn, a Swedish forestry statistician whom he credits with being the first, in Matérn (1960), to recommend this function for isotropic processes in an arbitrary dimension. The Matérn model, like the generalized Wendland model, has an additional parameter, ν, called the **smoothness parameter** because it controls the smoothness of realizations of the process. The exponential function is the special case of the Matérn function corresponding to $\nu = \frac{1}{2}$, and the Gaussian function is the limiting case as $\nu \to \infty$. The function $K_\nu(\cdot)$ in the definition of the Matérn covariance function is the modified Bessel function of the second kind, of order ν, one representation of which is given by

$$K_\nu(r) = \frac{\Gamma(\nu + \frac{1}{2})(2r)^\nu}{\sqrt{\pi}} \int_0^\infty \frac{\cos t}{(t^2 + r^2)^{\nu + 1/2}}\, dt.$$

Values of ν equal to $k + \frac{1}{2}$ for integer k yield simpler expressions for the Matérn function. We have already noted that the exponential model corresponds to $\nu = \frac{1}{2}$; furthermore, the cases $\nu = \frac{3}{2}$ and $\nu = \frac{5}{2}$ yield

$$\rho(r; \alpha) = [1 + (r/\alpha)] \exp(-r/\alpha)$$

and

$$\rho(r; \alpha) = \left(1 + \frac{r}{\alpha} + \frac{r^2}{3\alpha^2}\right) \exp(-r/\alpha),$$

respectively. These last two cases of the Matérn function coincide with Radon transforms of the exponential covariance function of orders 2 and 4 and therefore are given those names. Realizations of Gaussian processes with exponential, Gaussian, and these last two covariance functions are displayed in Figure 6.5. A progressive increase in the smoothness of the realizations with increasing ν is discernible.

The next five models in Table 6.2 all have a functional form related to the probability density function of a Cauchy distribution with location parameter 0, hence "Cauchy" is included in the name of the two most general models in this group. The gravity, rational quadratic, and magnetic models are special cases of the Cauchy model, which in turn is a special case of the generalized Cauchy model.

The Dagum family of correlation models, like the generalized Cauchy family, has three parameters. Necessary and sufficient conditions on its parameters for positive definiteness in Euclidean spaces of various dimensions are not yet known, but Table 6.2 gives sufficient conditions in $\mathbb{R}^3$ and $\mathbb{R}^d$ for all d provided by Porcu *et al.* (2007b) and Berg *et al.* (2008), respectively. The Dagum family includes the special case of the generalized Cauchy model with $\phi = 1$.

Note that the triangular and cosine correlation functions are positive definite in $\mathbb{R}$ but not in $\mathbb{R}^2$ or $\mathbb{R}^3$, and the circular correlation function is positive definite in $\mathbb{R}$ and $\mathbb{R}^2$ but not in $\mathbb{R}^3$. The remaining functions are positive definite in $\mathbb{R}^3$, and many are positive definite in an arbitrary number of dimensions. The zero-correlation model has no unknown parameters, whereas most of the remaining correlation functions listed have one parameter. The exceptions are the powered exponential, Matérn, and Cauchy correlation functions, which have two parameters, and the generalized Cauchy and Dagum functions, which have three.

More complicated isotropic covariance functions may be constructed from the correlation functions already presented, using the algebraic operations described in Section 6.1. An extremely important example of such a construction is obtained by adding a scaled no-correlation model to any isotropic mean-square continuous covariance function $C^*(r; \boldsymbol{\theta})$, yielding

$$C(r; \boldsymbol{\theta}) = \begin{cases} C^*(0; \boldsymbol{\theta}) + \sigma_0^2 & \text{for } r = 0, \\ C^*(r; \boldsymbol{\theta}) & \text{for } r > 0. \end{cases} \qquad (6.7)$$

(This is the isotropic case of a more general result first noted in Exercise 2.12.) If $C^*(r; \boldsymbol{\theta})$ is, for instance, the isotropic exponential covariance function, then (6.7) is called the isotropic exponential covariance function with nugget effect (σ_0^2 being the nugget effect), with a similar naming protocol when $C^*(r; \boldsymbol{\theta})$ is any other isotropic mean-square continuous covariance function. Such a function could result from homoscedastic, spatially uncorrelated measurement errors in the spatial linear model, or from a decomposition of $W(\mathbf{s})$ into the sum

$$W(\mathbf{s}) = W_c(\mathbf{s}) + W_{ms}(\mathbf{s}),$$

where $\{W_c(\mathbf{s}) : \mathbf{s} \in \mathcal{A}\}$ is an isotropic mean-square continuous process and $\{W_{ms}(\mathbf{s}) : \mathbf{s} \in \mathcal{A}\}$ is microscale variation (the latter was defined in Chapter 2). The corresponding covariance matrix $\boldsymbol{\Sigma}$ of the spatial linear model may be written as the sum of two matrices, one of which is a scalar multiple of the $n \times n$ identity matrix.

One feature that distinguishes isotropic covariance functions from one another is their limiting behavior as distance increases. With one exception, every covariance function included in Table 6.2 tends to zero as r increases. The exception is the cosine function, which is not positive definite in two or more dimensions, but in one dimension might be appropriate if the underlying process appears to exhibit periodic second-order dependence that does not damp out with increasing distance. The wave function is also nonmonotonic, but its oscillations damp out so that it tends to 0 as distance increases. It may be useful if the spatial dependence appears to vanish in the limit but some negative dependence is observed at intermediate distances. All of the remaining models in Table 6.2 conform strictly to Tobler's first law by decreasing monotonically to zero as distance increases.

Various other features or attributes, corresponding to particular elements of $\boldsymbol{\theta}$ or functions thereof, can be used to distinguish among isotropic covariance functions. These features include the following:

- *Variance.* For every random variable $Y(\mathbf{s})$ of the underlying process, $C(0; \boldsymbol{\theta})$ is its variance. As noted previously, an isotropic covariance model may be obtained from each correlation model in Tables 6.1 and 6.2 by simply multiplying the correlation model by a positive scale parameter σ^2.

- *Nugget effect.* The **nugget effect**, defined formally as $C(0; \boldsymbol{\theta}) - \lim_{r \to 0} C(r; \boldsymbol{\theta})$, is the "jump" in $C(r; \boldsymbol{\theta})$ at the origin, if such a jump exists. If no such jump exists, then the covariance function is continuous everywhere, and the underlying geostatistical process is mean-square continuous. The no-correlation model in Table 6.1 has a nugget effect but no dependence, and for this reason, it is also called the pure nugget model. None of the other models in the tables have a nugget effect, but they can easily be extended to include one via (6.7). The lower left panel in Figure 6.4 displays a simulated realization of a unit-variance Gaussian process with an isotropic spherical covariance function with nugget effect equal to 25% of the variance. When compared to the panel next to it, which displays a realization of a nuggetless Gaussian process with the same variance and same range parameter, the impact of the nugget effect on the smoothness of the realization is clearly discernible; in particular, the realization of the process with nugget effect appears to be more granular.

- *Partial sill.* The partial sill is defined as $\lim_{r \to 0} C(r; \boldsymbol{\theta})$; the rationale for the term will become clear in Section 6.5. For all of the models listed in Tables 6.1 and 6.2 except the no-correlation model, the partial sill is equal to the variance. However, for multi-component models that include a nugget effect, the partial sill does not coincide with the variance; rather, the variance is the sum of the nugget effect and partial sills of the remaining components.

- *Range.* The range is defined as the distance beyond which the covariance function is equal to 0, if such a distance exists. Thus, models with compact support have a range, which is denoted by α in Table 6.1; for globally supported covariance functions, the range does not exist. Because the covariance matrix of a spatial linear model corresponding to a compactly supported covariance function has some (perhaps many) null elements, such functions can have computational advantages for statistical inference, as was noted in Section 5.6. Note, by (6.3), that a valid compactly supported isotropic covariance function can be constructed as the product of one such function and an arbitrary isotropic covariance function.

- *Practical range.* For isotropic processes that do not have a range, the practical range is defined as the distance beyond which the covariance function is less than or equal to 5% of the variance, i.e., $0.05 \times C(0; \boldsymbol{\theta})$, if such a distance exists. Of the globally supported models listed in Table 6.2, all but the cosine model have a practical range. The practical range of a model must be determined on a case-by-case basis. For the exponential model, the practical range is approximately 3α. This is obtained by solving the equation

$$\sigma^2 \exp(-r/\alpha) = 0.05\sigma^2$$

for r, yielding $r = \alpha \ln 20 \doteq 3\alpha$. For the Gaussian model, a similar calculation reveals the practical range to be approximately equal to $\sqrt{3}\alpha$. Practical ranges for some other models are considered in Exercises 6.8 and 6.14–6.16.

- *Distance scale parameter.* For processes that have neither a range nor a practical range, there may nonetheless be a parameter that controls the rate at which the covariance changes with distance. For example, for the cosine model listed in Table 6.2, α controls that rate.

- *Smoothness parameter.* Some isotropic correlation models with two or more parameters have a parameter that controls the smoothness, i.e., differentiability, of the function at the origin. This attribute, in turn, has consequences for the mean-square differentiability of the corresponding Gaussian process and hence the smoothness of realizations of the process. In particular, a Matérn covariance function is $2k$ times differentiable, and a Gaussian process with that covariance function is k-times mean-square differentiable (in any direction), if and only if $\nu > k$ (Stein, 1999). Recall that Figure 6.5 illustrates the increasing smoothness of realizations of such a process as ν increases. Similarly, a Gaussian process with generalized Wendland covariance function is k-times mean-square differentiable (in any direction) if and only if $\nu > k - \frac{1}{2}$ (Bevilacqua *et al.*, 2022). In contrast, processes with powered exponential or Cauchy covariance functions are either infinitely mean-square differentiable (if $\theta = 2$) or not differentiable at all (if $\theta < 2$), with the consequence that these functions do not allow for modeling and estimating smoothness.

- *Long-range dependence parameter.* For all models save the generalized Cauchy and Dagum, the correlation vanishes at a rapid rate as distance increases. In order to allow for a slower rate (long-range dependence), the generalized Cauchy model has an additional parameter, ϕ (see Table 6.2). The smaller the value of ϕ, the stronger the long-range dependence. The parameter ϕ for the Dagum model appears to play a similar role.

6.3   Anisotropic covariance models

In practice, it is sometimes found, using directional semivariogram plots or the nonparametric tests for isotropy described in Chapter 3, that the covariance function for a particular dataset may be anisotropic. It is desirable, therefore, to expand our modeling possibilities to stationary covariances that are functions of direction as well as distance.

It has been stated (e.g., Clark, 1979, p.12) that the form of anisotropy observed most often in practice is a direction-dependent range (or practical range). If true, this is fortunate because it is relatively easy to construct valid covariance functions that are flexible with respect to the dependence of the range on direction. This dependence may be continuous or discontinuous (Allard *et al.*, 2016). One particularly useful type of continuous dependence of the range on direction is **geometric range anisotropy**. As an example of this type, consider a geostatistical process in $\mathbb{R}^2$ with an exponential covariance function whose range parameter varies with direction; more specifically, suppose that the practical range in the N-S (north-south) direction is a times as large as the practical range in the E-W (east-west) direction, and the practical ranges in the NE-SW and NW-SE directions are $a\sqrt{2/(1+a^2)}$ as large as the E-W practical range. Suppose further that the nugget effect (if any) and partial sill are direction-invariant. Then, the covariance function can be described well, at least in the four observed directions, by a single exponential model upon replacing Euclidean distance $r = \|\mathbf{h}\| = \|(h_u, h_v)^T\| = \sqrt{h_u^2 + h_v^2}$ with elliptical distance $\sqrt{h_u^2 + (h_v^2/a^2)}$. Loci of equal correlation for this covariance function are ellipses (in $\mathbb{R}^3$ they would be ellipsoids), with major axis aligned with the u-axis if $a < 1$ or with the v-axis if $a > 1$.

Any isotropic covariance function (not merely the exponential) in any dimension d can be generalized to make it geometrically anisotropic, merely by replacing $\|\mathbf{h}\|$ in the isotropic function with $\sqrt{\mathbf{h}^T \mathbf{B} \mathbf{h}}$, where $\mathbf{B}$ is a $d \times d$ positive definite matrix. In practice, some or all of the elements of $\mathbf{B}$ may be unknown, in which case it is sensible to regard them as additional model parameters. Thus, for example, any geometrically anisotropic version of the nuggetless exponential correlation function listed in Table 6.2 may be written, when $d = 2$, as

$$\rho(\mathbf{h}; \alpha, \eta, \xi) = \exp(-\sqrt{h_u^2 + 2\eta h_u h_v + \xi h_v^2}/\alpha)$$

for some parameters α, η, and ξ; the extension to $d = 3$ is straightforward but notationally more cumbersome. In the two-dimensional case, which we focus on exclusively in what follows, η and ξ are constrained to satisfy $\xi > \eta^2$ to ensure positive definiteness of $\mathbf{B}$, and isotropy corresponds to the special case of $(\eta, \xi) = (0, 1)$. An alternative parameterization of a geometrically anisotropic correlation function in two dimensions is in terms of an **anisotropy angle** (the direction of strongest spatial dependence) and an **anisotropy ratio** (the ratio of the smallest to largest ranges or practical ranges). This may be expressed mathematically as

$$\rho(\mathbf{h}; \alpha, \tau, \phi) = \rho(\|\mathbf{TRh}\|/\alpha),$$

$$\mathbf{T} = \begin{pmatrix} 1 & 0 \\ 0 & \tau \end{pmatrix}, \quad \mathbf{R} = \begin{pmatrix} \cos\phi & -\sin\phi \\ \sin\phi & \cos\phi \end{pmatrix}. \tag{6.8}$$

This is the parameterization that R packages `spmodel`, `gstat`, and `geoR` use, with ϕ representing the anisotropy angle and τ representing the anisotropy ratio. Its main advantage is greater interpretability of its parameters; a disadvantage is that the anisotropy angle is not well-defined when the anisotropy ratio is equal to one (i.e., when the correlation function is isotropic). Figure 6.6A displays a locus of equal correlation for a geometrically anisotropic correlation function with anisotropy angle equal to $\pi/3$ and anisotropy ratio equal to $1/3$. Furthermore, the first two columns of plots in Figure 6.7 display realizations of Gaussian

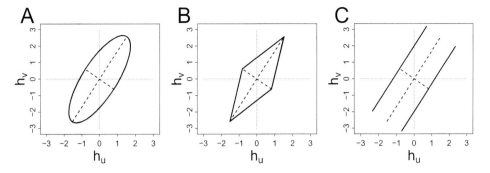

FIGURE 6.6
Loci of equal correlation for stationary correlation funtions with three types of anisotropy:
A) geometric anisotropy with anisotropy angle $\pi/3$ and anisotropy ratio $1/3$; B) product
anisotropy for an exponential correlation function with anisotropy angle $\pi/3$ and anisotropy
ratio $1/3$; C) zonal anisotropy with infinite range in direction $\pi/3$.

processes having geometrically anisotropic exponential correlation functions. Stronger spatial dependence in the direction of the anisotropy angle is clearly visible, particularly for
the realizations corresponding to the smaller (more extreme) anisotropy ratio.

The directional dependence of the range parameter of a geometrically anisotropic correlation function is highly structured, as it can be described by just two parameters: either
(η, ξ) or (ϕ, τ), depending on the parameterization. Geometric anisotropy and more general forms of the dependence of the range parameter on direction may be modeled, using a

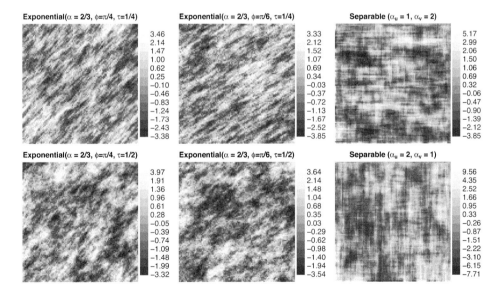

FIGURE 6.7
Simulations of some Gaussian processes with unit variance and anisotropic or separable
correlation functions.

polar-coordinates representation, as

$$\rho(\mathbf{h}; \boldsymbol{\theta}) = \rho(\|\mathbf{h}\|/\alpha(\phi)),$$

where we have ignored all parameters (such as smoothness parameters) except those pertaining to the directionally-varying range parameter, $\alpha(\phi)$. For a geometrically anisotropic model (in $\mathbb{R}^2$),

$$\alpha(\phi) = \frac{1}{\sqrt{(1/\alpha_1)^2 \cos^2(\phi - \phi_0) + (1/\alpha_2)^2 \sin^2(\phi - \phi_0)}},$$

where ϕ_0 is the anisotropy angle, α_1 is the range parameter in the ϕ_0-direction, and α_2 is the range parameter in the direction orthogonal to ϕ_0. Note that the parameterization in terms of the anisotropy angle and anisotropy ratio is related to this parameterization via $\phi = \phi_0$, $\alpha = \alpha_1$ and $\tau = \alpha_2/\alpha_1$.

Another continuous and highly structured form of range anisotropy occurs when the correlation function is stationary and separable, i.e., when it satisfies

$$\rho(\mathbf{h}; \boldsymbol{\theta}) = \prod_{i=1}^{d} \rho_i(h_i; \boldsymbol{\theta}_i),$$

where $\boldsymbol{\theta} = (\boldsymbol{\theta}_1^T, \ldots, \boldsymbol{\theta}_d^T)^T$ and each term in the product is a valid correlation function in $\mathbb{R}$. The validity of the product as a correlation function in $\mathbb{R}^d$ follows from (6.4). Thus, for example,

$$\begin{aligned}
\rho(\mathbf{h}; \alpha_u, \alpha_v) &= \exp[-(|h_u|/\alpha_u)] \exp[-(|h_v|/\alpha_v)] \\
&= \exp[-(|h_u|/\alpha_u) + (|h_v|/\alpha_v)]
\end{aligned} \tag{6.9}$$

is a range-anisotropic correlation function in $\mathbb{R}^2$. Its loci of equal correlation are rhombuses, and when the data locations have certain regular spatial configurations it has significant computational advantages over geometric range anisotropy because the covariance matrix of the corresponding spatial linear model can be written as the Kronecker product of two much smaller matrices; the advantages that occur in this situation were noted in Section 5.6. Realizations of Gaussian processes with this correlation function are displayed in the upper and lower right plots of Figure 6.7. Horizontal and vertical "striations" are discernible in these two plots, the former being more pronounced in the first plot and the latter more pronounced in the second.

A polar-coordinates representation of the dependence of the range parameter on direction for the covariance function given by (6.9) is

$$\alpha(\phi) = \frac{1}{(1/\alpha_1)|\cos\phi| + (1/\alpha_2)|\sin\phi|},$$

where $\alpha_1 = \alpha_u$ and $\alpha_2 = \alpha_v$. For this model, as for all separable models, the anisotropy is rather inflexible because it is constrained to align with the (Euclidean) coordinate system. It can be generalized by rotating the coordinate system by angle ϕ_0, whence the polar-coordinates representation of the range parameter is

$$\alpha(\phi) = \frac{1}{(1/\alpha_1)|\cos(\phi - \phi_0)| + (1/\alpha_2)|\sin(\phi - \phi_0)|}.$$

The equivalent representation in Euclidean coordinates, for any functional form of the correlation function, is

$$\rho(\mathbf{h}; \boldsymbol{\theta}) = \prod_{i=1}^{d} \rho_i(h_i^*; \boldsymbol{\theta}_i),$$

where h_i^* is the ith element of $\mathbf{TRh}$, with $\mathbf{T}$ and $\mathbf{R}$ defined as in (6.8). This type of anisotropy, called alternatively **componentwise anisotropy** (Allard *et al.*, 2016) or **product anisotropy** (Koch *et al.*, 2020), combines the relative flexibility of geometric anisotropy with respect to the direction of the anisotropy angle with the computational advantages of separability for gridded data. Surprisingly, the grid need not be aligned with the anisotropy angle for substantial computational advantages to be realized (Koch *et al.*, 2020).

Whereas the loci of equal correlation under geometric anisotropy are all the same shape (ellipses in $\mathbb{R}^2$ or ellipsoids in $\mathbb{R}^3$) regardless of the functional form (exponential, spherical, Cauchy, ...) of the correlation function, this is not so for product anisotropy. Figure 6.6B displays a locus of equal correlation for the special case of a product of two one-dimensional exponential correlation functions with ranges satisfying $\alpha_1 = 3\alpha_2$, rotated by an amount $\pi/3$. For this case, and for any rotated case of (6.9), the loci of equal correlation are rhombuses. For functional forms other than exponential, however, there are other types of superellipses, including astroids (e.g., for Matérn covariance functions with smoothness parameter less than $1/2$) and even ellipses (for the Gaussian correlation function), so that there is overlap of product anisotropy with geometric anisotropy. Ellipses occur only for the geometrically anisotropic Gaussian correlation function, however, the isotropic case of which is the only continuous correlation function that is simultaneously isotropic and separable (Zimmerman, 1986; De Iaco *et al.*, 2019).

Occasionally, directional sample semivariogram plots may suggest that the (practical) range varies with direction in a manner that does not conform to either geometric anisotropy or product anisotropy. One possibility that has been proposed for such a situation is a so-called **zonally range-anisotropic** correlation model, which in its simplest form is a degenerate case of a geometrically anisotropic model obtained by setting some of the diagonal elements of $\mathbf{T}$ in (6.8) equal to 0. Consider, for example, a zonally anisotropic exponential correlation function in $\mathbb{R}^2$ obtained by setting the second diagonal element of $\mathbf{T}$ equal to 0. This correlation function may be written as

$$\rho(r, \phi; \alpha, \theta_0) = \exp[-r/\alpha(\phi)],$$

where $\alpha(\phi) = \alpha/|\cos(\phi - \phi_0)|$. The practical range of this correlation function exists and is nonconstant in all directions except that which is orthogonal to ϕ_0 (see Figure 6.6C). In that singular direction, the practical range does not exist, and for convenience we say that the practical range is infinite in that direction. Furthermore, in that direction the correlation is identically equal to 1.0, and thus the underlying process itself is constant in that direction.

A pure zonally anisotropic model is seldom used in practice, but such a model sometimes is combined additively with other correlation functions to yield a model with more complicated anisotropies. For example, consider the following correlation model in $\mathbb{R}^2$:

$$\begin{aligned}
\rho(h_u, h_v; \alpha_{uv}, \alpha_u, \alpha_v) &= \frac{1}{3}\exp(-\sqrt{h_u^2 + h_v^2}/\alpha_{uv}) \\
&\quad + \frac{1}{3}\exp(-|h_u|/\alpha_u) + \frac{1}{3}\exp(-|h_v|/\alpha_v).
\end{aligned} \tag{6.10}$$

The first summand on the right-hand side of the equality in (6.10) is isotropic with a finite practical range, while the second summand is anisotropic with an infinite practical range in the v-direction. The final summand is anisotropic with infinite practical range in the u-direction. Overall, then, the model has practical ranges that vary with direction and are finite in all but the u- and v-directions. The first summand in the model may be modified to be anisotropic itself, and all three of them may be modified to have different functional forms. Furthermore, the model may be expanded to allow for infinite practical ranges in any finite number of directions by adding terms to the sum whose arguments are vectors in the orthogonal directions. However, an infinite practical range in even one direction seems unrealistic in practice, and is problematic theoretically. Adler (2010) showed that a sufficient condition for the parameters of the covariance function of a stationary Gaussian process to be estimated consistently from a single realization is for the correlation to tend to zero in all directions. Thus, for both practical and theoretical reasons, using a zonally range-anisotropic model or a linear combination of such models seems unwise. An alternative that has no theoretical shortcomings, but which is considerably more complicated, is the family of range-anisotropic models introduced by Allard *et al.* (2016), which are based on directional measures. This family includes the geometric and product range-anisotropic models as special cases, but also includes more general forms of range anisotropy while allowing the correlation to vanish in all directions.

As noted in Chapter 3, directional sample semivariogram plots sometimes reveal an anisotropy in the sill or nugget effect, rather than the range. Sill anisotropy is discussed in Section 6.5, but here we describe one practically important type of nugget anisotropy that can be modeled rather easily. Suppose that the data are obtained by extracting all material from vertical drillholes (or boreholes) sunk into a three-dimensional body of ore or soil, and that samples from the same drillhole are analyzed contemporaneously in the laboratory while samples from different drillholes are analyzed on different occasions, perhaps even by different individuals. For such data, it may be reasonable to assume that measurement errors from different drillholes are uncorrelated but measurement errors from the same drillhole are equally and positively correlated. This type of correlation structure may be imposed by adopting the spatial random effects model

$$Y(u, v, w) = \mu + W(u, v, w) + \delta_1(u, v) + \delta_2(u, v, w),$$

where $W(u, v, w)$ is mean-square continuous and isotropic with variance σ_W^2, and $\delta_1(\cdot)$ and $\delta_2(\cdot)$ are independent (both of each other and of $W(\cdot)$) white noise processes on $\mathbb{R}^2$ and $\mathbb{R}^3$ with variances σ_{01}^2 and σ_{02}^2, respectively. It can be shown (see Exercise 6.12) that under this model, measurement errors corresponding to measurements from the same drillhole have correlation $\sigma_{01}^2/(\sigma_{01}^2 + \sigma_{02}^2)$, and that the covariance function of $Y(\cdot)$ is

$$C_Y(h_u, h_v, h_w) = \begin{cases} \sigma_W^2 + \sigma_{01}^2 + \sigma_{02}^2 & \text{if } h_u = h_v = h_w = 0, \\ C_W(0, 0, h_w) + \sigma_{01}^2 & \text{if } h_u = h_v = 0, \ h_w \neq 0, \\ C_W(h_u, h_v, h_w) & \text{otherwise,} \end{cases} \tag{6.11}$$

where $C_W(h_u, h_v, h_w)$ is the covariance function of $\{W(\cdot)\}$. Hence, for this process the nugget effect is σ_{02}^2 in the vertical direction and $\sigma_{01}^2 + \sigma_{02}^2$ in all other directions.

The smoothness parameter of a covariance function may also be modeled as a function of direction, but the nature of its dependence on direction is more restricted than that of the range or nugget effect. If the smoothness parameter is a continuous function of direction, then for a very large class of covariance functions, including virtually all that are used in practice, it must be constant (Allard *et al.*, 2016). Furthermore, if the smoothness parameter of a covariance function in that same class varies discontinuously with direction, in $\mathbb{R}^2$ it may do so in at most one direction (which is a special case of zonal anisotropy), and in that direction, it must be smoother than in the other directions (Allard *et al.*, 2016). In $\mathbb{R}^3$ the greater smoothness must occur in either a single direction or a single plane of directions.

6.4 Nonstationary covariance models

An assumption of global second-order stationarity simplifies the modeling of the covariance structure of a geostatistical process, but may be unrealistic for some spatial datasets, especially if there are features such as roads or mountains in the spatial domain that could induce local dependencies. For this reason, a number of approaches have been developed for accommodating nonstationarity in spatial modeling; for a review of these, see Fouedjio (2017). Here we will describe the three approaches that fit most seamlessly within the framework of spatial linear models. One of these is to model dependence using generalized covariance functions, which is the topic of the next section. The other two construct explicitly nonstationary covariance functions.

The first explicit modeling approach uses auto-convolution in a manner similar to that described in Section 6.1, but with a spatially varying function $g(\cdot)$. This approach was first proposed by Higdon (1998) and Higdon *et al.* (1999), then developed further by Paciorek (2003), Stein (2005a), and Paciorek & Schervish (2006). It builds a valid nonstationary covariance function as

$$C(\mathbf{s}, \mathbf{s}') = \int_{\mathbb{R}^d} g_{\mathbf{s}}(\mathbf{x}) g_{\mathbf{s}'}(\mathbf{x}) \, d\mathbf{x}, \tag{6.12}$$

where $g_{\mathbf{s}}(\cdot)$ is a spatially varying square-integrable function on $\mathbb{R}^d$. For many choices of $g_{\mathbf{s}}(\cdot)$, the integral in (6.12) cannot be explicitly evaluated. A notable exception is

$$g_{\mathbf{s}}(\mathbf{x}) = (2\pi)^{-d/2} |\boldsymbol{\Omega}(\mathbf{s})|^{-1/2} \exp\{-(\mathbf{x} - \mathbf{s})^T [\boldsymbol{\Omega}(\mathbf{s})]^{-1} (\mathbf{x} - \mathbf{s})/2\}$$

(the probability density function of a d-dimensional normal distribution with mean vector $\mathbf{s}$ and covariance matrix $\boldsymbol{\Omega}(\mathbf{s})$), for which (6.12) takes the relatively simple form

$$C(\mathbf{s}, \mathbf{s}'; \boldsymbol{\theta}) = \sigma^2 |\boldsymbol{\Omega}(\mathbf{s})|^{1/4} |\boldsymbol{\Omega}(\mathbf{s}')|^{1/4} |(1/2)[\boldsymbol{\Omega}(\mathbf{s}) + \boldsymbol{\Omega}(\mathbf{s}')]|^{-1/2} \exp(-Q(\mathbf{s}, \mathbf{s}')), \tag{6.13}$$

where $Q(\mathbf{s}, \mathbf{s}') = 2(\mathbf{s} - \mathbf{s}')^T [\boldsymbol{\Omega}(\mathbf{s}) + \boldsymbol{\Omega}(\mathbf{s}')]^{-1} (\mathbf{s} - \mathbf{s}')$ and we have now written $C(\cdot, \cdot)$ as a parametric function and included a scale parameter σ^2. Observe that (6.13) has the form of a geometrically anisotropic Gaussian covariance function but with a spatially varying variance and a spatially varying anisotropy. Being a Gaussian covariance function, it has the undesirable property that realizations of a Gaussian process with this covariance function are infinitely differentiable (hence too smooth to be practically realistic). Therefore, as an extension of (6.13), Paciorek & Schervish (2006) proved that if $C_S(r)$ is *any* isotropic covariance function that is valid in an arbitrary dimension, then

$$C(\mathbf{s}, \mathbf{s}'; \boldsymbol{\theta}) = \sigma^2 |\boldsymbol{\Omega}(\mathbf{s})|^{1/4} |\boldsymbol{\Omega}(\mathbf{s}')|^{1/4} |(1/2)[\boldsymbol{\Omega}(\mathbf{s}) + \boldsymbol{\Omega}(\mathbf{s}')]|^{-1/2} C_S(\sqrt{Q(\mathbf{s}, \mathbf{s}')})$$

is a valid nonstationary covariance function in any dimension. From this general result, a nonstationary version of each isotropic covariance function listed in Table 6.2 that is valid in any dimension may be constructed. In particular, a nonstationary version of the Matérn covariance function is given by

$$\begin{aligned} C(\mathbf{s}, \mathbf{s}'; \boldsymbol{\theta}) = {} & \sigma^2 \frac{2^{1-\nu}}{\Gamma(\nu)} |\boldsymbol{\Omega}(\mathbf{s})|^{1/4} |\boldsymbol{\Omega}(\mathbf{s}')|^{1/4} |(1/2)[\boldsymbol{\Omega}(\mathbf{s}) + \boldsymbol{\Omega}(\mathbf{s}')]|^{-1/2} \\ & \times [(2\nu Q(\mathbf{s}, \mathbf{s}'))^{1/2}]^\nu K_\nu((2\nu Q(\mathbf{s}, \mathbf{s}'))^{1/2}), \end{aligned}$$

which includes a nonstationary version of the exponential covariance function as the special case $\nu = \frac{1}{2}$. A further extension of this function that allows the smoothness parameter, ν, to vary spatially was introduced by Stein (2005a).

Given a particular spatial dataset with spatial data locations $\mathbf{s}_1, \ldots, \mathbf{s}_n$, the parameters of the nonstationary covariance functions described in the previous paragraph that must be estimated include the $d \times d$ matrices $\boldsymbol{\Omega}(\mathbf{s}_1), \ldots, \boldsymbol{\Omega}(\mathbf{s}_n)$. This is generally too many parameters to estimate well unless further assumptions are made about how $\boldsymbol{\Omega}(\mathbf{s})$ varies spatially. Useful parsimonious models for $\boldsymbol{\Omega}(\mathbf{s})$ might assume that the anisotropy angle and the largest and smallest ranges (or practical ranges), suitably transformed, are low-order polynomial functions of spatial location or of distance from a particular topographic feature (e.g. a road or mountain range) that might affect the data's covariance structure.

The second explicit modeling approach considers nonstationary spatial dependence caused by the presence of a point source in the spatial domain. In this approach, the observed spatial process is regarded as being the result of a baseline stationary process that is modified by the action of the point source in such a way that observations equidistant from the point source are more highly correlated than they would be otherwise, regardless of how far apart the corresponding sites are. Several slightly different parametric nonstationary covariance functions result from different modes of modification by the point source; see Hughes-Oliver *et al.* (1998), Hughes-Oliver & González-Farias (1999), and Warren (2020). Here we describe the model proposed by Warren (2020), which is for use in $\mathbb{R}^2$. In this function, a site is represented in terms of its coordinates (h, ϕ) in a polar coordinate system centered at the point source. Thus, h is the Euclidean distance between the site and the point source, and ϕ is the angle of the vector (in radians) that extends from the point source to the site, relative to a specified reference line. It is assumed that there is no observation at the point source. The nonstationary covariance function is

$$
\begin{aligned}
C(\mathbf{s}, \mathbf{s}'; \boldsymbol{\theta}) &= C((h, \phi), (h', \phi'); \boldsymbol{\theta}) \\
&= \sigma^2 \Bigg((1-p)\rho_I(\sqrt{h^2 + h'^2 - 2hh'\cos(\phi - \phi')}; \boldsymbol{\xi}) \\
&\quad + p \exp\left\{ -\frac{|h - h'|}{\alpha_1} - \frac{\min[|\phi - \phi'|, 2\pi - \max(\phi, \phi') + \min(\phi, \phi')]}{\alpha_2} \right\} \Bigg).
\end{aligned}
$$

Here, $\rho_I(\cdot)$ represents any parametric isotropic correlation function of the Euclidean distance between sites (such as those listed in Tables 6.1 and 6.2), written in terms of their polar coordinates, with $\boldsymbol{\xi}$ denoting its parameters; the first term in the exponent causes the correlation function to decay as the distance of each site from the point source increases, with decay parameter α_1; the second term in the exponent causes the correlation function to decay as the difference in the directions (angles of incidence), modulo 2π, of the sites from the point source increases, with decay parameter α_2; $0 \leq p \leq 1$; and $\boldsymbol{\theta} = (\boldsymbol{\xi}^T, \alpha_1, \alpha_2, p)^T$. The positive definiteness of this model was established by Warren (2020). It and the point-source models in the other two references cited above may be generalized to accommodate multiple point sources and anisotropy around the point source (such as might be caused by a prevailing wind direction), though this can lead to model identifiability and fitting issues.

6.5 Semivariogram models and generalized covariance functions

The preceding development has focused entirely on covariance functions. For processes that are intrinsically stationary but not second-order stationary, models for the semivariogram are needed. As was noted in Section 2.2, the semivariogram $\gamma(\mathbf{h}; \boldsymbol{\theta})$ of an intrinsically stationary process is necessarily even and conditionally negative definite and satisfies $\gamma(\mathbf{0}; \boldsymbol{\theta}) = 0$. These properties are also sufficient for a real-valued function on $\mathbb{R}^d$ to be an intrinsically stationary semivariogram. In fact, a result analogous to Bochner's theorem holds for semivariograms (Gneiting *et al.*, 2001), though the characterization based on this result is not of a form as familiar to statisticians as characteristic functions so we do not describe it explicitly here. Although the semivariogram need not be bounded, its conditional

TABLE 6.3
Some parametric isotropic semivariogram models.

Semivariogram	Functional form	Parameter space
No-correlation	$\gamma(r; \sigma^2) = \begin{cases} 0 & \text{for } r = 0, \\ \sigma^2 & \text{for } r > 0 \end{cases}$	$\sigma^2 > 0$
Exponential	$\gamma(r; \sigma^2, \alpha) = \sigma^2[1 - \exp(-r/\alpha)]$	$\sigma^2 > 0, \alpha > 0$
Linear	$\gamma(r; \theta_1) = \theta_1 r$	$\theta_1 > 0$
Power	$\gamma(r; \theta_1, \theta_2) = \theta_1 r^{\theta_2}$	$\theta_1 > 0, 0 < \theta_2 < 2$

negative definiteness implies that it cannot grow too fast with increasing $\|\mathbf{h}\|$; in particular, as noted in Section 2.2.1, its growth must be sub-quadratic, meaning that $\gamma(\mathbf{h})/\|\mathbf{h}\|^2 \to 0$ as $\|\mathbf{h}\| \to \infty$. Furthermore, for a semivariogram to conform strictly to Tobler's first law, it must be monotone increasing in every direction.

For each of the covariance functions described earlier in this chapter there is a corresponding conditionally negative definite semivariogram, which can be obtained via the relation established in Section 2.2.1, which we write here as

$$\gamma(\mathbf{s}, \mathbf{s}'; \boldsymbol{\theta}) = \frac{1}{2} C(\mathbf{s}, \mathbf{s}; \boldsymbol{\theta}) + \frac{1}{2} C(\mathbf{s}', \mathbf{s}'; \boldsymbol{\theta}) - C(\mathbf{s}, \mathbf{s}'; \boldsymbol{\theta}).$$

This relation simplifies to

$$\gamma(\mathbf{h}; \boldsymbol{\theta}) = C(\mathbf{0}; \boldsymbol{\theta}) - C(\mathbf{h}; \boldsymbol{\theta})$$

when the process is second-order stationary, and to

$$\gamma(r; \boldsymbol{\theta}) = C(0; \boldsymbol{\theta}) - C(r; \boldsymbol{\theta}) \tag{6.14}$$

when the process is isotropic as well. Beyond semivariograms of this form, however, there are additional semivariograms that correspond to processes that are intrinsically stationary but not second-order stationary. Semivariograms of the former type are bounded, but those of the latter type are unbounded. Two examples of isotropic semivariograms of each type are listed in Table 6.3 and displayed in Figure 6.8. A bounded semivariogram is given the same name as its corresponding covariance function. The unbounded semivariograms in Table 6.3 are the linear semivariogram and the power semivariogram, whose names need no explanation. The linear model is a special case of the power model.

Isotropic semivariograms may have attributes that have some of the same names and meanings as they have for isotropic covariance functions, but others that are particular to semivariograms only. The **nugget effect** of an isotropic semivariogram is defined as $\lim_{r \to 0} \gamma(r; \boldsymbol{\theta})$. By (6.14), when the process is second-order stationary and isotropic, it is apparent that $\lim_{r \to 0} \gamma(r) = C(0) - \lim_{r \to 0} C(r)$, implying that the nugget effect of the semivariogram coincides with that of the covariance function in this case. In any case, a semivariogram with a nugget effect may be created from one without a nugget effect (such as the exponential, linear, or power semivariograms in Table 6.3) by adding the no-correlation semivariogram model to it. If the semivariogram is bounded, then the **sill** is defined as $\lim_{r \to \infty} \gamma(r; \boldsymbol{\theta})$, provided that this limit exists, and the **range** and **practical range** are, respectively, the distance beyond which the semivariogram is equal to its sill and the distance beyond which the semivariogram is greater than or equal to 95% of its sill, if these distances exist. For a bounded two-component semivariogram for which one of the components is a nugget effect, the **partial sill** is defined as the limit as $r \to \infty$ of the other component, if the limit exists. Thus, for such a semivariogram the sill is equal to the sum of the nugget effect and the partial sill (which explains the name of the latter).

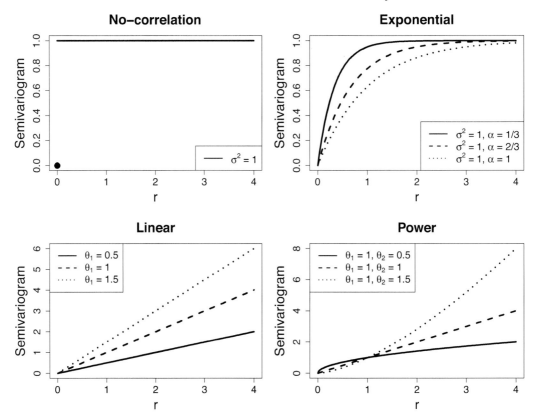

FIGURE 6.8
Some isotropic semivariogram models.

Using (6.14) again, it is apparent that the sill coincides with the process variance while the range, practical range, and partial sill of the semivariogram all coincide with those of the covariance function. Unbounded semivariograms may have some features for which there is no counterpart among covariance functions. Two of these are exemplified by the linear and power models in Table 6.3. Both models have a **scale parameter**, θ_1, and the power model also has a **growth parameter**, represented by θ_2 in Table 6.3.

Isotropic semivariogram models, like isotropic covariance models, may be modified to allow for anisotropy. Owing to the one-to-one relationship between covariance functions and bounded semivariograms for second-order stationary processes, anisotropy for the semivariogram of such a process may be modeled in exactly the same ways that it was for the covariance function. For unbounded semivariograms, the scale and growth parameters may be allowed to vary with direction, subject to certain restrictions. Allard *et al.* (2016) provide details and examples.

Recall from Section 2.2.2 that the notion of intrinsic stationarity may be generalized to the notion of intrinsic stationarity of order k ($k = 0$ being the intrinsically stationary case), with a corresponding extension of the (negative) semivariogram to a function called the generalized covariance function of order k (gcf-k). A gcf-k $G(\mathbf{h}; \boldsymbol{\theta})$ is necessarily k-conditionally positive definite and even; moreover, $G(\mathbf{0}; \boldsymbol{\theta}) = 0$ and $G(\mathbf{h}; \boldsymbol{\theta})/\|\mathbf{h}\|^{2k+2} \to 0$ as $\|\mathbf{h}\| \to \infty$ (Matheron, 1973). A particularly useful parametric family of generalized covariance functions for spatial linear models is the family of isotropic polynomial generalized covariance functions of order k, defined by

$$G_k(r;\boldsymbol{\theta}) = \sum_{l=0}^{k} (-1)^{l+1} \theta_l r^{2l+1}. \tag{6.15}$$

The elements of $\boldsymbol{\theta}$ in $G_k(r;\boldsymbol{\theta})$ must satisfy certain constraints, which were specified by Matheron (1973). When $k = 0$, (6.15) reduces to the negative of the linear semivariogram.

Because the spatial dependence of an IRF-k is modeled via a generalized covariance function rather than an ordinary covariance function, the covariance matrix of the corresponding spatial linear model is not specified (and need not even exist), and consequently the function's parameters cannot be estimated by the method of maximum likelihood. However, a variant known as residual maximum likelihood estimation is perfectly suited for such models. Details will be presented in Chapter 8.

6.6 Exercises

1. Show that if $C_1(\mathbf{s},\mathbf{s}';\boldsymbol{\theta}_1)$ and $C_2(\mathbf{s},\mathbf{s}';\boldsymbol{\theta}_2)$ are covariance functions in $\mathbb{R}^d \times \mathbb{R}^d$ for $\boldsymbol{\theta}_1 \in \Theta_1$ and $\boldsymbol{\theta}_2 \in \Theta_2$, any linear combination $C(\mathbf{s},\mathbf{s}';\boldsymbol{\theta}) = a_1 C_1(\mathbf{s},\mathbf{s}';\boldsymbol{\theta}_1) + a_2 C_2(\mathbf{s},\mathbf{s}';\boldsymbol{\theta}_2)$ with positive coefficients a_1 and a_2 is a covariance function in $\mathbb{R}^d \times \mathbb{R}^d$ for $\boldsymbol{\theta} \equiv (\boldsymbol{\theta}_1^T, \boldsymbol{\theta}_2^T)^T \in \Theta \equiv \Theta_1 \times \Theta_2$.

2. Using a scenario with just two sites, show that the cosine covariance function is not strictly positive definite in $\mathbb{R}$.

3. Show that the isotropic cosine covariance function listed in Table 6.2 is not positive definite in $\mathbb{R}^2$. (Hint: Consider three sites located at the corners of an equilateral triangle of length $\pi\alpha$.)

4. Consider the isotropic triangular correlation function, with range $\alpha > 0$, listed in Table 6.1. Use the following set-up to show that this function does not satisfy the positive definiteness property (6.1) in $\mathbb{R}^2$ and therefore cannot serve as a correlation function in $\mathbb{R}^2$. Set $n = 48$, take the sites $\mathbf{s}_1, \ldots, \mathbf{s}_{48}$ to form a 6×8 square grid with spacing $\alpha/\sqrt{2}$, and in (6.1) take $a_i = (-1)^{k+l}$ if site i lies in row k and column l of the grid.

5. Consider an isotropic second-order stationary geostatistical process in $\mathbb{R}^d$ with covariance function
$$C(r;\sigma^2,\kappa) = \begin{cases} \sigma^2 & \text{if } r = 0, \\ \sigma^2 \kappa & \text{if } r \neq 0, \end{cases}$$
where $\sigma^2 > 0$. Show that κ cannot be negative. (Hint: Consider the positive definiteness property (6.1), with $a_i = 1$ for every site i.)

6. Show that an isotropic correlation function in $\mathbb{R}^d$ is bounded below by $-1/d$. (Hint: Note that $\mathrm{E}[(\sum_{i=1}^{d+1} Y_i)^2] \geq 0$.)

7. Let $Y(\mathbf{s}) = W(\mathbf{s}) + Z(\mathbf{s})$, where $W(\mathbf{s}) \equiv W \sim N(0,\sigma_W^2)$, $\{Z(\mathbf{s}) : \mathbf{s} \in \mathbb{R}^d\}$ is a d-dimensional isotropic Gaussian process with exponential covariance function $\sigma_Z^2 \exp(-r/\alpha)$, and W and $\{Z(\mathbf{s}) : \mathbf{s} \in \mathbb{R}^d\}$ are independent. Obtain the covariance function of $\{Y(\mathbf{s}) : \mathbf{s} \in \mathbb{R}^d\}$ and show that it is isotropic but does not vanish to 0 as r tends to infinity.

8. Verify that the practical range of the isotropic Gaussian correlation function listed in Table 6.2 is approximately equal to $\sqrt{3}\alpha$, and derive the practical ranges of the wave, powered exponential, gravity, rational quadratic, magnetic, Cauchy, and generalized Cauchy correlation functions.

9. Graph the correlation functions listed in Tables 6.1 and 6.2 that are not displayed in Figures 6.2 and 6.3, for two or more choices of each parameter. Use distance increments of 0.1 ranging from 0 to 4, and pick appropriate parameter values so that the autocorrelation approaches 0 by distance 4.

10. For each of the correlation functions that you graphed in the previous exercise, plot a realization of a mean-zero, unit-variance Gaussian process on a 40 by 40 grid with grid spacing of 0.1. After generating the 1600×1600 covariance matrix, use the Cholesky decomposition for creating the realizations, and use the same set of independent normals when multiplying by the Cholesky factor across parameter values and models. How do the patterns change as α increases, and across different models? What happens to the range of realized values as α increases? Also, note that in some cases the covariance matrix may not be positive definite (make sure to include the J-Bessel function in your set of realizations). If this problem arises, how can you fix the problem and still create a realization using an eigen decomposition rather than a Cholesky decomposition?

11. Consider the Matérn covariance function with zero nugget, unit variance, and range parameter 0.4. For each smoothness parameter $\nu = 1/2$, $\nu = 3/2$, and $\nu = 5/2$, plot 5 realizations in 1 dimension for locations at 0.001 increments from 0 to 1. Describe how the smoothnesses of the realizations vary with the value of the smoothness parameter.

12. Verify (6.11).

13. Determine the left and right first and second derivatives of the isotropic triangular, spherical, and pentaspherical covariance function with respect to the range, α, at $r = \alpha$. What do you conclude about the existence of the first two derivatives of these covariance functions, with respect to the range, at $r = \alpha$?

14. An alternative parameterization of the Matérn correlation function is as follows: $\rho(r; \alpha, \nu) = \frac{1}{2^{\nu-1}\Gamma(\nu)} \left((2\nu)^{1/2} \frac{r}{\alpha}\right)^\nu K_\nu \left((2\nu)^{1/2} \frac{r}{\alpha}\right)$, where $\alpha > 0$ and $\nu > 0$. For values of ν equal to 0.25, 0.5, 1.0, 1.5, 2.0, 5.0, and 10.0, determine the practical range and plot it as a function of ν. Repeat for the same values of ν for the parameterization of the Matérn function given in Table 6.2. For which parameterization is the practical range less affected by ν?

15. Consider the covariance function

$$C(h_u, h_v; \alpha) = \sigma^2 \exp[-(|h_u| + |h_v|)/\alpha] \quad \text{for } -\infty < h_u < \infty, \ -\infty < h_v < \infty,$$

where $\alpha > 0$, which is a special case of (6.9).

(a) What is the shape of a locus of equal correlation for this function? (Describe as precisely as possible using terminology from planar geometry.)

(b) In terms of α, what is this correlation function's practical range in the NE-SW direction? How does it compare to the practical range in the N-S (or E-W) direction?

16. Consider the function

$$\rho(h_u, h_v; \alpha) = \exp(-|h_u|/\alpha) + \exp(-|h_v|/\alpha) + \exp[-(|h_u| + |h_v|)/\alpha],$$

defined for (h_u, h_v) in two-dimensional Euclidean space, where $\alpha > 0$.

(a) Is this function a valid correlation function for a two-dimensional geostatistical process? Explain why or why not.

(b) Assuming that the answer to part (a) is "Yes," what is the function's practical range in the NE-SW direction?

(c) Show that the practical range doesn't exist in the N-S direction.

17. Consider the function

$$\rho(h_1, \ldots, h_d; \alpha_1, \ldots, \alpha_d) = \exp\left(-\sum_{i=1}^{d} \frac{h_i^2}{\alpha_i}\right) \quad \text{for } -\infty < h_i < \infty \quad (i = 1, \ldots, d),$$

where $\alpha_i > 0$ for all i.

(a) Explain why this is a valid correlation function in $\mathbb{R}^d$.

(b) Consider the special case of this correlation function in which $d = 3$ and $\alpha_1 = \alpha_2 \neq \alpha_3$. What is the shape of a locus of equal correlation of this function? (Describe as precisely as possible using terminology from solid geometry.)

18. Consider the function

$$f(h_u, h_v; \alpha_u, \alpha_v) = \begin{cases} 1 - \frac{h_u}{\alpha_u} - \frac{h_v}{\alpha_v} + \frac{h_u h_v}{\alpha_u \alpha_v} & \text{for } 0 \leq h_u \leq \alpha_u, \ 0 \leq h_v \leq \alpha_v, \\ 0 & \text{otherwise,} \end{cases}$$

where $\alpha_u > 0$ and $\alpha_v > 0$. Could this function serve as a correlation function of a second-order stationary process in $\mathbb{R}^2$? Explain.

19. Show that the product of two anisotropic covariance functions can be isotropic. (Hint: Consider two geometrically anisotropic Gaussian covariance functions with anisotropy angles of 0 and 90 degrees, respectively.)

20. Let $\{Y(\mathbf{s}) : \mathbf{s} \in \mathbb{R}^d\}$ represent a geostatistical process with mean μ and covariance function

$$C(\mathbf{s}, \mathbf{s}'; \sigma^2, \theta) = \sigma^2 \left(\|\mathbf{s}\|^\theta + \|\mathbf{s}'\|^\theta - \|\mathbf{s} - \mathbf{s}'\|^\theta\right),$$

where $\sigma^2 > 0$ and $0 < \theta < 2$. Show that $\{Y(\mathbf{s}) : \mathbf{s} \in \mathbb{R}^d\}$ is intrinsically stationary and isotropic, and find its semivariogram.

7

Parametric Covariance Structures for Spatial-Weights Linear Models

Now we turn our attention to parametric covariance structures for areal data. As noted previously, models for the covariance structure of geostatistical data may be adapted for use with areal data simply by replacing (conceptually) each areal data site with its centroid or some other representative point, or by integrating over sites. In either case, the parametric covariance models for geostatistical data developed in the previous chapter may be applied to areal data. But as described in Section 2.3, another, quite different approach to modeling the covariance structure of areal data is through spatial weights matrices. Spatial weights matrices are based upon a specified neighborhood structure for the data sites which, when combined with a model such as the SAR, CAR, or SMA model, determine the data's covariance structure. This chapter expands upon Section 2.3 by modeling the elements of the spatial weights matrix (hence also modeling, albeit indirectly, the elements of the covariance matrix) as functions of one or more unknown parameters.

7.1 SAR and CAR models

7.1.1 General requirements of SAR models

Recall from Section 2.3 that the SAR model is given by

$$y_i = \mathbf{x}_i^T \boldsymbol{\beta} + \sum_{j \in \mathcal{N}_i} b_{ij}(y_j - \mathbf{x}_j^T \boldsymbol{\beta}) + d_i \quad (i = 1, \ldots, n), \tag{7.1}$$

or in vector form,

$$\mathbf{y} = \mathbf{X}\boldsymbol{\beta} + \mathbf{B}(\mathbf{y} - \mathbf{X}\boldsymbol{\beta}) + \mathbf{d}, \tag{7.2}$$

where $\mathbf{d}$ is a multivariate normal random vector with mean vector $\mathbf{0}$ and positive definite diagonal covariance matrix $\mathbf{K}_{\text{SAR}} = \text{diag}(\kappa_1^2, \ldots, \kappa_n^2)$. The b_{ij}'s, which are specified (perhaps up to some unknown parameters) by the modeler, relate each observation to the others, with b_{ii} taken to equal 0 for all i. Equation (7.1) may be rearranged to yield

$$(\mathbf{I} - \mathbf{B})(\mathbf{y} - \mathbf{X}\boldsymbol{\beta}) = \mathbf{d},$$

and provided that $\mathbf{I} - \mathbf{B}$ is nonsingular this equation may be manipulated [by premultiplying both sides by $(\mathbf{I} - \mathbf{B})^{-1}$] to obtain

$$\mathbf{y} = \mathbf{X}\boldsymbol{\beta} + (\mathbf{I} - \mathbf{B})^{-1}\mathbf{d}. \tag{7.3}$$

It follows easily that the covariance matrix of $\mathbf{y}$ is

$$\boldsymbol{\Sigma}_{\text{SAR}} = (\mathbf{I} - \mathbf{B})^{-1}\mathbf{K}_{\text{SAR}}(\mathbf{I} - \mathbf{B}^T)^{-1}. \tag{7.4}$$

DOI: 10.1201/9780429060878-7

Clearly this matrix is symmetric; by Theorem A.3 in Appendix A, it is also positive definite.

The SAR model is sometimes written in a slightly different way, as follows:

$$\mathbf{y} = \mathbf{X}\boldsymbol{\beta} + \mathbf{u}, \quad \mathbf{u} = \mathbf{B}\mathbf{u} + \mathbf{d}, \tag{7.5}$$

where $\mathbf{d}$ is assumed to be distributed as it was in (7.2). From the second equation in (7.5) we obtain $(\mathbf{I} - \mathbf{B})\mathbf{u} = \mathbf{d}$, yielding $\mathbf{u} = (\mathbf{I} - \mathbf{B})^{-1}\mathbf{d}$ (provided that $\mathbf{I} - \mathbf{B}$ is nonsingular). Substituting this expression for $\mathbf{u}$ into the first equation in (7.5) yields (7.3), thereby establishing that (7.5) is just another way to write the SAR model.

For future reference, let us summarize the conditions that must be satisfied for $\boldsymbol{\Sigma}_{\text{SAR}}$, as given by (7.4), to be a valid (symmetric positive definite) SAR covariance matrix. They are:

SAR1: $\mathbf{K}_{\text{SAR}}$ is diagonal with positive diagonal elements;

SAR2: $b_{ii} = 0$ for all i; and

SAR3: $\mathbf{I} - \mathbf{B}$ is nonsingular.

Note that $\mathbf{B}$ is not required to be symmetric, which allows the modeler to account for mechanisms thought to operate asymmetrically. For example, if the ith response is believed to have twice as much influence on the jth response as the jth has on the ith, then the modeler may take $b_{ji} = 2b_{ij}$. Also note that $\mathbf{B}$ and $\mathbf{K}_{\text{SAR}}$ are functionally independent, i.e., how the b_{ij}'s are specified in no way restricts how the κ_i^2's must be specified (or vice versa).

7.1.2 General requirements of CAR models

Recall, also from Section 2.3, that the CAR model arises by assuming that for $i = 1, \ldots, n$, the conditional distribution of y_i, given all other responses, is normal with mean

$$\text{E}(y_i|y_j, j \neq i) = \mathbf{x}_i^T \boldsymbol{\beta} + \sum_{j \in \mathcal{N}_i} c_{ij}(y_j - \mathbf{x}_j^T \boldsymbol{\beta}) \tag{7.6}$$

and positive variance κ_i^2, where $c_{ii} = 0$. Here the c_{ij}'s play a role analogous to the role that the b_{ij}'s play in the SAR model, except that they describe spatial dependence for conditional means rather than for the observations themselves. Nevertheless, under certain conditions, the conditional specification yields a model (a joint probability distribution) for the observations. Let $\mathbf{C} = (c_{ij})$ and $\mathbf{K}_{\text{CAR}} = \text{diag}(\kappa_1^2, \ldots, \kappa_n^2)$. Exercise 2.18 established that if $\mathbf{I} - \mathbf{C}$ is positive definite and $\mathbf{K}_{\text{CAR}}^{-1}(\mathbf{I} - \mathbf{C})$ is symmetric, then the covariance matrix of $\mathbf{y}$ is

$$\boldsymbol{\Sigma}_{\text{CAR}} = (\mathbf{I} - \mathbf{C})^{-1}\mathbf{K}_{\text{CAR}}, \tag{7.7}$$

and this matrix is positive definite.

A summary of the conditions required for $\boldsymbol{\Sigma}_{\text{CAR}}$ to be a valid covariance matrix is as follows:

CAR1: $\mathbf{K}_{\text{CAR}}$ is diagonal with positive diagonal elements;

CAR2: $c_{ii} = 0$ for all i;

CAR3: $\mathbf{I} - \mathbf{C}$ is positive definite; and

CAR4: $\mathbf{K}_{\text{CAR}}^{-1}(\mathbf{I} - \mathbf{C})$ is symmetric, or equivalently, $\frac{c_{ij}}{\kappa_i^2} = \frac{c_{ji}}{\kappa_j^2}$ for all i and j.

Note that conditions CAR1 and CAR2 are identical or completely analogous to SAR1 and SAR2. Condition CAR3, however, is stronger than SAR3, and the symmetry condition CAR4 has no counterpart for SAR models. Furthermore, CAR4 implies that $\mathbf{C}$ and $\mathbf{K}$ are not functionally independent. Thus, the requirements on $\mathbf{C}$ and $\mathbf{K}_{\mathrm{CAR}}$ to yield a valid covariance matrix in the CAR framework are more stringent than those on $\mathbf{B}$ and $\mathbf{K}_{\mathrm{SAR}}$ in the SAR framework. For example, by CAR1 and CAR4, c_{ij} and c_{ji} must have the same sign, whereas there is no such requirement for b_{ij} and b_{ji}. Furthermore, CAR4 implies that $\frac{c_{ij}}{c_{ji}} = \frac{c_{ik}}{c_{ki}} \cdot \frac{c_{kj}}{c_{jk}}$ for all i, j, k for which $c_{ij} \neq 0$, $c_{ji} \neq 0$, $c_{ik} \neq 0$, $c_{ki} \neq 0$, $c_{jk} \neq 0$, and $c_{kj} \neq 0$ (Exercise 7.1a). Nevertheless, $\mathbf{C}$ is not generally required to be symmetric.

7.1.3 Relationships to partial correlations

It was noted in Section 2.1 that the off-diagonal elements of the inverse of a covariance matrix $\boldsymbol{\Sigma}$ are closely related to the partial correlations between each pair of variables, conditioned on the remaining $n - 2$ variables. In fact, it turns out that each off-diagonal element of $\boldsymbol{\Sigma}^{-1}$, when divided by the square root of the product of the two corresponding diagonal elements, is the negative of the aforementioned partial correlation (Whittaker, 1990). That is,

$$\mathrm{corr}(y_i, y_j | y_k, k \neq i, j) = -\frac{\sigma^{ij}}{(\sigma^{ii}\sigma^{jj})^{1/2}}, \tag{7.8}$$

where $\boldsymbol{\Sigma}^{-1} = (\sigma^{ij})$. For a CAR model, (7.7) and CAR4 may be used to show that (7.8) reduces to

$$\mathrm{corr}(y_i, y_j | y_k, k \neq i, j) = \mathrm{sgn}(c_{ij})(c_{ij}c_{ji})^{1/2} \tag{7.9}$$

(see Exercise 7.2). By (7.9), c_{ij} and c_{ji} must satisfy $0 \leq c_{ij}c_{ji} \leq 1$ for all i and j. Moreover, (7.9) reveals that by specifying the elements of $\mathbf{C}$, the modeler is, in effect, specifying the partial correlations. In particular, specifying that c_{ij} equals zero specifies that the partial correlation between y_i and y_j, conditioned on the remaining observations, is zero. This provides a very useful interpretation of sparse neighbor weighting schemes under a CAR model, namely, that responses taken at sites that are not neighbors are conditionally independent, given the responses at all other sites. Furthermore, (7.9) reveals that if c_{ij} is taken to be a function of the distance between sites i and j for all i and j, then the partial correlation is also a function of that distance. This implies that it is possible to model the partial correlations as stationary, or even isotropic, using a CAR model.

However, similar statements cannot be made about the zero elements of $\mathbf{B}$ or stationary modeling of partial correlations via a SAR model because $\boldsymbol{\Sigma}_{\mathrm{SAR}}^{-1}$ is equal not to $\mathbf{K}_{\mathrm{SAR}}^{-1}(\mathbf{I}-\mathbf{B})$ but to $(\mathbf{I} - \mathbf{B})\mathbf{K}_{\mathrm{SAR}}^{-1}(\mathbf{I} - \mathbf{B}^T)$. Nevertheless, depending on how sparse $\mathbf{B}$ is, there may still be some pairs of responses that are conditionally independent. Consider, for example, a situation with four observations and a tridiagonal $\mathbf{B}$, i.e.,

$$\mathbf{B} = \begin{pmatrix} 0 & b_{12} & 0 & 0 \\ b_{21} & 0 & b_{23} & 0 \\ 0 & b_{32} & 0 & b_{34} \\ 0 & 0 & b_{43} & 0 \end{pmatrix}. \tag{7.10}$$

Then it may be verified (simply by carrying out the matrix multiplications; see Exercise 7.3) that the fourth element in the first row of $(\mathbf{I}-\mathbf{B})\mathbf{K}_{\mathrm{SAR}}^{-1}(\mathbf{I}-\mathbf{B}^T)$ is equal to zero, implying that the first and fourth observations are conditionally independent, given the second and third observations. However, the first and third observations are not conditionally independent (given the second and fourth observations), in contrast to what would be the case for a CAR with tridiagonal $\mathbf{C}$. For this reason, SAR models are sometimes said to be "less local" than

CAR models, meaning that any given response is influenced by more observations under a SAR model, given the same degree of sparsity among the spatial weights.

7.2 Parsimonious SAR and CAR models

When fully parameterized, $\mathbf{B}$ in the SAR model has $n^2 - n$ parameters, and $\mathbf{K}_{\mathrm{SAR}}$ has n parameters. Thus, the number of parameters in a SAR model can be as many as n^2. The number of free parameters in a fully parameterized CAR model is slightly less, due to the restrictions imposed by CAR4, but is still larger than n. It is not possible to sensibly estimate the parameters of such richly parameterized models from the n available observations. Even if many sites have relatively few neighbors, so that many of the elements of $\mathbf{B}$ or $\mathbf{C}$ are equal to zero, the data still may be insufficient to support the estimation of their nonzero elements, let alone the diagonal elements of $\mathbf{K}_{\mathrm{SAR}}$ or $\mathbf{K}_{\mathrm{CAR}}$. The solution to this problem is to parameterize $\mathbf{B}$ and $\mathbf{K}_{\mathrm{SAR}}$, or $\mathbf{C}$ and $\mathbf{K}_{\mathrm{CAR}}$, using a much smaller set of parameters.

By far the most commonly used and well-studied parameterizations of $\mathbf{B}$ and $\mathbf{C}$ are those given by the single multiplicative-parameter forms

$$\mathbf{B} = \rho_{\mathrm{SAR}}\mathbf{W} \quad \text{and} \quad \mathbf{C} = \rho_{\mathrm{CAR}}\mathbf{W}, \tag{7.11}$$

where $\mathbf{W} = (w_{ij})$ is a completely specified matrix such that $w_{ii} = 0$ for all i and w_{ij} $(i \neq j)$ represents the spatial proximity of site j to site i, and ρ_{SAR} and ρ_{CAR} are unknown real parameters. Despite the use of the same Greek symbol for these parameters as that commonly used to represent a population correlation, ρ_{SAR} and ρ_{CAR} are not correlations and the temptation to think of them as such must be resisted. We refer to them as the **spatial dependence parameters** of the SAR and CAR models, though this term also has the potential to mislead if one does not keep in mind that these parameters affect the spatial dependence only indirectly, through the spatial weights matrix.

The parameter spaces for the spatial dependence parameters, i.e., the values of ρ_{SAR} for which SAR3 holds and the values of ρ_{CAR} for which CAR3 and CAR4 hold, depend, not surprisingly, on $\mathbf{W}$. In particular, they can be expressed in terms of the eigenvalues of $\mathbf{W}$. Let $\{\lambda_i : i = 1, \ldots, n\}$ represent those eigenvalues, some of which could be complex. Henceforth assume, as is invariably the case in practice, that at least one site in the spatial dataset has at least one neighbor, so that $\mathbf{W} \neq \mathbf{0}$. Then at least one eigenvalue of $\mathbf{W}$ is nonzero, and, since (by Theorem A.7a in Appendix A) the sum of the eigenvalues of $\mathbf{W}$ is equal to the trace of $\mathbf{W}$, actually at least two eigenvalues are nonzero. Furthermore, if $\mathbf{W}$ is symmetric then all of its eigenvalues are real (by Theorem A.8 in Appendix A); if $\mathbf{W}$ is not symmetric, then some of its eigenvalues may be complex. In the symmetric case, then, at least one eigenvalue is positive and at least one is negative.

For a SAR model with $\mathbf{B} = \rho_{\mathrm{SAR}}\mathbf{W}$, SAR3 specializes to the condition that $\mathbf{I} - \rho_{\mathrm{SAR}}\mathbf{W}$ is nonsingular. It can be shown (see Exercise 7.4) that this condition is satisfied if and only if ρ_{SAR} is not the reciprocal of any real nonzero eigenvalue of $\mathbf{W}$. Thus, if A denotes the subset of eigenvalues of $\mathbf{W}$ that are real and nonzero, then the parameter space for ρ_{SAR} is $\Omega_{\mathrm{SAR}} \equiv \{\rho_{\mathrm{SAR}} : \rho_{\mathrm{SAR}} \neq \lambda_i^{-1}, \lambda_i \in A\}$. If all the eigenvalues of $\mathbf{W}$ are complex or zero, there are no constraints on ρ_{SAR}.

The parameter space for ρ_{CAR} can also be expressed in terms of the eigenvalues of $\mathbf{W}$, but due to CAR3 it is more restricted than that for ρ_{SAR}, and due to CAR4 it has implications for the main diagonal elements of $\mathbf{K}_{\mathrm{CAR}}$ (see below). Furthermore, the additional restrictions $\frac{w_{ij}}{w_{ji}} = \frac{w_{ik}}{w_{ki}} \cdot \frac{w_{kj}}{w_{jk}}$ for all i, j, k for which $w_{ij} \neq 0$, $w_{ji} \neq 0$, $w_{ik} \neq 0$, $w_{ki} \neq 0$, $w_{jk} \neq 0$, and

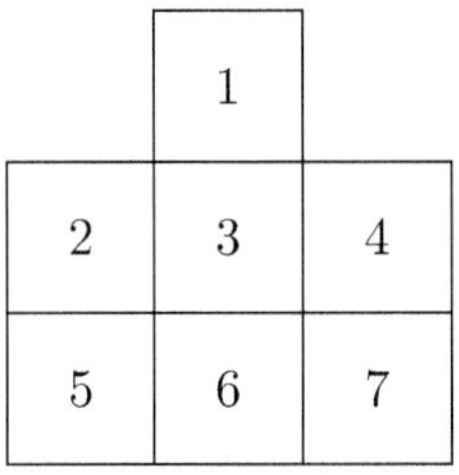

FIGURE 7.1
Layout of sites for the example of Section 7.2. Numbers within sites are the site indices.

$w_{kj} \neq 0$ must be placed on $\mathbf{W}$ for CAR4 to hold, although $\mathbf{W}$ generally need not be symmetric. Under CAR4, the eigenvalues of $\mathbf{W}$ are real and may be ordered as follows:

$$\lambda_1 \leq \lambda_2 \leq \cdots \leq \lambda_n,$$

where λ_1 is negative and λ_n is positive (Exercise 7.5a). CAR3 requires that all of the eigenvalues of $\mathbf{I} - \rho_{\text{CAR}}\mathbf{W}$ be positive, and Exercise 7.5b establishes that a necessary and sufficient condition for this to happen is $\lambda_1^{-1} < \rho_{\text{CAR}} < \lambda_n^{-1}$. Thus, the parameter space for ρ_{CAR} is $\Omega_{\text{CAR}} \equiv \{\rho_{\text{CAR}} : \lambda_1^{-1} < \rho_{\text{CAR}} < \lambda_n^{-1}\}$. Note that this parameter space includes 0.

The parameter spaces we have specified for ρ_{SAR} and ρ_{CAR} are the largest allowable, but in practice modelers sometimes use proper subsets of these spaces. For a SAR model in particular, if $\mathbf{W}$ is symmetric the parameter space for ρ_{SAR} is often taken to be identical to that for ρ_{CAR}, i.e., $(\lambda_1^{-1}, \lambda_n^{-1})$. This choice does exclude many valid SAR models, however.

For parameterizing $\mathbf{K}_{\text{SAR}}$ there is complete flexibility (subject to SAR1), owing to its functional independence from $\mathbf{B}$. One common parsimonious choice is $\mathbf{K}_{\text{SAR}} = \sigma^2 \mathbf{K}_0$ for some unknown parameter $\sigma^2 > 0$ and some fully specified diagonal matrix $\mathbf{K}_0 = \text{diag}(\kappa_1^2, \ldots, \kappa_n^2)$ with positive diagonal elements. If κ_i^2 is believed to be directly (or inversely) proportional to the area $|A_i|$ of site i for all i, then the modeler might take the diagonal elements of $\mathbf{K}_0$ to be those areas (or their reciprocals). In human population studies, a similar approach could be taken using not the area but the population size within the site. In the absence of such information, modelers often take $\mathbf{K}_0 = \mathbf{I}$. There is less flexibility in parameterizing $\mathbf{K}_{\text{CAR}}$. In fact, CAR4 completely determines $\mathbf{K}_{\text{CAR}}$, up to a scalar multiple, in terms of the elements of $\mathbf{C}$. Specifically, when $\mathbf{C} = \rho_{\text{CAR}}\mathbf{W}$, the κ_i^2's must satisfy $\kappa_i^2/\kappa_j^2 = w_{ij}/w_{ji}$ for all i and j such that $w_{ij} \neq 0$. Thus, if $\mathbf{W}$ is symmetric, then $\mathbf{K}_{\text{CAR}}$ must be of the form $\sigma^2 \mathbf{K}^* = \sigma^2 \text{diag}(\kappa_1^{*2}, \ldots, \kappa_n^{*2})$ for some $\sigma^2 > 0$, where $\kappa_i^{*2} = 1$ if $w_{ij} \neq 0$ for at least one j, and κ_i^{*2} is positive but otherwise arbitrary if $w_{ij} = 0$ for all j. The special case in which $\mathbf{K}_{\text{CAR}}$ has this form and the w_{ij}'s are binary spatial adjacency weights has been called the **homogeneous CAR model** by Cressie & Kapat (2008).

As an example of marginal variances and correlations corresponding to $\boldsymbol{\Sigma}_{\text{SAR}}$ and $\boldsymbol{\Sigma}_{\text{CAR}}$, consider the scenario depicted in Figure 7.1 in which there are seven sites (the number within the site is its index), and the spatial weights matrix has the single multiplicative-parameter form where $\mathbf{W}$ is simply a binary adjacency matrix. The corresponding spatial weights matrix is

$$\mathbf{W} = \begin{pmatrix} 0 & 0 & 1 & 0 & 0 & 0 & 0 \\ 0 & 0 & 1 & 0 & 1 & 0 & 0 \\ 1 & 1 & 0 & 1 & 0 & 1 & 0 \\ 0 & 0 & 1 & 0 & 0 & 0 & 1 \\ 0 & 1 & 0 & 0 & 0 & 1 & 0 \\ 0 & 0 & 1 & 0 & 1 & 0 & 1 \\ 0 & 0 & 0 & 1 & 0 & 1 & 0 \end{pmatrix}, \tag{7.12}$$

which is symmetric. The most extreme eigenvalues of $\mathbf{W}$ are approximately ± 2.524, so $\mathbf{I} - \rho\mathbf{W}$ is positive definite for $\rho \in (-0.396, 0.396)$. If we set $\rho_{\mathrm{SAR}} = 0.3$ and $\mathbf{K}_{\mathrm{SAR}} = \mathbf{I}$, then the matrix of marginal SAR variances (on the main diagonal) and correlations (on the off-diagonals) is

$$
\begin{pmatrix}
1.83 & 0.49 & 0.69 & 0.49 & 0.40 & 0.51 & 0.40 \\
 & 2.95 & 0.77 & 0.55 & 0.75 & 0.68 & 0.50 \\
 & & 5.57 & 0.77 & 0.66 & 0.81 & 0.66 \\
 & & & 2.95 & 0.50 & 0.68 & 0.75 \\
 & & & & 2.79 & 0.76 & 0.53 \\
 & & & & & 4.43 & 0.76 \\
 & & & & & & 2.79
\end{pmatrix}.
$$

Its CAR counterpart, likewise setting $\rho_{\mathrm{CAR}} = 0.3$ and $\mathbf{K}_{\mathrm{CAR}} = \mathbf{I}$, is

$$
\begin{pmatrix}
1.16 & 0.16 & 0.37 & 0.16 & 0.11 & 0.18 & 0.11 \\
 & 1.36 & 0.43 & 0.19 & 0.40 & 0.29 & 0.15 \\
 & & 1.81 & 0.43 & 0.29 & 0.48 & 0.29 \\
 & & & 1.36 & 0.15 & 0.29 & 0.40 \\
 & & & & 1.34 & 0.42 & 0.18 \\
 & & & & & 1.61 & 0.42 \\
 & & & & & & 1.34
\end{pmatrix}.
$$

We see that the variances and correlations under the SAR model are larger than those under the CAR model with the same value of the spatial dependence parameter. Moreover, although the diagonal elements of $\mathbf{K}_{\mathrm{SAR}}$ are equal and the same is true of the diagonal elements of $\mathbf{K}_{\mathrm{CAR}}$, the marginal variances under both models are heterogeneous; specifically, they are larger at sites that have more neighbors. This feature, called **topologically-induced heterogeneity** by Tiefelsdorf *et al.* (1999), is often unrealistic in practice. For a SAR model the heterogeneity among variances can be reduced substantially by an appropriate, non-identity choice of $\mathbf{K}_{\mathrm{SAR}}$. But owing to the restrictions imposed on $\mathbf{K}_{\mathrm{CAR}}$ by CAR4, this is not an option for a CAR model. Further discussion of this issue is deferred to the next section.

The top four plots in Figure 7.2 display realizations of observations following the single multiplicative-parameter SAR and CAR models described in this section, all with mean zero and $\mathbf{K}_{\mathrm{SAR}} = \mathbf{K}_{\mathrm{CAR}} = \mathbf{I}$. The spatial support for these displays is a 20×20 square grid of sites, with neighbors defined as adjacent sites within rows or columns. The top left plot is a collection of 400 independent and identically distributed standard normal random variables, which may be used as a baseline for comparison. The upper right plot corresponds to a SAR model with $\rho_{\mathrm{SAR}} = 0.2$ and shows discernible positive spatial correlation among the observations, but the correlation in the middle left plot, which corresponds to a SAR model with $\rho_{\mathrm{SAR}} = 0.25$, appears to be considerably stronger. The spatial correlation in the middle right plot, which corresponds to a CAR model with $\rho_{\mathrm{CAR}} = 0.25$, appears to be somewhat weaker than that in the SAR model with $\rho_{\mathrm{SAR}} = 0.25$, despite having the same value of the spatial dependence parameter. The parameter space for ρ_{CAR} in this case is $(-0.252824, 0.252824)$, so 0.25 is very close to the upper bound.

Non-multiplicative single-parameter parameterizations of the spatial weights matrix are possible. For a CAR model, Pettitt *et al.* (2002) took

$$
c_{ij} = \frac{\phi \gamma_{ij}}{1 + |\phi| \sum_{k \in \mathcal{N}_i} \gamma_{ik}} \text{ for } j \neq i, \quad \kappa_i^2 = \frac{\sigma^2}{1 + |\phi| \sum_{k \in \mathcal{N}_i} \gamma_{ik}}, \tag{7.13}
$$

where γ_{ij} is a fully specified function of the distance between sites i and j (hence $\gamma_{ij} = \gamma_{ji}$) such that $\gamma_{ii} = 0$ for all i, and $\gamma_{ij} \geq 0$ for all i and j; for example, γ_{ij} could be a binary

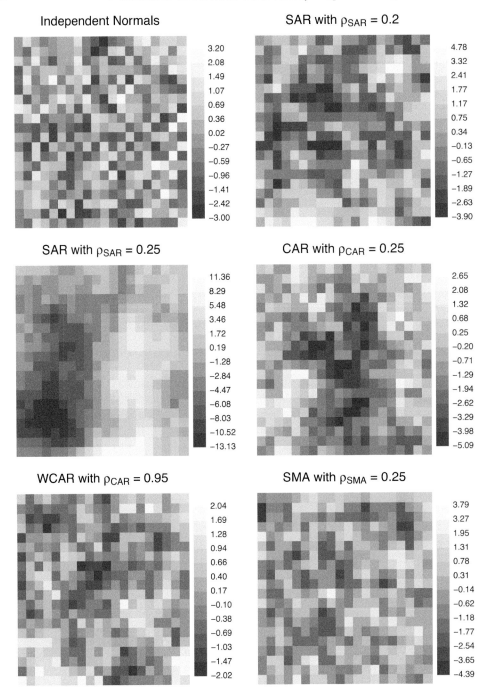

FIGURE 7.2
Realizations of some single-multiplicative-parameter SAR, CAR, and SMA processes on a
20 × 20 square grid. WCAR (the weighted CAR model) is row-standardized; the others are
not.

adjacency weight, or it could be the reciprocal of the distance between representative points within sites i and j. Here, ϕ is the single unknown parameter in the weights matrix. In contrast to the multiplicative parameterization in (7.11), ϕ is unconstrained. Therefore, determining the eigenvalues of the possibly asymmetric matrix $\mathbf{C}$ is unnecessary. It is left as an exercise (Exercise 7.6) to verify that CAR3 and CAR4 are satisfied for this model.

Occasionally, modelers may use additional parameters in the spatial weights matrix to allow for anisotropies or different strengths of dependence for neighbors of different orders. For example, to allow for anisotropy when sites lie on a regular rectangular grid, we could suppose that

$$\mathbf{\Omega} = \rho_1 \mathbf{W}_1 + \rho_2 \mathbf{W}_2,$$

where $\mathbf{W}_1$ and $\mathbf{W}_2$ are binary row-adjacency and column-adjacency matrices, respectively, and ρ_1 and ρ_2 are spatial dependence parameters corresponding to rows and columns. Of course, an alternative way to allow for anisotropy is to suppose that $\mathbf{\Omega} = \rho \mathbf{W}$ where the neighbor weights are equal to 1 within rows and equal to some user-specifed number (different than 1) within columns. An advantage of the first approach, relative to this alternative, is that it allows the data to inform the modeler about the degree of anisotropy. Two disadvantages are that the parameter space for a model with a multi-parameter weight matrix typically is considerably more difficult to specify than the parameter space for a single-parameter model, and the model is more challenging to estimate.

7.3 Standardization of spatial weights

For a binary adjacency spatial weights matrix, such as the one used in the seven-site example of the preceding section, $y_i - \mathbf{x}_i^T \boldsymbol{\beta}$ in (7.1) and $\mathrm{E}(y_i | y_j, j \neq i) - \mathbf{x}_i^T \boldsymbol{\beta}$ in (7.6) are (weighted) sums of neighboring random variables. It may seem more natural, however, to model these quantities as averages, rather than sums, of those random variables. This leads to the idea of **row standardization**. Most (if not all) applications of row standardization have been within the context of the single multiplicative-parameter weights matrices $\rho_{\mathrm{SAR}} \mathbf{W}$ and $\rho_{\mathrm{CAR}} \mathbf{W}$ defined in (7.11), with $\mathbf{W}$ symmetric besides, hence we limit our consideration to that context. Additional assumptions about the weights must be made to accommodate row standardization, which were not made for the general case. For starters, it must be assumed that $\mathbf{W} = (w_{ij})$ is nonnegative (a nonnegative matrix is a matrix whose elements are all nonnegative, which is not the same as a nonnegative definite matrix.) It must also be assumed that for each i, a j exists such that $w_{ij} > 0$, although we relax this assumption in the following section. Then, we are row-standardizing $\mathbf{W}$ when we divide the elements in the ith row by the corresponding row sum $w_{i\cdot} = \sum_{j \in \mathcal{N}_i} w_{ij}$ $(i = 1, \ldots, n)$. This results in a new, row-standardized spatial weights matrix $\overline{\mathbf{W}} = (\overline{w}_{ij})$, where

$$\overline{w}_{ij} = \frac{w_{ij}}{w_{i\cdot}}.$$

Define

$$\mathbf{D} = \mathrm{diag}\left(\frac{1}{w_{1\cdot}}, \frac{1}{w_{2\cdot}}, \ldots, \frac{1}{w_{n\cdot}}\right).$$

Then $\overline{\mathbf{W}}\mathbf{1} = \mathbf{1}$ and $\overline{\mathbf{W}} = \mathbf{D}\mathbf{W}$, as is easily verified. Note also that $\overline{\mathbf{W}}$ generally is asymmetric, even though $\mathbf{W}$ is symmetric (by assumption).

Exercise 7.7a establishes that if $\mathbf{W}$ and $\overline{\mathbf{W}}$ are as described in the previous paragraph, then the eigenvalues $\lambda_1, \lambda_2, \ldots, \lambda_n$ of $\overline{\mathbf{W}}$ (which, as it turns out, are real) satisfy $|\lambda_i| \leq 1$

for all i and $\lambda_i = 1$ for some i. Thus, the eigenvalues of $\overline{\mathbf{W}}$ may be ordered to satisfy

$$-1 \leq \lambda_1 \leq \lambda_2 \leq \cdots \leq \lambda_n = 1.$$

It follows from this result and the results of the previous section that the parameter space for ρ_{CAR} in this case is $(\lambda_1^{-1}, 1)$, where $\lambda_1^{-1} \leq -1$. For the sake of simplicity, and also to avoid some bizarre behavior (see Section 7.5) of the elements of the marginal covariance matrix when ρ_{SAR} and ρ_{CAR} are less than -1, modelers often take the parameter spaces for ρ_{SAR} and ρ_{CAR} to be $(-1, 1)$ when using a row-standardized spatial weights matrix. As for the κ_i^2's, for a row-standardized SAR they are unconstrained (apart from SAR1's positivity requirement), but for a row-standardized CAR of the form described in the previous paragraph, they are constrained by CAR4 to have form

$$\kappa_i^2 = \sigma^2/w_i. \tag{7.14}$$

for some $\sigma^2 > 0$ (Exercise 7.7b). It is customary to take these as the elements of $\mathbf{K}_{\mathrm{SAR}}$ for row-standardized SAR models also, though it is not necessary. The special case of a CAR model in which the κ_i^2's have this form and $\overline{\mathbf{W}}$ is obtained by row-standardizing a binary spatial adjacency weights matrix has been called the **weighted CAR model** by Cressie & Kapat (2008). The lower left plot in Figure 7.2 displays a realization of such a model, with $\rho_{\mathrm{CAR}} = 0.95$, on a 20×20 square grid. Some spatial correlation is evident in this realization, but there is little that distinguishes the pattern in this plot from that corresponding to a homogeneous CAR model (not shown).

By (7.7) and (7.14), the covariance matrix for the weighted CAR model may be written as

$$\boldsymbol{\Sigma} = \sigma^2(\mathbf{I} - \rho\overline{\mathbf{W}})^{-1}\mathbf{D}. \tag{7.15}$$

An equivalent expression for this covariance matrix is

$$\boldsymbol{\Sigma} = \sigma^2[\mathrm{diag}(\mathbf{W1}) - \rho\mathbf{W}]^{-1} \tag{7.16}$$

(see Exercise 7.9). This yields the following convenient computing formula for $\boldsymbol{\Sigma}^{-1}$:

$$\boldsymbol{\Sigma}^{-1} = (1/\sigma^2)[\mathrm{diag}(\mathbf{W1}) - \rho\mathbf{W}].$$

Recall from the previous section that when a binary adjacency spatial weights matrix is used and $\mathbf{K} = \mathbf{I}$, the marginal variances of observations under the SAR and CAR models tend to increase as the number of neighbors of the sites at which those observations are taken increases. Now, however, consider what happens to the marginal variances when we row-standardize, using the seven-site example introduced in the previous section. Row standardization of the $\mathbf{W}$-matrix given by (7.12) yields a new spatial weights matrix

$$\overline{\mathbf{W}} = \begin{pmatrix} 0 & 0 & 1 & 0 & 0 & 0 & 0 \\ 0 & 0 & \frac{1}{2} & 0 & \frac{1}{2} & 0 & 0 \\ \frac{1}{4} & \frac{1}{4} & 0 & \frac{1}{4} & 0 & \frac{1}{4} & 0 \\ 0 & 0 & \frac{1}{2} & 0 & 0 & 0 & \frac{1}{2} \\ 0 & \frac{1}{2} & 0 & 0 & 0 & \frac{1}{2} & 0 \\ 0 & 0 & \frac{1}{3} & 0 & \frac{1}{3} & 0 & \frac{1}{3} \\ 0 & 0 & 0 & \frac{1}{2} & 0 & \frac{1}{2} & 0 \end{pmatrix}.$$

It turns out that the most extreme eigenvalues of $\overline{\mathbf{W}}$ are ± 1.0. With this $\overline{\mathbf{W}}$ as our spatial weights matrix, $\rho_{\mathrm{SAR}} = 0.8$, and $\mathbf{K}_{\mathrm{SAR}}$ given by (7.14), the matrix of marginal variances (on the main diagonal) and correlations (on the off-diagonals) for the row-standardized SAR model is

$$
\begin{pmatrix}
2.75 & 0.60 & 0.82 & 0.60 & 0.53 & 0.61 & 0.53 \\
 & 2.12 & 0.80 & 0.61 & 0.85 & 0.74 & 0.59 \\
 & & 1.76 & 0.80 & 0.72 & 0.82 & 0.72 \\
 & & & 2.12 & 0.59 & 0.74 & 0.85 \\
 & & & & 2.17 & 0.82 & 0.62 \\
 & & & & & 1.82 & 0.82 \\
 & & & & & & 2.17
\end{pmatrix}.
$$

Its row-standardized CAR counterpart is

$$
\begin{pmatrix}
1.31 & 0.22 & 0.49 & 0.22 & 0.16 & 0.22 & 0.16 \\
 & 0.77 & 0.45 & 0.23 & 0.51 & 0.34 & 0.20 \\
 & & 0.49 & 0.45 & 0.34 & 0.46 & 0.34 \\
 & & & 0.77 & 0.20 & 0.34 & 0.51 \\
 & & & & 0.78 & 0.47 & 0.24 \\
 & & & & & 0.57 & 0.47 \\
 & & & & & & 0.78
\end{pmatrix}.
$$

We see, as we also saw when binary adjacency weights without row standardization were used, that the marginal variances and correlations are larger for the SAR model than for the CAR model. More importantly, however, and in contrast to the unstandardized case, row standardization yields marginal variances that are *inversely* related to the number of neighbors. In general, the marginal variances of row-standardized SAR and CAR models are largest at the margins of the study area and tend to decrease toward its interior. This feature is often as unrealistic as the aforementioned opposite situation that occurs when weights are not row-standardized. In most cases, the modeler would prefer to model the marginal variances as homogeneous, either because of a belief that the data satisfy (approximately) a second-order stationarity assumption, or simply as a result of a lack of prior knowledge that any observations are more (or less) reliable than any others. For this reason, Tiefelsdorf *et al.* (1999) introduced an alternative standardization scheme that stabilizes (approximately) the variances. The **variance-stabilizing spatial weights matrix**, denoted by $\widetilde{\mathbf{W}} = (\widetilde{w}_{ij})$, is a spatial weights matrix for which

$$
\widetilde{w}_{ij} = \frac{w_{ij}}{\left(\sum_{j\in\mathcal{N}_i} w_{ij}^2\right)^{1/2}}. \tag{7.17}
$$

CAR4 requires the corresponding matrix $\mathbf{K}_{\mathrm{CAR}}$ to have ith diagonal element

$$
\frac{\sigma^2}{\left(\sum_{j\in\mathcal{N}_i} w_{ij}^2\right)^{1/2}}
$$

for some $\sigma^2 > 0$.

   To illustrate, consider yet again the seven-site example, for which the most extreme eigenvalues of $\widetilde{\mathbf{W}}$ are ± 1.577957 and the allowable parameter space for ρ_{CAR} is therefore $(-0.63373, 0.63373)$. The marginal correlations of SAR and CAR models with $\rho_{\mathrm{SAR}} = \rho_{\mathrm{CAR}} = 0.5$ are similar to those of their row-standardized counterparts, so we consider those cases, for which the matrices of marginal variances and correlations are as follows:

$$
\begin{pmatrix}
2.33 & 0.58 & 0.77 & 0.58 & 0.50 & 0.59 & 0.50 \\
 & 2.65 & 0.80 & 0.61 & 0.82 & 0.73 & 0.57 \\
 & & 3.37 & 0.80 & 0.71 & 0.83 & 0.71 \\
 & & & 2.65 & 0.57 & 0.73 & 0.82 \\
 & & & & 2.61 & 0.81 & 0.60 \\
 & & & & & 3.07 & 0.81 \\
 & & & & & & 2.61
\end{pmatrix},
$$

$$
\begin{pmatrix}
1.24 & 0.20 & 0.44 & 0.20 & 0.14 & 0.21 & 0.14 \\
 & 1.04 & 0.46 & 0.22 & 0.46 & 0.33 & 0.19 \\
 & & 0.97 & 0.46 & 0.33 & 0.48 & 0.33 \\
 & & & 1.04 & 0.19 & 0.33 & 0.46 \\
 & & & & 1.04 & 0.46 & 0.22 \\
 & & & & & 0.98 & 0.46 \\
 & & & & & & 1.04
\end{pmatrix}.
$$

We see that these marginal variances, especially those for the CAR model, are somewhat more homogeneous than the marginal variances for models with either no standardization or row standardization. Even better variance stabilization seems to occur if the divisor in the weights given by (7.17) is modified from $\left(\sum_{j \in \mathcal{N}_i} w_{ij}^2\right)^{1/2}$ to $\left(\sum_{j \in \mathcal{N}_i} w_{ij}^\alpha\right)^{1/\alpha}$, where α is slightly less than 2 for SAR models and slightly greater than 2 for CAR models; see Exercise 7.10.

There is another parametric CAR model, which is related to the weighted CAR model, called the **autocorrelated CAR model** (Cressie & Kapat, 2008). For this model, $\mathbf{C}$ has the single multiplicative-parameter form $\mathbf{C} = \rho_{\mathrm{CAR}} \mathbf{W}^*$, where $\mathbf{W}^* = (w_{ij}^*)$ and w_{ij}^* is equal to $\sqrt{w_{j\cdot}}/\sqrt{w_{i\cdot}}$ if $j \in \mathcal{N}_i$ and is equal to 0 otherwise, for some weighted CAR model with $\mathbf{W} = (w_{ij})$, and $\kappa_i^2 = \sigma^2/w_{i\cdot}$ (where it is assumed that $w_{i\cdot} > 0$ for all i). The most interesting aspect of this model is that the squared partial correlation of each pair of observations, given the rest, is constant across pairs and equal to ρ_{CAR}^2. As a consequence, $-1 < \rho_{\mathrm{CAR}} < 1$ for this model (Exercise 7.11a), but there is a further restriction, as the parameter space for ρ_{CAR} under this model is identical to what it is under the homogeneous CAR model (Exercise 7.11b).

7.4   Islands

Islands are sites that have no neighbors. If site i is an island, then the ith rows of $\mathbf{B}$ and $\mathbf{C}$ consist of all zeros, which can cause problems if not dealt with appropriately. For example, the row sum corresponding to an island is zero, which precludes row standardization of the spatial weights matrix for such sites. One possibility for dealing with an island is to simply amalgamate it with the nearest areal unit and combine responses on those units by averaging or summing. But there is some loss of information and potentially some arbitrariness in this approach, as the "nearest" areal unit may not be well defined. Furthermore, combining the islands' responses with those at other sites may result in a probabilistic structure for the observations not consistent with that corresponding to the originally specified neighborhood structure (an exception being when islands are combined only with other islands). Ver Hoef *et al.* (2018b) suggested a more satisfactory approach, which is to take the marginal covariance matrix of the observations to be

$$\Sigma = \left(\begin{array}{cc} \sigma_{\mathrm{I}}^2 \mathbf{I} & \mathbf{0} \\ \mathbf{0} & \Sigma_{\mathrm{M}} \end{array} \right),$$

where the observations are ordered in $\mathbf{y}$ so that all islands appear first. Here $\sigma_{\mathrm{I}}^2 > 0$ is the assumed constant variance of responses at islands and Σ_{M} is the SAR or CAR covariance matrix for "mainland" sites, i.e., sites that have neighbors, given by expressions analogous to (7.4) or (7.7). Note that the common variance of responses at islands in this formulation is permitted to be unequal to the variance(s) in Σ_{M}. A slightly more general formulation would allow for heterogeneous variances at islands by assuming that the covariance matrix of responses at islands is $\sigma^2 \mathbf{K}_{\mathrm{I}}$ where $\mathbf{K}_{\mathrm{I}}$ is a specified diagonal matrix with positive diagonal elements. This formulation is consistent with what was described previously in Section 7.2 for the parameterizations of $\mathbf{K}_{\mathrm{SAR}}$ and $\mathbf{K}_{\mathrm{CAR}}$.

7.5 Relationships between spatial weights and marginal correlations

As noted previously, modeling the spatial correlation of areal data using spatial weights matrices is rather more indirect than using a geostatistical model. Although (7.4) and (7.7) express the marginal covariance matrices among the observations for the SAR and CAR models as functions of the spatial weights matrices and $\mathbf{K}$, because those expressions involve matrix inverses it is not obvious how the marginal correlations among observations are related to the spatial weights, even when those weights are parameterized parsimoniously. In this section, we study this relationship for several scenarios.

Let us consider first the important case where the spatial weights matrix is given by $\rho \mathbf{W}$, where $\mathbf{W}$ is a binary adjacency matrix (hence symmetric), ρ is either ρ_{SAR} or ρ_{CAR}, and $\mathbf{K} = \sigma^2 \mathbf{I}$ for some $\sigma^2 > 0$. Figure 7.3 displays the relationship between marginal correlations of neighbors (using the rook's definition) and ρ_{SAR} or ρ_{CAR}, for observations on $\sqrt{n} \times \sqrt{n}$ square lattices, over the interval $(\lambda_1^{-1}, \lambda_n^{-1})$ for which the CAR covariance matrix is positive definite. (Although SAR models are valid for some values of ρ_{SAR} outside this interval, consideration of the marginal correlations for such models, which behave somewhat erratically, is relegated to Exercise 7.12a.) The upper left panel of the figure indicates that, for the SAR model on a 3×3 lattice, the relationship is monotonic, symmetric about 0, nearly linear, and such that the marginal correlation tends to 1 or -1 as ρ_{SAR} tends to the right or left endpoint of the interval. Only two curves are apparent in this case because there are only two distinct marginal correlations among neighbors for a 3×3 lattice: one corresponding to a site-pair in which one site is a corner site, and the other corresponding to a site-pair in which one site is the interior site. The upper right panel of the figure reveals that, for the CAR model on a 3×3 lattice, the relationship between the marginal correlations and ρ_{CAR} is similar to what it is for the SAR model but not quite as linear; in particular, the marginal correlations increase more slowly with ρ_{CAR} than with ρ_{SAR} over the central portion of the interval, resulting in stronger curvature near the interval's boundaries. Thus, a somewhat larger ρ_{CAR} than ρ_{SAR} is needed to yield an appreciable marginal correlation. The bottom pair of panels in Figure 7.3 correspond to a 6×6 lattice. When compared to the top pair, they suggest that regardless of the size of the lattice, the relationship between the marginal correlations and ρ_{SAR} (or ρ_{CAR}) is qualitatively similar to that seen in the 3×3 case, but the curvature near the boundaries of $(\lambda_1^{-1}, \lambda_n^{-1})$ becomes more pronounced as the size of the lattice increases.

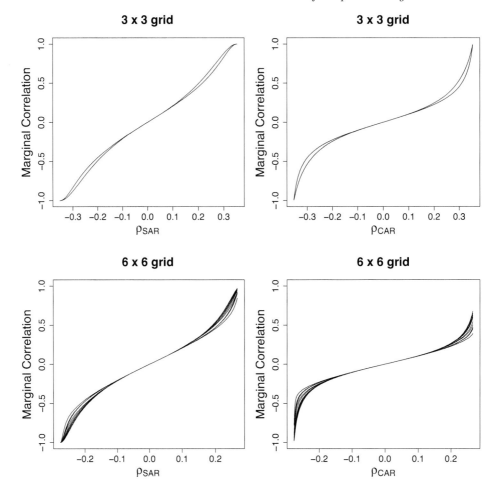

FIGURE 7.3
Relationship between spatial dependence parameter and marginal correlations of neighbors, under SAR and CAR models, for 3×3 and 6×6 square grids with binary spatial weights.

  The next figure, Figure 7.4, displays the relationships between the same quantities for scenarios that are exactly the same as those just described except that the original weights have been row-standardized. Accordingly, the parameter space for ρ_{CAR} is $(\lambda_1^{-1}, 1)$, where it turns out that $\lambda_1^{-1} = -1$ for both lattices, and the diagonal elements of $\mathbf{K}$ are given by (7.14). The relationships are qualitatively very similar to what they were when weights were not row-standardized. There is a feature, however, that is not easily discernible in the figure and was not present previously: some of the curves cross at values of ρ other than zero. That is, the ranking (from smallest to largest) of the marginal correlations is not consistent over the entire interval. But while this feature is non-intuitive, it is too small to be of much concern to a modeler.

  Some additional scenarios with square lattice spatial configurations yield similar results, so they are not displayed here. These scenarios include using anisotropic weights or the queen's definition of neighbors (so that diagonal squares, not merely adjacent squares, are considered neighbors).

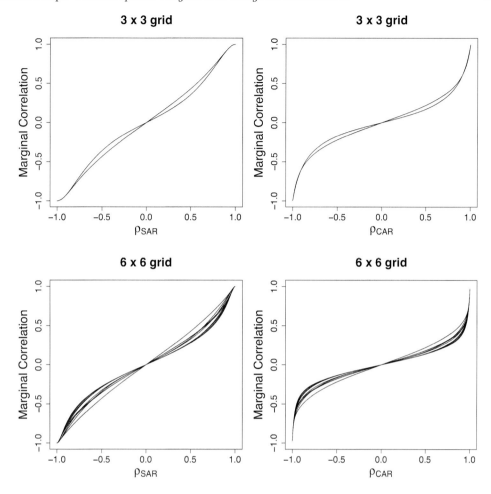

FIGURE 7.4
Relationship between spatial dependence parameter and marginal correlations of neighbors for 3×3 and 6×6 square grids and row-standardized spatial weights.

Thus, for observations on a square lattice, the dependence of the marginal correlations among neighboring observations on ρ_{SAR} (or ρ_{CAR}), though nonlinear, at least appears to be monotonic over the interval on which $\mathbf{I} - \rho_{CAR}\mathbf{W}$ is positive definite. But is the same true for observations on irregular lattices, for which some sites have many more neighbors than others? We answer this question by considering another scenario using the irregular lattice of the 48 contiguous states of the United States of America plus the District of Columbia, which is very similar to a scenario first studied by Wall (2004). Figure 7.5 shows for this lattice the relationship between marginal correlations among neighbors and ρ_{SAR} (and ρ_{CAR}), again over the interval for which the CAR covariance matrix is positive definite. Spatial weights for the top two panels were binary adjacency weights, for which the relevant interval is $(-0.3489, 0.1847)$; weights for the bottom two panels were row-standardized, for which the relevant interval is $(\lambda_1^{-1} = -1.3924, 1)$. The figure indicates that regardless of whether or not the weights are row-standardized, when ρ is positive all of the marginal correlations exhibit a nonlinear behavior over the relevant interval similar to that observed for the square lattices, but when ρ is negative some behave quite differently. In fact, when ρ

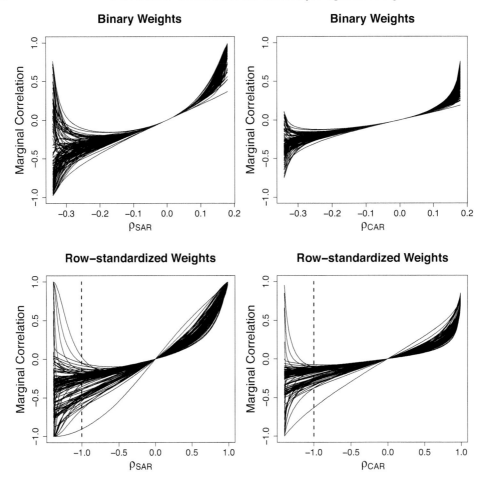

FIGURE 7.5
Relationship between spatial dependence parameter and marginal correlations of neighbors for states in the contiguous United States of America and the District of Columbia, under SAR and CAR models using binary spatial adjacency weights and row-standardized versions of those weights. The dotted vertical line in the bottom two plots marks the left endpoint of the parameter space (-1) used in practice by many modelers.

is negative, some marginal correlations are positive and decrease as ρ increases! The state-pairs corresponding to the marginal correlations exhibiting this bizarre behavior are not geographically clustered or otherwise discernibly different, so there is no ready explanation for this phenomenon. In the row-standardized case, the phenomenon is especially evident over the subinterval $(-1.3924, -1)$. Thus, a modeler could reduce, though not altogether eliminate, such behavior by taking the parameter space for ρ to be $(-1, 1)$ rather than $(\lambda_1^{-1}, 1)$. This practice, as noted previously, is often followed.

Assunção & Krainski (2009) also studied the behavior of marginal correlations under row-standardized CAR models. They proved that when $\rho_{CAR} > 0$, the marginal correlations corresponding to all site-pairs increase monotonically with ρ_{CAR}, but at pair-specific rates, as suggested empirically by Figure 7.4. The rates depend on the complete neighborhood structure (second-order neighbors, third-order neighbors, etc.) of the sites in the site-pairs, not merely their first-order neighborhood structure, especially when ρ_{CAR} is close to 1. In fact, the closer that ρ_{CAR} is to 1, the greater the impact of higher-order neighborhoods on the marginal correlations. When the sites in one site-pair have considerably more higher-order neighbors than the sites in another, the ranks of their marginal correlations can switch as ρ_{CAR} increases, as was shown in Figure 7.4.

7.6 Variants of SAR and CAR models

The SAR model we have considered so far is but one of several variants and extensions of SAR models that have been proposed in the literature. Furthermore, there is a variant of the CAR model that has received considerable attention. This section briefly describes these variants.

All of the proposed SAR variants are special cases of the **general nested SAR model** given by

$$\mathbf{y} = \mathbf{X}\boldsymbol{\beta} + \mathbf{B}_{\text{Lag}}\mathbf{y} + \mathbf{B}_{\text{Reg}}\mathbf{Z}\boldsymbol{\gamma} + \mathbf{u}, \quad \mathbf{u} = \mathbf{B}_{\text{Err}}\mathbf{u} + \mathbf{d},$$

where $\mathbf{Z}$, like $\mathbf{X}$, is a specified matrix of regressors; $\boldsymbol{\gamma}$ is a vector of unknown parameters; $\mathbf{B}_{\text{Lag}}$, $\mathbf{B}_{\text{Reg}}$, and $\mathbf{B}_{\text{Err}}$ are spatial weights matrices that incorporate spatial dependence associated with the response, the regressors, and the errors, respectively; and $\mathbf{d}$ is a vector whose elements are independent and identically distributed $\text{N}(0, \sigma^2)$ random variables. A special, but still quite general, case of this model that includes all of the SAR variants proposed to date is obtained by parameterizing each spatial weights matrix using a single multiplicative parameter, yielding

$$\mathbf{y} = \mathbf{X}\boldsymbol{\beta} + \rho_{\text{Lag}}\mathbf{W}\mathbf{y} + \rho_{\text{Reg}}\mathbf{W}\mathbf{Z}\boldsymbol{\gamma} + \rho_{\text{Err}}\mathbf{W}\mathbf{u} + \mathbf{d}.$$

Some special cases of this model that have names are as follows:

- SAR error model ($\rho_{\text{Lag}} = \rho_{\text{Reg}} = 0$):

$$\mathbf{y} = \mathbf{X}\boldsymbol{\beta} + \mathbf{u}, \quad \mathbf{u} = \rho_{\text{Err}}\mathbf{W}\mathbf{u} + \mathbf{d}.$$

  This is the classical SAR model featured earlier in this chapter.

- SAR lag model ($\rho_{\text{Reg}} = \rho_{\text{Err}} = 0$):

$$\mathbf{y} = \mathbf{X}\boldsymbol{\beta} + \rho_{\text{Lag}}\mathbf{W}\mathbf{y} + \mathbf{d}$$

  (Lesage & Pace, 2009).

- SAR lagged X model ($\rho_{\text{Lag}} = \rho_{\text{Err}} = 0, \rho_{\text{Reg}} = 1, \mathbf{Z} = \mathbf{X}$):

$$\mathbf{y} = \mathbf{X}\boldsymbol{\beta} + \mathbf{W}\mathbf{X}\boldsymbol{\gamma} + \mathbf{d}$$

  (Vega & Elhorst, 2015).

- Durbin lag model ($\rho_{\text{Err}} = 0, \rho_{\text{Reg}} = 1$):

$$\mathbf{y} = \mathbf{X}\boldsymbol{\beta} + \rho_{\text{Lag}}\mathbf{W}\mathbf{y} + \mathbf{W}\mathbf{Z}\boldsymbol{\gamma} + \mathbf{d}$$

  (Lesage & Pace, 2009). The special case that results upon setting $\mathbf{Z} = \mathbf{X}$ has been called the SAR mixed model (Kissling & Carl, 2008).

- Durbin error model ($\rho_{\text{Lag}} = 0, \rho_{\text{Reg}} = 1$):

$$\mathbf{y} = \mathbf{X}\boldsymbol{\beta} + \mathbf{WZ}\boldsymbol{\gamma} + \mathbf{u}, \quad \mathbf{u} = \rho_{\text{Err}}\mathbf{Wu} + \mathbf{d}$$

(Lesage & Pace, 2009).

- Double spatial coefficient model ($\rho_{\text{Reg}} = 0$):

$$\mathbf{y} = \mathbf{X}\boldsymbol{\beta} + \rho_{\text{Lag}}\mathbf{Wy} + \mathbf{u}, \quad \mathbf{u} = \rho_{\text{Err}}\mathbf{Wu} + \mathbf{d}.$$

Of the SAR variants, the SAR lag model, Durbin lag model, and double spatial coefficient model allow the spatial weights matrix to smooth the explanatory variables in the model matrix in addition to creating spatial correlation among the responses. As a consequence, those three models are not linear models, i.e., the expectations of the responses are not linear functions of the unknown model parameters. Thus, they fall outside the purview of this book. Furthermore, the SAR lag model and SAR mixed model performed poorly in an empirical study, relative to the SAR error model (Kissling & Carl, 2008), which led Ver Hoef *et al.* (2018b) to discourage their use. The SAR lagged X model is merely a Gauss-Markov linear model with model matrix $(\mathbf{X}|\mathbf{WX})$, so it regards the observations as independent and can be estimated by ordinary least squares. The Durbin error model is a mixed linear model with the same parameter space for ρ_{Err} as the SAR error model, but with model matrix $(\mathbf{X}|\mathbf{WZ})$ rather than $\mathbf{X}$.

The conditional formulation (7.6) does not allow for the same types of extensions of the CAR model that were just described for the SAR model. Nevertheless, there is one variant of a CAR model, due to Besag & Kooperberg (1995) and called the **intrinsic CAR**, or **ICAR**, model, which has been used very often within a Bayesian hierarchical modeling paradigm. In an ICAR model, $c_{ij} = w_{ij}/w_{i\cdot}$ in (7.6) and $\kappa_i^2 = \sigma^2/w_{i\cdot}$. In the case of binary neighbor weights, for example,

$$c_{ij} = \frac{1}{|\mathcal{N}_i|} \quad \text{and} \quad \kappa_i^2 = \frac{\sigma^2}{|\mathcal{N}_i|} \quad (i = 1, \ldots, n).$$

That is, the conditional mean of the ith response, given the rest, is the average response of its neighbors, and the ith conditional variance is inversely proportional to the number of neighbors of site i. Thus, the ICAR model can be viewed as a limiting case, as $\rho_{\text{CAR}} \to 1$, of a single multiplicative-parameter CAR model for which the weights have been row-standardized. However, the ICAR model does not specify a proper joint distribution for $\mathbf{y}$ because $(\mathbf{I} - 1 \cdot \mathbf{W})\mathbf{1} = \mathbf{0}$, i.e., the precision matrix is singular. Equivalently, the covariance matrix of $\mathbf{y}$ does not exist. Consequently, the ICAR model has been relegated for use as an improper prior for spatial random effects in a hierarchical Bayesian model. One reason it is very attractive for that purpose is that its only unknown parameter is σ^2. Of course, this also reduces its flexibility.

7.7   Spatial moving average models

As introduced in Section 2.3, a **spatial moving average (SMA) model** for $\mathbf{y}$ is given by

$$y_i = \mathbf{x}_i^T \boldsymbol{\beta} + \sum_{j \in \mathcal{N}_i} m_{ij} d_j + d_i,$$

where the spatial weights matrix is denoted by $\mathbf{M} = (m_{ij})$ and satisfies $m_{ii} = 0$ for all i, and the d_i's are independent and identically distributed normal variables with mean zero and positive variance κ_i^2. In vector form, the model is

$$\mathbf{y} = \mathbf{X}\boldsymbol{\beta} + (\mathbf{M} + \mathbf{I})\mathbf{d},$$

where $\mathbf{d}$ is a multivariate normal random vector with mean vector $\mathbf{0}$ and positive definite diagonal covariance matrix $\mathbf{K}_{\text{SMA}} = \text{diag}(\kappa_1^2, \ldots, \kappa_n^2)$. The covariance matrix of the observations is

$$\boldsymbol{\Sigma}_{\text{SMA}} = (\mathbf{M} + \mathbf{I})\mathbf{K}_{\text{SMA}}(\mathbf{M}^T + \mathbf{I}), \tag{7.18}$$

which is positive definite if and only if $\mathbf{M} + \mathbf{I}$ is nonsingular (by Theorem A.3 in Appendix A).

A summary of the conditions required for $\boldsymbol{\Sigma}_{\text{SMA}}$ to be a valid covariance matrix is as follows:

SMA1: $\mathbf{K}_{\text{SMA}}$ is diagonal with positive diagonal elements;

SMA2: $m_{ii} = 0$ for all i;

SMA3: $\mathbf{M} + \mathbf{I}$ is nonsingular.

As was true for $\mathbf{B}$ and $\mathbf{C}$, $\mathbf{M}$ generally has too many parameters to be estimated from the available data, so it is usually parameterized much more parsimoniously. The single multiplicative-parameter model, $\mathbf{M} = \rho_{\text{SMA}}\mathbf{W}$, is most commonly used. The parameter space for ρ_{SMA} is the set of values for which $\rho_{\text{SMA}}\mathbf{W} + \mathbf{I}$ is nonsingular, which may be obtained in a fashion rather similar to how the parameter spaces for ρ_{SAR} and ρ_{CAR} were found. The parameter space so obtained is $\Omega_{\text{SMA}} = \{\rho_{\text{SMA}} : \rho_{\text{SMA}} \neq -\lambda_i^{-1}, \lambda_i \in A\}$ where A was defined in the description of Ω_{SAR}.

To illustrate, the matrix of marginal variances and correlations for the seven-site example introduced in Section 7.2 following a single multiplicative-parameter SMA model with $\rho_{\text{SMA}} = 0.3$ is

$$\begin{pmatrix} 1.09 & 0.08 & 0.49 & 0.08 & 0 & 0.08 & 0 \\ & 1.18 & 0.47 & 0.08 & 0.51 & 0.15 & 0 \\ & & 1.36 & 0.47 & 0.14 & 0.46 & 0.14 \\ & & & 1.18 & 0 & 0.15 & 0.51 \\ & & & & 1.18 & 0.49 & 0.08 \\ & & & & & 1.27 & 0.49 \\ & & & & & & 1.18 \end{pmatrix}.$$

Most of the marginal correlations are small, or even zero. The only appreciable correlations are those between adjacent sites. Furthermore, the marginal variances are relatively homogeneous in comparison to the marginal variances of the SAR and CAR models without some type of standardization.

There is a notable difference in the manner in which the covariance matrix of a SMA depends on its spatial weights matrix $\mathbf{M}$, as given by (7.18), and the manners in which the covariance matrices of a SAR or CAR depend on their spatial weights matrices $\mathbf{B}$ and $\mathbf{C}$, as given by (7.4) and (7.7). Specifically, $\boldsymbol{\Sigma}_{\text{SMA}}$ depends on the spatial weights directly, while $\boldsymbol{\Sigma}_{\text{SAR}}$ and $\boldsymbol{\Sigma}_{\text{CAR}}$ depend on the spatial weights through the inverses of matrices involving those weights. Put another way, the SMA is fundamentally a model for the covariance matrix, while the SAR and CAR are fundamentally models for the precision matrix. It should not surprise us, therefore, if the behavior of the SMA model is somewhat different than that of the SAR and CAR models.

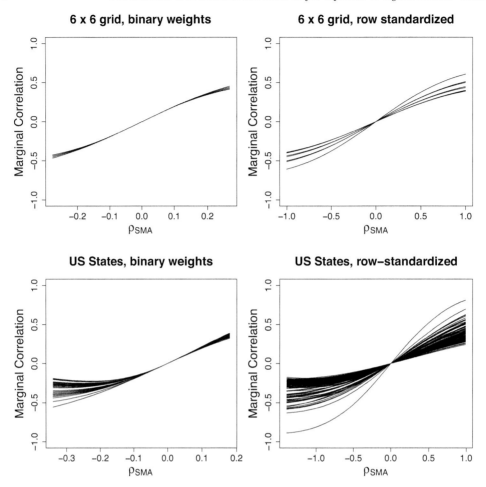

FIGURE 7.6
Relationship between spatial dependence parameter and marginal correlations of neighbors for a 6×6 square grid and for states in the contiguous United States of America and the District of Columbia, under SMA models using binary and row-standardized weights.

Figure 7.6 shows the relationship between ρ_{SMA} and the marginal correlations among neighbors for sites on a 6×6 lattice (top two panels) and the contiguous United States plus the District of Columbia (bottom two panels), using binary adjacency weights and their row-standardized counterparts. In contrast to what was seen previously for the SAR and CAR models, the marginal correlations are well-behaved. However, they are rather limited in their extent, as most of their absolute values are less than 0.5 over the entire parameter space. Thus, very strong marginal correlations may not be adequately modeled via an SMA model.

The lower right panel of Figure 7.2 displays a realization of observations on a 20×20 square grid following the single-parameter SMA model described in this section, with $\rho_{SMA} = 0.25$. The spatial correlation among the observations induced by the model is discernible, but relatively weak, reinforcing what was seen in Figure 7.6.

From the foregoing, it seems that a SMA model with nonzero weights for adjacent sites only is relatively limited in its ability to create large autocorrelations. One way to allow for the possibility of stronger correlations is to create more neighbors by expanding the neighborhoods; see Exercise 7.14.

SMA models may be combined with SAR models to form so-called **SARMA models**. For example, a SARMA error model is of the form

$$y_i = \mathbf{x}_i^T \boldsymbol{\beta} + \sum_{j \in \mathcal{N}_i} b_{ij}(y_j - \mathbf{x}_j^T \boldsymbol{\beta}) + \sum_{j \in \mathcal{N}_i} m_{ij} d_j + d_i, \tag{7.19}$$

where $\mathbf{d} = (d_i)$ is distributed as in the SMA model. Provided that $\mathbf{I} - \mathbf{B}$ is nonsingular, the covariance matrix of the observations under this model is $(\mathbf{I} - \mathbf{B})^{-1}(\mathbf{M} + \mathbf{I})\mathbf{K}_{\text{SMA}}(\mathbf{M}^T + \mathbf{I})(\mathbf{I} - \mathbf{B}^T)^{-1}$; furthermore, this covariance matrix is positive definite if $\mathbf{M} + \mathbf{I}$ is nonsingular (Exercise 7.15). Parameterizing $\mathbf{B}$ and $\mathbf{M}$ as $\rho_{\text{SAR}}\mathbf{W}$ and $\rho_{\text{SMA}}\mathbf{W}$, respectively, yields a parsimonious SARMA error model with covariance matrix

$$\boldsymbol{\Sigma}_{\text{SARMA}} = (\mathbf{I} - \rho_{\text{SAR}}\mathbf{W})^{-1}(\rho_{\text{SMA}}\mathbf{W} + \mathbf{I})\mathbf{K}_{\text{SMA}}(\rho_{\text{SMA}}\mathbf{W}^T + \mathbf{I})(\mathbf{I} - \rho_{\text{SAR}}\mathbf{W}^T)^{-1}. \tag{7.20}$$

Although the temporal antecedents of SARMA models, known as ARMA models, are very commonly used for time series data, SARMA models appear to have been used rather sparingly for spatial data.

7.8 Relationships between models

Observe that if $\mathbf{K}_{\text{SAR}} = \mathbf{I}$ in (7.4), then $\boldsymbol{\Sigma}_{\text{SAR}} = (\mathbf{I} - \mathbf{B})^{-1}[(\mathbf{I} - \mathbf{B})^{-1}]^T = [(\mathbf{I} - \mathbf{B}^T)(\mathbf{I} - \mathbf{B})]^{-1} = (\mathbf{I} - \mathbf{B} - \mathbf{B}^T + \mathbf{B}^T\mathbf{B})^{-1}$. Some authors, upon comparing this last expression to (7.7), have claimed that any SAR model with spatial weights matrix $\mathbf{B}$ and $\mathbf{K}_{\text{SAR}} = \mathbf{I}$ is a CAR model with (symmetric) spatial weights matrix $\mathbf{B} + \mathbf{B}^T - \mathbf{B}^T\mathbf{B}$ and $\mathbf{K}_{\text{CAR}} = \mathbf{I}$. By the same token, upon comparing (7.4) to (7.18), one might claim that a SAR model with spatial weights matrix $\mathbf{B}$ and $\mathbf{K}_{\text{SAR}} = \mathbf{I}$ is equivalent to a SMA model with $\mathbf{M} = (\mathbf{I} - \mathbf{B})^{-1} - \mathbf{I}$ and $\mathbf{K}_{\text{SMA}} = \mathbf{I}$. Both of these claims are false; see Exercise 7.17. However, it is true that every SAR model is a CAR model with some spatial weights matrix, and every CAR model is a SAR model with some spatial weights matrix; see Exercise 7.18a,c. In fact, Ver Hoef *et al.* (2018a) proved the following even stronger results:

- Any positive definite covariance matrix may be expressed as the SAR covariance matrix $(\mathbf{I} - \mathbf{B})^{-1}\mathbf{K}_{\text{SAR}}(\mathbf{I} - \mathbf{B}^T)^{-1}$ for an infinite number of pairs of matrices $\mathbf{B}$ and $\mathbf{K}_{\text{SAR}}$;

- Any positive definite covariance matrix may be expressed as the CAR covariance matrix $(\mathbf{I} - \mathbf{C})^{-1}\mathbf{K}_{\text{CAR}}$ for a unique pair of matrices $\mathbf{C}$ and $\mathbf{K}_{\text{CAR}}$.

Using similar techniques it can be demonstrated that any positive definite covariance matrix may also be expressed as the SMA covariance matrix $(\mathbf{M} + \mathbf{I})\mathbf{K}_{\text{SMA}}(\mathbf{M}^T + \mathbf{I})$ for an infinite number of pairs of matrices $\mathbf{M}$ and $\mathbf{K}_{\text{SMA}}$ (see Exercise 7.18b). Thus, any one of the SAR, CAR, and SMA models can be written as a model of either of the other two types. Moreover, any geostatistical covariance structure given by one of the models described in Chapter 6 can be obtained via an appropriately constructed SAR, CAR, or SMA model.

Typically, modelers take $\mathbf{B}$, $\mathbf{C}$, and $\mathbf{M}$ in their SAR, CAR, and SMA models to be sparse, containing mostly zeros. Although any of these models can be expressed as any other, it is not clear whether a sparse-matrix case of such a model can be obtained from a sparse-matrix case of any of the others. However, Ver Hoef *et al.* (2018b) argued that a SAR model with sparse $\mathbf{B}$ can indeed be obtained from a CAR model with sparse $\mathbf{C}$. A case of the opposite was shown previously in Section 7.1.3, and another, due to Besag (1974), is considered in Exercise 7.19.

Similarly, it is not immediately clear whether geostatistical covariance structures can be expressed as sparse-matrix cases of a spatial-weights model. However, Rue & Tjelmeland (2002) showed that Gaussian processes with several commonly used geostatistical covariance functions (exponential, spherical, Gaussian, Matérn) can be well-approximated by sparse CARs when the observations lie on a regular lattice. This connection can be exploited to reduce the computations associated with spatial prediction (Hartman & Hössjer, 2008). Lindgren *et al.* (2011) carried this idea further by establishing that a Gaussian process with a Matérn covariance function having integer (in $\mathbb{R}^2$) or half-integer (in $\mathbb{R}$) smoothness parameter may be closely approximated by a sparse SAR, regardless of the spatial configuration of data locations. Again, the utility of such an approximation is a huge reduction in the computational burden of fitting the geostatistical model and performing inferences, a topic that will be taken up again in the next chapter.

7.9 Exercises

1. (a) Show that CAR4 implies that $\frac{c_{ij}}{c_{ji}} = \frac{c_{ik}}{c_{ki}} \cdot \frac{c_{kj}}{c_{jk}}$ for all i, j, k for which $c_{ij} \neq 0$, $c_{ji} \neq 0$, $c_{ik} \neq 0$, $c_{ki} \neq 0$, $c_{jk} \neq 0$, and $c_{kj} \neq 0$.

   (b) Use the result of part (a) to give an example of a matrix $\mathbf{C}$ that satisfies CAR2 and CAR3 but does not satisfy CAR4 for any $\mathbf{K}_{\text{CAR}}$.

2. Use (7.7) and CAR4 to show that (7.8) reduces to (7.9) for a CAR model.

3. Consider a SAR model with $\mathbf{B}$ given by (7.10) and any $\mathbf{K}_{\text{SAR}}$ satisfying SAR1. Show that the fourth element in the first row of $\boldsymbol{\Sigma}_{\text{SAR}}^{-1}$ is equal to zero, implying that the first and fourth observations are conditionally independent, given the second and third observations.

4. For a SAR model with the single multiplicative-parameter spatial weights matrix $\rho_{\text{SAR}}\mathbf{W}$, show that SAR3 is satisfied if and only if ρ_{SAR} is not the reciprocal of any real nonzero eigenvalue of $\mathbf{W}$. (This is an extension of Proposition 2(i) of Ver Hoef *et al.* (2018b).)

5. For a CAR model with single multiplicative-parameter spatial weights matrix $\rho_{\text{CAR}}\mathbf{W}$, show that:

   (a) CAR4 implies that the eigenvalues of $\mathbf{W}$ are real, and if they are denoted in ordered fashion as $\lambda_1 \leq \lambda_2 \leq \cdots \leq \lambda_n$, then λ_1 is negative and λ_n is positive;

   (b) CAR3 is satisfied if and only if $\lambda_1^{-1} < \rho_{\text{CAR}} < \lambda_n^{-1}$.

6. Show that CAR3 and CAR4 are satisfied for the unconstrained CAR model given by (7.13).

7. Let $\mathbf{W}$ and $\overline{\mathbf{W}}$ be spatial weights matrices of the single multiplicative-parameter forms described in Section 7.2 (thus $\mathbf{W}$ is symmetric), and let $\lambda_1 \leq \lambda_2 \leq \ldots \leq \lambda_n$ be the ordered eigenvalues of $\overline{\mathbf{W}}$.

   (a) Show that
   $$-1 \leq \lambda_1 \leq \lambda_2 \leq \cdots \leq \lambda_n = 1.$$

(b) Show that the diagonal elements of the matrix $\mathbf{K}_{\mathrm{CAR}}$ corresponding to $\overline{\mathbf{W}}$ are constrained by CAR4 to have form $\kappa_i^2 = \sigma^2/w_i$. ($i = 1, \ldots, n$) for some $\sigma^2 > 0$.

8. Determine the parameter space for ρ in single multiplicative-parameter SAR, CAR, and SMA models for data for which the spatial configuration is as displayed in Figure 2.4 and the weights matrix ($\mathbf{B}$, $\mathbf{C}$, or $\mathbf{M}$) is as prescribed in Section 2.3.2.

9. Show that (7.15) and (7.16) are equivalent.

10. For the seven-site example considered in this chapter, show that:

    (a) the marginal variances of the SAR model with $\rho_{\mathrm{SAR}} = 0.5$ are better stabilized when the divisor of (7.17) is replaced with $\left(\sum_{j \in \mathcal{N}_i} w_{ij}^{5/3} \right)^{3/5}$;

    (b) the marginal variances of the CAR model with $\rho_{\mathrm{CAR}} = 0.5$ are better stabilized when the divisor of (7.17) is replaced with $\left(\sum_{j \in \mathcal{N}_i} w_{ij}^{20/9} \right)^{9/20}$.

11. For the autocorrelated CAR model described in Section 7.3:

    (a) Show that $-1 < \rho_{\mathrm{CAR}} < 1$. Hint: use (7.9).
    (b) Show that CAR3 and CAR4 are satisfied if and only if $\lambda_1^{-1} < \rho_{\mathrm{CAR}} < \lambda_n^{-1}$.

12. Construct figures analogous to Figure 7.3, showing the marginal correlations corresponding to adjacent site-pairs in 3×3 and 6×6 square lattices, for each of the following, and comment on each figure:

    (a) SAR models with $\mathbf{W}$ a binary adjacency matrix and $-1 \le \rho_{\mathrm{SAR}} \le 1$;
    (b) the unconstrained CAR model with elements of $\mathbf{C}$ and $\mathbf{K}$ given by (7.13), where $-20 < \phi < 20$ and γ_{ij} is simply the adjacency indicator variable evaluated at sites i and j;
    (c) the autocorrelated CAR model with $\mathbf{W}$ a binary adjacency matrix and $\lambda_1^{-1} < \rho_{\mathrm{CAR}} < \lambda_n^{-1}$.

13. Explain why the SAR lag model, Durbin lag model, and double spatial coefficient model are not linear models.

14. Suppose that the neighborhoods in the seven-site example of Section 7.2 are expanded by using the queen's rather than rook's definition of neighbor. Find the matrix of marginal variances and correlations for a single-multiplicative-parameter SMA model with $\rho_{\mathrm{SMA}} = 0.3$ and $\mathbf{K}_{\mathrm{SAR}} = \mathbf{I}$. How do the marginal variances and correlations compare to those based on the rook's definition of neighbor, which are displayed in Section 7.7?

15. Derive the covariance matrix for the SARMA model given by (7.19) assuming that $\mathbf{I} - \mathbf{B}$ is nonsingular, and establish that it is positive definite if, in addition, $\mathbf{M} + \mathbf{I}$ is nonsingular.

16. What model does the SARMA model with covariance matrix (7.20) reduce to when $\rho_{\mathrm{SAR}} = -\rho_{\mathrm{SMA}}$?

17. Explain the fallacy of each of the following claims:

    (a) Any SAR model with spatial weights matrix $\mathbf{B}$ and $\mathbf{K}_{\mathrm{SAR}} = \mathbf{I}$ is a CAR model with spatial weights matrix $\mathbf{B} + \mathbf{B}^T - \mathbf{B}^T\mathbf{B}$ and $\mathbf{K}_{\mathrm{CAR}} = \mathbf{I}$.
    (b) Any SAR model with spatial weights matrix $\mathbf{B}$ and $\mathbf{K}_{\mathrm{SAR}} = \mathbf{I}$ also a SMA model with spatial weights matrix $\mathbf{M} = (\mathbf{I} - \mathbf{B})^{-1} - \mathbf{I}$ and $\mathbf{K}_{\mathrm{SMA}} = \mathbf{I}$.

18. The following lemma is a straightforward consequence of matrix multiplication:
    Lemma. If $\mathbf{D}$ is an $n \times n$ diagonal matrix and $\mathbf{Q}$ is an $n \times n$ square matrix with all zeros on the diagonal, then $\mathbf{DQ}$ and $\mathbf{QD}$ have all zeros on the diagonal.

    (a) Use this lemma to show that any positive definite covariance matrix $\boldsymbol{\Sigma}$ can be expressed as the covariance matrix of a SAR model for a non-unique pair of matrices $\mathbf{B}$ and $\mathbf{K}_{\text{SAR}}$.

    (b) Similarly, show that any positive definite covariance matrix $\boldsymbol{\Sigma}$ can be expressed as the covariance matrix of a SMA model for a non-unique pair of matrices $\mathbf{M}$ and $\mathbf{K}_{\text{SMA}}$.

    (c) Show that any positive definite covariance matrix $\boldsymbol{\Sigma}$ can be expressed as the covariance matrix of a CAR model for a unique pair of matrices $\mathbf{C}$ and $\mathbf{K}_{\text{CAR}}$.

19. Consider observations on a rectangular lattice wrapped on a torus, so that sites in the top row of the lattice are neighbors of sites in the bottom row (and vice versa) and sites in the leftmost column are neighbors with sites in the rightmost column (and vice versa). Suppose that y_{ij}, the random variable in the ith row and jth column of the lattice, follows the SAR model

$$y_{ij} = \beta_1 y_{i-1,j} + \beta_2 y_{i+1,j} + \beta_3 y_{i,j-1} + \beta_4 y_{i,j+1} + d_{ij},$$

where $\text{Var}(d_{ij}) = \sigma^2$, for all i and j, for some $\sigma^2 > 0$. Show that the conditional mean of y_{ij} given the remaining variables is

$$
\begin{aligned}
\text{E}(y_{ij}&|\{y_{kl}, (k,l) \neq (i,j)\}) \\
= \quad &(1 + \beta_1^2 + \beta_2^2 + \beta_3^2 + \beta_4^2)^{-1}\{(\beta_1 + \beta_2)(y_{i-1,j} + y_{i+1,j}) \\
&+ (\beta_3 + \beta_4)(y_{i,j-1} + y_{i,j+1}) - (\beta_1\beta_4 + \beta_2\beta_3)(y_{i-1,j-1} + y_{i+1,j+1}) \\
&- (\beta_1\beta_3 + \beta_2\beta_4)(y_{i-1,j+1} + y_{i+1,j-1}) - \beta_1\beta_2(y_{i-2,j} + y_{i+2,j}) \\
&- \beta_3\beta_4(y_{i,j-2} + y_{i,j+2})\},
\end{aligned}
$$

and thus the specified SAR model for which only adjacent sites are neighbors leads to a CAR model with additional sites as neighbors.

8

Likelihood-Based Inference

Chapters 4 and 5 considered the estimation of a spatial linear model's mean structure by methods that either did not take the covariance structure into account (Chapter 4) or accounted for it but unrealistically assumed it to be known up to the value of a scalar multiplier (Chapter 5). Those estimation methods—ordinary and generalized least squares estimation—are not optimal unless the assumed covariance structure is, in fact, the actual covariance structure of the data. A more flexible approach, which is likely to perform better, is to assume that the covariance structure is that which is prescribed by one of the parametric families of covariance or spatial weighting functions described in Chapters 6 and 7, and to estimate the parameters of that function prior to or simultaneously with the model's mean parameters. This chapter describes methods for accomplishing that task based on maximizing the likelihood function, or a variant known as the residual likelihood function, of the data. It is assumed throughout that the observations have a multivariate normal distribution, which is guaranteed if the underlying spatial process is Gaussian. Interestingly, it turns out that likelihood-based estimation of the mean structure under this assumption coincides with generalized least squares estimation using the estimated covariance parameters in place of their unknown values. Consequently, what was presented in Chapter 5 is highly relevant.

8.1 Maximum likelihood estimation

Consider once again the spatial general linear model defined and developed in Sections 1.3 and 2.4, and suppose that the model's errors have a multivariate normal distribution. This model specifies that

$$\mathbf{y} = \mathbf{X}\boldsymbol{\beta} + \mathbf{e},$$

where $\mathbf{e}$ has mean vector $\mathbf{0}$ and covariance matrix $\boldsymbol{\Sigma}(\boldsymbol{\theta})$ whose elements are known positive definite functions of a parameter vector $\boldsymbol{\theta}$. The joint parameter space for $\boldsymbol{\beta}$ and $\boldsymbol{\theta}$ is taken to be the Cartesian product of $\mathbb{R}^p$ and Θ, where Θ is the set of m-vectors $\boldsymbol{\theta}$ for which $\boldsymbol{\Sigma}(\boldsymbol{\theta})$ is positive definite.

Because the joint distribution of $\mathbf{y}$ is multivariate normal, the likelihood function is given by

$$\ell(\boldsymbol{\beta}, \boldsymbol{\theta}; \mathbf{y}) = (2\pi)^{-n/2} |\boldsymbol{\Sigma}(\boldsymbol{\theta})|^{-1/2} \exp\{-(\mathbf{y} - \mathbf{X}\boldsymbol{\beta})^T [\boldsymbol{\Sigma}(\boldsymbol{\theta})]^{-1} (\mathbf{y} - \mathbf{X}\boldsymbol{\beta})/2\}.$$

Maximum likelihood estimates (MLEs) of $\boldsymbol{\beta}$ and $\boldsymbol{\theta}$ are defined as any values of these parameters (in $\mathbb{R}^p$ and Θ, respectively) that maximize $\ell(\boldsymbol{\beta}, \boldsymbol{\theta}; \mathbf{y})$. Because the log function is monotone increasing, any such values also maximize the **log-likelihood function**,

$$\log \ell(\boldsymbol{\beta}, \boldsymbol{\theta}; \mathbf{y}) = -\frac{n}{2} \log(2\pi) - \frac{1}{2} \log |\boldsymbol{\Sigma}(\boldsymbol{\theta})| - \frac{1}{2}(\mathbf{y} - \mathbf{X}\boldsymbol{\beta})^T [\boldsymbol{\Sigma}(\boldsymbol{\theta})]^{-1} (\mathbf{y} - \mathbf{X}\boldsymbol{\beta}).$$

DOI: 10.1201/9780429060878-8

Furthermore, for any fixed $\boldsymbol{\theta} \in \Theta$, say $\boldsymbol{\theta}_0$, the unique value of $\boldsymbol{\beta}$ that maximizes $\log \ell(\boldsymbol{\beta}, \boldsymbol{\theta}_0; \mathbf{y})$ is that which minimizes

$$(\mathbf{y} - \mathbf{X}\boldsymbol{\beta})^T [\boldsymbol{\Sigma}(\boldsymbol{\theta}_0)]^{-1} (\mathbf{y} - \mathbf{X}\boldsymbol{\beta}).$$

It follows that $\tilde{\boldsymbol{\theta}}$ and $\tilde{\boldsymbol{\beta}}$ are MLEs of $\boldsymbol{\theta}$ and $\boldsymbol{\beta}$ if (and only if) $\tilde{\boldsymbol{\theta}}$ is any value of $\boldsymbol{\theta} \in \Theta$ that maximizes the so-called **profile log-likelihood function**

$$L(\boldsymbol{\theta}; \mathbf{y}) = -\frac{n}{2} \log(2\pi) - \frac{1}{2} \log |\boldsymbol{\Sigma}(\boldsymbol{\theta})| - \frac{1}{2} \mathbf{y}^T \mathbf{E}(\boldsymbol{\theta}) \mathbf{y}, \tag{8.1}$$

and $\tilde{\boldsymbol{\beta}} = \{\mathbf{X}^T [\boldsymbol{\Sigma}(\tilde{\boldsymbol{\theta}})]^{-1} \mathbf{X}\}^{-1} \mathbf{X}^T [\boldsymbol{\Sigma}(\tilde{\boldsymbol{\theta}})]^{-1} \mathbf{y}$. Here,

$$\mathbf{E}(\boldsymbol{\theta}) = [\boldsymbol{\Sigma}(\boldsymbol{\theta})]^{-1} - [\boldsymbol{\Sigma}(\boldsymbol{\theta})]^{-1} \mathbf{X} \{\mathbf{X}^T [\boldsymbol{\Sigma}(\boldsymbol{\theta})]^{-1} \mathbf{X}\}^{-1} \mathbf{X}^T [\boldsymbol{\Sigma}(\boldsymbol{\theta})]^{-1} \tag{8.2}$$

and $L(\boldsymbol{\theta}; \mathbf{y}) = \log \ell(\tilde{\boldsymbol{\beta}}, \boldsymbol{\theta}; \mathbf{y})$. Observe that the MLE of $\boldsymbol{\beta}$ has the form of the GLS estimator, with the unknown parameter $\boldsymbol{\theta}$ replaced by its MLE. It is thus an instance of an empirical GLS estimator and could be represented alternatively as $\hat{\boldsymbol{\beta}}_{EGLS}$.

For all practically useful spatial general linear models, one of the parameters of the covariance structure is a scale parameter, say σ^2, and it is possible to express the covariance matrix as a multiple of this parameter and a matrix that is free of this parameter. That is, the covariance matrix may be written as

$$\boldsymbol{\Sigma}(\boldsymbol{\theta}) = \sigma^2 \mathbf{R}(\boldsymbol{\theta}_{-1}) \tag{8.3}$$

for some matrix $\mathbf{R}(\boldsymbol{\theta}_{-1})$, where $\boldsymbol{\theta} = (\sigma^2, \boldsymbol{\theta}_{-1}^T)^T$, $\boldsymbol{\theta}_{-1} = (\theta_2, \theta_3, \ldots, \theta_m)^T \in \Theta_{-1}$ and Θ_{-1} is defined such that $\Theta = \{\sigma^2 > 0, \boldsymbol{\theta}_{-1} \in \Theta_{-1}\}$. One important case of this occurs, for example, when the process is stationary, for then the observations are homoscedastic and $\boldsymbol{\Sigma}(\boldsymbol{\theta})$ can be expressed in the aforementioned way with σ^2 being the common variance (or equivalently the semivariogram sill) of the observations and $\mathbf{R}(\boldsymbol{\theta}_{-1})$ being a correlation matrix. But in general, the process need not be stationary and $\mathbf{R}(\boldsymbol{\theta}_{-1})$ need not be a correlation matrix for (8.3) to hold. In any case where it does hold, it may be shown that $L(\boldsymbol{\theta}; \mathbf{y})$ assumes its maximum value at $(\tilde{\sigma}^2, \tilde{\boldsymbol{\theta}}_{-1}^T)^T$, where

$$\tilde{\sigma}^2 = \mathbf{y}^T \mathbf{Q}(\tilde{\boldsymbol{\theta}}_{-1}) \mathbf{y} / n$$

and $\tilde{\boldsymbol{\theta}}_{-1}$ is any value of $\boldsymbol{\theta}_{-1} \in \Theta_{-1}$ that maximizes

$$L_{-1}(\boldsymbol{\theta}_{-1}; \mathbf{y}) = -\frac{n}{2} \log(2\pi) - \frac{1}{2} \log |\mathbf{R}(\boldsymbol{\theta}_{-1})| - \frac{n}{2} \log[\mathbf{y}^T \mathbf{Q}(\boldsymbol{\theta}_{-1}) \mathbf{y}]. \tag{8.4}$$

Here

$$\mathbf{Q}(\boldsymbol{\theta}_{-1}) = \sigma^2 \mathbf{E}(\boldsymbol{\theta}) = [\mathbf{R}(\boldsymbol{\theta}_{-1})]^{-1} - [\mathbf{R}(\boldsymbol{\theta}_{-1})]^{-1} \mathbf{X} \{\mathbf{X}^T [\mathbf{R}(\boldsymbol{\theta}_{-1})]^{-1} \mathbf{X}\}^{-1} \mathbf{X}^T [\mathbf{R}(\boldsymbol{\theta}_{-1})]^{-1}.$$

This is an additional instance of profiling, applied in this case to the scale parameter σ^2. The advantage of maximizing $L_{-1}(\boldsymbol{\theta}_{-1}; \mathbf{y})$ rather than $L(\boldsymbol{\theta}; \mathbf{y})$ is that the dimensionality of the maximization problem is reduced from m to $m - 1$. This can yield substantial savings in computation. It is easy to show that $\tilde{\sigma}^2$ is equal to $[(n - p)/n]\hat{\sigma}_{EGLS}^2$ where $\mathbf{R}(\tilde{\boldsymbol{\theta}}_{-1})$ is substituted for $\mathbf{R}(\tilde{\boldsymbol{\theta}})$ in expression (5.12) for $\hat{\sigma}_{EGLS}^2$ (see Exercise 8.1).

The problem of maximizing $L(\boldsymbol{\theta}; \mathbf{y})$ for $\boldsymbol{\theta} \in \Theta$ (or $L_{-1}(\boldsymbol{\theta}_{-1}; \mathbf{y})$ for $\boldsymbol{\theta}_{-1} \in \Theta_{-1}$) is a constrained nonlinear optimization problem for which a closed-form solution exists only in very special cases. In general, therefore, ML estimates must be obtained numerically. Most data analysts will use existing software packages for this purpose. Software packages in R that

obtain maximum likelihood estimates of parameters of several stationary Gaussian geostatistical models include `spmodel` (the `splm` function), `geoR` (the `likfit` function), `fields`
(various functions), and `RandomFields` (the `RFfit` function). Packages that obtain maximum likelihood estimates for Gaussian geostatistical models with convolution-based nonstationary covariance functions (in $\mathbb{R}^2$ only) or with big data, respectively, are `convo-SPAT`
and `ExaGeoStatR`. Spatial autoregressive and moving average models may be estimated via
maximum likelihood using functions `spautolm` and `errorsarlm` in package `spatialreg` or
function `spautor` in `spmodel`.

For those data analysts who opt to (or must) write their own code, the remainder of
this section describes some algorithms for the numerical maximization of the spatial general
linear model's log-likelihood function and many issues associated with that maximization.
Even those analysts who use existing software stand to benefit from a cursory understanding
of these algorithms and issues, if for no other reason than to gain a healthy degree of
skepticism of the results produced by the software. Our description is mostly in terms of
$L(\boldsymbol{\theta}; \mathbf{y})$ but applies equally well to $L(\boldsymbol{\theta}_{-1}; \mathbf{y})$.

Perhaps the simplest (conceptually) numerical approach for maximizing the profile log-
likelihood function is a brute-force **grid search**. This amounts to evaluating the function
at a grid of points overlaid upon the parameter space, Θ. This is often effective when Θ
is low-dimensional, but it becomes computationally prohibitive as the number of parameters increases or as the analyst uses a finer grid mesh. The computational burden arises
primarily because evaluating the profile log-likelihood at the values of $\boldsymbol{\theta}$ on the grid requires repeatedly forming and inverting the covariance matrix (or some related matrix
arising from its Cholesky or singular value decomposition) and evaluating its determinant. In general, these operations require $O(n^3)$ computations, but as noted in Section
5.6, in some practical cases (e.g. a CAR or SAR model, or a geostatistical model with
a separable or compactly supported covariance function) it may be possible to perform
the matrix inversions relatively efficiently. In many of the same cases, the determinant
may be evaluated efficiently also. For example, in the case of a SAR model with co-
variance matrix $\sigma^2(\mathbf{I} - \rho\mathbf{W})^{-1}(\mathbf{I} - \rho\mathbf{W}^T)^{-1}$, the matrix denoted generically by $\mathbf{R}(\boldsymbol{\theta}_{-1})$
in (8.4) may be written as $\mathbf{R}(\rho) = (\mathbf{I} - \rho\mathbf{W})^{-1}(\mathbf{I} - \rho\mathbf{W}^T)^{-1}$, which has determinant
$|\mathbf{I} - \rho\mathbf{W}|^{-2} = [\prod_{i=1}^{n}(1 - \rho\lambda_i)]^{-2}$ where $\lambda_1, \ldots, \lambda_n$ are the eigenvalues of $\mathbf{W}$. The $\{\lambda_i\}$
may be determined once and for all, so that the determinant of $\mathbf{R}(\rho)$ for each value of ρ
encountered in a grid search (or any other algorithm) may be evaluated by squaring and
reciprocating a simple n-fold product of scalars (Ord, 1975).

The primary alternative to a grid search is an **iterative algorithm**, in which a guess
("iterate") of the parameter is repeatedly updated until some stopping criterion is satisfied,
with the final iterate taken as the estimate. One important class of iterative algorithms is
gradient algorithms. In a gradient algorithm, the $(k+1)$st iterate $\boldsymbol{\theta}^{(k+1)}$ is computed by
updating the kth iterate $\boldsymbol{\theta}^{(k)}$ according to the equation

$$\boldsymbol{\theta}^{(k+1)} = \boldsymbol{\theta}^{(k)} + \rho^{(k)}\mathbf{M}^{(k)}\mathbf{g}^{(k)},$$

where $\rho^{(k)}$ is a scalar, $\mathbf{M}^{(k)}$ is an $m \times m$ matrix, and $\mathbf{g}^{(k)}$ is the gradient of $L(\boldsymbol{\theta}; \mathbf{y})$ evaluated
at $\boldsymbol{\theta} = \boldsymbol{\theta}^{(k)}$, i.e. $\mathbf{g}^{(k)} = \partial L(\boldsymbol{\theta}; \mathbf{y})/\partial\boldsymbol{\theta}|_{\boldsymbol{\theta}=\boldsymbol{\theta}^{(k)}}$. The matrix product of $\mathbf{M}^{(k)}$ and $\mathbf{g}^{(k)}$ can be
thought of as defining the search direction for the $(k+1)$st iterate (relative to the kth
iterate), while $\rho^{(k)}$ defines the size of the step to be taken in that direction.

Three gradient algorithms commonly used in conjunction with maximizing a profile log-
likelihood function are the methods of **steepest ascent**, **Newton-Raphson**, and **Fisher
scoring**. In the method of steepest ascent, $\mathbf{M}^{(k)} = \mathbf{I}$ (the $m \times m$ identity matrix) for all
k, implying that each new iterate is obtained by moving away from the previous one in the
direction of greatest increase in the function. In the Newton-Raphson procedure, $\mathbf{M}^{(k)}$ is

the inverse of the $m \times m$ matrix whose (i, j)th element is $-\partial^2 L(\boldsymbol{\theta}; \mathbf{y})/\partial\theta_i\partial\theta_j|_{\boldsymbol{\theta}=\boldsymbol{\theta}^{(k)}}$. In the Fisher scoring algorithm, $\mathbf{M}^{(k)} = (\mathcal{I}^{(k)})^{-1}$ where $\mathcal{I}^{(k)}$ is the **Fisher information matrix** associated with $L(\boldsymbol{\theta}; \mathbf{y})$ evaluated at $\boldsymbol{\theta}^{(k)}$, i.e. $\mathcal{I}^{(k)}$ is the $m \times m$ matrix whose (i, j)th element is $\mathrm{E}[-\partial^2 L(\boldsymbol{\theta}; \mathbf{y})/\partial\theta_i\partial\theta_j|_{\boldsymbol{\theta}=\boldsymbol{\theta}^{(k)}}]$. Thus, Fisher scoring is identical to Newton-Raphson except that the second-order partial derivatives are replaced by their expectations. Expressions for both the second-order partial derivatives and their expectations may be found in Section 8.3. For all three algorithms, the default step size is $\rho^{(k)} = 1$.

In order for a gradient algorithm to be applicable, the first-order partial derivatives of the profile log-likelihood function with respect to the elements of $\boldsymbol{\theta}$ must exist at $\boldsymbol{\theta}^{(0)}$ and at all iterates produced by the algorithm. Moreover, in order to employ the Newton-Raphson or Fisher scoring methods in particular, the second-order partial derivatives must exist at the same values. This is equivalent to requiring that the second-order partial derivatives of the covariance function with respect to the elements of $\boldsymbol{\theta}$ exist at these values. This requirement is ensured for many, but not all, covariance functions; the spherical covariance function provides a counterexample (see Exercise 6.13).

Which gradient algorithm converges most rapidly to a maximum likelihood estimate depends on the situation and is not easy to predict. Because the method of steepest ascent does not require the evaluation of any second-order derivatives or their expectations, it is not as demanding computationally as the other two methods on a per-iteration basis. However, it may require many more iterations to converge than the others. Furthermore, not all potentially useful iterative algorithms are gradient algorithms. A golden section search (Kiefer, 1953) or the Nelder-Mead simplex algorithm (Nelder & Mead, 1965), for example, may also be effective.

Several practical decisions must be made to implement any iterative algorithm. These include choices of starting value(s) for $\boldsymbol{\theta}$, parameterization of the covariance function, stopping rule, and methods for accommodating the constraints on $\boldsymbol{\theta}$, which typically are linear inequality constraints (also known as "box constraints"). For a stationary geostatistical covariance model, one sensible choice of starting value is an estimate of $\boldsymbol{\theta}$ obtained as a result of parametrically smoothing the semivariogram using the method described in Section 3.6.2. For a CAR model with single dependence parameter ρ and binary non-standardized weights, a reasonable starting value for ρ might be the correlation coefficient between neighbor residuals obtained by regressing responses at each pair of neighboring sites on those at the $n - 2$ remaining sites. As a general rule, the closer the starting value is to the maximum of the profile log-likelihood function (assuming such a maximum exists), the faster the algorithm will converge. On the other hand, there are benefits to starting at several widely dispersed values, as described further below.

For some models, it may be possible to improve the performance of an iterative algorithm by reparameterizing the covariance function. Gradient algorithms, in particular, generally perform best when the profile log-likelihood function is approximately a quadratic function of the parameters. Thus, any transformation of the parameters that renders the profile log-likelihood function more nearly quadratic is likely to be helpful. This possibility does not appear to have been fully exploited by existing software packages for likelihood-based estimation of spatial general linear models.

Every iterative algorithm requires a rule for determining when the algorithm should be stopped. A rule based on the relative change in the profile log-likelihood function from one iteration to the next generally is preferable to a rule based on the absolute change in that function, or to a ceiling on the number of iterations. A rule used by some software packages is to stop when $|L(\boldsymbol{\theta}^{(k+1)}; \mathbf{y}) - L(\boldsymbol{\theta}^{(k)}; \mathbf{y})|/|L(\boldsymbol{\theta}^{(k)}; \mathbf{y})| < 1.0 \times 10^{-8}$.

In general, an iterative algorithm can produce iterates (and a final estimate) that lie outside the allowable parameter space. This can cause the algorithm to fail. However, several procedures are available for ensuring that the parameter constraints are respected on each iteration. The simplest but also least widely applicable of these is to transform the parameter(s) so that one or more of its constraints is eliminated. For example, the enforcement of the lower bound on a correlation parameter required to lie in (0,1) may be avoided by representing the same parameter as $\exp(-\theta)$ with new parameter space $(0, \infty)$ (for θ). Two more general approaches for enforcing constraints are interior penalty and partial stepping techniques. The interior penalty method (Carroll, 1961) is applicable when the constraints are of the form $\theta_i[>, \geq]0$ $(i = 1, \ldots, m)$, where $[>, \geq]$ is used to indicate that the equality may or may not be strict. In this case, the constrained maximization of $L(\boldsymbol{\theta}; \mathbf{y})$ is replaced by the unconstrained maximization of

$$L^*(\boldsymbol{\theta}; \mathbf{y}) \equiv L(\boldsymbol{\theta}; \mathbf{y}) - \sum_{i=1}^{m} \phi_i/\theta_i,$$

where $\phi_1, \ldots, \phi_m$ are small positive constants. The idea is that the modified function $L^*(\cdot; \mathbf{y})$ is close to $L(\cdot; \mathbf{y})$ except near the boundary of the parameter space where it assumes very large negative values, which serves to keep iterates from getting too close to the boundary. A partial stepping technique replaces the constant stepsize ($\rho^{(k)} \equiv 1$) with a stepsize sufficiently small that the next iterate remains in the parameter space. Perhaps the simplest form of this is step halving (Jennrich & Sampson, 1976), which consists of using the first step in the sequence $1, 1/2, 1/4, \ldots$ for which the next iterate remains in the parameter space and increases the profile log-likelihood function.

In many standard statistical problems, a unique MLE exists. For spatial linear models, however, there is no guarantee of either existence or uniqueness. Furthermore, the profile log-likelihood function can have many local maxima, any of which an iterative algorithm may converge to instead of the global maximum. Figure 8.1 displays the profile log-likelihood function $L_{-1}(\theta; \mathbf{y})$ for nine randomly generated realizations of $\mathbf{y}$ on an 8×8 square grid with unit spacing, from three constant-mean, unit-variance isotropic Gaussian processes. The covariance functions of all three processes are parameterized in such a way that θ represents the range (for the spherical case) or the practical range (for the other two cases), to facilitate a more direct comparison. The top row corresponds to three realizations of a process with spherical correlation function with a range of 3.0, and the remaining rows correspond to three realizations each from processes with exponential and Gaussian correlation functions for which the practical range is 3.0. All three profile log-likelihoods for the process with spherical covariance function have multiple, widely separated modes, which is consistent with results reported by Warnes & Ripley (1987) and Mardia & Watkins (1989). The profile log-likelihoods corresponding to the other two processes are unimodal, and in each case the mode is not far from the true value of the practical range. These results and the experience of the authors suggest that multiple modes are uncommon in practice for isotropic covariance functions within the Matérn class when the dataset is of a sufficiently large size typical of most applications and the dependence is not too strong relative to the dataset's spatial extent. Not much is known about the existence and extent of multimodality of the profile log-likelihood function for processes with covariance functions other than the spherical or Matérn. In any case, a reasonable practical strategy for determining whether a local maximum obtained by an iterative algorithm is likely to be the unique global maximum is to repeat the algorithm from multiple widely dispersed starting values. If the algorithm obtains the same local maximum regardless of starting value, then the analyst can be reasonably assured (though never 100% certain) that the local maximum is a global maximum.

The computations required to carry out maximum likelihood estimation of the parameters of a spatial general linear model can be quite burdensome. This is especially so when the model has an aggregate-geostatistical covariance structure, for which the elements of

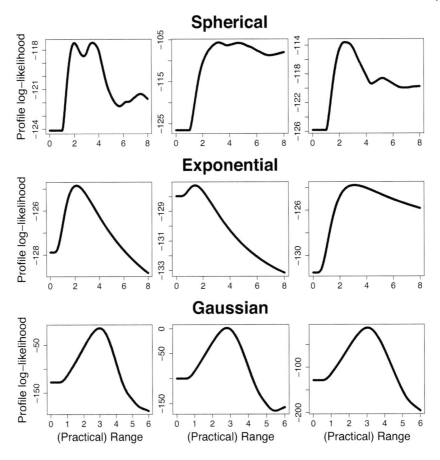

FIGURE 8.1
Profile log-likelihood functions of the range or practical range for realizations (three each) from isotropic, constant-mean, unit-variance Gaussian processes with three correlation functions.

$\mathbf{X}$ and $\boldsymbol{\Sigma}$ in (8.1) must be obtained by integrating the mean and covariance function over data sites, rather than merely evaluating those functions at data sites. In practice, the integration must be done numerically. For obtaining $\boldsymbol{\Sigma}$, the numerical integration must be repeated for each parameter combination in the (grid search or iterative) maximization procedure, whereas for $\mathbf{X}$ it may be carried out just once, due to the linearity of the mean function. Some general strategies for reducing the computational burden of likelihood-based estimation of spatial general linear models are described in Section 8.7.

8.2   REML estimation

Although MLEs of the spatial covariance parameters comprising $\boldsymbol{\theta}$ have several desirable properties, they also have a well-known shortcoming: they are biased as a consequence of the "loss in degrees of freedom" from estimating $\boldsymbol{\beta}$ (Harville, 1977). This bias may be substantial even for moderately sized samples if either the spatial correlation is strong or p (the dimensionality of $\boldsymbol{\beta}$) is large. However, the bias can be reduced substantially, and in some special cases eliminated completely, by employing a variant of maximum likelihood estimation known as **residual** (or restricted) **maximum likelihood (REML) estimation**. REML was originally proposed for use with components-of-variance models by Patterson & Thompson (1971, 1974), then later for use with spatial models by Kitanidis (1983). For both types of models, it is now as popular as maximum likelihood estimation, if not more so.

In REML estimation, the likelihood (or equivalently the log-likelihood) function associated with any set of $n - p$ linearly independent linear combinations of the observations known as **error contrasts**, rather than the likelihood function associated with the observations, is maximized. An error contrast is a linear combination of the observations, i.e. $\mathbf{a}^T\mathbf{y}$, that has expectation zero for all $\boldsymbol{\beta}$ and all $\boldsymbol{\theta} \in \Theta$. For example, if the geostatistical process has constant mean, then the difference of any two observations, or more generally, any linear combination of the observations whose coefficients sum to zero, is an error contrast. Two error contrasts $\mathbf{a}^T\mathbf{y}$ and $\mathbf{b}^T\mathbf{y}$ are said to be linearly independent if $\mathbf{a}$ and $\mathbf{b}$ are linearly independent vectors. Any set of $n - p$ linearly independent elements of $(\mathbf{I} - \mathbf{P_X})\mathbf{y}$ may serve as the required error contrasts for the residual likelihood. The log-likelihood function associated with any such set differs by at most an additive constant (which does not depend on $\boldsymbol{\beta}$ or $\boldsymbol{\theta}$) from the function

$$L_R(\boldsymbol{\theta}; \mathbf{y}) = -\frac{n-p}{2}\log(2\pi) - \frac{1}{2}\log|\boldsymbol{\Sigma}(\boldsymbol{\theta})| - \frac{1}{2}\log|\mathbf{X}^T[\boldsymbol{\Sigma}(\boldsymbol{\theta})]^{-1}\mathbf{X}| - \frac{1}{2}\mathbf{y}^T\mathbf{E}(\boldsymbol{\theta})\mathbf{y}, \quad (8.5)$$

where $\mathbf{E}(\boldsymbol{\theta})$ was defined by (8.2). Observe that $L_R(\boldsymbol{\theta}; \mathbf{y})$ differs from the profile log-likelihood function $L(\boldsymbol{\theta}; \mathbf{y})$ given by (8.1) only additively by an extra term, $-\frac{1}{2}\log|\mathbf{X}^T[\boldsymbol{\Sigma}(\boldsymbol{\theta})]^{-1}\mathbf{X}|$, and by an additive constant that does not depend on $\boldsymbol{\theta}$. A REML estimate (REMLE) of $\boldsymbol{\theta}$ is any value $\tilde{\boldsymbol{\theta}}_R \in \Theta$ at which L_R attains its maximum. This estimate generally must be obtained via the same kinds of numerical procedures used to obtain a MLE, and multimodality is again an issue (Dietrich & Osborne, 1991). Once a REMLE of $\boldsymbol{\theta}$ is obtained, the corresponding estimator of $\boldsymbol{\beta}$ is obtained as the empirical generalized least squares estimator evaluated at $\boldsymbol{\theta} = \tilde{\boldsymbol{\theta}}_R$, i.e.

$$\tilde{\boldsymbol{\beta}}_R = \{\mathbf{X}^T[\boldsymbol{\Sigma}(\tilde{\boldsymbol{\theta}}_R)]^{-1}\mathbf{X}\}^{-1}\mathbf{X}^T[\boldsymbol{\Sigma}(\tilde{\boldsymbol{\theta}}_R)]^{-1}\mathbf{y}.$$

Subsequently, we refer to this estimator of $\boldsymbol{\beta}$ as the REML-EGLS estimator. Also, similar to the situation with the profile log-likelihood function $L(\boldsymbol{\theta}; \mathbf{y})$, when $\boldsymbol{\Sigma}(\boldsymbol{\theta}) = \sigma^2\mathbf{R}(\boldsymbol{\theta}_{-1})$ the residual log-likelihood is maximized at $(\tilde{\sigma}_R^2, \tilde{\boldsymbol{\theta}}_{-1,R}^T)^T$, where

$$\tilde{\sigma}_R^2 = \mathbf{y}^T\mathbf{Q}(\tilde{\boldsymbol{\theta}}_{-1,R})\mathbf{y}/(n-p)$$

and $\tilde{\boldsymbol{\theta}}_{-1,R}$ is any value of $\boldsymbol{\theta}_{-1} \in \Theta_{-1}$ that maximizes

$$\begin{aligned}L_{-1,R}(\boldsymbol{\theta}_{-1}; \mathbf{y}) &= -\frac{n-p}{2}\log(2\pi) - \frac{1}{2}\log|\mathbf{R}(\boldsymbol{\theta}_{-1})| - \frac{1}{2}\log|\mathbf{X}^T[\mathbf{R}(\boldsymbol{\theta}_{-1})]^{-1}\mathbf{X}| \\ &\quad - \frac{n-p}{2}\log(\mathbf{y}^T\mathbf{Q}(\boldsymbol{\theta}_{-1})\mathbf{y}).\end{aligned}$$

The divisor $(n - p)$ in the REML estimator of σ^2 differs from that (n) in the MLE. This reveals one way in which REML estimation accounts for the loss of p degrees of freedom incurred by estimating the mean structure. Moreover, it is clear that $\tilde{\sigma}_R^2$ coincides with $\hat{\sigma}_{EGLS}^2$ where $\mathbf{R}(\tilde{\boldsymbol{\theta}}_{-1,R})$ is substituted for $\mathbf{R}(\tilde{\boldsymbol{\theta}})$ in (5.12).

TABLE 8.1
Empirical bias and mean-squared error of maximum likelihood (ML) and residual
maximum likelihood (REML) estimators of the variance (σ^2) and correlation
parameter (ρ) of an exponential covariance function (parameterized in
power-correlation form), for three mean structures: an intercept-only model, row effects
only, and both row and column effects.

Mean effects	bias($\tilde{\sigma}^2$) ML	REML	bias($\tilde{\rho}$) ML	REML	MSE($\tilde{\sigma}^2$) ML	REML	MSE($\tilde{\rho}$) ML	REML
constant	−0.049	0.007	−0.038	−0.014	0.041	0.054	0.009	0.009
row	−0.204	0.030	−0.096	−0.014	0.068	0.164	0.019	0.012
row-column	−0.360	0.618	−0.183	−0.012	0.146	295.725	0.045	0.018

Most R packages that can be used to obtain MLEs of spatial linear models may also be
used to obtain REMLEs, a notable exception being `spatialreg`.

Does REML effectively reduce the bias incurred by maximum likelihood estimation of
the covariance parameters of spatial linear models? And does it improve covariance parameter estimation overall, in terms of mean squared error? Two published simulation studies
have compared the performance of ML and REML estimators of covariance parameters for
such models. An early study by Zimmerman & Zimmerman (1991) for isotropic Gaussian
processes with constant mean and nuggetless exponential covariance function parameterized
as

$$C(r; \sigma^2, \rho) = \sigma^2 \rho^r$$

indicated that in this case the MLE of the correlation decay parameter (ρ) was negatively
biased, while the REMLE was nearly unbiased. Furthermore, the MLE and REMLE of the
sill (σ^2) were both biased, but in opposite directions, and the positive bias of the REMLE
was much more extreme than the negative bias of the MLE. In fact, the distribution of
the sill's REMLE had a quite heavy upper tail. In a subsequent study, Irvine *et al.* (2007)
compared the performance of MLEs and REMLEs for isotropic Gaussian processes with
constant mean and exponential-plus-nugget covariance functions parameterized in terms of
a sill, nugget-to-sill ratio, and practical range. For the practical range (results on the sill,
nugget effect, or their ratio were not included) they obtained results broadly similar to those
of Zimmerman & Zimmerman (1991) but somewhat less favorable to REML; for example,
the upper tail of the empirical distribution of the REMLE of the practical range was, in many
cases, so heavy that this estimator's mean squared error was orders of magnitude larger
than that of the MLE. The heavy-upper-tail problem became worse as either the practical
range or the nugget-to-sill ratio increased. Thus, the REMLEs of one or more covariance
parameters in the models considered by these studies had a heavy upper tail. On the other
hand, it must be noted that both studies considered only processes with constant means,
i.e. $p = 1$. In light of the aforementioned rationale for REML, we might expect the relative
performance of REML to ML estimation to improve as the dimensionality of $\boldsymbol{\beta}$ increases.
Table 8.1 contains results of a small simulation-based investigation of this question for an
example of an isotropic unit-variance Gaussian process that has a nuggetless exponential
correlation function, which is parameterized in the same way as in the Zimmerman &
Zimmerman (1991) study, with $\rho = 0.5$. The process was observed on a 12×12 square grid
with unit spacing. Three mean structures were considered: the first was, as in the previous
studies, merely an overall constant (for which $p = 1$), the second consisted of row effects (for
which $p = 12$) and the third was an additive row-column coordinate effects structure (for
which $p = 23$). REMLEs and MLEs were obtained using functions in the `spmodel` package.
The table gives the empirical biases and mean squared errors of the MLEs and REMLEs of

TABLE 8.2
Empirical average length, length[CI($\boldsymbol{\beta}$)], and coverage, cover[CI($\boldsymbol{\beta}$)], of nominal 90% confidence intervals for the elements of $\boldsymbol{\beta}$ when using maximum likelihood (ML) and residual maximum likelihood (REML) estimators for an exponential covariance function (parameterized in power-correlation form), for the same three mean structures listed in Table 8.1.

	length[CI(β)]		cover[CI(β)]	
Mean effects	**ML**	**REML**	**ML**	**REML**
constant	0.741	0.805	0.823	0.850
row	1.609	1.928	0.822	0.887
row-column	1.384	2.011	0.750	0.884

σ^2 and ρ over 1000 simulations for each mean structure. These results indicate that in this case REML is indeed superior to ML for estimating ρ, but contrary to expectation, REML is inferior to ML for estimating σ^2 (with respect to both bias and mean squared error), drastically so for the model with row and column effects. If ρ is changed to $\sqrt{0.5}$ (so that the correlation is 0.5 at distance 2.0 rather than 1.0), REML is inferior to ML for estimating both σ^2 and ρ. On the basis of these results, it appears that, insofar as the estimation of covariance parameters is concerned, REML is *not* generally superior to ML.

Of course, the parameters of the covariance function are not the only objects of inferential interest for a spatial linear model. For example, the regression coefficients will generally also be of interest, indeed often more so. Both ML and REML estimators of the regression coefficients are unbiased (Kackar & Harville, 1981). However, ML- and REML-based confidence intervals for the regression coefficients might differ. Table 8.2 gives the average lengths and empirical coverage probabilities of two-sided nominal 90% confidence intervals for the regression coefficients, obtained by "plugging" the MLE and REMLE of $\boldsymbol{\theta}$ into $\hat{\sigma}^2_{GLS}$ and $\mathbf{R}$ in (5.5), using the same simulations that were used to produce the results for Table 8.1. These results clearly show that the the plug-in intervals using the MLEs are often too narrow and therefore undercover. The plug-in intervals using the REMLE frequently are too narrow also, but they are wider than their MLE-based counterparts and have better coverage probabilities overall, as the large positive bias of the REMLE of σ^2 helpfully counteracts the inherent underestimation of the variance of the empirical GLS estimator by the plug-in estimator. Thus, insofar as inference on the regression coefficients is concerned, REML estimation appears to perform better than ML estimation. The reader is encouraged to do some additional investigations of the relative performance of ML and REML estimation of covariance and regression parameters in Exercise 8.4. A comparison of the quality of predictive inferences based on ML and REML estimation is also important, but is deferred to Chapter 9.

Although the performance of REML estimation for the estimation of the variance in spatial general linear models is disappointing, it does have one distinct advantage over ML estimation: it may be used to estimate parameters of Gaussian intrinsic random fields of order k (IRF-k's). Thus, in particular, it may be used to estimate the parameters of an unbounded semivariogram model. Recall from Section 2.2 that an IRF-k is characterized by the probabilistic structure of certain linear combinations known as generalized increments; more specifically, the kth-order generalized increments of a Gaussian IRF-k are jointly Gaussian with mean zero and a generalized covariance function. The generalized covariance function does not specify a covariance structure for the observations themselves, but only for the generalized increments. It turns out that generalized increments for any Gaussian IRF-k coincide with error contrasts for a Gaussian process with kth-order polynomial mean structure, so that REML estimation may proceed for such processes; owing to the lack of a fully specified covariance structure for the observations, however, ML estimation cannot. For example, REML, but not ML, may be used to estimate the power semivariogram (listed in Table 6.3) of an isotropic IRF-0. The REML log-likelihood in this case is

$$L_R(\boldsymbol{\theta}; \mathbf{y}) = -\frac{n-1}{2}\log(2\pi) - \frac{1}{2}\log|\mathbf{A}^T\boldsymbol{\Gamma}(\boldsymbol{\theta})\mathbf{A}| - \frac{1}{2}\mathbf{y}^T\mathbf{A}[\mathbf{A}^T\boldsymbol{\Gamma}(\boldsymbol{\theta})\mathbf{A}]^{-1}\mathbf{A}^T\mathbf{y},$$

where $\mathbf{A}$ is the $(n-1) \times n$ matrix whose ith row has ith element 1, $(i+1)$st element -1, and zeroes elsewhere; and $\boldsymbol{\Gamma}(\boldsymbol{\theta})$ is the $n \times n$ matrix with (i,j)th element $\gamma(\mathbf{s}_i - \mathbf{s}_j; \boldsymbol{\theta}) = \sigma^2 \|\mathbf{s}_i - \mathbf{s}_j\|^{\theta_2}$.

In general, REML estimation of a generalized covariance function, like ML estimation of a covariance function, is a challenging computing problem. Also like ML estimation, however, a significant reduction in computation can be achieved in some practically important special cases. Barendregt (1987) and Zimmerman (1989a) described some of those cases and the computational reductions possible for each. The power semivariogram and other members of the class of polynomial generalized covariance functions given by (6.15) are among those cases.

8.3 Asymptotic regimes and results

The finite-sample distributions of ML and REML estimators of the parameters of spatial linear models generally are intractable. Therefore, in order to construct confidence intervals and make other inferences about the parameters, we resort to using the asymptotic distributions of the estimators. For continuous Gaussian spatial processes, there are two (at least) quite different "regimes" under which asymptotic properties of the ML and REML estimators could be considered: **increasing-domain asymptotics**, in which the minimum distance between data locations is bounded away from zero and thus the spatial domain of observation is unbounded in the limit, and **fixed-domain** (also called infill) **asymptotics**, in which observations are taken ever more densely within a fixed and bounded domain (Figure 8.2). The average distance between sites tends to increase under the increasing-domain regime but decrease under the fixed-domain regime, with implications for the degree of spatial dependence, in some aggregate sense, that exists among the observations. "Mixed" regimes in which the minimum intersite distance decreases to zero and the domain of observation increases without bound (possibly at different rates) could also be imagined, but most investigations of asymptotic properties of spatial statistics have assumed either the increasing-domain or fixed-domain regime.

For an areal model, it is possible to imagine analogs of the two main regimes: an increasing-domain regime in which more sites of roughly the same size are added, thereby expanding their total spatial coverage without bound, and a fixed-domain regime in which the existing sites are repeatedly partitioned in such a way that their maximum size decreases to zero. However, if spatial weights are based solely upon contiguity, these two analogs could yield essentially the same spatial dependence structure. For example, if the sites form a square grid and the spatial weights matrix $\mathbf{W}$ is that which corresponds to the rook's definition of contiguity, the SAR model's covariance matrix is invariant to the size of the grid cells and thus as the number of grid cells increases the degree of aggregate spatial dependence is reduced (hence aligning more with the increasing-domain regime), regardless of whether the additional grid cells are the result of adding more of the same size or partitioning existing ones. It is only when the spatial weights are at least partly distance-based that the two regimes are distinct.

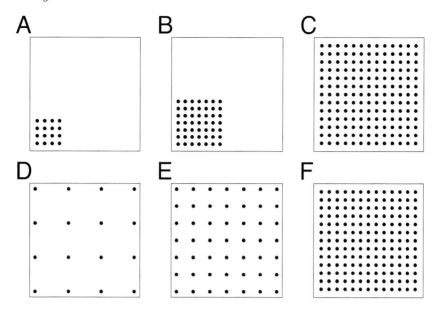

FIGURE 8.2
Depiction of two asymptotic regimes for geostatistical data: A,B,C) increasing-domain regime; D,E,F) fixed-domain regime. The squares containing the data locations represent a spatial domain of constant area.

Within the increasing-domain framework, it is known that, under certain regularity conditions, the MLEs of parameters $\boldsymbol{\beta}$ and $\boldsymbol{\theta}$ of geostatistical models are consistent (i.e., they converge in probability to the corresponding true values) and asymptotically normally distributed, with covariance matrices given by $\{\mathbf{X}^T[\boldsymbol{\Sigma}(\boldsymbol{\theta})]^{-1}\mathbf{X}\}^{-1}$ and $[\mathcal{I}(\boldsymbol{\theta})]^{-1}$, respectively, where $\mathcal{I}(\boldsymbol{\theta})$ is the **(expected) Fisher information matrix** for the covariance parameters, defined by

$$
\begin{aligned}
\mathcal{I}(\boldsymbol{\theta}) &= \left(-\mathrm{E}\left(\frac{\partial^2 L(\boldsymbol{\theta};\mathbf{y})}{\partial\theta_i\,\partial\theta_j}\right)\right) \\
&= \left(\frac{1}{2}\mathrm{tr}\left(\boldsymbol{\Sigma}^{-1}\frac{\partial\boldsymbol{\Sigma}}{\partial\theta_i}\boldsymbol{\Sigma}^{-1}\frac{\partial\boldsymbol{\Sigma}}{\partial\theta_j}\right)\right)
\end{aligned}
\tag{8.6}
$$

(Mardia & Marshall, 1984). Furthermore, the MLEs of $\boldsymbol{\beta}$ and $\boldsymbol{\theta}$ are asymptotically independent. Thus, the asymptotic covariance matrix of the MLE of $\boldsymbol{\beta}$ is the same as the finite-sample covariance matrix of the GLS estimator, while that of the MLE of $\boldsymbol{\theta}$ is the same matrix that multiplies the gradient vector in the method of scoring as it applies to the profile log-likelihood function. Both are relatively easy to evaluate for all of the covariance functions in Tables 6.1 and 6.2 except the generalized Wendland and Matérn models; however, the results of De Oliveira & Han (2022) make the task surmountable even for the Matérn model. To illustrate, in the case of an exponential model with nugget effect, $\boldsymbol{\Sigma} = \sigma^2\mathbf{R}(\alpha,\theta_0) = \sigma^2[\mathbf{A}(\alpha) + \theta_0\mathbf{I}]$, where the (i,j)th element of $\mathbf{A}(\alpha)$ is $\exp(-r_{ij}/\alpha)$, and thus $\mathcal{I}(\boldsymbol{\theta})$ is given by (8.6) where

$$
\begin{aligned}
\partial\boldsymbol{\Sigma}/\partial\sigma^2 &= \mathbf{R}(\alpha,\theta_0), \\
\partial\boldsymbol{\Sigma}/\partial\alpha &= \sigma^2\frac{\partial}{\partial\alpha}\mathbf{A}(\alpha) = \sigma^2\mathbf{D}\circ\mathbf{A}(\alpha)/\alpha^2, \\
\partial\boldsymbol{\Sigma}/\partial\theta_0 &= \sigma^2\mathbf{I},
\end{aligned}
$$

$\mathbf{D}$ is the $n\times n$ matrix of intersite distances, and $\circ$ is the Hadamard (elementwise) product.

The expressions for the elements of the expected information matrix given by (8.6) are considerably simpler than those for the elements of the observed information matrix,

$$\mathcal{I}_O(\boldsymbol{\theta}) = -\left(\frac{\partial^2 L(\boldsymbol{\theta}; \mathbf{y})}{\partial \theta_i \, \partial \theta_j}\right),$$

which we do not provide here. We use only the former matrix in what follows, although the latter matrix is asymptotically equivalent.

Under the same framework and similar regularity conditions, the REML estimator of $\boldsymbol{\theta}$ is also consistent and asymptotically normal with covariance matrix given by $[\mathcal{I}_R(\boldsymbol{\theta})]^{-1}$, where

$$\begin{aligned}\mathcal{I}_R(\boldsymbol{\theta}) &= \left(-\mathrm{E}\left(\frac{\partial^2 L_R(\boldsymbol{\theta}; \mathbf{y})}{\partial \theta_i \, \partial \theta_j}\right)\right) \\ &= \left(\frac{1}{2}\mathrm{tr}\left(\mathbf{E}\frac{\partial \boldsymbol{\Sigma}}{\partial \theta_i}\mathbf{E}\frac{\partial \boldsymbol{\Sigma}}{\partial \theta_j}\right)\right)\end{aligned} \qquad (8.7)$$

(Cressie & Lahiri, 1996), where $\mathbf{E}$ was defined in (8.2). Again, this information matrix is relatively easy to evaluate for most covariance functions.

Analogous results hold, again under suitable regularity conditions, for the ML and REML estimators of parameters of areal models (Ord, 1975; Lee, 2004); see Exercises 8.2 and 8.3.

The aforementioned asymptotic results may be used to construct (increasing-domain) asymptotically valid confidence intervals for covariance parameters having any desired coverage probability $1 - \epsilon$. The confidence intervals are of the form

$$\text{Estimator} \pm [z_{\epsilon/2} \times (\text{Estimated asymptotic variance of estimator})^{1/2}],$$

where $z_{\epsilon/2}$ is the upper $\epsilon/2$ percentage point of the standard normal distribution and the estimated asymptotic variance of the estimator is the appropriate main diagonal element of $[\mathcal{I}(\boldsymbol{\theta})]^{-1}$ evaluated at the MLE (or of $[\mathcal{I}_R(\boldsymbol{\theta})]^{-1}$ evaluated at the REMLE).

The regularity conditions required for the aforementioned asymptotic properties to hold are rather technical and will not be given here, but it is worth noting that they require the covariance function to tend to 0 as distance increases in every direction (for a geostatistical model) or the sum of spatial weights for a site to be uniformly bounded across sites (for an areal model), among other requirements. All but one of the covariance functions listed in Tables 6.1 and 6.2 satisfy the first of these conditions; the lone exception is the cosine function. Another regularity condition for geostatistical models requires the second-order partial derivatives of the covariance function to exist and be continuous over the entire parameter space for $\boldsymbol{\theta}$ — a property that holds for the Matérn class (and many others) but not for the triangular, circular, and spherical covariance functions.

The available results under fixed-domain asymptotics are considerably more limited. Moreover, they suggest that the asymptotic behavior of MLEs can be quite different in this framework than in the increasing-domain framework. For example, for a one-dimensional, zero-mean, stationary and isotropic Gaussian process with nuggetless exponential covariance function, the MLEs of the process variance (σ^2) and range parameter (α) are not consistent under fixed-domain asymptotics, although their quotient is consistent for the quotient of those two parameters (σ^2/α) (Ying, 1991). This result extends to the entire class of Matérn covariance functions in dimensions 1, 2, or 3 (for the quantity $\sigma^2/\alpha^{2\nu}$) (Zhang, 2004) and also to the generalized Wendland family of functions (for a similar quantity) (Bevilacqua *et al.*, 2019). A simulation-based demonstration of this phenomenon was given by Zhang (2004), and another is provided in Table 8.3. The demonstration features a one-dimensional

TABLE 8.3

Empirical mean-squared errors of maximum likelihood estimates of parameters of a spatial linear model under fixed-domain and increasing-domain asymptotic regimes.

	Fixed-domain			Increasing-domain		
n	σ^2	α	σ^2/α	σ^2	α	σ^2/α
50	0.280	0.0128	1.45	0.261	0.0130	3.28
250	0.243	0.0099	0.20	0.071	0.0032	0.25
1000	0.256	0.0102	0.05	0.020	0.0009	0.06

stationary Gaussian process with known mean 0 and nuggetless exponential covariance function

$$C(r; \sigma^2, \alpha) = \sigma^2 \exp(-r/\alpha) \quad \text{for} \quad r \geq 0,$$

simulated at points $1/(n+1), 2/(n+1), \ldots, n/(n+1)$ on $(0,1)$, where $n \in \{50, 250, 1000\}$, $\sigma^2 = 1$, and $\alpha = 0.2$. The data locations for different n conform to a fixed-domain asymptotic regime. MLEs and REMLEs of each parameter and their quotient, σ^2/α, were computed for each of 1000 simulated realizations of this process, and the empirical mean squared error was determined for each parameter estimate. Furthermore, the entire experiment was repeated at the 250 points $1/51, 2/51, \ldots, 250/51$ and the 1000 points $1/51, 2/51, \ldots, 1000/51$, which are representative of an increasing-domain asymptotic regime. Only the results for the MLEs are shown in Table 8.3, as the REMLEs exhibited very similar asymptotic behavior. The results comport with the aforementioned fact that σ^2/α is consistently estimable under both asymptotic regimes while σ^2 and α individually are consistently estimable under only the increasing-domain regime.

Typically in practice, spatial data are observed at a finite number of sites (points or subregions) with no intention or possibility of taking more observations, and it is not clear which asymptotic regime is the most appropriate one to appeal to for asymptotic standard errors of the parameter estimates. One rationale for choosing between the two regimes is how well the asymptotic distributions of the MLEs or REMLEs of parameters of interest approximate the finite-sample distributions of those estimators. For the same process as the one featured in Table 8.3 and for one with a nugget effect but otherwise identical, Zhang & Zimmerman (2005) found that the approximations to the MLE's finite-sample distribution provided by the two asymptotic frameworks perform about equally well for those parameters for which the MLE is consistent under both frameworks, but the finite-sample approximation provided by the fixed-domain framework performs better for those parameters that cannot be estimated consistently under fixed-domain asymptotics. It seems, therefore, that it would be best to base inferences on fixed-domain asymptotic results, if they were available. But unfortunately, such results are not available for most important practical cases, such as models in two or more dimensions with unknown mean. Estimated standard errors and confidence limits of MLEs and REMLEs provided by existing software packages are based on increasing-domain asymptotics, and should therefore be "taken with a grain of salt."

8.4 Information content and interplay among parameter estimates

In principle, the (expected) Fisher information matrix can reveal how much information the data have for estimating some model parameters, relative to others. Specifically, the

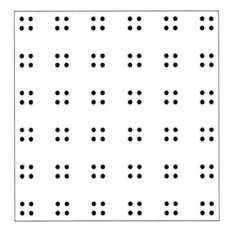

FIGURE 8.3
Spatial configuration of data sites for Case IV of the illustration of information content in Section 8.4.

reciprocals of the elements along the main diagonal of the inverse information matrix are the inverses of the Cramér-Rao lower bound for the variances of unbiased estimators of the corresponding parameters, and may thus be regarded as measures of the information content that the data have about the corresponding parameters (De Oliveira & Han, 2022). These quantities may be compared to each other, but the comparisons are meaningful only for parameters that are commensurate (measured in the same units) or unitless.

To illustrate, consider the information matrix associated with maximum likelihood estimation of a geostatistical model in $\mathbb{R}^2$ with constant mean and isotropic exponential covariance function with nugget effect,

$$C(r; \sigma_1^2, \sigma_0^2, \alpha) = \begin{cases} \sigma_1^2 + \sigma_0^2 & \text{if } r = 0, \\ \sigma_1^2 \exp(-r/\alpha) & \text{otherwise.} \end{cases}$$

Here σ_1^2 is the partial sill, σ_0^2 is the nugget effect, and α is the range parameter. The units of the two variance components, σ_1^2 and σ_0^2, are the squared units of the observations, and the units for α are the units of distance. Since the variance components are commensurate, the reciprocals of the main diagonal elements of the inverse of the information matrix corresponding to them may be compared meaningfully. Of course, the results of these comparisons may depend on the spatial configuration of the data sites, the range parameter, and the values of the variance components themselves.

Table 8.4 gives numerical values of the information content of the MLEs of σ_1^2 and σ_0^2 for a benchmark case (Case I) of data on a 12×12 square grid with unit spacing, with parameters set to $\sigma_1^2 = 0.75$, $\sigma_0^2 = 0.25$, and $\alpha = 2$. Values of the same quantities are also shown for three other cases: Case II is identical to Case I except that the values of σ_1^2 and σ_0^2 are interchanged; Case III is identical to Case I except that $\alpha = 3$; and Case IV is identical to Case I except for the spatial configuration of data sites, which are obtained by shifting each site in the 12×12 grid by $1/4$ unit to the left or right and by $1/4$ unit up or down, as shown in Figure 8.3, to create a more clustered pattern of sites.

The cases presented in Table 8.4 are not sufficiently extensive to draw many strong general conclusions, except possibly for the conclusion that if changing the value of a parameter while keeping the spatial configuration unchanged results in greater information content for one of the variance components, then the information content for the other tends to decrease.

TABLE 8.4

Information content on, and asymptotic correlations between, maximum likelihood estimators of covariance parameters, for four cases of the example described in Section 8.4.

	Info. content		Asymptotic correlations		
Case	σ_1^2	σ_0^2	(σ_1^2, σ_0^2)	(σ_1^2, α)	(σ_0^2, α)
I	16.7	43.0	-0.51	0.06	0.71
II	31.6	32.2	-0.72	-0.52	0.68
III	12.5	94.1	-0.24	0.47	0.60
IV	18.2	105.2	-0.34	0.32	0.60

From the comparison of Case IV to Case I, it also appears that a more clustered spatial configuration leads to an increase in the information content for both variance components.

For the more general Matérn model with unknown smoothness parameter, De Oliveira & Han (2022) made similar comparisons of the information content on the smoothness parameter (which is unitless) and a scaled, unitless version of the range parameter. Their key finding was that the information about the smoothness parameter can be large, even larger than that about the range parameter, thus contradicting unfounded claims in the literature that geostatistical data provide very little information about the smoothness parameter. This suggests that the practice of fixing the smoothness parameter and estimating only the remaining parameters, which has been recommended by some authors, is unwarranted.

The information matrix is useful not only for what it can tell us about relative information content on the parameters of a spatial linear model, but also for what it reveals about the interplay between estimates of different parameters. Specifically, this interplay may be quantified by the correlations in the asymptotic correlation matrix among the parameter estimates, obtained from the inverse of the information matrix by dividing each of its elements by the square root of the product of the appropriate main diagonal elements. To illustrate, consider again the example with four cases defined above. Table 8.4 gives the asymptotic correlations between MLEs of each pair of parameter estimates. Again, owing to the limited breadth of the cases considered, definitive conclusions are probably not justified, but in all four cases the partial sill and nugget effect were negatively related, while the range and nugget effect were positively related. These relationships agree with intuition. Because the variance of each observation is $\sigma_1^2 + \sigma_0^2$ and this quantity should be estimated well by the sample variance of the observations, if one variance component is overestimated it is natural to expect the other to be underestimated. And, because the correlation between observations at a fixed distance can be estimated reasonably well by the sample correlation between such observations, if the range parameter is overestimated (implying stronger correlation) it is reasonable to expect the nugget effect to compensate by also being overestimated (implying weaker correlation).

De Oliveira & Han (2022) conducted a similar study of the interplay between parameter estimates for the more general Matérn model with unknown smoothness parameter. Perhaps the most interesting of their findings was the strong positive association between the estimates of the nugget effect and the smoothness parameter; that is, the more smooth the process is estimated to be, the larger the estimated nugget effect. This makes sense intuitively; as stated by De Oliveira & Han (2022), "Discordant observations collected at close by locations might be explained either as coming from a non-smooth random field model or due to the presence of measurement error. If the smoothness were to be fixed at a value that is higher than the one supported by the data, an overestimation of the nugget effect would result to compensate. The opposite effect is expected when the smoothness is fixed at a value that is too low." In the example presented by those authors, there was also a strong association, this one negative, between the estimates of the range and smoothness parameters. The intuition for this is similar, albeit in the opposite direction, as discordant observations at proximate locations could also be explained by a small range. If the smoothness were fixed at a value too large for the data, the estimate of the range parameter would need to be smaller than it should be, in order to compensate.

8.5   Hypothesis testing and model selection

Typically, there is not just one, but many spatial linear models under consideration for the data at hand. For example, there may be several explanatory variables that have been measured, and one wishes to determine which of them, if any, are important in explaining the response. While such a determination could be (and often has been) made in the context of a nonspatial model using ordinary least squares regression methods, ignoring spatial correlation to make this determination is unwise as it may lead to the selection of too few explanatory variables and larger prediction errors (Hoeting *et al.*, 2006). It is generally better to determine which explanatory variables are important in the context of a model with spatial dependence (assuming, of course, that indeed there is residual spatial dependence in the data). There may be several candidate models for the residual spatial dependence as well. For example, in a geostatistical situation where the spatial dependence appears to be stationary, isotropic, and monotone decreasing with increasing distance, a method for choosing among various reasonable covariance functions (spherical, exponential, Gaussian, etc.) would be useful. Or, for areal data one may wish to compare a CAR with first- and second-order spatial neighborhoods to one with merely first-order neighborhoods. So methods are needed for comparing models on the basis of both their mean structure and covariance structure. Depending on the nature of the models being compared and the goals of the investigator, the comparisons may take the form of a statistical hypothesis test, or they may be based on less formal criteria.

When two models are nested, i.e., when one model is a special case of the other, the models may be compared formally using a **likelihood ratio test**. Typically, this amounts to comparing twice the difference in the maximized log-likelihood functions corresponding to the two models to percentiles of the chi-square distribution with q degrees of freedom, where q is the difference in dimensionality of the parameter spaces for the two models. If the test statistic exceeds the $100(1 - \epsilon)$ percentile of the appropriate chi-square distribution, then the P-value for the test of the smaller, "reduced" model against the larger, "full" model is less than ϵ, and the smaller model is rejected in favor of the larger model at the ϵ level of significance. The success of this testing procedure is predicated on the assumption that twice the negative log of the likelihood ratio test statistic is asymptotically distributed, under the null hypothesis, as a chi-squared random variable with q degrees of freedom. This assumption is justified under increasing-domain asymptotics and appropriate regularity conditions, but perhaps not otherwise.

The likelihood ratio testing approach just described may be used to compare nested models that differ with respect to their mean structures, their covariance structures, or both their mean and covariance structures. A variant based on twice the difference in the maximized REML log-likelihoods is also asymptotically valid (subject to the caveat on asymptotic regime and regularity conditions), but only for comparing nested models with different covariance structures. Nested models with different mean structures should not be compared using this variant because error contrasts for the two models are different functions of the data.

Any number of non-nested spatial linear models may be compared informally within the (standard) likelihood framework using a **penalized likelihood criterion**. There are several such criteria, of which the two most frequently used are Akaike's Information Criterion

$$AIC = -2\log \ell(\tilde{\boldsymbol{\beta}}, \tilde{\boldsymbol{\theta}}; \mathbf{y}) + 2(p+m),$$

and Schwartz's Bayesian Information Criterion,

$$BIC = -2\log \ell(\tilde{\boldsymbol{\beta}}, \tilde{\boldsymbol{\theta}}; \mathbf{y}) + (p+m)\log n.$$

A penalized likelihood criterion balances model fit, as measured by minus twice the maximized log-likelihood function, against model complexity, as measured by a penalty term [$2(p+m)$ in AIC and $(p+m)\log n$ in BIC]. The model with the smallest value of the criterion is judged to be best. Because $\log n > 2$ when $n > 8$, in many practical cases BIC tends to favor more parsimonious models than AIC. Models with different mean structures or different covariance structures, or both, may be compared, though it is not universally accepted that penalizing mean and covariance parameters equally (as these criteria do) is appropriate. Furthermore, for spatial linear models the use of penalized likelihood criteria, like likelihood ratio testing, is perhaps only justifiable within an increasing-domain framework (Huang & Chen, 2007).

Still another issue pertains to BIC only. The multiplier, $\log n$, in the penalty term of BIC can be justified theoretically only for independent observations. In more general settings, including correlated data, the appropriate penalty term for BIC appears to be the log of the determinant of the observed Fisher information matrix (Kass & Raftery, 1995; Neath & Cavanaugh, 2012). However, simulation studies of the choice of BIC's penalty term for various spatial linear models are not (yet) available, so the "jury is still out" on this issue.

Penalized likelihood criteria may also be used to compare non-nested models within a REML framework. Analogs of AIC and BIC for this framework are

$$AIC_R = -2L_R(\tilde{\boldsymbol{\theta}}_R) + 2m$$

and

$$BIC_R = -2L_R(\tilde{\boldsymbol{\theta}}_R) + m\log n.$$

Many authors have argued that these criteria, like the formal likelihood-based testing procedure for comparing nested models within a REML framework, should be used only to compare models that differ with respect to their covariance structures. Surprisingly, however, Gurka (2006) demonstrated that AIC_R and BIC_R, with m replaced by $m+p$, may be reasonably effective for comparing non-nested models with different covariance structures *and* different mean structures. Nevertheless, because none of Gurka's covariance structures were spatial, we do not consider the issue to have been settled for spatial models, so we recommend continuing the standard practice of comparing models with different mean structures using ML estimation only until stronger evidence to suggest otherwise becomes available.

For areal data, SAR, CAR, and SMA models with specified spatial weights matrices may be compared using penalized likelihood criteria. Within any one of these models, the criteria may also be used to compare different neighborhood structures, different methods for standardizing the weights, etc.

Of the software packages in R that perform likelihood-based estimation of spatial linear models, `spmodel` provides AIC and AIC_R, but also $\log \ell(\tilde{\boldsymbol{\beta}}, \tilde{\boldsymbol{\theta}}; \mathbf{y})$ and $\log L_R(\tilde{\boldsymbol{\theta}}_R; \mathbf{y})$, from which it is easy to compute BIC and BIC_R. Another penalized likelihood criterion, known as corrected AIC, is also available in `spmodel`. Furthermore, `likfit` in `geoR` provides AIC and BIC; `RFfit` in `RandomFields` provides those two plus the corrected AIC; and `spautolm` and `errorsarlm` in `spatialreg` provide AIC (only).

There are many ways to try to create a parsimonious model when there are multiple explanatory variables for the mean structure. One way is to add each explanatory variable, one at a time, and then use P-values, *AIC*, *BIC*, a prediction-based criterion called leave-one-out cross-validation (LOOCV) to be introduced in Section 9.10, or any of several other criteria available in the literature to decide which term to keep. Then, we can consider adding a second term, removing the one already added from the candidate pool. This process can proceed through the interaction terms as well, and is known as a forward selection procedure.

We could also take the opposite approach, and start with all terms in the model, then remove the one with the highest P-value, or use *AIC*, *BIC*, LOOCV, etc., to eliminate the worst model after removing each term one at a time. This process can proceed, generally removing the higher-order interaction terms first, and continuing to the main effects. This is called a backward selection procedure. Various modifications are possible, such as reinstating some previously eliminated terms, one at a time, at each step, because P-values and likelihood-based criteria can change depending on the particular set of terms in the model. For example, it is entirely possible for a term to be removed at one stage of the model-building process, only to find that, if it re-enters the model later, it has a significant P-value that suggests it should be included in the model.

Forward and backward selection procedures do not examine all possible combinations of explanatory variables, and the order of selection may in fact miss the single best model. Thus, rather then going through a series of models sequentially, another approach is to simply try all possible combinations of explanatory variables, and again use a whole-model criteria, such as *AIC*, *BIC*, LOOCV, etc, to choose the single best model. The problem with this approach is that it can be computationally expensive, especially for spatial models that may take extra time to fit when sample sizes are large. It is also possible to try all possible spatial correlation models in Tables 6.1 and 6.2, but doing so would require a lot of model-fitting, especially if taken in combination with every possible mean structure.

Yet another idea is to keep several more-or-less equivalent models rather than selecting a single one, which has been called "multi-model inference" (Burnham & Anderson, 2002). This could be especially useful for prediction, where the predictions from each model can be averaged. This has been called "model averaging" (e.g. Hoeting *et al.*, 1999). The use of multiple models make the predictions robust to model misspecification and is intuitively appealing. On the other hand, the use of multiple models can be confusing if we are interested in fixed effects and covariance parameters, as it may be difficult to summarize their meaning across models, and fitting many models can be very costly in terms of time, especially if one wants to do model diagnostics, etc., which require some human interpretation (Ver Hoef & Boveng, 2015).

8.6   Robustness to model misspecification

How robust are the likelihood-based estimators of spatial linear model parameters to model misspecification? More specifically, how robust are estimators of the mean structure parameters to misspecification of the covariance structure, and how robust are estimators of the covariance structure parameters to misspecification of the mean structure? Let us address the second question first. The toy examples displayed in Figures 3.13 and 4.3, and the data analyses presented later in this chapter, indicate that the best-fitting covariance structure for a dataset can change dramatically when, for example, the fitted mean structure is changed from a polynomial trend surface of one order to a polynomial trend surface of another order. This suggests that estimators of the covariance structure (and therefore also

TABLE 8.5

Empirical average length, length[CI(β)], and empirical coverage probability, coverage[CI(β)], of nominal 90% confidence intervals for the elements of β for simulated data having three mean structures and three covariance models, which are estimated using REML-EGLS, acting as though each of the covariance models is the true model, and using OLS (fitted model "nocorr"). Columns under length[CI(β)] and cover[CI(β)] correspond to the three true models: spherical ("sph"), exponential ("exp"), and Gaussian ("Gau"). The fitted models have the same designations, apart from "nocorr."

Mean effects	Fitted model	length[CI(β)]			coverage[CI(β)]		
		sph	exp	Gau	sph	exp	Gau
constant	sph	0.606	0.692	0.542	0.906	0.808	0.879
	exp	0.980	0.798	1.010	0.982	0.867	0.990
	Gau	0.698	0.678	0.732	0.954	0.820	0.967
	nocorr	0.270	0.264	0.271	0.542	0.379	0.579
row	sph	1.855	1.990	1.690	0.890	0.898	0.861
	exp	2.242	1.922	2.335	0.927	0.890	0.950
	Gau	2.278	2.231	2.285	0.951	0.940	0.965
	nocorr	1.238	1.202	1.245	0.727	0.699	0.739
row-column	sph	1.874	1.903	1.684	0.890	0.867	0.856
	exp	2.636	2.025	2.935	0.944	0.879	0.964
	Gau	2.216	2.144	2.228	0.943	0.924	0.951
	nocorr	1.198	1.154	1.214	0.705	0.669	0.726

conclusions drawn about the nature and strength of the spatial dependence) are not very robust to mean structure misspecification. To address the first question, a simulation study similar to the one presented in Section 8.2 was conducted. In this study, data were simulated at the nodes of a 12×12 square, unit-spaced grid in $\mathbb{R}^2$ from models with constant, row-effects, and additive row-column-effects mean structures and three isotropic nuggetless covariance functions having variance 1.0. The three covariance functions were the spherical, exponential, and Gaussian models. To make the strength of correlation comparable across models, the range parameters were set to values that yield a correlation of 0.5 at unit distance. One thousand data sets were simulated from each model. REML-EGLS estimates of mean parameters were obtained for each simulated dataset under the covariance model that generated it and under the two other covariance models, with empirical average length and coverage probability of the nominal 90% confidence interval for the elements of β (over the 1000 simulations) used to quantify the quality of estimation (as was done previously in Table 8.2). For comparison, OLS estimates were also obtained (corresponding to a model with no correlation). Results are given in Table 8.5. These show that the confidence intervals for the elements of β, when obtained from fits of the spatially correlated models, are usually narrowest for the fitted spherical model, widest for the fitted exponential model, and intermediate for the fitted Gaussian model, regardless of the true model and the mean structure. The coverage probabilities of these intervals mostly follow this same ordering, ranging from roughly 78% to 99%. The intervals obtained via OLS are much narrower than even those for the spherical model, and the corresponding coverage probabilities are much smaller than those corresponding to the spatially correlated models. On the whole, these results suggest that while it is probably a stretch to say that REML-EGLS inference for mean parameters is robust to covariance model misspecification, the quality of those inferences is not terrible, provided that a model allowing for positive spatial correlation is fitted. In contrast, results obtained using OLS are really quite bad.

TABLE 8.6
Empirical bias of maximum likelihood estimators of the variance (σ^2) and unit-distance correlation (UDC) for simulated data having three mean structures and three covariance models, which are estimated acting as though each of the covariance models is the true model. Models are designated as described in the caption of Figure 8.5.

Mean effects	Fitted model	bias($\tilde{\sigma}^2$)			bias($\widetilde{UDC}$)		
		sph	exp	Gau	sph	exp	Gau
constant	sph	0.015	0.178	−0.083	0.006	0.014	−0.045
	exp	0.043	−0.054	0.063	0.024	−0.039	0.029
	Gau	0.796	0.821	0.591	−0.029	−0.060	0.051
row	sph	−0.089	0.032	−0.178	−0.007	−0.020	−0.057
	exp	−0.113	−0.202	−0.087	−0.017	−0.096	−0.002
	Gau	0.661	0.688	0.447	−0.037	−0.074	0.043
row-column	sph	−0.197	0.032	−0.279	−0.028	−0.020	−0.073
	exp	−0.273	−0.358	−0.235	−0.080	−0.185	−0.048
	Gau	0.529	0.525	0.318	−0.052	−0.100	0.035

The same nontrivial covariance models fitted to obtain Table 8.5 may also be used to investigate the robustness of the estimation of selected covariance attributes to covariance model misspecification. Specifically, for those same models, maximum likelihood estimates of the process variance and the unit-distance correlation were obtained for each simulated dataset under the model that generated it and also under the two other models, with empirical bias and mean squared error of the estimates (over the 1000 simulations) used to quantify the quality of estimation (as was done previously in Table 8.1). Maximum likelihood, rather than REML, estimation was used here because of the large MSEs for REML estimates of σ^2, even when obtained using the true model (see Section 8.2). Results are given in Tables 8.6 and 8.7. For purposes of interpretation of these results, it is useful to note that the spherical and exponential models are more similar to each other than either is to the Gaussian model, as neither is differentiable at the origin while the Gaussian is infinitely differentiable there. Perhaps not surprisingly, therefore, the estimation bias is quite similar, for both parameters, when the true covariance function is spherical and the fitted covariance function is exponential, or vice versa. When the fitted model is Gaussian, however, the bias of the MLE of σ^2 is much larger. Interestingly, the bias of the MLE of σ^2 is much smaller when the fitted model is spherical or exponential than when it is Gaussian, even when the true model is Gaussian. Taken together, these results suggest that maximum likelihood estimation of important attributes of an isotropic covariance structure may be quite robust to covariance model misspecification, provided that the smoothness of the model at the origin is not grossly overspecified.

Koch *et al.* (2020) conducted an analogous simulation study to compare the robustness of isotropic and product-anisotropic covariance functions. Interestingly, they found that a product-anisotropic Matérn covariance function fit much better to data generated from a process with isotropic Matérn covariance function than an isotropic Matérn covariance function fit to data generated from a process with a product-anisotropic Matérn covariance function. This finding, plus the greater computational efficiency of the product-anisotropic functions for gridded data, led them to argue for the general use of product-anisotropic functions rather than isotropic and geometrically anisotropic functions when the data lie on a grid.

TABLE 8.7
Empirical mean-squared error (MSE) of maximum likelihood estimators of the variance (σ^2) and unit-distance correlation (*UDC*) for simulated data having three mean structures and three covariance models, which are estimated acting as though each of the covariance models is the true model. Models are designated as described in the caption of Figure 8.5.

Mean effects	Fitted model	MSE($\tilde{\sigma}^2$)			MSE($\widetilde{UDC}$)		
		sph	exp	Gau	sph	exp	Gau
constant	sph	0.036	0.127	0.074	0.004	0.013	0.009
	exp	0.052	0.038	0.049	0.006	0.008	0.005
	Gau	0.735	0.777	0.437	0.002	0.007	0.003
row	sph	0.039	0.094	0.104	0.005	0.018	0.012
	exp	0.049	0.069	0.042	0.007	0.018	0.005
	Gau	0.534	0.566	0.293	0.003	0.009	0.003
row-column	sph	0.068	0.094	0.145	0.007	0.018	0.016
	exp	0.100	0.147	0.085	0.015	0.045	0.009
	Gau	0.361	0.352	0.188	0.006	0.015	0.002

8.7 Estimation for large spatial data sets

Environmental datasets with thousands, even millions, of observations are becoming more common. The computations required to carry out exact likelihood-based estimation of spatial linear models for datasets this large can be prohibitively burdensome, regardless of whether a grid search or an iterative algorithm is used to maximize the likelihood function. Most of the burden, as noted previously, is associated with evaluating (repeatedly) the inverse and determinant of the model's covariance matrix (or matrices derived from it), which is of order n^3. A few scenarios in which these evaluations can be performed more efficiently were described in Section 5.6. Those scenarios involve either highly structured spatial configurations of sites (e.g., nodes of a square grid), certain properties of the covariance function (e.g., separability or compact support), or very specific models (e.g., CAR models). This section describes some additional strategies for reducing computations, all of which are based on approximations to the likelihood function that are more easily maximized than the likelihood itself. Our description is not exhaustive, however, and the development of new strategies is an active area of research.

A number of strategies for reducing computations may be gathered under the heading of **data partitioning**. As the name implies, these strategies are based on a partitioning of the observation vector $\mathbf{y}$ into subvectors $\mathbf{y}_1, \ldots, \mathbf{y}_b$. Letting $\mathbf{y}_{(j)} = (\mathbf{y}_1^T, \ldots, \mathbf{y}_j^T)^T$, $p(\mathbf{y}; \boldsymbol{\beta}, \boldsymbol{\theta})$ denote the joint density of $\mathbf{y}$, and $p(\mathbf{y}_j | \mathbf{y}_{(j-1)}; \boldsymbol{\beta}, \boldsymbol{\theta})$ denote the conditional density of $\mathbf{y}_j$ given $\mathbf{y}_{(j-1)}$, the (exact) likelihood may be written as

$$p(\mathbf{y}; \boldsymbol{\beta}, \boldsymbol{\theta}) = p(\mathbf{y}_1; \boldsymbol{\beta}, \boldsymbol{\theta}) \prod_{j=2}^{b} p(\mathbf{y}_j | \mathbf{y}_{(j-1)}; \boldsymbol{\beta}, \boldsymbol{\theta}). \tag{8.8}$$

Vecchia (1988) proposed to approximate (8.8) by replacing each complete conditioning vector $\mathbf{y}_{(j-1)}$ with a subvector $\mathbf{u}_{(j-1)}$ of $\mathbf{y}_{(j-1)}$ so that the matrices whose inverses and determinants must be evaluated to compute the conditional densities are much smaller. Since there is no unique way to partition $\mathbf{y}$ in this scheme, the question arises as to how $\mathbf{y}_j$ and $\mathbf{u}_{(j-1)}$ should be chosen. Vecchia considered vectors $\mathbf{y}_j$ of length one only, ordered by the values of either of the two coordinate axes of data locations, and he recommended choosing $\mathbf{u}_{(j-1)}$ to consist of the q most proximate observations to $\mathbf{y}_j$, where q is much smaller than n. The smaller the value of q, of course, the smaller the computational burden but the cruder the approximation to the true likelihood. For datasets of size $n = 100$ simulated from several Gaussian processes in $\mathbb{R}^2$, Vecchia showed that taking $q = 10$ resulted in good approximations; in practice, of course, an appropriate choice of q will depend on the range of spatial dependence relative to the distance between observations.

Stein *et al.* (2004) extended Vecchia's original proposal by allowing the $\mathbf{y}_j$'s to be of non-uniform (and non-unity) lengths, and applied it to the residual likelihood function rather than the ordinary likelihood function. For simulated Gaussian data of size $n = 1000$ in $\mathbb{R}^2$, they found that taking $q = 16$ was sufficient for the (statistical) efficiencies of the approximate REML estimators, relative to the exact REML estimators, to be in the range 80–95%. In contrast to Vecchia, however, they found that it was sometimes advantageous to include some observations in $\mathbf{u}_{(j-1)}$ that were rather distant (spatially) from $\mathbf{y}_j$. This was particularly true when the spatial dependence was strong relative to the extent of the spatial domain of observation, a situation that was not represented in any of Vecchia's examples. Both Vecchia's and Stein et al.'s methods can reduce the computational burden from $O(n^3)$ to $O(n)$ if the conditioning sets are chosen to be sufficiently small.

Caragea & Smith (2007) proposed three additional variations on data partitioning. In the "big blocks" method, the likelihood is approximated by the joint density of the b block mean responses. This aggregates all the information in a block to a single variate, but allows for dependence among the block means. In the "small blocks" method, more widely known as the **independent blocks method**, the likelihood is approximated by the b-fold product of joint k-dimensional densities for the observations in a block, thus treating the blocks as if they were independent (and the covariance matrix as if it were block diagonal). Finally, the "hybrid" method approximates the likelihood by the big blocks likelihood multiplied by a b-fold product of densities, where each term in the product is the joint density of k observations in a block, conditional on the block mean. The approximate likelihoods produced by all three of these methods are computable in $O(n^2)$ computations if b is chosen appropriately [specifically, if it grows at a rate between $O(n^{1/2})$ and $O(n^{2/3})$]. On the basis of running times and simulation studies, Caragea & Smith (2008) recommended the hybrid method over the other two. They also found the hybrid method to perform better than Vecchia's method and, in some cases, Stein et al.'s method.

The independent-blocks method is merely one instance of a general approximate likelihood-based estimation strategy known as **composite likelihood** (Besag, 1975; Lindsay, 1988). Curriero & Lele (1999) considered a particular composite likelihood estimation scheme. For an intrinsically stationary Gaussian process, they formed a composite log-likelihood function from the marginal densities of all pairwise differences among the observations. Ignoring constant terms, this function is

$$CL(\boldsymbol{\theta}; \mathbf{y}) = -\frac{1}{2} \sum_{i=1}^{n-1} \sum_{j>i} \left\{ \log[\gamma(\mathbf{s}_i - \mathbf{s}_j; \boldsymbol{\theta})] + \frac{(y_i - y_j)^2}{2\gamma(\mathbf{s}_i - \mathbf{s}_j; \boldsymbol{\theta})} \right\}, \tag{8.9}$$

where $\gamma(\cdot; \boldsymbol{\theta})$ is the semivariogram of the process. Maximization of $CL(\boldsymbol{\theta}; \mathbf{y})$ with respect to $\boldsymbol{\theta}$ yields a composite MLE. Note that by considering only pairwise differences, the unknown constant mean parameter is eliminated from this composite log-likelihood, which is similar in spirit to REML. The composite MLE is consistent and asymptotically normal under certain regularity conditions within an increasing-domain asymptotic framework, and it turns out that this consistency holds even if the process is not Gaussian. This idea can be extended to Gaussian IRF-ks also.

Bevilacqua *et al.* (2012) proposed a weighted version of Curriero and Lele's composite likelihood criterion function, which downweights or removes altogether those summands in (8.9) for which $\|\mathbf{s}_i - \mathbf{s}_j\|$ is sufficiently large. They also introduced a second type of weighted estimator called a weighted composite score estimator and showed that for large datasets both weighted estimation methods provide an excellent tradeoff between statistical and computational efficiency when compared to REML estimation.

Ver Hoef *et al.* (2023) proposed a variation on the independent blocking idea. They maximized the likelihood for a model with an independent-blocks form of the covariance matrix, but for estimating the covariance matrix of the MLE of $\boldsymbol{\beta}$ based on that likelihood they used the actual covariance matrix of $\mathbf{y}$ rather than its block diagonal approximation, thereby accounting for the correlations between blocks for the estimation of uncertainty.

A quite different strategy for approximating the likelihood is **covariance tapering** (Kaufman *et al.*, 2008). In this approach, elements of the covariance matrix corresponding to spatially distant pairs of observations are deliberately set to zero so that algorithms for sparse matrix inversion and determinant evaluation may be used. This cannot be done arbitrarily, however, or else it could result in a covariance function that is not positive definite. Suppose that $Y(\cdot)$ is isotropic with covariance function $C_0(r; \boldsymbol{\theta})$, and let $\rho_T(r; \tau)$, where τ is specified, be a "tapering function," i.e. an isotropic continuous correlation function that is identically zero whenever $r \geq \tau$. Then define

$$C_1(r; \boldsymbol{\theta}, \tau) = C_0(r; \boldsymbol{\theta})\rho_T(r; \tau) \quad \text{for} \quad r \geq 0.$$

Plainly, $C_1(\cdot)$ has the same variance as $C_0(\cdot)$ and preserves some of its shape; moreover, by (6.3), $C_1(\cdot)$ is a valid (positive definite) covariance function. Furthermore, Kaufman et al. showed that if $C_0(\cdot)$ belongs to the Matérn family of covariance functions, then a tapering function may be found for which $C_1(\cdot)$ has the same behavior near the origin as $C_0(\cdot)$. Now, if $C_1(\cdot)$ was actually the covariance function of $Y(\cdot)$, then the log-likelihood function corresponding to the vector of observations $\mathbf{y}$ would be

$$\log \ell_{1T}(\boldsymbol{\beta}, \boldsymbol{\theta}; \mathbf{y}) = -\frac{n}{2}\log(2\pi) - \frac{1}{2}\log|\boldsymbol{\Sigma}(\boldsymbol{\theta}) \circ \mathbf{T}(\tau)| - \frac{1}{2}(\mathbf{y} - \mathbf{X}\boldsymbol{\beta})^T[\boldsymbol{\Sigma}(\boldsymbol{\theta}) \circ \mathbf{T}(\tau)]^{-1}(\mathbf{y} - \mathbf{X}\boldsymbol{\beta}),$$

where $\mathbf{T}(\tau)$ is the matrix with (i, j)th element $\rho_T(\|\mathbf{s}_i - \mathbf{s}_j\|; \tau)$ and "$\circ$" refers to the Hadamard (elementwise) product of two matrices. Kaufman et al. considered a so-called "one-taper estimator" obtained by maximizing this approximation to the actual log-likelihood, and showed via simulation that it performs quite well, though it is slightly biased. As a less biased alternative, they proposed another estimator, the "two-taper estimator," which is based on the theory of unbiased estimating equations. The approximation to the log-likelihood function that yields this estimator is

$$\log \ell_{2T}(\boldsymbol{\beta}, \boldsymbol{\theta}; \mathbf{y}) = -\frac{n}{2}\log(2\pi) - \frac{1}{2}\log|\boldsymbol{\Sigma}(\boldsymbol{\theta}) \circ \mathbf{T}(\tau)| - \frac{1}{2}(\mathbf{y} - \mathbf{X}\boldsymbol{\beta})^T\{[\boldsymbol{\Sigma}(\boldsymbol{\theta}) \circ \mathbf{T}(\tau)]^{-1} \circ \mathbf{T}\}(\mathbf{y} - \mathbf{X}\boldsymbol{\beta}).$$

Again using simulation, Kaufman et al. showed that the two-taper estimator performs even better (is less biased) than the one-taper estimator, especially when the degree of tapering is severe (i.e., when τ is small); however, it is not as computationally efficient. Larger values of τ result in smaller bias and variance for both estimators, so as a practical matter it is best to choose τ to be the largest possible value for which the computations are feasible. Also, it turns out that, under a Matérn model at least, the variances of the tapered estimators are on a par with that of the ordinary MLE, and information-based estimation of these variances performs about as well as it does for the ordinary MLE.

Another distinct class of methods for dealing with large spatial datasets approximates the covariance matrix of the spatial process by that of a random effects model

$\mathbf{y} = \mathbf{X}\boldsymbol{\beta} + \mathbf{Z}\mathbf{b} + \mathbf{e}$, i.e., as

$$\boldsymbol{\Sigma}(\boldsymbol{\theta}) = \mathbf{Z}\mathbf{G}(\boldsymbol{\theta})\mathbf{Z}^T + \mathbf{D}(\boldsymbol{\theta}),$$

where $\mathbf{Z}$ is an $n \times r$ matrix, $\mathbf{G}(\boldsymbol{\theta})$ is an $r \times r$ nonnegative definite matrix, and $\mathbf{D}(\boldsymbol{\theta})$ is a positive definite diagonal matrix for all $\boldsymbol{\theta} \in \Theta$. Here r is relatively small and fixed as n is allowed to increase, hence such methods are called **low-rank process approximation methods**. They include fixed-rank kriging (FRK) (Cressie & Johannesson, 2008), Gaussian predictive process modeling (Banerjee *et al.*, 2008), and multiresolution Gaussian process modeling (Nychka *et al.*, 2015), which differ with respect to how $\mathbf{Z}$ and $\mathbf{G}$ are modeled. FRK models the ith row of $\mathbf{Z}$ as a vector of r basis functions evaluated at $\mathbf{s}_i$ $(i = 1, \ldots, n)$, which attempt to capture different scales of spatial dependence, and imposes no structure other than positive definiteness on $\mathbf{G}$. Gaussian predictive process modeling takes $\mathbf{G}$ to be the covariance matrix of the (partially) observed process, but evaluated at a fixed set of r points, called knots, where the process may or may not be observed, and then takes the ith row of $\mathbf{Z}$ to be the vector whose inner product with the values of the process at the knots yields the conditional mean of y_i given those values. The Gaussian predictive process is therefore an orthogonal projection of the original process onto an r-dimensional subspace. The multiresolution Gaussian process model takes the rows of $\mathbf{Z}$ to be vectors of basis functions, similar to FRK, and takes $\mathbf{G}$ to be the covariance matrix of a Gaussian SAR model. The covariance matrix corresponding to each low-rank approximation method may be inverted efficiently using (5.11), and the determinant of this covariance matrix may also be efficiently evaluated using the matrix determinant lemma,

$$|\mathbf{Z}\mathbf{G}(\boldsymbol{\theta})\mathbf{Z}^T + \mathbf{D}(\boldsymbol{\theta})| = |\mathbf{G}(\boldsymbol{\theta})||\mathbf{D}(\boldsymbol{\theta})||[\mathbf{G}(\boldsymbol{\theta})]^{-1} + \mathbf{Z}^T[\mathbf{D}(\boldsymbol{\theta})]^{-1}\mathbf{Z}|.$$

There are many other methods for dealing with large spatial datasets. We will merely list three more, with references: Markov random field approximations (Rue & Tjelmeland, 2002; Lindgren *et al.*, 2011), spectral approximations (Fuentes, 2007), and nearest-neighbor Gaussian process approximations (Datta *et al.*, 2016a,b).

Given the large (possibly overwhelming!) variety of approaches for reducing the computational burden of likelihood-based estimation of spatial linear models, which should a data analyst use? Each method sacrifices statistical efficiency for computational feasibility: data partitioning does so by ignoring information about the covariances between some observations, while covariance tapering and low-rank process approximation methods do so by adopting slightly or substantially different models for the observations. Stein (2014) showed that there is a class of processes for which any low-rank process approximation method yields much smaller expected log-likelihoods than, and is thus inferior to, the independent blocks approximation method. Heaton *et al.* (2019), in a comparative study of many methods, also found the low-rank process approximation methods to perform less well, although he did not account for computing time in the comparisons. Ver Hoef *et al.* (2023) demonstrated that the independent blocks approximation method with variance adjustment performed as well as a nearest-neighbor Gaussian process approximation approach, both in terms of computing time and estimation/prediction accuracy. There being no evidence, then, that the independent blocks approximation method is non-competitive, it is currently our recommended choice, and it is available in `spmodel`.

8.8   Examples

8.8.1   The wet sulfate deposition data

We now illustrate the methodology presented in this chapter by fitting spatial general linear models to each of the four example data sets introduced in Chapter 1. We begin with the

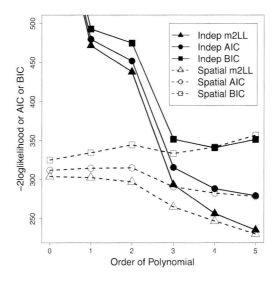

FIGURE 8.4

Two times the minimized negative log-likelihood (m2LL, triangles), *AIC* (circles), and *BIC* (squares) for polynomial models of the spatial coordinates, for both independent error models (solid lines and solid shapes) and spatial models (dashed line and open shapes), for the clean wet sulfate deposition data.

wet sulfate deposition data. Exploratory data analyses scattered throughout Chapter 3 suggested that it would be wise to transform these data to the square-root scale and to set aside three outliers, yielding a modified dataset that was called the clean wet sulfate deposition data. Polynomials of orders zero through five were fitted to these modified data by OLS in Section 4.5.1, and summary statistics from those fits suggested that polynomials up to third-order (and possibly higher) were worthy of further consideration. The analyses presented in Section 4.5.1 also suggested that spatial correlation exists and is isotropic in the residuals from a quadratic fit, but perhaps does not exist in the residuals from cubic and higher-order fits.

Here, we fit polynomial models to the clean wet sulfate deposition data once again, but this time using likelihood-based methods. Models having polynomial mean structures up to (and including) order five and having several geostatistical covariance functions were estimated by ML and REML, using the R package `spmodel`. This package also returns *AIC* and the minimized negative log-likelihood, from which we computed *BIC*. For example, Figure 8.4 shows $-2\log\ell(\tilde{\boldsymbol{\beta}},\tilde{\boldsymbol{\theta}})$ (which we denote as m2LL), *AIC*, and *BIC* computed for all polynomial models on the two spatial coordinates, from the constant mean model up to a fifth-order polynomial, for both independent error models and spatially-correlated error models, where the spatial correlation is modeled by an isotropic exponential covariance function with nugget effect. Note that m2LL decreases as the order of the polynomial increases, as it must, because the models are nested. *AIC* and *BIC* are greater than m2LL due to the penalties they impose on the number of parameters, which is readily seen in Figure 8.4. We generated a similar figure using the complete dataset (not shown). This figure was quite similar overall, except that all of the values were approximately 50 larger on the vertical axis. The negative log-likelihoods were much larger because the three outliers add a lot of variance to the overall fits. For example, REML fits of the third-order polynomial for the complete data resulted in a partial sill estimate of 1.693, while for the cleaned data it was only 0.463. Similarly, the estimated nugget effect for the complete data was 0.220,

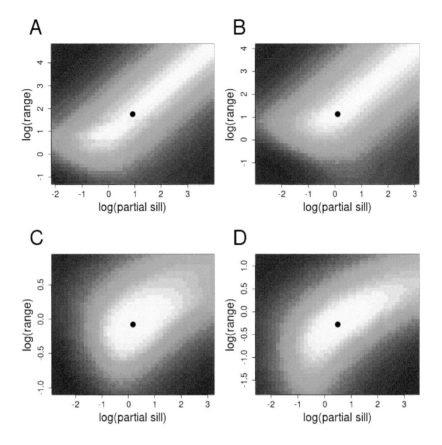

FIGURE 8.5
Residual log-likelihood surface for the logarithm of the partial sill, $\log(\sigma^2 - \sigma_0^2)$, by the logarithm of the range parameters, $\log \alpha$, while holding the nugget effect, σ_0^2, constant at its REML estimate, for the clean wet sulfate deposition data. A) Circular model; B) spherical model; C) Gaussian model; D) gravity model. Yellower colors correspond to larger likelihood values, and bluer colors correspond to smaller likelihood values. All models had constant mean. The REML estimates are shown as solid black circles.

while for the cleaned data it was 0.118. The estimated range was also smaller for the cleaned data. Finally, the spatial R^2 for the complete data was 0.123, while it was 0.176 for the cleaned data. The better fit of the cleaned data, and smaller variances, result in more precise estimates and prediction, so we will continue to use the cleaned data for the rest of this example, and also when we return to it in Chapter 9 for prediction.

For both the independent models and spatial models, *AIC* generally decreases most rapidly from the second- to the third-order polynomial, and continues to decrease, but more slowly, for higher-order polynomials. Because *AIC* has a reputation for favoring overly complex models, we also consider *BIC*. While *AIC* suggests that the fifth-order polynomial is best for both the independence and spatial models, *BIC* suggests that either a fourth-order polynomial is best if the errors are assumed to be independent, or a constant-mean model is best (the third-order model is a close second) if the errors are spatially correlated. We will consider model selection for these data in more detail after we introduce prediction in Chapter 9.

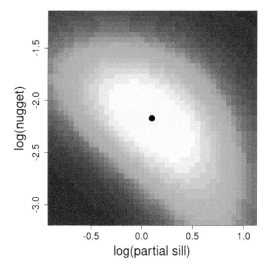

FIGURE 8.6
Residual log-likelihood surface for the logarithm of the partial sill, $\log(\sigma^2 - \sigma_0^2)$, by the logarithm of the nugget effect, $\log(\sigma_0^2)$, while holding the range parameter constant at its REML estimate for a constant-mean model with spherical covariance function, for the clean wet sulfate deposition data. The REML estimate is shown as a solid black circle.

Fitting spatial models is more complicated than fitting models with independent errors, of course. One feature of almost all covariance models listed in Tables 6.1 and 6.2 is that there is "correlation" in the likelihood between the partial sill and the range parameter. Figure 8.5 demonstrates this for the residual log-likelihood corresponding to a constant-mean model with circular, spherical, Gaussian, and gravity covariance models. In fact, recall from Section 8.3 that, for the case of Matérn models at least, maximum likelihood estimators of the process variance, σ^2, and range parameter, α, are inconsistent under fixed-domain asymptotics, but the MLE of their ratio, σ^2/α, *is* consistent. This difficulty in estimating the variance and range manifests most vividly in the long ridges seen in Figure 8.5AB. Values of the residual log-likelihood are nearly equal along these ridges, and, in fact, practitioners of geostatistics are well aware that optimization algorithms do not always converge as the range and partial sill both increase or both decrease along such a ridge. However, as will be noted in the next chapter, predictions are largely insensitive to the values of σ^2 and α themselves, but are sensitive to their ratio. Thus, it matters little where estimates settle on ridges such as those seen in Figure 8.5AB because the ratio remains relatively constant along them. Practitioners of geostatistics often set a bound on one of these parameters to force convergence, realizing that little changes in the log-likelihood by letting the range and partial sill continue to change during optimization.

It is also instructive to examine how the likelihood changes as the partial sill and nugget effect vary, while holding the range constant. The residual log-likelihood surface corresponding to the spherical model, holding the range constant, is given in Figure 8.6. Notice that there is an inverse relationship between the estimated partial sill and nugget effect. This makes sense because the overall variation in the data is constant, regardless of how much is attributed to the nugget effect and how much to the partial sill. Thus, as one goes up the other must go down, as expressed in the surface shown in Figure 8.6. This phenomenon was also discussed in Section 8.4.

TABLE 8.8
Estimated covariance parameters using REML, and corresponding
estimated standard errors obtained using the Fisher information
matrix, for the clean wet sulfate deposition data.

Model	Estimate			Standard error		
	psill	nugget	range	psill	nugget	range
exponential	2.653	0.111	4.598	6.127	0.027	11.030
spherical	1.091	0.115	3.085	0.134	0.039	0.458
gaussian	1.197	0.190	0.920	0.437	0.061	0.346
gravity	1.624	0.170	0.761	0.797	0.021	0.170
rquad	1.285	0.173	0.960	0.593	0.021	0.184
magnetic	1.205	0.175	1.140	0.541	0.021	0.200
circular	2.405	0.113	5.453	5.117	0.026	11.508

The Fisher information matrix associated with REML may be used to construct confidence intervals for the covariance parameters, as suggested in Section 8.3. We fitted constant mean models by REML estimation, and the estimated values of covariance parameters for various covariance models along with their estimated standard errors based on the Fisher information matrix are given in Table 8.8.

It is interesting to examine not only the asymptotic variances of the elements of the REMLE of $\boldsymbol{\theta}$, but also the asymptotic correlations among them. For the circular model in Table 8.8, the asymptotic correlation matrix, with parameters in the order partial sill, nugget effect, and range, is

$$\begin{pmatrix} 1.0000 & -0.0758 & 0.9932 \\ -0.0758 & 1.0000 & -0.0020 \\ 0.9932 & -0.0020 & 1.0000 \end{pmatrix}.$$

Here, as in Figure 8.5A, we can see the strong correlation between the partial sill and the range, and, as in Figure 8.6, the weaker, but still evident, negative correlation between the partial sill and the nugget effect.

The preceding has demonstrated the value of examining the residual log-likelihood surface by plotting two parameters at a time to look for multimodality and noteworthy relationships. The asymptotic covariance matrix of the covariance parameter estimates revealed the same features and can be used to construct confidence intervals. Another way to assess covariance parameter estimation is to use profile likelihood. Consider the residual log-likelihood function $L_{-i,R}(\theta_i; \tilde{\boldsymbol{\theta}}_{-i,R}, \tilde{\boldsymbol{\beta}}_R, \mathbf{y})$, where the ith component of $\boldsymbol{\theta}$ has been held constant at θ_i, and the function has been maximized over all other parameters, whose maximizing values are denoted as $\tilde{\boldsymbol{\theta}}_{-i,R}$ and $\tilde{\boldsymbol{\beta}}_R$. Then a profile likelihood plot for the ith component of $\boldsymbol{\theta}$ is one which plots $L_{-i,R}(\theta_i; \tilde{\boldsymbol{\theta}}_{-i,R}, \tilde{\boldsymbol{\beta}}_R, \mathbf{y})$ for various values of θ_i.

Let us consider the Matérn covariance function, which is very popular because of its additional parameter that controls smoothness and differentiability of the associated Gaussian process. In Table 6.2, this parameter was denoted as ν, and it was called the smoothness parameter. Recall that if $\nu = 1/2$, the Matérn model reduces to the exponential model; if $\nu = 1$ it reduces to what has been called the Whittle model; if $\nu = 3/2$ it is the Radon model of order 2; if $\nu = 5/2$ it is the Radon model of order 4; and the limiting case, as $\nu \to \infty$, is the Gaussian model. In `spmodel`, the smoothness parameter is allowed to range from $1/5$ to 5. For the Matérn model, Figure 8.7 shows the profile likelihood decreasing slowly as values for the partial sill and range parameter increase beyond their REMLEs. The nugget effect has a pronounced peak in the profile likelihood. As we mentioned in Section 8.4, some authors claim that the smoothness parameter for the Matérn model is difficult to estimate,

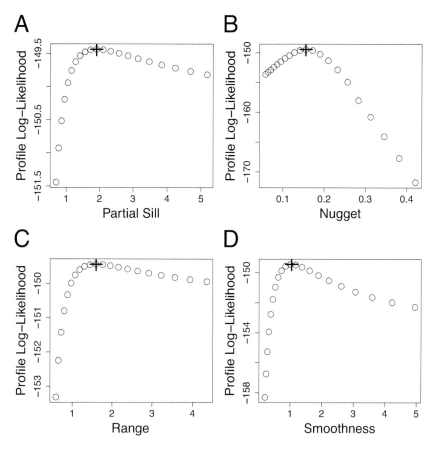

FIGURE 8.7
Profile likelihood for parameters in the Matérn model fitted to the clean wet sulfate deposition data with a constant mean: A) Partial sill; B) nugget effect; C) range parameter; D) smoothness parameter. Each parameter's REML estimate is shown by a plus symbol at the corresponding value of the profile likelihood.

and they suggest either setting it equal to a single value (say 1), or trying a small number of values and choosing the one that optimizes the likelihood or some other model selection criteria. However, for these data, the smoothness parameter has a fairly well-defined maximum, and optimization had no issues.

8.8.2 The harbor seal trends data

Next we analyze the harbor seal trends data, which are areal. Hence, we consider models based on neighbor relationships (Figure 1.4), i.e., spatial-weights linear models (Chapter 7). Many models of this type are amenable to fast computing because they are specified through the inverse of the covariance matrix, e.g., $\mathbf{\Sigma}^{-1} = \mathbf{K}_{\mathrm{CAR}}^{-1}(\mathbf{I} - \rho_{\mathrm{CAR}}\mathbf{W})$ (which is (2.37) using the parsimonious model for $\mathbf{C}$ given by (7.11)). It is the inverse of the covariance matrix that is used in the likelihoods (e.g, equations (8.1) and (8.4)), the evaluation of which can be computationally demanding for large data sets if the covariance structure of the model is specified directly through the covariance matrix, as it is for a geostatistical model (Chapter 6). Both spatial-weights and geostatistical models also require $|\mathbf{\Sigma}^{-1}|$ in the likelihood, and there are special algorithms for determining this for sparse matrices, such as the inverse covariance matrices of CAR (especially) and SAR models. Hence CAR and SAR models, which merely require an inverse of a diagonal matrix and a determinant of a sparse matrix, would seem to have a computational advantage over pseudo-geostatistical and aggregate-geostatistical models for the harbor seal trends data.

However, one interesting "wrinkle" of these data is that some of the responses are missing (unobserved). Ultimately, we will want to predict the responses at the locations where they are unobserved, so as noted in Section 2.3.2, the spatial-weights models must be formulated using a neighborhood structure that includes those locations. Recall that for the likelihood evaluations and optimization, we need the inverse of the covariance matrix for *only* the observed data. This mismatch between neighborhood structure formulations needed for estimation and prediction results in a situation where the computational advantages of the spatial-weights models can vanish. To make this clear, let us order the locations so that all those where the response was observed are first, followed by those where the response is unobserved. Then the partitioned covariance matrix (for observed and unobserved responses) and its inverse are

$$\boldsymbol{\Sigma} = \begin{bmatrix} \boldsymbol{\Sigma}_{oo} & \boldsymbol{\Sigma}_{ou} \\ \boldsymbol{\Sigma}_{uo} & \boldsymbol{\Sigma}_{uu} \end{bmatrix} \text{ and } \boldsymbol{\Sigma}^{-1} = \begin{bmatrix} \boldsymbol{\Sigma}^{oo} & \boldsymbol{\Sigma}^{ou} \\ \boldsymbol{\Sigma}^{uo} & \boldsymbol{\Sigma}^{uu} \end{bmatrix},$$

where the subscripts and superscripts o and u label blocks of the matrices corresponding to the observed and unobserved responses, respectively. Unfortunately, neither $\boldsymbol{\Sigma}^{oo}$ nor $(\boldsymbol{\Sigma}^{oo})^{-1}$ is equal to the matrix needed for likelihood-based estimation, $\boldsymbol{\Sigma}_{oo}^{-1}$. For the spatial-weights models, the most straightforward way to obtain $\boldsymbol{\Sigma}_{oo}^{-1}$ would seem to be to first compute $\boldsymbol{\Sigma} = (\boldsymbol{\Sigma}^{-1})^{-1}$, then carve out and invert its upper left block $\boldsymbol{\Sigma}_{oo}$. This requires two inversions. However, there is a potentially faster way. Recall from Section 5.6 that if we have already obtained $\boldsymbol{\Sigma}^{-1}$, then we can obtain $\boldsymbol{\Sigma}_{oo}^{-1}$ as

$$\boldsymbol{\Sigma}_{oo}^{-1} = \boldsymbol{\Sigma}^{oo} - \boldsymbol{\Sigma}^{ou}(\boldsymbol{\Sigma}^{uu})^{-1}\boldsymbol{\Sigma}^{uo}, \tag{8.10}$$

which requires a single inverse, $(\boldsymbol{\Sigma}^{uu})^{-1}$, that has dimensions less than those of the complete covariance matrix $(\boldsymbol{\Sigma}^{-1})^{-1}$, followed by some matrix multiplication and addition. If the dimensions of $\boldsymbol{\Sigma}_{uu}$ are (much) smaller than $\boldsymbol{\Sigma}_{oo}$, then this strategy will require (much) less computation. Conversely, if the dimensions of $\boldsymbol{\Sigma}_{uu}$ are larger than $\boldsymbol{\Sigma}_{oo}$, then spatial-weights models are more costly for estimation, in terms of the computational demand required for matrix inversion, than geostatistical models. For the harbor seal trends example, there are 306 and 157 locations where the response was observed and unobserved, respectively, so the inversion of $\boldsymbol{\Sigma}_{oo}$ using (8.10) is more efficient in this case.

We fit spatial-weights models with three distinct orders of neighbors: first-order, second-order (defined as sites that are not first-order neighbors but have a first-order neighbor in common), and fourth-order (defined as sites that have a second-order neighbor, but no first-order neighbor, in common). Figure 1.4 shows the neighborhood relationships of these orders for the 463 sites. The corresponding neighbor weights matrices are denoted by $\mathbf{W}_1$, $\mathbf{W}_2$, and $\mathbf{W}_4$ and were constructed using the method described in Section 2.3.1, with $\mathbf{W}_2$ inclusive of first-order neighbors and $\mathbf{W}_4$ inclusive of first-, second-, and third-order neighbors in this case. Also, recall that for these data we have information on a categorical (factor) variable indicating membership in one of five different genetic stocks (Figure 1.2), hence we can include stock effects in the mean structure of the model.

For these data, we will consider a variety of models that allow us to determine whether a spatially correlated error structure is even necessary, and if so, whether a CAR or SAR model is more appropriate. For the spatially correlated models, we can consider the first-, second-, and fourth-order neighborhood models as described above, and models with and without

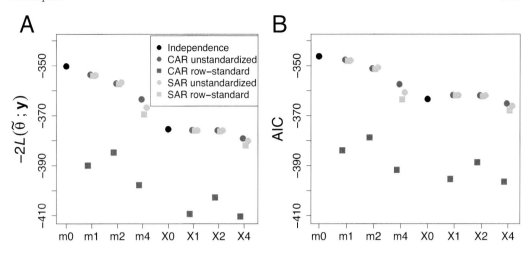

FIGURE 8.8
Log-likelihood and *AIC* for a variety of spatial-weights models for the harbor seal trends data: A) Minus two times the log-likelihood at the MLE; B) *AIC* for the same models. A model label beginning with "m" is a constant-mean model, while one beginning with "X" has a separate mean for each genetic stock. The number following these letters indicates the neighborhood order for the model.

row standardization of the binary weight matrices. Figure 8.8 shows twice the negative of the maximized log-likelihood of $\boldsymbol{\beta}$ and $\boldsymbol{\theta}$, and *AIC*, for all of these models. Generally, we see that models with a stock-effects mean structure fit better than those with constant mean. Interestingly, for the models with stock effects, the quality of the fits of first- and second-order CAR and SAR models with unstandardized weights, and SAR models of the same two orders with standardized weights, are almost identical to the quality of fit of the independence model, and so *AIC* favors the independence model because the spatial models have one more parameter (here, the spatial covariance matrix is $\boldsymbol{\Sigma} = \sigma^2 \mathbf{R}_\rho$, where σ^2 is an overall variance parameter and $\mathbf{R}_\rho$ is a CAR or SAR covariance matrix that depends on the single spatial dependence parameter ρ). The most dramatic feature of Figure 8.8 is the much improved fit when using row standardization with the CAR models. The best overall models suggested by *AIC* were the first- and fourth-order, row-standardized CAR models with a stock-effects mean structure. Recall that possible outliers were identified in Section 3.7.4, and in particular, one trend value (0.835) was considerably higher than the others. Here, we removed that value and refit several of the models in Figure 8.8, and the results changed very little. Therefore, we use the full dataset for all further analyses.

Because the models fitted have only two covariance parameters, visualization of their residual log-likelihood surfaces is straightforward. Consider the model with stock-effects mean structure and CAR covariance matrix

$$\mathrm{Var}(\mathbf{e}) = \sigma^2 \mathrm{diag}(\mathbf{W}_4 \mathbf{1} - \rho \mathbf{W}_4)^{-1}.$$

Note, by (7.16), that this is an equivalent way to write the CAR model with fourth-order neighbors and row standardization. The residual log-likelihood surface corresponding to this model is shown in Figure 8.9. The REML estimate is the value that maximizes this surface, shown as a black circle, where $\tilde{\rho} = 0.762$ and $\tilde{\sigma}^2 = 0.277$. There are several ways to make inferences on the covariance parameters, including profile likelihood and the Fisher information matrix.

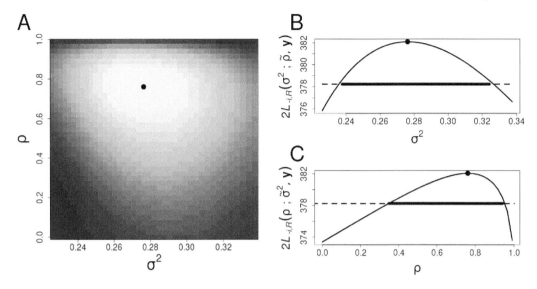

FIGURE 8.9

Residual log-likelihood and related quantities for inference for the harbor seal trends data: A) The residual log-likelihood surface for ρ and σ^2. Yellower colors are higher values, and bluer colors are lower values. The solid black circle is the maximum of the surface, yielding the REML estimate. B) Profile residual likelihood confidence interval for σ^2. The curve is twice the residual log-likelihood optimized for all parameters except σ^2, which is held constant at the value given on the horizontal axis. The solid black circle is the REML estimate, $\tilde{\sigma}_R^2$, and the dashed line is $2L_{-\sigma^2,R}(\tilde{\sigma}_R^2; \tilde{\rho}_R, \mathbf{y}) - \chi_{\epsilon,1}^2$ for $\epsilon = 0.05$. The horizontal black line forms the 95% confidence interval. C) Profile residual likelihood confidence interval for ρ with the same features as described for B.

    Beginning with the profile likelihood method, consider the profile residual log-likelihood functions for the two covariance parameters, which are denoted here as $L_{-i,R}(\theta_i; \tilde{\boldsymbol{\theta}}_{-i,R}, \mathbf{y})$ $(i = 1, 2)$, where the ith element of $\boldsymbol{\theta}$ has been held constant at θ_i and the function has been maximized over all other parameters, whose values are denoted as $\tilde{\boldsymbol{\theta}}_{-i,R}$. Then an asymptotically valid (under increasing-domain asymptotics) $100(1-\epsilon)\%$ confidence interval for θ_i may be obtained as the set of all values of θ_i for which $2L_{-i,R}(\theta_i; \tilde{\boldsymbol{\theta}}_{-i,R}, \mathbf{y})$ is greater than $2L_{-i,R}(\tilde{\theta}_{i,R}; \tilde{\boldsymbol{\theta}}_{-i,R}, \mathbf{y}) - \chi_{\epsilon,1}^2$, where $\chi_{\epsilon,\nu}^2$ is the $100(1-\epsilon)$th percentile of the chi-squared distribution with ν degrees of freedom. Using $\epsilon = 0.05$ leads to a 95% confidence interval and the well-known value $\chi_{0.05,1}^2 = 3.841$, and $2L_{-i,R}(\tilde{\theta}_{i,R}; \tilde{\boldsymbol{\theta}}_{-i,R}, \mathbf{y}) - 3.841$ is shown by the horizontal line in each of Figure 8.9B,C. The confidence interval is given by the solid portion of the horizontal line, above which all values of $2L_{-i,R}(\theta_i; \tilde{\boldsymbol{\theta}}_{-i,R}, \mathbf{y})$ are greater than $2L_{-i,R}(\tilde{\theta}_{i,R}; \tilde{\boldsymbol{\theta}}_{-i,R}, \mathbf{y}) - 3.841$. For σ^2, the 95% confidence interval is $[0.238, 0.324]$, and for ρ it is $[0.347, 0.941]$.

    A second approach to finding a confidence interval, as suggested in Section 8.3, is based on the Fisher information matrix (under increasing-domain asymptotics) associated with REML estimation. For our model, this matrix is given by

$$\mathcal{I}_R(\boldsymbol{\theta}) = \left(\frac{1}{2} \text{tr} \left(\mathbf{E} \frac{\partial \boldsymbol{\Sigma}}{\partial \theta_i} \mathbf{E} \frac{\partial \boldsymbol{\Sigma}}{\partial \theta_j} \right) \right),$$

where $\mathbf{E} = \mathbf{\Sigma}^{-1} - \mathbf{\Sigma}^{-1}\mathbf{X}(\mathbf{X}^T\mathbf{\Sigma}^{-1}\mathbf{X})^{-1}\mathbf{X}^T\mathbf{\Sigma}^{-1}$ as in (8.2), $\mathbf{\Sigma}^{-1} = (\mathbf{D} - \rho\mathbf{W}_4)/\sigma^2$, and

$$\frac{\partial\mathbf{\Sigma}}{\partial\sigma^2} = (\mathbf{D} - \rho\mathbf{W}_4)^{-1},$$

$$\frac{\partial\mathbf{\Sigma}}{\partial\rho} = \sigma^2(\mathbf{D} - \rho\mathbf{W}_4)^{-1}\mathbf{W}_4(\mathbf{D} - \rho\mathbf{W}_4)^{-1}.$$

Estimates of the asymptotic standard errors of the elements of $\tilde{\boldsymbol{\theta}}_R$ are given by the square roots of the elements on the main diagonal of $[\mathcal{I}_R(\boldsymbol{\theta})]^{-1}$, evaluated at $\boldsymbol{\theta} = \tilde{\boldsymbol{\theta}}_R$, which we denote as $[\mathcal{I}_R(\tilde{\boldsymbol{\theta}}_R)]^{-1}$. Then an approximate $100(1-\epsilon)\%$ confidence interval for θ_i is

$$\tilde{\theta}_{i,R} \pm z_{\epsilon/2}\sqrt{\{[\mathcal{I}_R(\tilde{\boldsymbol{\theta}}_R)]^{-1}\}_{ii}}.$$

For these data, the estimated asymptotic standard errors for $\tilde{\sigma}_R^2$ and $\tilde{\rho}_R$ were 0.018 and 0.124, respectively, so the 95% confidence interval for σ^2 was $[0.240, 0.313]$ and for ρ it was $[0.519, 1.006]$, which may be compared to those obtained previously using the profile likelihood. Observe that the asymptotic method based on the Fisher information matrix gives an upper confidence limit for ρ that lies slightly outside of its parameter space, which is a drawback of this method compared to profile likelihood. In practice, this situation is dealt with by moving the confidence limit to the nearest point on the boundary of the parameter space.

It is also interesting to examine the asymptotic correlations among the parameter estimates, as revealed by the Fisher information matrix. Let $\mathbf{S}(\boldsymbol{\theta})$ be a diagonal matrix containing reciprocals of the square roots from the main diagonal of $[\mathcal{I}_R(\boldsymbol{\theta})]^{-1}$. Then an estimate of the asymptotic correlation matrix of σ^2 and ρ is

$$\mathbf{S}(\boldsymbol{\theta})[\mathcal{I}_R(\boldsymbol{\theta})]^{-1}\mathbf{S}(\boldsymbol{\theta})\Big|_{\boldsymbol{\theta}=\tilde{\boldsymbol{\theta}}_R} = \begin{pmatrix} 1.000 & 0.141 \\ 0.141 & 1.000 \end{pmatrix},$$

which shows little correlation between the estimates of ρ and σ^2, and this is also evident in the shape of the residual log-likelihood surface (Figure 8.9A).

One feature of a spatial-weights model is that it is nonstationary, and it is interesting to investigate this property for our data. First, consider two CAR models with the stock-effects mean structure and a fourth-order neighborhood structure: one where the neighbor weights matrix is row-standardized, i.e., $\text{Var}(\mathbf{e}) = \mathbf{\Sigma}_{rs} = \sigma_{rs}^2(\mathbf{I} - \rho_{rs}\overline{\mathbf{W}}_4)^{-1}\mathbf{K}_{rs}$, and another where it is not, i.e., $\text{Var}(\mathbf{e}) = \mathbf{\Sigma}_{un} = \sigma_{un}^2(\mathbf{I} - \rho_{un}\mathbf{W}_4)^{-1}\mathbf{K}_{un}$. The diagonal elements of $\mathbf{\Sigma}_{rs}$, plotted as a function of the number of neighbors, are shown in Figure 8.10A. Note that the marginal variances decrease with increasing numbers of neighbors. This is reasonable because the spatial weights are equal to 1/(number of neighbors) when they are row-standardized, so we are *averaging* over neighboring values, and the variance of an average decreases as sample size increases. On the other hand, the marginal variances of $\mathbf{\Sigma}_{un}$ are shown in Figure 8.10B, where the spatial weights all equal 1, so we are *summing* over neighboring values, and the variance of a sum increases with sample size. Moreover, even for a fixed number of neighbors, the variance is not constant because (7.6) specifies a conditional expectation and the marginal variances involve the inverse of the matrix specified by these weights. In summary, both of these models exhibit nonstationary variances, in contrast to classical geostatistical models, for which variances are constant at all spatial locations.

Similarly, we can study autocorrelation as a function of the order of the neighbor. Let $\mathcal{N}_i^{[1]}$ be the set of all first-order neighbors of site i (sites sharing a boundary with site i), $\mathcal{N}_i^{[2]}$ be the set of all neighbors of $\mathcal{N}_i^{[1]}$, exclusive of any already in $\mathcal{N}_i^{[1]}$, $\mathcal{N}_i^{[3]}$ be the set of all neighbors of $\mathcal{N}_i^{[2]}$, exclusive of any already in $\mathcal{N}_i^{[1]}$ or $\mathcal{N}_i^{[2]}$, etc., up to sixth-order

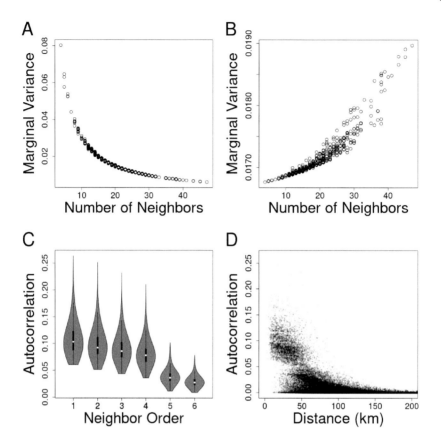

FIGURE 8.10
Nonstationarity in spatial-weights models for the harbor seal trends data. A) Marginal variance as a function of number of neighbors for a row-standardized CAR model; B) Marginal variance as a function of number of neighbors for a binary weights CAR model; C) Violin plots of autocorrelation as a function of neighbor order for a row-standardized CAR model; D) Scatterplot of all pairwise correlations as a function of centroid distance for a row-standardized CAR model.

neighbors. Then pairwise autocorrelations taken from Σ_{rs}, as a function of neighbor order, are plotted as violin plots in Figure 8.10C. In general, autocorrelation decreases as a function of neighbor order, but there is wide variation for a given order. It is also interesting to see that there is a sudden decrease in autocorrelation, on average, after order four. Nonetheless, Figure 8.10C also shows that autocorrelation exists in Σ_{rs} beyond order four, even though $\overline{\mathbf{W}}_4$ contains all zeros beyond order four. Like nonconstant variances, this is due to the fact that Σ_{rs} is obtained through an inverse involving $\overline{\mathbf{W}}_4$, rather than from $\overline{\mathbf{W}}_4$ directly.

Although distance is often not well-defined for spatial-weights models, such as when sites are polygons as in this example, one unique definition of distance is obtained by associating a centroid with each polygon, and then measuring Euclidean distances between centroids (the pseudo-geostatistical approach). Plotting autocorrelation as a function of between-centroid distance shows that, in general, autocorrelation decreases with distance (Figure 8.10D), just as it did for neighbor order (8.10C). Again, this situation may be compared to an

TABLE 8.9
Maximum likelihood (empirical GLS) estimates of fixed effects using a
row-standardized CAR model with provision for islands, as given by the `spmodel`
package in `R`, for the harbor seal trends data. Entries in the second column are
the MLEs of the elements of $\boldsymbol{\beta}$, the third column gives the corresponding
estimated standard errors, the fourth column is a computed z-value obtained by
dividing the entry in the second column by the corresponding entry in the third
column, and the fifth column is the estimated probability of obtaining the
z-value under the null hypothesis that the effect was zero.

Coefficients	Estimate	Std. Error	z-value	Pr($> |z|$)
stock Clarence Strait	0.0070	0.0127	0.546	0.585
stock Dixon/Cape Decision	0.0518	0.0155	3.340	0.001
stock Glacier Bay/Icy Strait	-0.0716	0.0233	-3.075	0.002
stock Lynn Canal/Stephens	-0.0264	0.0207	-1.279	0.201
stock Sitka/Chatham	0.0198	0.0169	1.177	0.239

isotropic geostatistical model, where autocorrelation is fixed for any given distance between
two locations.

Before moving on to estimation of fixed effects, we consider a few more models. One
very handy feature of the `spmodel` software is that it allows the estimation of a separate
variance parameter for islands (isolated sites), as discussed in Section 7.4. By default, the
software uses the geometry in `sf` objects and determines neighbors as any polygons that
share a common boundary. Fitting a CAR model with the stock-effects mean structure
in this way, and ordering the data so that the isolated polygons are listed first, the ML
estimated covariance matrix for the row-standardized model is

$$\boldsymbol{\Sigma}(\tilde{\boldsymbol{\theta}}) = \begin{pmatrix} 0.043\mathbf{I} & \mathbf{0} \\ \mathbf{0} & 0.032(\mathbf{I} - 0.270\overline{\mathbf{W}})^{-1}\overline{\mathbf{K}} \end{pmatrix},$$

and $-2\log\ell(\tilde{\boldsymbol{\beta}}, \tilde{\boldsymbol{\theta}}; \mathbf{y}) = -418.99$. With five fixed effects and three covariance parameters,
$AIC = $ -402.99 and, even though the model has an extra covariance parameter, it is better
than all of those in Figure 8.8. Using between-centroid distances again, the best pseudo-
geostatistical model with the stock-effects mean structure, using MLE, is a circular model,
which has $-2\log\ell(\tilde{\boldsymbol{\beta}}, \tilde{\boldsymbol{\theta}}; \mathbf{y}) = -390.66$, and, again with three covariance parameters, AIC is
-374.66. This is considerably worse than the row-standardized CAR model with provision
for islands. Finally, the variance-stabilizing weights of Section 7.3 were used in a CAR model
(without provision for islands) with stock effects, yielding $-2\log\ell(\tilde{\boldsymbol{\beta}}, \tilde{\boldsymbol{\theta}}; \mathbf{y}) = -399.93$, and
AIC was -385.93.

Any choice of a covariance model, and fitting it, leads to the next topic — inference
about the fixed effects — which was covered broadly in Chapter 5. For the remainder of
this subsection, let $\boldsymbol{\Sigma}(\boldsymbol{\theta})$ be estimated by ML using a row-standardized CAR model with
provision for islands, as described in the previous paragraph, and with a stock-effects mean
structure, as this appeared to be the best model according to AIC. Then the empirical GLS
estimate of stock effects, $\tilde{\boldsymbol{\beta}} = \{\mathbf{X}^T[\boldsymbol{\Sigma}(\tilde{\boldsymbol{\theta}})]^{-1}\mathbf{X}\}^{-1}\mathbf{X}^T[\boldsymbol{\Sigma}(\tilde{\boldsymbol{\theta}})]^{-1}\mathbf{y}$, is given in Table 8.9.

Standard errors for $\tilde{\beta}_i$, the ith element of $\tilde{\boldsymbol{\beta}}$, are estimated as $\tilde{se}(\tilde{\beta}_i) = \sqrt{(\{\mathbf{X}^T[\boldsymbol{\Sigma}(\tilde{\boldsymbol{\theta}})]^{-1}\mathbf{X}\}^{-1})_{ii}}$. As discussed in Section 5.7, the distributional properties of empirical generalized least squares are complicated due to the fact that, from (8.3), $\boldsymbol{\Sigma}(\tilde{\boldsymbol{\theta}}) = \tilde{\sigma}^2\mathbf{R}(\tilde{\boldsymbol{\theta}}_{-1})$. Nevertheless, assuming that $\tilde{\boldsymbol{\beta}}$ is approximately normal, an approximate z-value
can be formed as $\tilde{\beta}_i/\tilde{se}(\tilde{\beta}_i)$, the fourth column in Table 8.9. From there, approximate

P-values measuring the strength of evidence against the null hypothesis that $\beta_i = 0$ (against a two-sided alternative) may be computed and are given in the fifth column in Table 8.9.

It is also possible to pivot on the estimated z-value to obtain confidence intervals for β_i. Specifically, a $100(1 - \epsilon)\%$ confidence interval for β_i is $\tilde{\beta}_i \pm z_{\epsilon/2}\widetilde{se}(\tilde{\beta}_i)$. For example, a 95% confidence interval for the log-trend of the Dixon/Cape Decision stock is $0.0518 \pm (1.96)(0.0155) = [0.0214, 0.0822]$, meaning that we can be approximately 95% certain that this stock is growing by a rate somewhere between 2% to 8% per year.

In addition to estimates of stock trends, we are often interested in specific contrasts among the trends. Figure 1.2 shows stocks labeled 8 and 9 as the two northern stocks, while stocks 10, 11, and 12 are southern stocks. We would like to estimate the difference in average trend between the northern and southern stocks. Stock 8 is Glacier Bay/Icy Strait, and stock 9 is Lynn Canal/Stephens, so the contrast of interest is $(\beta_1 + \beta_2 + \beta_5)/3 - (\beta_3 + \beta_4)/2 = \boldsymbol{\ell}^T\tilde{\boldsymbol{\beta}}$, where $\boldsymbol{\ell}^T = (1/3, 1/3, -1/2, -1/2, 1/3)$. The estimate of this contrast is 0.0752, with standard error 0.0178. Thus, an approximate 95% confidence interval of the difference in log-trend between the northern and southern stocks is $[0.0403, 0.1102]$, or in other words, we can be about 95% certain that the average trend of the southern stocks is somewhere between 4% and 11% higher than that of the northern stocks.

In addition to estimating individual effects and contrasts or other linear combinations thereof, there may be interest in whether stock effects as a whole should be included in the model, or whether a constant mean would suffice. We will now describe two ways to investigate this. In Section 5.1, an F-test for the joint linear hypothesis $\mathbf{L}^T\boldsymbol{\beta} = \boldsymbol{\ell}_0$, where $\mathbf{L}^T$ is a specified $q \times p$ matrix with linearly independent rows and $\boldsymbol{\ell}_0$ is a specified q-vector, was presented for the case where $\mathbf{R}$ is known. For our example, the hypothesis of interest is of this form with

$$\mathbf{L}^T = \begin{pmatrix} 1 & -1 & 0 & 0 & 0 \\ 0 & 1 & -1 & 0 & 0 \\ 0 & 0 & 1 & -1 & 0 \\ 0 & 0 & 0 & 1 & -1 \end{pmatrix}$$

and $\boldsymbol{\ell}_0 = \mathbf{0}_4$ (a vector of four zeros), and the test statistic given by (5.6) with $\mathbf{R}(\tilde{\boldsymbol{\theta}}_{-1})$ substituted for $\mathbf{R}$, is

$$\frac{(\mathbf{L}^T\tilde{\boldsymbol{\beta}})^T\{\mathbf{L}^T[\mathbf{X}^T(\mathbf{R}(\tilde{\boldsymbol{\theta}}_{-1}))^{-1}\mathbf{X}]^{-1}\mathbf{L}\}^{-1}\mathbf{L}^T\tilde{\boldsymbol{\beta}}}{q\tilde{\sigma}^2} = 136.27.$$

The corresponding P-value is 0.000110. Thus, the statistical evidence for including stock effects in the mean structure is very strong.

An alternative way to determine the effect of genetic stock, after accounting for the overall mean effect, is by a likelihood ratio test, as described in Section 8.5. Recall that $-2\log \ell(\tilde{\boldsymbol{\beta}}, \tilde{\boldsymbol{\theta}}; \mathbf{y}, \mathbf{X}_{\text{stock}}) = -418.99$ for the row-standardized CAR model with provision for islands and with a stock-effects mean structure. For the model with the same covariance structure but a constant mean, $-2\log \ell(\tilde{\boldsymbol{\beta}}, \tilde{\boldsymbol{\theta}}; \mathbf{y}, \mathbf{X}_{\text{constant}}) = -398.74$. The difference in these two values is 20.24, and under increasing-domain aymptotics this is approximately distributed as a chi-squared random variable with four degrees of freedom. The corresponding P-value is 0.00045, which leads to the same conclusion as above. These results were obtained using the `anova` function in `spmodel`.

8.8.3 The caribou forage data

Spatial models for the designed experiment that produced the caribou forage data (Figure 1.5) may be based on neighbor relationships, which lead to linear models using spatial

TABLE 8.10

Model comparisons among covariance structures using MLE based on twice the negative maximized log-likelihood function ($-2\log\ell(\tilde{\boldsymbol{\beta}}, \tilde{\boldsymbol{\theta}}; \mathbf{y})$), *AIC*, *BIC*, and LOOCV, for the caribou forage data. The basic mean structure was $\boldsymbol{\mu} = \mu + \alpha_i + \gamma_j + \tau_{ij}$, where μ is an overall effect, α_i is the effect of the ith level of Water, γ_j is the effect of the jth level of Tarp, and τ_{ij} is their interaction effect. In the penultimate row of the table, $\mathrm{sph}(\mathbf{D}; \rho) = [\mathbf{I} - (1.5/\rho)\mathbf{D} + (0.5/\rho^3)\mathbf{D}^3]I(\mathbf{D} < \rho)$.

mean	$\boldsymbol{\Sigma}$	$-2\log\ell(\tilde{\boldsymbol{\beta}}, \tilde{\boldsymbol{\theta}}; \mathbf{y})$	*AIC*	*BIC*	LOOCV
μ	$\sigma^2\mathbf{I}$	-20.3	-6.29	3.52	0.0465
μ+row+col	$\sigma^2\mathbf{I}$	-42.9	-10.92	11.50	0.0511
μ	$\sigma^2(\mathbf{I} - \rho\mathbf{W}_1)^{-1}\mathbf{K}_1$	-22.6	-6.63	4.58	0.0396
μ	$\sigma^2(\mathbf{I} - \rho\overline{\mathbf{W}}_1)^{-1}\overline{\mathbf{K}}_1$	-24.7	-8.69	2.52	0.0373
μ	$\sigma^2(\mathbf{I} - \rho\mathbf{W}_2)^{-1}\mathbf{K}_2$	-20.9	-4.94	6.27	0.0396
μ	$\sigma^2(\mathbf{I} - \rho\overline{\mathbf{W}}_2)^{-1}\overline{\mathbf{K}}_2$	-23.1	-7.14	4.07	0.0418
μ	$\sigma^2(\mathbf{I} - \rho\mathbf{W}_1)^{-1}(\mathbf{I} - \rho\mathbf{W}_1^T)^{-1}$	-22.8	-6.84	4.37	0.0394
μ	$\sigma^2(\mathbf{I} - \rho\overline{\mathbf{W}}_1)^{-1}(\mathbf{I} - \rho\overline{\mathbf{W}}_1^T)^{-1}$	-24.0	-8.02	3.19	0.0370
μ	$\sigma^2(\mathbf{I} - \rho\mathbf{W}_2)^{-1}(\mathbf{I} - \rho\mathbf{W}_2^T)^{-1}$	-20.9	-4.88	6.33	0.0394
μ	$\sigma^2(\mathbf{I} - \rho\overline{\mathbf{W}}_2)^{-1}(\mathbf{I} - \rho\overline{\mathbf{W}}_2^T)^{-1}$	-21.0	-4.99	6.22	0.0430
μ	$\sigma^2\mathrm{sph}(\mathbf{D}; \rho) + \delta^2\mathbf{I}$	-22.7	-4.69	7.92	0.0381
μ	$\sigma^2\exp(-\mathbf{D}/\rho) + \delta^2\mathbf{I}$	-23.3	-5.30	7.31	0.0389

weights (Chapter 7), or on distances between plot centroids (which in this case are grid-points), leading to pseudo-geostatistical models (Chapter 6). Here, we investigate both approaches and compare them.

We fit a variety of models by maximizing the log-likelihood. For spatial weights models, we define neighbors based on the "rook's move," and we denote the corresponding binary spatial weights matrix by $\mathbf{W}_1$. If this matrix is row-standardized, it is denoted by $\overline{\mathbf{W}}_1$. The corresponding diagonal matrix under a CAR model is denoted by $\overline{\mathbf{K}}_1$. An alternative spatial weights matrix that includes first-order and second-order neighbors (the latter defined as sites having a common first-order neighbor) is denoted by $\mathbf{W}_2$ and is constructed according to the procedure described in Section 2.3.1. The corresponding diagonal matrix under a CAR model is denoted by $\mathbf{K}_2$. If row standardization is applied to $\mathbf{W}_2$, the resulting matrix is denoted by $\overline{\mathbf{W}}_2$, with corresponding diagonal matrix $\overline{\mathbf{K}}_2$.

In Section 4.5.3 we carried out two classical analyses of this designed experiment with uncorrelated errors: one where the water and tarp main effects and their interaction comprised the mean structure, and another with the same mean effects but that also included row and column effects to account for spatial trends. Table 8.10 gives fitting results for those two models and for a variety of CAR and SAR models with first and second-order neighbors, with and without row standardization, and for the spherical and exponential geostatistical models with nugget effects. Note that we have denoted the range parameter of the pseudo-geostatistical models in the table by ρ (rather than the usual α) to avoid confusion with the main effect of water. Measures of fit included in Table 8.10 are $-2\log\ell(\tilde{\boldsymbol{\beta}}, \tilde{\boldsymbol{\theta}}; \mathbf{y})$, *AIC*, *BIC*, and LOOCV. While it may seem strange to include a metric of predictive performance such as LOOCV for these data in light of the fact that no prediction is to be performed, LOOCV is a valuable model diagnostic even when interest is centered on estimation of parameters because it evaluates the fit of both the mean structure and covariance structure.

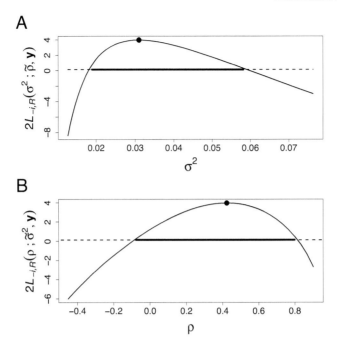

FIGURE 8.11

A) Profile likelihood-based confidence interval for σ^2 in the spatial linear model for the caribou forage data, using REML. The curve is twice the residual log-likelihood optimized for all covariance parameters except σ^2, which is held constant at the value given along the horizontal axis. The abscissa of the closed circle is the REMLE, $\tilde{\sigma}_R^2$, and the dashed line is $2L_{-\sigma^2,R}(\tilde{\sigma}_R^2; \tilde{\rho}_R, \mathbf{y}) - \chi_{\epsilon,1}^2$ for $\epsilon = 0.05$. The horizontal black line forms the 95% confidence interval. B) Profile likelihood-based confidence interval for ρ, obtained in the same way as that for σ^2.

Table 8.10 shows that the uncorrelated-errors model with row and column effects minimizes $-2\log\ell(\tilde{\boldsymbol{\beta}}, \boldsymbol{\theta}; \mathbf{y})$ by a large margin among all models, but at the cost of 15 mean structure parameters (9 more than the other models). *AIC* also suggests that this is the best model, even when penalizing for the extra mean parameters, but on the basis of *BIC* and LOOCV it compared poorly to all other models. Among the spatial-weights models, the first-order neighbor models with row standardization were best according to all three criteria, with little difference between the CAR and SAR models. They also appear to be slightly better than both of the pseudo-geostatistical models.

We proceed with the row-standardized SAR model as the best model, partly based on the results in Table 8.10 and partly because we already examined a CAR model for the harbor seal trends data. As we did for those data, here we use profile likelihood (Figure 8.11) to make inferences about the covariance parameters. The REML estimate of ρ is 0.424, with a 95% confidence interval from -0.079 to 0.799, and the REML estimate of σ^2 is 0.0310, with a 95% confidence interval from 0.0189 to 0.0582.

Figure 8.12 compares the REML surfaces among row-standardized first-order SAR and CAR models and an exponential pseudo-geostatistical model, where Figure 8.12A corresponds to the residual log-likelihood surface that was used to construct the profile-likelihood confidence intervals shown in Figure 8.11. The row-standardized first-order CAR model (Figure 8.12B) has $\tilde{\rho}_R = 0.695$ and $\tilde{\sigma}_R^2 = 0.0934$, both substantially larger than their counterparts in the SAR model. The resulting correlations among observations, plotted for all

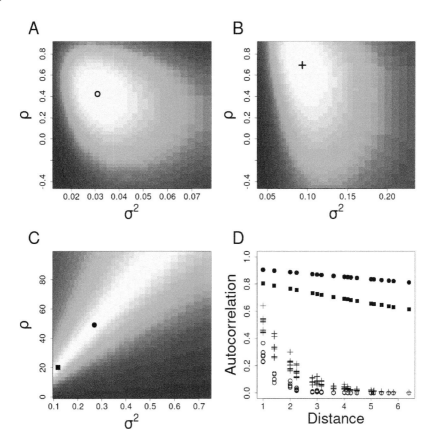

FIGURE 8.12

REML surfaces for models fitted to the caribou forage data, with mean structures consisting of water and tarp main effects and their interaction: A) Row-standardized, first-order SAR covariance structure (the open circle shows the REMLE); B) Row-standardized, first-order CAR covariance structure (the plus symbol shows the REMLE); C) Pseudo-geostatistical exponential covariance structure (the closed circle shows the REMLE and the black square shows the REMLE of σ^2 with ρ fixed at 20); D) Autocorrelation as a function of distance between centroids, where the symbols used correspond to those used for the REML estimators in A, B, and C.

possible pairwise distances in Figure 8.12D, are higher for the CAR model than for the SAR model, and both are nonstationary. The estimated pseudo-geostatistical exponential model has a nugget effect of 0.0222, with a partial sill of 0.269 and a range parameter (ρ in the last line of Table 8.10) of 49.2. When the range parameter is fixed at 20, the REML estimate of the nugget effect is 0.0218 and the partial sill is 0.118. Part of the reason to hold the range parameter at 20 was to observe how the residual likelihood is still optimized along the ridge of higher values, and to see how that affects the parameter estimates. The resulting autocorrelations, plotted for all possible pairwise distances in Figure 8.12D, are stationary for both estimated exponential models and much higher than those of the SAR and CAR models. Of course, the model with range parameter 49.2 has higher autocorrelations than the model with range parameter 20. We will consider further inferences for these data in Chapter 11.

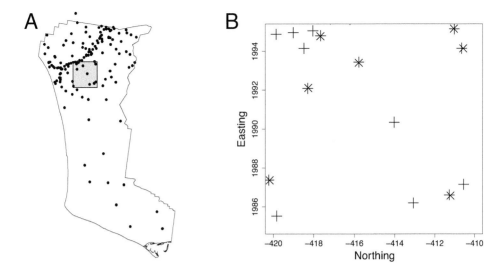

FIGURE 8.13

Sampling locations for the moss heavy metals data from 2001: A) Cape Krusenstern study area; B) An expanded view of the shaded rectangular box in the study area, showing primary sampling locations (+) and autocorrelation sampling locations (×).

8.8.4   The moss heavy metals data

In Section 4.5.4, we used a linear model for the natural logarithm of lead concentration in moss tissue samples that included explanatory variables for 1) year of sample (2001 or 2006), 2) the natural logarithm of distance-from-road, and 3) side-of-road (north or south), and their two-way and three-way interactions. Here we will allow for spatially correlated errors and incorporate random effects that accommodate the special sampling design of this study.

The spatial sampling design affects how efficiently we estimate spatial correlation, how well we estimate fixed effects, and how precisely we predict at unsampled locations. These issues will be discussed in greater detail in Chapter 10. However, intuitively, it would seem that in order to estimate the spatial correlation well, we should have some pairs of sample locations that are very close together, as that will help anchor the fitted spatial covariance function at short distances. This notion guided the selection of sample sites for the moss heavy metals study, although it is not apparent from the earlier figures (e.g., Figure 1.8). Figure 8.13A shows the Cape Krusenstern study area with all samples from 2001, and the shaded rectangular box within the study area is expanded in Figure 8.13B. The primary sample units are shown as + symbols, which were targeted for a certain vegetation class and then randomized within strata, where strata were determined by distance from the haul road. At some of the primary sites, "autocorrelation sites" were established, which were sampled in random directions at distances between 10 and 20 meters from the primary sites. The autocorrelation sites are shown with × symbols in Figure 8.13B.

At some locations, duplicate samples were taken. This was done because the investigators were interested in the microscale variation due to grabbing one handful of moss versus reaching over and grabbing a different handful of moss. In terms of our spatial correlation matrix, these samples will have a distance of (essentially) zero, which results in a correlation of 1.0 for the corresponding off-diagonal elements of the covariance matrix. These off-diagonal 1's can cause that matrix to fail to be positive definite, and cause computer algorithms for inverses, determinants, etc., to be unstable. However, if we add a random effect for location to the model for repeatedly sampling at a single location, which adds a variance term to the main diagonal elements of the covariance matrix, these issues are resolved. So let $\mathbf{Z}_1$ represent a design matrix of dummy variables (ones and zeros) that indicate the location at which a sample occurs.

Additionally, moss samples are sent to a laboratory for analysis of lead concentration. The process of extracting the heavy metal, and then measuring it by a machine, may have what is sometimes called "measurement error." In order to assess measurement error, after moss samples were homogenized some of the homogenized samples were split into two, and then both were run through the heavy metal extraction and concentration measurement process. Let $\mathbf{Z}_2$ represent a design matrix of dummy variables that label laboratory replicates that are nested within both location and (possibly) a duplicate sample for a location, as described in the previous paragraph.

The covariance matrix resulting from the sampling design above can be illustrated with a simple example of three locations, where location 1 is far from locations 2 and 3 (and its autocorrelation with them is 0.01), but location 3 is an autocorrelation location for location 2 (and their autocorrelation is 0.99). Additionally, suppose that location 2 had two duplicate samples, and for each duplicate sample, there were two laboratory replicate samples, for a grand total of 6 samples across locations. Then

$$
\mathbf{Z}_1 = \begin{pmatrix} 1 & 0 & 0 \\ 0 & 1 & 0 \\ 0 & 1 & 0 \\ 0 & 1 & 0 \\ 0 & 1 & 0 \\ 0 & 0 & 1 \end{pmatrix} \quad \text{and} \quad \mathbf{Z}_2 = \begin{pmatrix} 1 & 0 & 0 & 0 \\ 0 & 1 & 0 & 0 \\ 0 & 1 & 0 & 0 \\ 0 & 0 & 1 & 0 \\ 0 & 0 & 1 & 0 \\ 0 & 0 & 0 & 1 \end{pmatrix}.
$$

Notice that the laboratory replicate effect, nested within location and duplicate samples, is confounded with the nugget effect, which is appropriate, because for this model we consider the nugget effect to be measurement error. Let σ_{PS}^2 be the variance (partial sill) associated with the spatial process, let σ_{LO}^2 be the variance of the random location effects, let σ_{DU}^2 be the variance of the random effects for samples within location, and let σ_{ME}^2 be the variance of laboratory measurement errors, which in this context may be taken as the nugget effect. The resulting covariance matrix is

$$
\mathbf{\Sigma} = \sigma_{PS}^2 \mathbf{R} + \sigma_{LO}^2 \mathbf{Z}_1 \mathbf{Z}_1^T + \sigma_{DU}^2 \mathbf{Z}_2 \mathbf{Z}_2^T + \sigma_{ME}^2 \mathbf{I},
$$

and for the simple example with 3 locations and 6 samples,

$$
\mathbf{\Sigma} = \sigma_{PS}^2 \begin{pmatrix} 1 & 0.01 & 0.01 & 0.01 & 0.01 & 0.01 \\ 0.01 & 1 & 1 & 1 & 1 & 0.99 \\ 0.01 & 1 & 1 & 1 & 1 & 0.99 \\ 0.01 & 1 & 1 & 1 & 1 & 0.99 \\ 0.01 & 1 & 1 & 1 & 1 & 0.99 \\ 0.01 & 0.99 & 0.99 & 0.99 & 0.99 & 1 \end{pmatrix} + \sigma_{LO}^2 \begin{pmatrix} 1 & 0 & 0 & 0 & 0 & 0 \\ 0 & 1 & 1 & 1 & 1 & 0 \\ 0 & 1 & 1 & 1 & 1 & 0 \\ 0 & 1 & 1 & 1 & 1 & 0 \\ 0 & 1 & 1 & 1 & 1 & 0 \\ 0 & 0 & 0 & 0 & 0 & 1 \end{pmatrix}
$$

$$
+ \sigma_{DU}^2 \begin{pmatrix} 1 & 0 & 0 & 0 & 0 & 0 \\ 0 & 1 & 1 & 0 & 0 & 0 \\ 0 & 1 & 1 & 0 & 0 & 0 \\ 0 & 0 & 0 & 1 & 1 & 0 \\ 0 & 0 & 0 & 1 & 1 & 0 \\ 0 & 0 & 0 & 0 & 0 & 1 \end{pmatrix} + \sigma_{ME}^2 \begin{pmatrix} 1 & 0 & 0 & 0 & 0 & 0 \\ 0 & 1 & 0 & 0 & 0 & 0 \\ 0 & 0 & 1 & 0 & 0 & 0 \\ 0 & 0 & 0 & 1 & 0 & 0 \\ 0 & 0 & 0 & 0 & 1 & 0 \\ 0 & 0 & 0 & 0 & 0 & 1 \end{pmatrix}.
$$

TABLE 8.11
REML-EGLS estimates of mean parameters for model selected
by backward elimination, for the moss heavy metals data.

Parameter	Estimate	Std. Error	z-value	P-value
μ	8.07345	0.22059	36.599	< 0.0001
$\alpha_{i=2006}$	-0.40732	0.26060	-1.563	0.118
β	-0.57895	0.01880	-30.791	< 0.0001
$\tau_{j=\text{South}}$	-0.11134	0.01229	-9.059	< 0.0001

In practice, the autocorrelation will not be known as supposed above; instead, it will be estimated, along with the four variance parameters. We now suppose that the spatial autocorrelation is isotropic and exponential, hence given by the model $\rho(r; \alpha) = \exp(-r/\alpha)$. Thus there are five covariance parameters comprising $\boldsymbol{\theta}$, i.e., $\boldsymbol{\theta} = (\sigma_{PS}^2, \alpha, \sigma_{LO}^2, \sigma_{DU}^2, \sigma_{ME}^2)$. An additional feature of the covariance function is that we assume that observations taken in different years are independent, but share the same set of covariance parameters.

For this example, we will not explore other covariance functions, as that was covered in previous examples. Instead, we will illustrate the search for a parsimonious mean structure. Seven explanatory variables, namely year, side-of-road, distance-from-road, and their three two-way interactions and single three-way interaction, are candidates for inclusion.

A backward elimination process for model selection, with the P-value as the criterion for elimination, was used to create a parsimonious mean structure for the data. We first fitted the complete model using REML-EGLS. We removed the term with the highest P-value, then refitted the resulting seven-variable model (including the intercept as a variable) and removed the term with the highest P-value, etc. Our cutoff for removal was $P > 0.15$. We prefer to be conservative here in removing terms; by retaining terms with somewhat weak evidence, we allow readers to assess the importance of retained explanatory variables for themselves. We removed terms in the following order, with P-value for removal in parentheses: 1) side-of-road ($P = 0.715$), 2) year × side-of-road ($P = 0.715$), 3) year × distance-from-road × side-of-road ($P = 0.768$), and 4) year × distance-from-road ($P = 0.382$). The REML-EGLS estimates of mean parameters produced by `spmodel` for the final model are given in Table 8.11; the notation used for the parameters in the table is the same as that used in model (4.11) and Table 4.8. This model is non-hierarchical, i.e., it includes an interaction between two variables without including both of the individual variables. Usually, non-hierarchical models should be avoided, but in this case such a model is scientifically reasonable, since we would expect the intercepts on the two sides of the road to be identical, as the pollutant is produced at the road, but prevailing winds could cause differences in the decay in concentration (i.e., different slopes) on the two sides.

The estimated range parameter in the exponential correlation model was 11.125, implying that the estimated practical range (the distance beyond which the autocorrelation is less than 0.05), is about 33 km.

The empirical Aitken model version of R^2, as described in Section 5.7, is

$$R^2(\tilde{\boldsymbol{\Sigma}}_R) = \frac{(\mathbf{X}\tilde{\boldsymbol{\beta}}_R - \mathbf{1}\tilde{\beta}_{0,R})^T \tilde{\boldsymbol{\Sigma}}_R^{-1} (\mathbf{X}\tilde{\boldsymbol{\beta}}_R - \mathbf{1}\tilde{\beta}_{0,R})}{(\mathbf{y} - \mathbf{1}\tilde{\beta}_{0,R})^T \tilde{\boldsymbol{\Sigma}}_R^{-1} (\mathbf{y} - \mathbf{1}\tilde{\beta}_{0,R})}.$$

$R^2(\tilde{\boldsymbol{\Sigma}}_R)$ may be interpreted as the proportion of the total variation of the data explained by the fitted mean structure, or equivalently by the fixed effects. Hence $1 - R^2(\tilde{\boldsymbol{\Sigma}}_R)$ is the proportion due to the random effects. This latter proportion may, in turn, be apportioned to the four variance components. For example, the proportion of the total variation that is

TABLE 8.12

REML estimates of variance components for the moss heavy metals data.

Variance Component	Symbol	Estimate	Proportion
Fixed Effects	R^2	0.8120	0.8120
Partial Sill	σ^2_{PS}	0.2016	0.1284
Locations	σ^2_{LO}	0.0640	0.0408
Duplicates	σ^2_{DU}	0.0267	0.0170
Measurement Error	σ^2_{ME}	0.0028	0.0018

explained by the spatial component of the model, the partial sill, is

$$(1 - R^2(\tilde{\Sigma}_R)) \times \frac{\tilde{\sigma}^2_{PS,R}}{\tilde{\sigma}^2_{PS,R} + \tilde{\sigma}^2_{LO,R} + \tilde{\sigma}^2_{DU,R} + \tilde{\sigma}^2_{ME,R}}.$$

The complete apportionment of variability for the moss heavy metals data is given in Table 8.12, which includes the estimated variance for each component.

Also, recall that samples had laboratory replicates, and replicate error was confounded with the nugget effect as measurement error. One might ask, what is a separate estimate of variance based on the replicate samples, where we use the average squared deviation from the mean of each pair of replicate samples, with the standard bias correction? That is, what is

$$\overline{\sigma}^2_{ME} = \frac{1}{k-1} \sum_{i=1}^{k} \sum_{j=1}^{2} (y_{ij} - \overline{y}_{i\cdot})^2,$$

where $\overline{y}_{i\cdot}$ is the mean for the ith pair of replicates $(i = 1, \ldots, k)$. For these data, $\overline{\sigma}^2_{ME} = 0.00282$, which is virtually identical to the REML estimate of measurement error variance listed in Table 8.12.

The fitted model is superimposed upon the data in Figure 8.14, which reveals several interesting features. First, note that the fitted intercept and lines for 2006 are lower than 2001. Recall that some mitigation measures were put in place to lower the amount of ore dust escaping from the back of trucks transporting ore on the haul road between 2001 and 2006. That appears to have had the effect of lowering lead concentrations in moss, although under the spatial model this effect is not statistically significant (the year effect for the fitted model had a P-value of 0.118, see above) and, as noted in Section 4.5.4, factors other than mitigation measures could have played a role in the difference between years. Visually, just looking at the raw data, it seems that there should be stronger evidence of the effect. For example, for a given distance from the haul road, almost all of the values north of the road in 2006 are lower than those in 2001. It is often difficult to make strong inferences about mean structure parameters in spatial models, as we discuss further in Chapter 11. However, in Chapter 9, we will use a prediction procedure called conditional simulation to compare estimated averages of realized surfaces, rather than mean parameters from a model, and our inferences will be much stronger.

Another noteworthy feature of Figure 8.14 is that the fitted line for 2006 on the north side of the road does not seem to fit very well, as it lies above most of the raw data. What is going on here? This is a good example of how estimation of the mean structure under a spatial model may behave differently than it does under non-spatial models, which was first illustrated in Section 5.2. Recall the second toy example of that section, in which we simulated responses with a bowl-shaped pattern at values of a single x-coordinate. If we

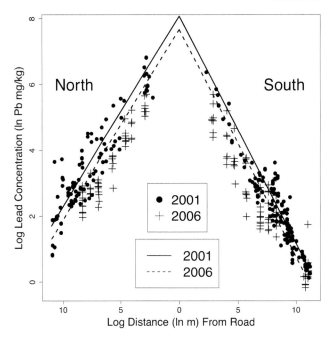

FIGURE 8.14
Log concentrations of lead in moss tissues as a function of year, side of the road (numbers to the left of 0 on the horizontal axis are north while numbers to the right are south), and log of the distance from the road, with the REML-EGLS fit of the model described in Section 8.8.4 superimposed.

fit a spatial model with constant mean to these data, using REML with an exponential covariance function, we obtain estimates of 86.9 for the range, 121.2 for the partial sill, and 0.215 for the nugget effect. These correspond to very strong spatial correlation, as the range and partial sill are far beyond the bounds of the x-coordinate and response, respectively. This is not unusual, as we have seen the strong ridge in the likelihood between the partial sill and range parameter before (e.g., Figures 8.5A,B and 8.12C), and an optimum that is well beyond the extremes of the response. Also recall that for that same toy example, the generalized least squares estimate of the mean was larger than all but a few observations, so that the generalized residuals did not sum to zero. It seems that the fitted line for 2006 on the north side of the road is behaving similarly, and the same appears to be true of the fitted line for 2006 on the south side of the road. The same is not true for the data from 2001. One might argue that the intercept for the north side of the road is tied to the intercept for the south side of the road, which complicates the interpretation. This is true, but the model could have allowed for separate intercepts and slopes for all year by side-of-road combinations. We leave it as an exercise to determine if the apparent fits in 2006 improve under those conditions (Exercise 8.6). Another possibility would be to add a quadratic term in distance-from-road, but it appears that doing so might improve only the fit to the 2006 data on the north side of the road, at the expense of greater model complexity. The current fitted model is perfectly capable and reasonable for the observed data, especially once we understand the behavior of the model-fitting as just described.

8.9 Exercises

1. Verify that $\tilde{\sigma}^2$ is equal to $[(n-p)/n]\hat{\sigma}^2_{EGLS}$ where $\mathbf{R}(\tilde{\boldsymbol{\theta}}_{-1})$ is substituted for $\mathbf{R}(\tilde{\boldsymbol{\theta}})$ in expression (5.12) for $\hat{\sigma}^2_{EGLS}$.

2. Consider the Gaussian SAR model

$$\mathbf{y} = \mathbf{X}\boldsymbol{\beta} + \mathbf{e}, \quad \mathbf{e} \sim \mathrm{N}(\mathbf{0}, \sigma^2(\mathbf{I} - \rho\mathbf{W})^{-1}(\mathbf{I} - \rho\mathbf{W}^T)^{-1}),$$

   with parameter space $\{\boldsymbol{\beta} \in \mathbb{R}^p, \sigma^2 > 0, \rho \in \Omega_{SAR}\}$, where Ω_{SAR} was defined in Section 7.2. The information matrix for this model (under increasing-domain asymptotics) is

$$\mathcal{I}(\boldsymbol{\beta}, \sigma^2, \rho) = (1/\sigma^4) \begin{pmatrix} \sigma^2\mathbf{X}^T\mathbf{F}^T\mathbf{F}\mathbf{X} & \mathbf{0} & \mathbf{0} \\ \mathbf{0} & n/2 & \sigma^2\mathrm{tr}(\mathbf{G}) \\ \mathbf{0} & \sigma^2\mathrm{tr}(\mathbf{G}) & \sigma^4[\mathrm{tr}(\mathbf{G}^T\mathbf{G}) + \xi] \end{pmatrix},$$

   where $\mathbf{F} = \mathbf{I} - \rho\mathbf{W}$, $\mathbf{G} = \mathbf{W}\mathbf{F}^{-1}$, $\xi = \sum_{i=1}^{n} \frac{\lambda_i^2}{(1-\rho\lambda_i)^2}$, and $\lambda_1, \ldots, \lambda_n$ are the eigenvalues of $\mathbf{W}$ (Ord, 1975; Cliff & Ord, 1981; Waller & Gotway, 2004). Furthermore, consider the special case of this model that is observed on a 5×5 square lattice, with constant mean β, the rook's definition of contiguity, and binary weights in $\mathbf{W}$.

   (a) Obtain approximate standard errors of the MLEs of β, σ^2, and ρ when $\sigma^2 = 1$ and $\rho = 0.15$.

   (b) Use the results of part (a) to give expressions for asymptotically valid $100(1 - \epsilon)\%$ confidence intervals for β, σ^2, and ρ.

3. Repeat the activities described in the previous problem, but this time for the Gaussian CAR model

$$\mathbf{y} = \mathbf{X}\boldsymbol{\beta} + \mathbf{e}, \quad \mathbf{e} \sim \mathrm{N}(\mathbf{0}, \sigma^2(\mathbf{I} - \rho\mathbf{W})^{-1}),$$

   with parameter space $\{\boldsymbol{\beta} \in \mathbb{R}^p, \sigma^2 > 0, \rho \in \Omega_{CAR}\}$, where Ω_{CAR} was defined in Section 7.2. The information matrix for this model (under increasing-domain asymptotics) is

$$\mathcal{I}(\boldsymbol{\beta}, \sigma^2, \rho) = (1/\sigma^4) \begin{pmatrix} \sigma^2\mathbf{X}^T\mathbf{F}\mathbf{X} & \mathbf{0} & \mathbf{0} \\ \mathbf{0} & n/2 & (\sigma^2/2)\mathrm{tr}(\mathbf{G}) \\ \mathbf{0} & (\sigma^2/2)\mathrm{tr}(\mathbf{G}) & \sigma^4\xi/2 \end{pmatrix},$$

   where $\mathbf{F}$, $\mathbf{G}$, ξ, and $\lambda_1, \ldots, \lambda_n$ are defined as in the previous exercise (Cliff & Ord, 1981; Waller & Gotway, 2004).

4. Repeat the simulation study of the performance of ML and REML estimation presented in Tables 8.1 and 8.2. Also repeat using two other covariance functions (e.g., spherical and Gaussian) of your choice.

5. Repeat the analysis of information content and interplay among parameter estimates presented in Section 8.4, but this time for the wave model, Gaussian model, and order-two Radon transform of the exponential model. Comment on any notable differences from the results in Table 8.4.

6. Fit a mixed linear model to the caribou forage data that has a mean structure consisting of fixed effects for water and tarp and their interaction, and covariance structure consisting of independent random row, column, and residual effects. Compare the fit of this model to those listed in Table 8.10 using the same criteria used in that table.

7. Fit a model to the moss heavy metals data that allows for year-specific intercepts and year-specific slopes on each side of the road. Does the fit of the model for 2006 improve?

8. Consider the pdf of an n-variate normal distribution for a general linear model with positive definite covariance matrix, i.e.

$$f(\mathbf{y}; \boldsymbol{\beta}, \boldsymbol{\theta}) = \frac{\exp\left(-\frac{1}{2}(\mathbf{y} - \mathbf{X}\boldsymbol{\beta})^T \boldsymbol{\Sigma}^{-1}(\mathbf{y} - \mathbf{X}\boldsymbol{\beta})\right)}{(2\pi)^{n/2}|\boldsymbol{\Sigma}|^{1/2}},$$

where the $n \times p$ matrix $\mathbf{X}$ has full column rank. By carrying out the following steps, show that the residual log-likelihood function $L_R(\boldsymbol{\theta}; \mathbf{y})$ displayed in (8.5) may be derived by p-dimensional integration of this pdf over $\boldsymbol{\beta}$ and taking the natural logarithm of the result.

- Add and subtract $\mathbf{X}\hat{\boldsymbol{\beta}}_{GLS}$ into $\mathbf{y} - \mathbf{X}\boldsymbol{\beta}$ in the expression for the pdf given above.
- Prove that $2(\mathbf{y} - \mathbf{X}\hat{\boldsymbol{\beta}}_{GLS})^T \boldsymbol{\Sigma}^{-1}(\mathbf{X}\hat{\boldsymbol{\beta}}_{GLS} - \mathbf{X}\boldsymbol{\beta}) = 0$.
- Isolate only those terms that contain $\boldsymbol{\beta}$.
- Use the fact that

$$\int_{-\infty}^{\infty} \cdots \int_{-\infty}^{\infty} \exp\left(-\frac{1}{2}(\mathbf{x} - \boldsymbol{\mu})^T \mathbf{A}^{-1}(\mathbf{x} - \boldsymbol{\mu})\right) d\mathbf{x} = (2\pi)^{m/2}|\mathbf{A}|^{1/2}$$

for an $m \times 1$ vector $\mathbf{x}$, $m \times 1$ constant vector $\boldsymbol{\mu}$, and positive definite $m \times m$ matrix $\mathbf{A}$. (Notice that this is how the normalizing constant of a multivariate normal distribution may be obtained).

9

Spatial Prediction

In previous chapters, we described two main objectives of spatial data analysis and methods for meeting those objectives using a spatial general linear model. Those objectives were 1) characterization of spatial dependence in the response variable, and 2) estimating the effects of spatial location and other explanatory variables on the response in the presence of spatial dependence. In some analyses of spatial data, there is a third objective, which may even be the primary objective: spatial prediction, i.e., the prediction of the response variable at some, perhaps all, locations where it is not observed. For example, as part of the analysis of the harbor seal trends data we may wish to predict the trends in harbor seal numbers in the polygons in which no observations were taken. Or, based upon the wet sulfate deposition observations at weather stations, we may wish to predict wet sulfate deposition at every site in the contiguous United States, in effect creating a map of predictions. In both cases, accompanying the predictions with estimates of their uncertainties is highly desirable.

Methods dedicated to spatial prediction that are based on a spatial linear model are the topic of this chapter. Collectively they are often called **kriging**, having been named after the South African mining engineer, D. G. Krige, who was among the first to develop and apply them to spatial data, though similar methods had been used previously in other contexts. The type of inference is referred to as prediction, rather than estimation, to emphasize the fact that the objects of inferential interest are regarded under the model as random variables rather than fixed unknown parameters.

Our presentation begins with a review of theory and methods for best linear unbiased prediction in the context of a (not necessarily spatial) Aitken model. Then we consider applications of this theory and methodology to several spatial prediction objectives.

9.1 Classical theory and methodology of best linear unbiased prediction

Recall the Aitken linear model given by

$$\mathbf{y} = \mathbf{X}\boldsymbol{\beta} + \mathbf{e},$$

where $\mathbf{e}$ is a random vector with mean $\mathbf{0}$ and covariance matrix $\sigma^2\mathbf{R}$. This is a model for the observed response vector $\mathbf{y}$, the purpose of which is to allow us to estimate the effects (if any) of the explanatory variables in $\mathbf{X}$ on the response. Now let us augment this model with a model for an unobserved predictand u, the purpose of which is to allow us to predict the predictand by exploiting its relationship with $\mathbf{y}$. Specifically, the joint model for $\mathbf{y}$ and u after augmentation is

$$\begin{pmatrix} \mathbf{y} \\ u \end{pmatrix} = \begin{pmatrix} \mathbf{X} \\ \mathbf{x}_u^T \end{pmatrix} \boldsymbol{\beta} + \begin{pmatrix} \mathbf{e} \\ e_u \end{pmatrix},$$

DOI: 10.1201/9780429060878-9

where the joint distribution of $\mathbf{e}$ and e_u is multivariate normal with mean vector $\mathbf{0}_{n+1}$ and covariance matrix

$$\mathrm{Var}\left(\begin{array}{c}\mathbf{e}\\ e_u\end{array}\right)=\sigma^2\left(\begin{array}{cc}\mathbf{R}&\mathbf{r}_{\mathbf{y}u}\\ \mathbf{r}_{\mathbf{y}u}^T&r_{uu}\end{array}\right).$$

Here $\sigma^2\mathbf{r}_{\mathbf{y}u}$ specifies the covariances between the elements of $\mathbf{y}$ and u, and $\sigma^2 r_{uu}$ specifies the variance of u. Observe, crucially, that the means of u and $\mathbf{y}$ under this model are linear functions of the same parameter vector ($\boldsymbol{\beta}$). Also, assume that $\mathbf{r}_{\mathbf{y}u}$ and r_{uu}, like $\mathbf{R}$, are known; in a later section we will relax this assumption.

The method of **best linear unbiased prediction** applied to this augmented model yields the best linear unbiased predictor (BLUP) of u. Here, by "linear" we mean that the predictor is a linear combination of the observed responses; that is, the predictor must be of the form $\boldsymbol{\lambda}^T\mathbf{y}$ for some n-vector $\boldsymbol{\lambda}=(\lambda_i)$ that does not depend on $\mathbf{y}$. By "unbiased" we mean that the expectation of the predictor matches that of the predictand; this requires that $\boldsymbol{\lambda}^T\mathbf{X}\boldsymbol{\beta}=\mathbf{x}_u^T\boldsymbol{\beta}$ for all p-vectors $\boldsymbol{\beta}$ or equivalently that

$$\boldsymbol{\lambda}^T\mathbf{X}=\mathbf{x}_u^T.\tag{9.1}$$

And, by "best" we mean that the variance of the difference between the predictor and the predictand, i.e, $\mathrm{Var}(\boldsymbol{\lambda}^T\mathbf{y}-u)$, is minimized. (Note that it is the variance of the prediction error, not the variance of the predictor, that is minimized.) Thus, the BLUP of u is the linear predictor $\boldsymbol{\lambda}^T\mathbf{y}$ that minimizes $\mathrm{Var}(\boldsymbol{\lambda}^T\mathbf{y}-u)$ subject to the constraint specified by equation (9.1). Methods of multivariate calculus may be used to solve this problem (see, for example, Goldberger (1962)), yielding the following predictor as the BLUP of u:

$$\hat{u}=\mathbf{x}_u^T\hat{\boldsymbol{\beta}}_{GLS}+\mathbf{r}_{\mathbf{y}u}^T\mathbf{R}^{-1}(\mathbf{y}-\mathbf{X}\hat{\boldsymbol{\beta}}_{GLS}),\tag{9.2}$$

where $\hat{\boldsymbol{\beta}}_{GLS}$ is the generalized least squares estimator of $\boldsymbol{\beta}$ that was first introduced in Chapter 5. Since the elements of $\hat{\boldsymbol{\beta}}_{GLS}$ are linear functions of $\mathbf{y}$, $\hat{u}$ is a linear predictor, and it may be verified that the corresponding $\boldsymbol{\lambda}$ satisfies (9.1); see Exercise 9.1. Furthermore, the minimized prediction error variance is given by

$$\mathrm{Var}(\hat{u}-u)=\sigma^2[r_{uu}-\mathbf{r}_{\mathbf{y}u}^T\mathbf{R}^{-1}\mathbf{r}_{\mathbf{y}u}+(\mathbf{x}_u^T-\mathbf{r}_{\mathbf{y}u}^T\mathbf{R}^{-1}\mathbf{X})(\mathbf{X}^T\mathbf{R}^{-1}\mathbf{X})^{-1}(\mathbf{x}_u-\mathbf{X}^T\mathbf{R}^{-1}\mathbf{r}_{\mathbf{y}u})]$$
$$\equiv\sigma^2 V.\tag{9.3}$$

Using this prediction error variance, a $100(1-\epsilon)\%$ prediction interval for u may be constructed as follows:

$$\hat{u}\pm t_{\epsilon/2,n-p}\hat{\sigma}_{GLS}\sqrt{V},$$

where $\hat{\sigma}_{GLS}^2$ is the generalized residual mean square that, like $\hat{\boldsymbol{\beta}}_{GLS}$, was first introduced in Chapter 5.

It is illuminating to compare expression (9.2) for the BLUP to the conditional mean of u given $\mathbf{y}$ when $\boldsymbol{\beta}$ is known and the joint distribution of u and $\mathbf{y}$ is multivariate normal:

$$\mathbf{x}_u^T\boldsymbol{\beta}+\mathbf{r}_{\mathbf{y}u}^T\mathbf{R}^{-1}(\mathbf{y}-\mathbf{X}\boldsymbol{\beta}).$$

Plainly, the BLUP has the same form as the conditional mean, with $\hat{\boldsymbol{\beta}}_{GLS}$ substituted for $\boldsymbol{\beta}$. Furthermore, the prediction error variance for the conditional mean is

$$\sigma^2(r_{uu}-\mathbf{r}_{\mathbf{y}u}^T\mathbf{R}^{-1}\mathbf{r}_{\mathbf{y}u}).$$

Upon comparison with (9.3), this shows that the term

$$\sigma^2(\mathbf{x}_u^T-\mathbf{r}_{\mathbf{y}u}^T\mathbf{R}^{-1}\mathbf{X})(\mathbf{X}^T\mathbf{R}^{-1}\mathbf{X})^{-1}(\mathbf{x}_u-\mathbf{X}^T\mathbf{R}^{-1}\mathbf{r}_{\mathbf{y}u})$$

in the BLUP's prediction error variance, which is always nonnegative, may be interpreted as the penalty for having to estimate $\boldsymbol{\beta}$.

The preceding has supposed that only one predictand is of interest, whereas actually it is often the case that the prediction of a vector of random variables $\mathbf{u}$ is desired. Expressions (9.2) and (9.3) for the BLUP and its prediction error variance may be generalized straightforwardly for this objective. For example, the vector of BLUPs of $\mathbf{u}$ is

$$\hat{\mathbf{u}} = \mathbf{X_u}^T \hat{\boldsymbol{\beta}}_{GLS} + \mathbf{R}_{\mathbf{yu}}^T \mathbf{R}^{-1}(\mathbf{y} - \mathbf{X}\hat{\boldsymbol{\beta}}_{GLS}), \tag{9.4}$$

which has prediction error covariance matrix

$$\begin{aligned}
\mathrm{Var}(\hat{\mathbf{u}} - \mathbf{u}) &= \sigma^2[\mathbf{R}_{\mathbf{uu}} - \mathbf{R}_{\mathbf{yu}}^T \mathbf{R}^{-1} \mathbf{R}_{\mathbf{yu}} \\
&\quad + (\mathbf{X_u}^T - \mathbf{R}_{\mathbf{yu}}^T \mathbf{R}^{-1} \mathbf{X})(\mathbf{X}^T \mathbf{R}^{-1} \mathbf{X})^{-1}(\mathbf{X_u} - \mathbf{X}^T \mathbf{R}^{-1} \mathbf{R}_{\mathbf{yu}})] \\
&\equiv \sigma^2 \mathbf{V}, \tag{9.5}
\end{aligned}$$

where the meanings of the symbols $\mathbf{R}_{\mathbf{uu}}$, $\mathbf{R}_{\mathbf{yu}}$, and $\mathbf{X_u}^T$ are obvious by analogy with (9.3). The main diagonal elements of $\sigma^2 \mathbf{V}$ are the prediction error variances of the individual BLUPs, and the off-diagonal elements of $\sigma^2 \mathbf{V}$ are their prediction error covariances. The latter are not needed for constructing marginal prediction intervals for the predictands, but are needed for constructing joint prediction regions and some types of simultaneous prediction intervals for those quantities.

Now let us specialize the preceding results to best linear unbiased prediction under spatial random and mixed effects models. Recall, from Section 5.4, that a spatial mixed effects model has the form

$$\mathbf{y} = \mathbf{X}\boldsymbol{\beta} + \mathbf{Z}\mathbf{b} + \mathbf{d},$$

where $\mathbf{y}$, $\mathbf{X}$, and $\boldsymbol{\beta}$ are defined as for any spatial linear model, $\mathbf{Z}$ is a specified $n \times q$ matrix, $\mathbf{b}$ is a q-vector of zero-mean random effects with positive definite covariance matrix $\sigma^2 \mathbf{G}$, $\mathbf{d}$ is an n-vector of zero-mean random variables with positive definite diagonal covariance matrix $\sigma^2 \mathbf{D}$, and $\mathbf{b}$ and $\mathbf{d}$ are uncorrelated. Consider the problem of predicting an arbitrary vector of linear combinations of fixed and random effects, $\mathbf{X_u}^T \boldsymbol{\beta} + \mathbf{Z_u}^T \mathbf{b}$. This can be seen as a special case of the prediction problem described in the previous paragraph, with

$$\mathbf{R} = (1/\sigma^2)\mathrm{Var}(\mathbf{Z}\mathbf{b} + \mathbf{e}) = \mathbf{Z}\mathbf{G}\mathbf{Z}^T + \mathbf{D} \quad \text{and} \quad \mathbf{R}_{\mathbf{yu}} = (1/\sigma^2)\mathrm{Cov}(\mathbf{Z}\mathbf{b} + \mathbf{e}, \mathbf{Z_u}^T \mathbf{b}) = \mathbf{Z}\mathbf{G}\mathbf{Z_u}.$$

Thus, specializing (9.4), the BLUP of $\mathbf{X_u}^T \boldsymbol{\beta} + \mathbf{Z_u}^T \mathbf{b}$ is

$$\mathbf{X_u}^T \hat{\boldsymbol{\beta}}_{GLS} + \mathbf{Z_u}^T \mathbf{G}\mathbf{Z}^T (\mathbf{Z}\mathbf{G}\mathbf{Z}^T + \mathbf{D})^{-1}(\mathbf{y} - \mathbf{X}\hat{\boldsymbol{\beta}}_{GLS}).$$

The covariance matrix of prediction errors associated with this predictor is $\sigma^2 \mathbf{V}$, where in this case, specializing (9.5) yields

$$\begin{aligned}
\mathbf{V} &= \mathbf{Z_u}^T \mathbf{G}\mathbf{Z_u} - \mathbf{Z_u}^T \mathbf{G}\mathbf{Z}^T (\mathbf{Z}\mathbf{G}\mathbf{Z}^T + \mathbf{D})^{-1} \mathbf{Z}\mathbf{G}\mathbf{Z_u} + \{[\mathbf{X_u}^T - \mathbf{Z_u}^T \mathbf{G}\mathbf{Z}^T (\mathbf{Z}\mathbf{G}\mathbf{Z}^T + \mathbf{D})^{-1}\mathbf{X}] \\
&\quad \times [\mathbf{X}^T (\mathbf{Z}\mathbf{G}\mathbf{Z}^T + \mathbf{D})^{-1}\mathbf{X}]^{-1}[\mathbf{X_u}^T - \mathbf{Z_u}^T \mathbf{G}\mathbf{Z}^T (\mathbf{Z}\mathbf{G}\mathbf{Z}^T + \mathbf{D})^{-1}\mathbf{X}]^T\}.
\end{aligned}$$

Computationally efficient inversion of $\mathbf{Z}\mathbf{G}\mathbf{Z}^T + \mathbf{D}$ in these expressions may be accomplished using (5.11).

9.2 Ordinary kriging

Now we consider kriging, i.e., applications of best linear unbiased prediction to spatial linear models. For our first application, suppose that the model is a geostatistical Aitken model with covariance matrix $\sigma^2 \mathbf{R} = (C(\mathbf{s}_i, \mathbf{s}_j))$ and that we wish to predict a response $y_0 \equiv Y(\mathbf{s}_0)$ where $\mathbf{s}_0$ denotes an arbitrary point in $\mathcal{A}$. Usually $\mathbf{s}_0$ will be unsampled, but the methodology is flexible enough to allow it to be a data site. Suppose further that the mean of $Y(\cdot)$ is constant, and write this constant as β_0. Then $\mathrm{E}(y_0) = \beta_0$ and $\mathrm{E}(\mathbf{y}) = \mathbf{1}\beta_0$. Write the GLS estimator of β_0 as $\hat{\beta}_{0,GLS}$. Let $r_{00} = (1/\sigma^2)\mathrm{Var}(y_0)$, and let $\mathbf{r}_0$ denote the n-vector whose ith element is $(1/\sigma^2)\mathrm{Cov}(y_i, y_0)$. Then, specializing (9.2) to this setup, the BLUP of y_0 is

$$\hat{y}_0 = \hat{\beta}_{0,GLS} + \mathbf{r}_0^T \mathbf{R}^{-1}(\mathbf{y} - \hat{\beta}_{0,GLS}\mathbf{1}). \tag{9.6}$$

In this context, the BLUP is called the **ordinary kriging (OK) predictor** of y_0 and the elements of the vector $\boldsymbol{\lambda}$ defined implicitly in (9.6) are called the **OK weights** corresponding to the given covariance function and given spatial configuration of data and prediction sites. The OK weights sum to one by (9.1), but are otherwise unrestricted; importantly, they do not depend on $\mathbf{y}$. Furthermore, specializing (9.3) to this setup yields the OK predictor's prediction error variance, also called the **OK variance**:

$$
\begin{aligned}
\sigma_{OK}^2(\mathbf{s}_0) &\equiv \mathrm{Var}(\hat{y}_0 - y_0) \\
&= \sigma^2[r_{00} - \mathbf{r}_0^T \mathbf{R}^{-1}\mathbf{r}_0 + (1 - \mathbf{r}_0^T \mathbf{R}^{-1}\mathbf{1})(\mathbf{1}^T \mathbf{R}^{-1}\mathbf{1})^{-1}(1 - \mathbf{1}^T \mathbf{R}^{-1}\mathbf{r}_0)] \\
&\equiv \sigma^2 v_0.
\end{aligned} \tag{9.7}
$$

Thus, in the trivial case of observations and predictand that are uncorrelated and homoscedastic with variance σ^2, the OK weights are all equal to $\frac{1}{n}$ (so the OK predictor is merely the sample mean) and the OK variance is $\sigma^2(1 + \frac{1}{n})$. In general, a $100(1 - \epsilon)\%$ prediction interval for y_0 is given by

$$\hat{y}_0 \pm t_{\epsilon/2, n-1} \hat{\sigma}_{GLS} \sqrt{v_0}.$$

A nontrivial toy example will serve to illustrate the calculation of heterogeneous OK weights and the OK variance, and the effects of the spatial configuration of the data and prediction sites on those quantities. The spatial configuration of five data sites and three prediction sites is displayed in Figure 9.1. The data sites are identical to those for the toy example of Sections 4.2 and 5.2, and the prediction sites are $\mathbf{s}_{01} = (2, 4)$, $\mathbf{s}_{02} = (4, 2)$, and $\mathbf{s}_{03} = (4, 4)$. The responses and predictands are assumed to have constant mean β_0 and covariance function identical to that of the aforementioned toy example, i.e., the isotropic spherical covariance function

$$C(r) = \begin{cases} \sigma^2 \left(1 - \frac{3r}{8} + \frac{r^3}{128}\right) & \text{for } 0 \leq r \leq 4, \\ 0 & \text{for } r > 4. \end{cases}$$

Then

$$\mathbf{R} = \begin{pmatrix} 1.000 & 0.633 & 0.492 & 0.249 & 0 \\ & 1.000 & 0.633 & 0.492 & 0 \\ & & 1.000 & 0.633 & 0.061 \\ & & & 1.000 & 0.014 \\ & & & & 1.000 \end{pmatrix},$$

$r_{00} = 1$, and

$$\mathbf{r}_0 = \begin{pmatrix} 0.633 \\ 0.492 \\ 0.633 \\ 0.313 \\ 0.086 \end{pmatrix}, \quad \begin{pmatrix} 0.014 \\ 0.061 \\ 0.249 \\ 0.313 \\ 0.249 \end{pmatrix}, \quad \text{or} \quad \begin{pmatrix} 0.086 \\ 0.061 \\ 0.249 \\ 0.116 \\ 0.633 \end{pmatrix},$$

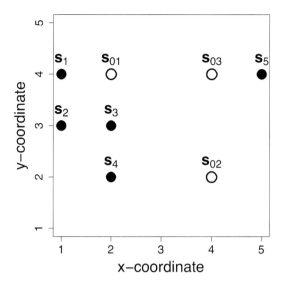

FIGURE 9.1
Spatial configuration of data sites (closed circles) and prediction sites (open circles) for the first toy example of Section 9.2.

depending on whether the prediction site is s_{01}, s_{02}, or s_{03}. The OK weights and OK variance in this scenario may be determined using (9.6) and (9.7) and are given in the first three columns of Table 9.1 (one column for each prediction site). For each prediction site, it can be seen that the weights sum to 1.0 (allowing for roundoff error) and that some of them are negative; negative weights are discussed further in Section 9.8. Broadly, the OK weights are inversely related to the distance from the data sites to the prediction site. For instance, the observation that receives by far the most weight for the prediction of $y_{03} \equiv Y(s_{03})$ is y_5, as s_5 is much closer to s_{03} than any other data site. The relationship between the weights and distances from the prediction site is not perfect, however. For example, the first and third data sites are equidistant from s_{01}, but the corresponding weights are not equal. Also of interest, the spatial weights corresponding to data sites separated from the prediction site by more than the range are not zero. The OK variance is largest at s_{02}, which seems reasonable because s_{02} is furthest from any data site and is therefore not as reliably predicted.

TABLE 9.1
OK weights and OK variances for the first toy example introduced in Section 9.2 (first three columns), and for two variations (remaining six columns). The variations take the covariance function to be exponential and Gaussian, respectively, rather than spherical.

	Spherical			Exponential			Gaussian		
	s_{01}	s_{02}	s_{03}	s_{01}	s_{02}	s_{03}	s_{01}	s_{02}	s_{03}
λ_1	0.488	0.094	0.088	0.396	0.129	0.121	0.692	0.365	0.060
λ_2	-0.061	-0.129	-0.118	0.058	0.054	0.032	-0.537	-0.672	-0.370
λ_3	0.511	0.184	0.296	0.370	0.159	0.161	1.043	0.084	0.712
λ_4	-0.048	0.425	0.037	0.038	0.298	0.107	-0.266	0.795	-0.177
λ_5	0.111	0.426	0.696	0.139	0.360	0.578	0.069	0.429	0.775
$\sigma^2_{OK}(s_{0i})$	0.459	0.917	0.563	0.688	1.039	0.804	0.078	0.605	0.191

Why are the OK weights not perfectly related to distance, and why do observations at data sites farther than the range contribute to the OK predictor? There are two main reasons. First, (9.6) indicates that the weights do not depend on $\mathbf{r}_0$ directly, but on $\mathbf{R}^{-1}\mathbf{r}_0$. Thus, unless $\mathbf{R}$ is diagonal, the weight associated with the ith observation is affected by spatial correlations between the predictand and observations at sites other than site i. Second, the OK predictor includes the GLS estimator of the unknown constant mean, to which all observations contribute. Recall from Sections 5.2 and 5.3, however, that the contributions of each observation to the GLS estimator of the mean typically are not equal, but instead are affected by the spatial configuration of data sites and in particular by how spatially isolated a data site is from the others. Thus, for example, if three data sites are equidistant from a prediction site but two of the data sites are much closer to each other than to the third data site, the weights for those two data sites usually will be smaller than that for the third site.

To see how much of an effect the functional form of the covariance function might have on ordinary kriging, the previous analysis was repeated using the isotropic exponential covariance function $C(r) = \exp(-3r/4)$ and the isotropic Gaussian covariance function $C(r) = \exp(-3r^2/16)$. The practical range of both these covariance functions is 4.0, the same as the range of the previous spherical covariance function, so in this sense the spatial correlation structures prescribed by the three models are of comparable strength. Results are given in Table 9.1. The OK weights and OK variances under the exponential model are rather similar to those under the spherical model, but not identical. There are no negative weights under the exponential model, for example. The OK weights under the Gaussian model are rather more different, however, and the OK variances are much smaller than their counterparts under the other two models. The explanation for this is that the correlation at small distances is much stronger under the Gaussian model than it is under a spherical or exponential model with the same (effective) range, and as will be seen in Section 9.5, stronger spatial correlation generally results in more precise prediction.

A second toy example will serve to illustrate some additional features of ordinary kriging. Consider a scenario in which there are 10 data sites spaced haphazardly within the unit interval in one dimension, and suppose that prediction is desired at every point in the interval. A one-dimensional interval rather than a planar region is chosen for this example to enhance visualization of the predictions. The process $Y(\cdot)$ is assumed to be second-order stationary with exponential covariance function $C(r) = \exp(-20r)$, so that the practical correlation range is 0.15. Figure 9.2A plots the observations and their locations, and Figure 9.2B displays the OK "surface" (the curve connecting OK predictors at all points) and corresponding OK variances for this scenario. The OK surface gives some idea of the nature of the interpolation performed by ordinary kriging. The interpolation obviously is not linear, and the OK surface is not very smooth, in the sense that it appears to be nondifferentiable at data sites. But the interpolation is *perfect*, i.e., the OK predictor at any data site is equal to the observation at that site. Correspondingly, the OK variance at any data site is zero. In fact, this is true of OK prediction universally, not merely for this toy example. To see this, consider the OK predictor at an arbitrary data site $\mathbf{s}_i$, where the data sites and spatial linear model are arbitrary except that the mean is constant. The observation at $\mathbf{s}_i$, $y_i \equiv Y(\mathbf{s}_i)$, is trivially a linear unbiased combination of the observations (the OK weight for the ith observation is 1 and the OK weights for the other observations are all zeros), and $\text{Var}(\hat{y}_0 - y_i) = \text{Var}(0) = 0$ when $\hat{y}_0 = y_i$, thus minimizing the prediction error variance. Many kriging practitioners have viewed perfect interpolation favorably, as it fully "honors the data" by reproducing observations at data sites. However, as will be described in Section 9.4, it can lead to some undesirable features. A final noteworthy feature in Figure 9.2B is that the OK variances tend to increase nonlinearly with distance from the prediction site to the nearest data site.

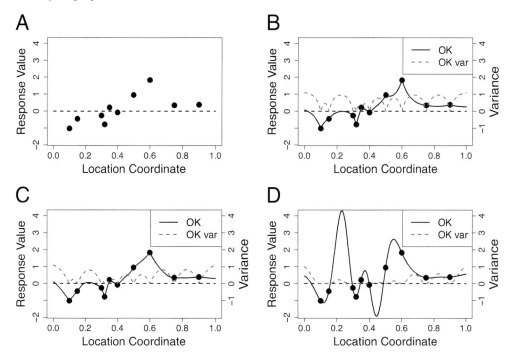

FIGURE 9.2
Displays pertaining to the second toy example of Section 9.2. The spatial domain is a unit interval. A) The observations at their spatial locations; B) the OK surface and the OK variances when the covariance function is exponential with practical range 0.15; C–D) the same quantities as in B when the covariance function is spherical and Gaussian, respectively, with the same range or practical range.

Figures 9.2C,D display the OK predictions and OK variances for the same scenario as the previous one, except that the exponential covariance function is replaced by the spherical and Gaussian covariance functions, respectively, with the same range or practical range as the exponential. The results for the spherical case are not much different from those from the exponential case, but the results for the Gaussian case are quite different. For the latter, the OK surface at data sites appears to be much smoother, and predictions away from data sites stray, in some cases, much beyond the range of the observations. In addition, there is less spatial variation in the OK variances under the Gaussian model, and they are smaller.

Although $Y(\cdot)$ for the toy examples presented in this section was assumed to be second-order stationary (even isotropic), this is not generally necessary. Expressions (9.6) and (9.7) for the OK predictor and OK variance are valid for any constant-mean nonstationary process with finite variances. Moreover, those expressions can be extended for use with a constant-mean process that lacks finite variances but has finite semivariances. Let $\mathbf{\Gamma} = (\gamma(\mathbf{s}_i, \mathbf{s}_j))$ be the semivariance matrix for the responses at data locations, and let $\boldsymbol{\gamma}_0 = (\gamma(\mathbf{s}_i, \mathbf{s}_0))$ be the vector of semivariances of the (data site, prediction site) pairs of responses. The extended expressions for the OK predictor and OK variance are

$$\hat{y}_0 = (\mathbf{1}^T \mathbf{\Gamma}^{-1} \mathbf{1})^{-1} \mathbf{1}^T \mathbf{\Gamma}^{-1} \mathbf{y} + \boldsymbol{\gamma}_0^T \mathbf{\Gamma}^{-1} [\mathbf{y} - \mathbf{1}(\mathbf{1}^T \mathbf{\Gamma}^{-1} \mathbf{1})^{-1} \mathbf{1}^T \mathbf{\Gamma}^{-1} \mathbf{y}] \qquad (9.8)$$

and

$$\mathrm{Var}(\hat{y}_0 - y_0) = \boldsymbol{\gamma}_0^T \mathbf{\Gamma}^{-1} \boldsymbol{\gamma}_0 - (1 - \boldsymbol{\gamma}_0^T \mathbf{\Gamma}^{-1} \mathbf{1})(\mathbf{1}^T \mathbf{\Gamma}^{-1} \mathbf{1})^{-1}(1 - \mathbf{1}^T \mathbf{\Gamma}^{-1} \boldsymbol{\gamma}_0). \qquad (9.9)$$

TABLE 9.2

OK weights and OK variances for prediction
at three sites for the areal data example
introduced in Section 9.2.

	A_1	A_3	A_7
λ_1	—	0.167	0.143
λ_2	0.215	0.417	0.064
λ_3	0.270	—	−0.288
λ_4	0.215	0.417	0.596
λ_5	0.268	−0.335	−0.329
λ_6	−0.236	0.667	0.517
λ_7	0.268	−0.335	—
$\sigma^2_{OK}(A_i)$	1.665	1.033	0.887

Expressions (9.8) and (9.9) coincide with (9.6) and (9.7) when the process has finite variances (Exercise 9.11). Extensions of the expressions can also be made for prediction when $Y(\cdot)$ is a higher-order intrinsic random function.

Ordinary kriging in the context of a spatial-weights linear model for areal data is achieved using the same expressions used for the predictor and prediction error variance in the geostatistical context, but with the elements of $\mathbf{R}$, $\mathbf{r}_0$ and r_{00} determined by the spatial-weights model, of course. As an example, Table 9.2 gives the OK weights and OK variance for a scenario identical to that of the example in Section 7.2, except that either y_1, y_3, or y_7 is predicted from the other observations. The model for the observations and predictands has constant mean, the spatial weights matrix is that for a single multiplicative-parameter SAR with $\rho_{SAR} = 0.3$, and $\mathbf{K}_{SAR} = \mathbf{I}$. The SAR model is formulated in the context of the complete seven-site configuration. The OK weights tend to be largest for the most proximate data sites to the prediction site, although they are also affected by the spatial configuration of the sites as a whole. In these respects, the results are similar to what they were for the toy examples of ordinary kriging under geostatistical models, which were presented earlier in this section. Somewhat surprisingly, however, the OK variance is not smallest when predicting at the interior site (A_3). Counterparts of the entries in Table 9.2 for some other spatial-weights models are requested in Exercises 9.9 and 9.10.

9.3   Universal kriging

The optimality of ordinary kriging is limited to a model in which the observations and predictands have constant mean. This is too restrictive for many applications. Accordingly, our second application of the theory and methodology of best linear unbiased prediction is to the prediction of y_0 under a spatial Aitken model with completely general linear mean structure, i.e., $E(\mathbf{y}) = \mathbf{X}\boldsymbol{\beta}$ for a full column rank (but otherwise arbitrary) $n \times p$ matrix $\mathbf{X}$ of explanatory variables observed at data locations, and $E(y_0) = \mathbf{x}_0^T \boldsymbol{\beta}$ for some p-vector $\mathbf{x}_0$ whose elements are the values of those same explanatory variables at $\mathbf{s}_0$. Specializing (9.2) to this more general setup, the BLUP of y_0, called the **universal kriging (UK) predictor** in this context, is

$$\hat{y}_{0,UK} = \mathbf{x}_0^T \hat{\boldsymbol{\beta}}_{GLS} + \mathbf{r}_0^T \mathbf{R}^{-1}(\mathbf{y} - \mathbf{X}\hat{\boldsymbol{\beta}}_{GLS}), \qquad (9.10)$$

and the elements of the vector $\boldsymbol{\lambda}$ defined implicitly in (9.10) are called the **UK weights** corresponding to the given mean and covariance functions and the given spatial configuration

TABLE 9.3

UK weights and UK variances for the first toy example introduced in Section 9.2, assuming a planar mean function.

	$\mathbf{s}_{01}$	$\mathbf{s}_{02}$	$\mathbf{s}_{03}$
λ_1	0.616	−0.494	0.217
λ_2	−0.116	−0.085	−0.135
λ_3	0.551	0.125	0.313
λ_4	−0.217	0.980	−0.089
λ_5	0.167	0.474	0.694
$\sigma^2_{UK}(\mathbf{s}_{0i})$	0.495	1.397	0.586

of data sites and prediction sites. Specializing (9.3) to this setup yields the UK predictor's prediction error variance, also called the **UK variance**:

$$
\begin{aligned}
\sigma^2_{UK}(\mathbf{s}_0) &\equiv \text{Var}(\hat{y}_{0,UK} - y_0) \\
&= \sigma^2[r_{00} - \mathbf{r}_0^T \mathbf{R}^{-1} \mathbf{r}_0 \\
&\quad + (\mathbf{x}_0^T - \mathbf{r}_0^T \mathbf{R}^{-1} \mathbf{X})(\mathbf{X}^T \mathbf{R}^{-1} \mathbf{X})^{-1}(\mathbf{x}_0 - \mathbf{X}^T \mathbf{R}^{-1} \mathbf{r}_0)] \\
&\equiv \sigma^2 v_{0,UK}.
\end{aligned}
\tag{9.11}
$$

A $100(1 - \epsilon)\%$ prediction interval for y_0 is given by

$$
\hat{y}_{0,UK} \pm t_{\epsilon/2, n-p} \hat{\sigma}_{GLS} \sqrt{v_{0,UK}}.
$$

All of these expressions reduce to those given previously for ordinary kriging when $\mathbf{X} = \mathbf{1}$. Furthermore, expressions analogous to (9.10) and (9.11) pertaining to a process with finite semivariances may be obtained by replacing 1 and $\mathbf{1}$ in expressions (9.8) and (9.9) with $\mathbf{x}_0$ and $\mathbf{X}$, respectively.

To contrast OK and UK, we modify the first toy example of Section 9.2 by supposing that the mean function of $Y(\cdot)$ is given by the planar surface $\text{E}(Y(u, v)) = \beta_{00} + \beta_{10}u + \beta_{01}v$, while retaining everything else. Thus, the ith row of $\mathbf{X}$ is $(1, u_i, v_i)$, and $\mathbf{x}_0^T = (1, u_0, v_0)$. The UK weights and UK variances for this modified example are given in Table 9.3. It may be verified (Exercise 9.12) that the UK weights in this table not only sum to one, but also satisfy

$$
\sum_{i=1}^n \lambda_i u_i = u_0 \quad \text{and} \quad \sum_{i=1}^n \lambda_i v_i = v_0,
\tag{9.12}
$$

as dictated by (9.1). Compared to the OK weights in Table 9.1, the UK weights are more dispersed. Furthermore, the UK variances are larger than their OK counterparts computed under the same spherical covariance model. This last feature is intuitively reasonable because a model with nonconstant mean structure requires that more parameters be estimated, thereby increasing prediction uncertainty. A more rigorous argument is that the UK predictor satisfies the lone unbiasedness constraint of ordinary kriging (provided that the mean function includes an unknown additive constant) and is therefore unbiased under a constant-mean model, but because there are additional constraints that the UK weights must satisfy, the UK predictor does not generally coincide with the linear unbiased predictor that minimizes the prediction error variance under that model, namely the OK predictor. Of course, the OK predictor generally is not unbiased under the planar mean model.

For further insights, we also modify the second toy example of Section 9.2 by supposing that the mean is not constant, but is instead given by the linear trend function $\text{E}(Y(s)) = \beta_0 + \beta_1 s$. All other aspects of the example are retained as they were. Figure 9.3B plots the

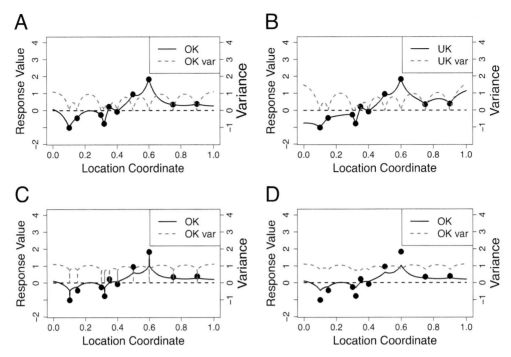

FIGURE 9.3
Displays pertaining to setups similar to the second toy example of Section 9.2. A) The OK surface and OK variances when the covariance function is exponential with practical range 0.05; B) the UK surface and UK variances corresponding to a model with linear trend but the same exponential covariance function; C) the OK surface and OK variances when the covariance function is exponential with the same sill and practical range as before but a 50% nugget attributable to measurement error; D) the surface of noiseless predictors and their prediction error variances when the covariance function is the same as in C.

UK surface and UK variances for this modified example. The surface suggests that the UK predictor, like the OK predictor, is a perfect interpolator; this can indeed be established in a manner similar to how it was established for the OK predictor. However, except at data sites, the UK surface is different than the OK surface shown in Figure 9.3A, most noticeably near the endpoints of the unit interval. At the endpoints, the OK predictor is close to the overall mean of the observations. Not so for the UK predictor, which, as a consequence of the positive fitted trend, lies well below the OK predictor at and near 0 and well above the OK predictor at and near 1. A final noteworthy feature is that the UK variance tends to be as large or larger than the OK variance over the entire unit interval, except at data sites.

A universal kriging scenario in the context of a spatial-weights linear model for areal data is considered in Exercise 9.13.

9.4   Noiseless prediction

As noted previously, the OK predictor is a perfect interpolator, regardless of the nature of the covariance function. However, this property has unexpected consequences when the covariance function has a nugget effect attributable to microscale variation, measurement error, or some combination of the two. To illustrate, Figure 9.3C displays the OK surface for a scenario identical to that of the second toy example of Section 9.2, except that the covariance function is modified to have a 50% nugget effect but the same sill, so that

$$C(r) = \begin{cases} 0.50 \exp(-20r) & \text{if } r > 0, \\ 1.00 & \text{if } r = 0. \end{cases}$$

The OK surface still honors the data, but does so by making the interpolator discontinuous at data locations. This seems unrealistic, hence undesirable.

When the nugget effect is attributable, at least in part, to measurement error, a case can be made that the predictand of primary inferential interest should be not y_0 but a "noiseless version" of y_0, i.e., y_0 minus its measurement error. This is because, quoting Cressie (1991), p. 128, "when predicting at location $\mathbf{s}_0$, one wants to know what actually exists at that location and not the value distorted by, for example, laboratory measurement error." It turns out that in such a case, the BLUP of the noiseless response coincides with the OK predictor of the noisy response at locations that are not data sites, but it is different at data sites. In fact, if the nugget effect is entirely attributable to measurement error, the BLUP of the noiseless response is such that the overall prediction surface is continuous. Figure 9.3D displays this phenomenon; the jumps at data locations that were present in Figure 9.3C have disappeared. Thus, the BLUP of a noiseless response is not necessarily a perfect interpolator. Figure 9.3D also plots the prediction error variance of the BLUP of the noiseless response as a function of the prediction site. Appropriately, these prediction error variances are not equal to zero at data sites.

9.5 Effects of covariance function parameters on kriging

The values of covariance function parameters can affect kriging, both ordinary and universal. Here we consider only the effects of the sill, range, and nugget on the OK weights and OK variances. The effect of the sill is easy to describe: it has no effect whatsoever on the OK weights, but it affects the OK variances multiplicatively. That is, if one process has twice as large a sill as another but is identical otherwise, then the OK variances for the one will be twice as large as those for the other. To study the effects of the nugget and range on ordinary kriging, let us re-use the spatial configuration of the first toy example in Section 9.2, in conjunction with six models (covariance functions). The models are all exponential with a sill of 1.0. Models 1, 2, and 3 are isotropic and nuggetless and have practical ranges equal to 2, 4, and 8, respectively; Models 4 and 5 are isotropic with practical range equal to 4 and nuggets equal to 0.25 and 0.50, respectively; and Model 6 is geometrically anisotropic with ranges 4 and 8 in the N–S and E–W directions, respectively, and zero nugget. OK weights and OK variances for prediction at $\mathbf{s}_{01}$ under the six models are given in Table 9.4. Three key findings emerge. First, as the practical range increases, i.e., as the spatial correlation gets stronger, the weights become more heterogeneous and the variances decrease. The heterogeneity among the weights is such that those that were already larger than average (0.2 in this scenario) when the practical range is 2 get even larger as the practical range increases, and those that were smaller than average get even smaller. Second, as the nugget increases (with the sill held constant), the weights become more homogeneous, resulting in more smoothing and less interpolation of the data, and the variances increase. Third, as the geometric anisotropy becomes more pronounced, the weights corresponding to data sites aligned in the direction of strongest correlation with the prediction site increase, at

TABLE 9.4

OK weights and OK variances under six models (first six columns), and inverse distance weights and inverse distance-squared weights (last two columns), for a scenario with spatial configuration identical to the first toy example of Section 9.2. Models 1–6 are defined in Section 9.5. Prediction error variances of the IDW and IDSW predictors, denoted as σ^2_{IDW} and σ^2_{IDSW}, respectively, are given in the last two rows; each is computed under the model corresponding to the given column.

	Model 1	Model 2	Model 3	Model 4	Model 5	Model 6	IDW	IDSW
λ_1	0.311	0.396	0.446	0.331	0.280	0.594	0.282	0.350
λ_2	0.128	0.058	0.006	0.125	0.163	0.027	0.200	0.175
λ_3	0.285	0.370	0.428	0.253	0.253	0.188	0.282	0.350
λ_4	0.114	0.038	-0.015	0.135	0.135	0.011	0.141	0.087
λ_5	0.162	0.139	0.135	0.169	0.169	0.180	0.094	0.039
$\sigma^2_{OK}(\mathbf{s}_{01})$	0.999	0.688	0.397	0.834	0.965	0.474	—	—
$\sigma^2_{IDW}(\mathbf{s}_{01})$	1.010	0.718	0.437	0.846	0.973	0.592	—	—
$\sigma^2_{IDSW}(\mathbf{s}_{01})$	1.024	0.711	0.427	0.854	0.998	0.577	—	—

the expense of weights corresponding to data sites aligned with the prediction site in the direction of weakest correlation.

9.6   Kriging versus inverse distance weighting

Kriging, in all of its forms, is merely one family of methods for spatial interpolation. Another family used by many environmental scientists is **inverse distance weighting** (IDW). In this approach, the predicted value of a response at any location is a weighted linear combination of the observed data, with weights inversely related to the distances from the data sites to the prediction site. The most commonly used form of inverse distance weighting uses the normalized (to sum to one) inverse distances themselves as weights, so that the predictor is

$$\breve{y}_0 = \frac{\sum_{i=1}^{n} \|\mathbf{s}_i - \mathbf{s}_0\|^{-1} y_i}{\sum_{i=1}^{n} \|\mathbf{s}_i - \mathbf{s}_0\|^{-1}},$$

where, if $\mathbf{s}_0 = \mathbf{s}_i$ for some i, then $\|\mathbf{s}_i - \mathbf{s}_0\|^{-1}$ is defined to equal 1, and $\|\mathbf{s}_j - \mathbf{s}_0\|^{-1}$ for $j \neq i$ is defined to equal 0. Another form uses normalized inverse squared distances as weights, and the general form is

$$\breve{y}_0 = \frac{\sum_{i=1}^{n} \|\mathbf{s}_i - \mathbf{s}_0\|^{-\alpha} y_i}{\sum_{i=1}^{n} \|\mathbf{s}_i - \mathbf{s}_0\|^{-\alpha}},$$

where α is any positive constant.

Clearly, an IDW predictor is linear, and it is a perfect interpolator. Because the weights are inherently nonnegative, IDW predictors are constrained to the range of observations. Furthermore, because the weights are normalized to sum to one, an IDW predictor is unbiased when the observations have constant mean, but is generally biased otherwise. Since ordinary kriging yields the best linear unbiased predictor under a constant-mean model, the theory of best linear unbiased prediction tells us that the IDW predictor has larger prediction error variance than the OK predictor. Prediction error variances associated with

IDW predictors traditionally have not been reported, but they could be, by applying (2.20) to $(\mathbf{a}^T, -1) \begin{pmatrix} \mathbf{y} \\ y_0 \end{pmatrix}$ where $\mathbf{a}$ is the vector of inverse-distance weights.

How much worse is IDW prediction relative to ordinary kriging? The answer depends on the spatial configuration of data and prediction sites and the strength of the spatial dependence. The last two columns of Table 9.4 present the weights corresponding to inverse distance weighting with $\alpha = 1$ (IDW) and $\alpha = 2$ (IDSW) for prediction at $\mathbf{s}_{01}$ in the first toy example of Section 9.2. The prediction error variances for the two inverse distance weighting predictors under the six covariance models considered in Section 9.5 are also given in the last two rows of the table. For most of the models, the inverse distance weighting predictors perform only a few percentage points worse than the OK predictors. The decay in relative performance is worse when the spatial dependence is strong than when it is weak (Model 3 versus Model 1), and is quite substantial (about 20%) when the spatial correlation is geometrically anisotropic (Model 6). Many authors have carried out more extensive comparative studies of kriging and IDW, using real and simulated data, with varying results. In a comprehensive numerical experiment, Zimmerman *et al.* (1999) found that the empirical ordinary and universal kriging variances were consistently smaller (by 20-30% on the log scale) than the empirical prediction error variances of inverse distance weighting, though the magnitude of the difference depended to some degree on the mean function, sampling pattern, strength of spatial correlation, and level of noise. Some other studies report less of a difference. A major weakness of inverse distance weighting relative to kriging is that it cannot adapt to the nature, strength, anisotropy, or nonstationarity of the spatial correlation, or to nonstationarity of the variance. Another weakness of inverse distance weighting is that it does not adapt as easily to the change-of-support prediction problems described in the next section. An advantage of IDW prediction is, of course, that it requires much less computation than kriging, as no parameter estimation (hence no matrix inversion) is involved.

9.7 Change-of-support prediction problems

Implicit in our presentations of ordinary and universal kriging to this point was the supposition that the predictands have the same support as the observations. More specifically, expressions (9.6) and (9.7) for the OK predictor and OK variance, and expressions (9.10) and (9.11) for their UK counterparts, are appropriate in either of two settings. The first setting is one in which the data and spatial linear model are geostatistical (hence the data locations are points and the model includes a covariance function that models the covariance between responses at points) and the problem is to predict responses at points—a problem known as **point-to-point kriging**. The second setting is one in which the data and spatial linear model are areal (hence the data locations are subregions and the model includes a spatial weighting matrix to indirectly model the covariance between subregions) and the problem is to predict responses corresponding either to data locations or to subregions that do not overlap with any data locations. Sometimes, in the geostatistical context, prediction of the average value $\overline{Y}(B) \equiv \int_B Y(\mathbf{s}) d\mathbf{s}/|B|$ over a region (block) $B \subset D$ of positive d-dimensional volume $|B|$ is desired, rather than prediction of $Y(\cdot)$ at one or more points. Historically, for example, mining engineers were interested in this type of prediction problem because the economics of mining required the extraction of material in relatively large blocks. This type of kriging is called **block kriging**. Expressions for the universal block kriging predictor of $\overline{Y}(B)$ and its associated kriging variance are identical to

(9.10) and (9.11), respectively, but with revised definitions $r_{00} = (1/\sigma^2) \int_B \int_B C(\mathbf{s}, \mathbf{t}) \, d\mathbf{s} \, d\mathbf{t}$, $\mathbf{r}_0 = (1/\sigma^2)[C(B, \mathbf{s}_1), \dots, C(B, \mathbf{s}_n)]^T$ where $C(B, \mathbf{s}_i) = |B|^{-1} \int_B C(\mathbf{s} - \mathbf{s}_i) \, d\mathbf{s}$, and $\mathbf{x}_0 = [\overline{X}_1(B), \dots, \overline{X}_p(B)]^T$ where $\overline{X}_j(B) = |B|^{-1} \int_B X_j(\mathbf{s}) \, d\mathbf{s}$ and p is the number of explanatory variables. In practice, these integrals may not be available in closed form, in which case they may be approximated using numerical integration methods, provided that $Y(\cdot)$ is mean-square continuous; continuity of the covariance function at the origin suffices if $Y(\cdot)$ is stationary (Stein, 1999).

A version of block kriging suitable for prediction of an average or total of realized response values within a finite-population areal sampling framework was proposed by Ver Hoef (2008) and has been used successfully to estimate and monitor moose populations in Alaska and northwest Canada. Let $A_1, \dots, A_N$ denote a population of areal sites, of which only $n < N$ are actually sampled. Without loss of generality let $A_1, \dots, A_n$ denote the sampled sites. Furthermore, let $\mathbf{y}_o$ and $\mathbf{y}_u$ ("o" for observed and "u" for unobserved) denote the n-vector of observations at sampled sites and the $(N - n)$-vector of response values at unsampled sites. We wish to use the observational vector $\mathbf{y}_o$ to predict the finite-population average

$$\overline{Y} = \frac{1}{N} \mathbf{1}_N^T \begin{pmatrix} \mathbf{y}_o \\ \mathbf{y}_u \end{pmatrix} = (\mathbf{b}_o^T, \mathbf{b}_u^T) \begin{pmatrix} \mathbf{y}_o \\ \mathbf{y}_u \end{pmatrix},$$

where $\mathbf{b}_o = (1/N)\mathbf{1}_n$ and $\mathbf{b}_u = (1/N)\mathbf{1}_{N-n}$ (for predicting the finite-population total we merely omit the factor of $1/N$). The prediction is based on the following spatial linear model for the observations and unobserved responses:

$$\begin{pmatrix} \mathbf{y}_o \\ \mathbf{y}_u \end{pmatrix} = \begin{pmatrix} \mathbf{X}_o \\ \mathbf{X}_u \end{pmatrix} \boldsymbol{\beta} + \begin{pmatrix} \mathbf{e}_o \\ \mathbf{e}_u \end{pmatrix},$$

where $\mathbf{X}_o$ and $\mathbf{X}_u$ are the matrices of explanatory variables for the observed and unobserved responses, respectively, and $\mathbf{e}_o$ and $\mathbf{e}_u$ are the corresponding vectors of model errors. Denote the covariance matrix of the model errors, in partitioned form, as

$$\mathrm{Var}\begin{pmatrix} \mathbf{e}_o \\ \mathbf{e}_u \end{pmatrix} = \begin{pmatrix} \boldsymbol{\Sigma}_{oo} & \boldsymbol{\Sigma}_{ou} \\ \boldsymbol{\Sigma}_{uo} & \boldsymbol{\Sigma}_{uu} \end{pmatrix}.$$

Then the best linear unbiased predictor of $\overline{Y}$ is

$$\hat{\overline{Y}} = \mathbf{b}_o^T \mathbf{y}_o + \mathbf{b}_u^T \mathbf{X}_u \hat{\boldsymbol{\beta}}_{GLS} + \mathbf{b}_u^T \boldsymbol{\Sigma}_{uo} \boldsymbol{\Sigma}_{oo}^{-1} (\mathbf{y}_o - \mathbf{X}_o \hat{\boldsymbol{\beta}}_{GLS}),$$

where $\hat{\boldsymbol{\beta}}_{GLS}$ is the generalized least squares estimator of $\boldsymbol{\beta}$ based on the observed data, i.e., $\hat{\boldsymbol{\beta}}_{GLS} = (\mathbf{X}_o^T \boldsymbol{\Sigma}_{oo}^{-1} \mathbf{X}_o)^{-1} \mathbf{X}_o^T \boldsymbol{\Sigma}_{oo}^{-1} \mathbf{y}_o$. The associated prediction error variance is

$$\begin{aligned} \mathrm{Var}(\hat{\overline{Y}} - \overline{Y}) = \ & \mathbf{b}_u^T [\boldsymbol{\Sigma}_{uu} - \boldsymbol{\Sigma}_{uo} \boldsymbol{\Sigma}_{oo}^{-1} \boldsymbol{\Sigma}_{ou} \\ & + (\mathbf{X}_u^T - \mathbf{X}_o^T \boldsymbol{\Sigma}_{oo}^{-1} \boldsymbol{\Sigma}_{ou})^T (\mathbf{X}_o^T \boldsymbol{\Sigma}_{oo}^{-1} \mathbf{X}_o)^{-1} (\mathbf{X}_u^T - \mathbf{X}_o^T \boldsymbol{\Sigma}_{oo}^{-1} \boldsymbol{\Sigma}_{ou})] \mathbf{b}_u. \end{aligned}$$

Block kriging is an example of a change-of-support prediction problem, i.e., a problem in which the predictands and observations have different support. In particular, block kriging is a solution to an *upscaling* problem, which is inference at a cruder level of support than the observations. The opposite type of problem is *downscaling*, or inference at a finer level of support than the observations. Prediction at points on the basis of areal data is a problem of this type, and **area-to-point (ATP) kriging** (Kyriakidis, 2004) is a solution within the geostatistical linear model framework. With the predictand in this case denoted once again by $y_0 = Y(\mathbf{s}_0)$ and the data denoted by $Y(B_1), Y(B_2), \dots, Y(B_n)$, ATP kriging is merely BLUP with $r_{00} = (1/\sigma^2)\mathrm{Var}(y_0)$, $\mathbf{r}_0 = (1/\sigma^2)[C(B_1, \mathbf{s}_0), \dots, C(B_n, \mathbf{s}_0)]^T$ where $C(B_i, \mathbf{s}_0) =$

$|B|_i^{-1} \int_{B_i} C(\mathbf{s}, \mathbf{s}_0) \, d\mathbf{s}$, and $\mathbf{R} = (1/\sigma^2)(C(B_i, B_j))$ where $C(B_i, B_j) = \int_{B_i} \int_{B_j} C(\mathbf{s}, \mathbf{t}) \, d\mathbf{s} \, d\mathbf{t}$. The integrals in these expressions may be approximated in the same way they are for block kriging. A very pleasing property of ATP kriging is that it produces *coherent* predictions, i.e., point predictions whose average over any areal data location B_i is equal to $Y(B_i)$. A slight variation is area-and-point-to-point kriging (Goovaerts, 2010), for use when the data have both areal and point support.

Yet another change-of-support prediction problem is prediction from areal data to a block that is neither a data site nor the union of data sites. This is neither an upscaling nor downscaling problem, but a spatial misalignment problem. Nevertheless, best linear unbiased prediction of blocks under spatial misalignment can be dealt with in a geostatistical context using an approach similar to block and ATP kriging called **area-to-area (ATA) kriging** (Goovaerts, 2008). A relevant `R` package for both ATP and ATA kriging is `atakrig`. Alternative Bayesian approaches for handling spatial misalignment have also been proposed; see Mugglin *et al.* (2000), Gelfand *et al.* (2001), and Gotway & Young (2002). Some of these approaches can deal with prediction of a misaligned areal predictand within the context of a spatial-weights model for areal data.

9.8 The screening effect and negative kriging weights

The **screening effect** refers to a decrease in the magnitude of the OK weight for a data site from what it would otherwise be were there not another data site in or approximately in a position between it and the prediction site, "screening" it, as it were, from the prediction site. Figure 9.4 and the accompanying Table 9.5 depict the screening effect for a simple scenario with two data sites, one fixed and the other transient, and a prediction site, under models with three covariance functions: spherical with range 4, exponential with practical range 4, and Gaussian with practical range 4. The OK weight assigned to the fixed data site located at $(1.1, 0)$, which is slightly farther than the transient data site from the prediction site at $(0, 0)$, is almost as large as the weight assigned to the transient data site when the two data sites lie on opposite sides of the prediction site, but it decreases as the transient site moves along the perimeter of the unit circle into an intervening position. When the covariance function is spherical or exponential, the OK weights of the datum at the fixed site are positive and decrease toward zero as the transient site intervenes, and the fixed-site datum therefore has less influence on the prediction. However, when the covariance function is Gaussian, the weight at the fixed data site decreases well beyond zero, becoming negative and large in magnitude. This implies that the screened datum will have a large negative effect on the prediction. Thus, the existence of a screening effect may depend on the covariance function. Of some additional interest, for the spherical and exponential covariance functions the OK variance increases monotonically as the transient data site moves into the intervening position, but this is not so for the Gaussian covariance function.

The screening effect has been invoked as an informal justification for the common practice of basing spatial prediction on a small number of observations at sites close to the prediction site, rather than on all observations, in order to reduce computations for large data sets or to protect against model misspecification at large spatial scales. Further details on this practice are given in Section 9.9, but here we describe a more rigorous justification for it in the context of at least some covariance functions. This justification was established by Stein (2002, 2011, 2015), who gave the screening effect a more formal definition in terms of a limit of the ratio of two OK variances, for prediction at an unobserved site, under a constant-mean, second-order stationary, d-dimensional and mean-square

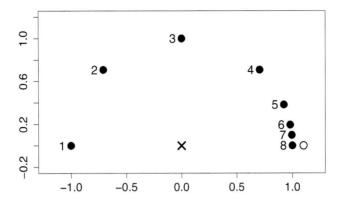

FIGURE 9.4
Two data sites and a prediction site for the demonstration of the screening effect. The prediction site ($\times$) and one data site (open circle) are fixed at the origin and $(1.1, 0)$, respectively, and the other data site (closed circle) occupies eight positions on the perimeter of the unit circle, beginning at $(-1, 0)$ and moving clockwise to $(1, 0)$.

continuous geostatistical process. The two OK variances correspond to the OK predictors based on observations at a set of data sites "near" in some sense to the prediction site, and observations taken on the union of that set and another set of more distant data sites. Both sets are indexed by a parameter, ϵ, controlling the observation locations, that tends to zero. For example, the near sites might be a fixed number of sites that exclude the origin and shrink toward the prediction site as $\epsilon \downarrow 0$ while maintaining the same relative spatial configuration, and the more distant sites might also be fixed in number and exclude the origin, but shrink toward another non-origin site while maintaining their relative spatial configuration. If the limit as $\epsilon \downarrow 0$ of the ratio of the two OK variances is 1.0 for a given covariance function, then an asymptotic screening effect is said to hold for that covariance function. Stein (2011) showed that an asymptotic screening effect, so defined, holds for

TABLE 9.5
The OK weight at the fixed data site located at $(1.1, 0)$, and the OK variance at the prediction site located at $(0, 0)$, for the spatial configurations displayed in Figure 9.4 under the spherical, exponential, and Gaussian covariance functions. The screening effect of the transient data site on the fixed data site is demonstrated as the former moves clockwise along the perimeter of the unit circle from $(-1, 0)$ to $(1, 0)$.

Site	Spherical OK weight	Spherical OK variance	Exponential OK weight	Exponential OK variance	Gaussian OK weight	Gaussian OK variance
1	0.476	0.411	0.478	0.692	0.472	0.092
2	0.474	0.433	0.478	0.705	0.468	0.120
3	0.467	0.502	0.475	0.752	0.453	0.203
4	0.442	0.618	0.462	0.861	0.361	0.312
5	0.389	0.687	0.437	0.952	0.011	0.342
6	0.296	0.719	0.392	1.007	−1.141	0.317
7	0.175	0.731	0.333	1.033	−3.653	0.239
8	0.034	0.734	0.264	1.045	−8.044	0.100

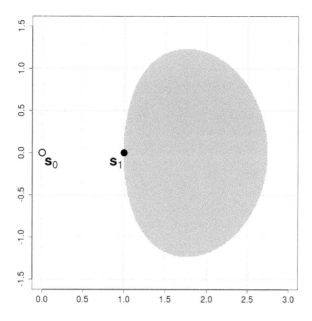

FIGURE 9.5
The negative-weights rainshadow for a second data site, corresponding to ordinary kriging at the given prediction site ($\mathbf{s}_0$) based on observations at the given data site ($\mathbf{s}_1$) and a second data site, under a Gaussian covariance function.

some covariance functions, but not others; in particular, it holds for the Matérn family with finite smoothness parameters, but does not hold for the triangular, spherical, and Gaussian covariance functions or for any separable covariance function. Later, Stein (2015) considered a somewhat different definition of an asymptotic screening effect but found that the covariance functions that satisfy this definition were similar to those that satisfy the earlier definition. Building off of Stein's work, Porcu *et al.* (2020b) argued that an asymptotic screening effect also exists for certain classes of compactly supported covariance functions. It is worth noting that there appears to be no strong connection between the finite-sample screening effect exemplified in Figure 9.4 and its asymptotic counterparts.

In several scenarios presented in this chapter, negative weights have occurred, most often at data locations that are screened (partially or completely) by another data location. For example, in the first toy example of Section 9.2, when predicting at $\mathbf{s}_{01}$, the second and fourth data locations were assigned negative weights. The second data location lies farther than the first and third data locations from the prediction site, in a direction not too different from either, and the fourth data location is completely screened by the third. Another example is displayed in Figure 9.5. For a scenario with a single data site, a single prediction site, and the Gaussian covariance function $C(r) = \exp(-r^2/3)$, the figure depicts the data site's "negative-weights rainshadow," i.e., the region within which a second data site would be assigned a negative weight. The rainshadow's area is substantial.

There is a divergence of opinion on the level of concern that should be accorded to negative weights. On one hand, if the actual numerical value of the OK predictor is reasonable despite the presence of some negative weights, there seems to be little reason for concern, especially since any adjustments made to the weights will only increase the prediction error variance. Furthermore, negative weights allow the OK predictor to take on values outside the range of the data, which is reasonable because the actual response at a prediction site could be more extreme than the most extreme observations. On the other hand, if the variable being measured is inherently positive, then for just the right set of weights, some of which are negative, the OK predictor can be negative, which obviously is undesirable. Barnes & Johnson (1984), Szidarovszky *et al.* (1987) and Deutsch (1996) proposed methods for imposing nonnegativity on the weights. Deutsch's method is the simplest of the three, as it merely involves setting the negative weights and some related small positive weights to zero, and then renormalizing the remaining weights to sum to one. All three approaches sacrifice optimality of the predictor for nonnegativity of the weights, but they retain the properties of linearity and unbiasedness.

9.9   Kriging in practice

In practice, two modifications are often made to the ordinary and universal kriging procedures described in this chapter. The first reckons with the reality that the parameter vector $\boldsymbol{\theta}$ of the covariance function, which heretofore in this chapter was assumed to be known, is actually unknown. Thus, as was the case for the GLS estimator of $\boldsymbol{\beta}$, the OK and UK predictors and their prediction error variances as they stand cannot actually be computed by practitioners. The conventional approach for dealing with the lack of knowledge of $\boldsymbol{\theta}$ is to substitute an estimate of $\boldsymbol{\theta}$, such as the maximum likelihood estimator $\tilde{\boldsymbol{\theta}}$, for $\boldsymbol{\theta}$ wherever it appears (implicitly) in expressions (9.6) and (9.10) for the OK and UK predictors. Analogous to empirical GLS as introduced in Section 5.7, these predictors are called the **empirical OK and UK predictors**. Thus, for example, the empirical UK predictor of y_0 (using the MLE of $\boldsymbol{\theta}$) is given by

$$\hat{\tilde{y}}_{0,UK} = \mathbf{x}_0^T \hat{\boldsymbol{\beta}}_{EGLS} + \tilde{\mathbf{r}}_0^T \tilde{\mathbf{R}}^{-1}(\mathbf{y} - \mathbf{X}\hat{\boldsymbol{\beta}}_{EGLS}), \tag{9.13}$$

where $\tilde{\mathbf{r}}_0 = \mathbf{r}_0(\tilde{\boldsymbol{\theta}})$, $\tilde{\mathbf{R}} = \mathbf{R}(\tilde{\boldsymbol{\theta}})$, and $\hat{\boldsymbol{\beta}}_{EGLS} = \hat{\boldsymbol{\beta}}_{GLS}(\tilde{\boldsymbol{\theta}})$ (the latter is the EGLS estimator of $\boldsymbol{\beta}$ introduced in Chapter 5). Furthermore, the conventional approach for estimating the prediction error variance of the empirical OK and UK predictors is to substitute $\tilde{\boldsymbol{\theta}}$ for $\boldsymbol{\theta}$ wherever it appears in expressions (9.7) and (9.11). For example, $\text{Var}(\hat{\tilde{y}}_{0,UK} - y_0)$ is estimated by

$$\begin{aligned}
&\widetilde{\text{Var}}(\hat{\tilde{y}}_{0,UK} - y_0) \\
={} & \hat{\sigma}_{EGLS}^2 [\tilde{r}_{00} - \tilde{\mathbf{r}}_0^T \tilde{\mathbf{R}}^{-1} \tilde{\mathbf{r}}_0 \\
& + (\mathbf{x}_0^T - \tilde{\mathbf{r}}_0^T \tilde{\mathbf{R}}^{-1}\mathbf{X})(\mathbf{X}^T \tilde{\mathbf{R}}^{-1}\mathbf{X})^{-1}(\mathbf{x}_0 - \mathbf{X}^T \tilde{\mathbf{R}}^{-1}\tilde{\mathbf{r}}_0)] \\
\equiv{} & \hat{\sigma}_{EGLS}^2 \tilde{v}_{0,UK},
\end{aligned} \tag{9.14}$$

where $\tilde{r}_{00} = r_{00}(\tilde{\boldsymbol{\theta}})$ and $\hat{\sigma}_{EGLS}^2$ is the estimated generalized mean square associated with the EGLS estimator of $\boldsymbol{\beta}$. The conventional approximate $100(1-\epsilon)\%$ prediction interval for y_0, in this empirical universal kriging setting, is

$$\hat{\tilde{y}}_{0,UK} \pm t_{\epsilon/2,n-p}\hat{\sigma}_{EGLS}\sqrt{\tilde{v}_{0,UK}}.$$

Although this approach to dealing with the lack of knowledge of $\boldsymbol{\theta}$ is reasonable, the empirical OK and UK predictors generally are not linear functions of the observations, nor are they "best" in any known sense. Like the EGLS estimators of regression coefficients,

TABLE 9.6

Performance measures of empirical OK prediction (using ML and REML to estimate the covariance parameters) and OK prediction (using the true covariance parameters) at two sites under a spatial linear model with constant mean and exponential covariance function.

Performance measure	Estimation method	$\rho = 0.3$		$\rho = 0.7$	
		Site I	Site II	Site I	Site II
RMSPE	ML	0.815	0.873	0.550	0.632
RMSPE	REML	0.814	0.873	0.551	0.633
RMSPE	True	0.803	0.863	0.546	0.629
Width	ML	2.139	2.277	1.394	1.585
Width	REML	2.140	2.283	1.401	1.592
Width	True	2.082	2.258	1.413	1.609
Coverage	ML	0.808	0.810	0.789	0.789
Coverage	REML	0.812	0.811	0.789	0.791
Coverage	True	0.805	0.813	0.799	0.801

however, they are unbiased under rather weak conditions (Kackar & Harville, 1981). Furthermore, the prediction error variances of the empirical predictors generally are intractable and not actually equal to (9.14); however, it is known that when the distribution of **y** is multivariate normal and $\boldsymbol{\theta}$ is estimated by ML or REML, the empirical UK predictor's prediction error variance is at least as large as that of the UK predictor when $\boldsymbol{\theta}$ is known (Zimmerman & Cressie, 1992). (A similar statement applies to the empirical OK predictor.) This is intuitively reasonable because (9.14) does not account for the additional prediction error incurred by estimating $\boldsymbol{\theta}$. In fact, empirical studies have shown that the empirical UK predictor's prediction error variance is often considerably larger than that of the UK predictor, which in turn also tends to be larger than (9.14). Zimmerman & Cressie (1992) gave a modified estimator of the prediction error variance of the empirical UK predictor, which performs reasonably well when the spatial dependence is not too strong. Another possibility is to estimate the prediction error variance via a parametric bootstrap (Sjöstedt de Luna & Young, 2003). However, because these alternatives are considerably more complicated, most practitioners obtain prediction intervals by the conventional approach.

Table 9.6 presents results of a small simulation study of the conventional prediction interval when the parameters are estimated by ML and REML. Observations following a spatial linear model with constant mean and isotropic exponential covariance function, the latter parameterized as $C(r) = \sigma^2 \rho^r$, were simulated on an 12×12 square grid with unit spacing, and at two prediction sites. Prediction Site I is located at the grid's center, which has coordinates (6.5,6.5) in a coordinate system in which the abscissas of columns and ordinates of rows are $1, 2, \ldots, 12$. Prediction Site II is located at (6.5,0.5), which is near the midpoint of the first row. True values of the parameters were $\beta_0 = 0$, $\sigma^2 = 1$, and $\rho = 0.3$ or 0.7. Parameters were estimated from the 144 observations on the grid, after which empirical ordinary kriging was used to obtain conventional 80% prediction intervals for the responses at the two prediction locations. For comparison, ordinary kriging using the true covariance parameters was carried out as well. Performance was measured by the root mean-squared prediction error (RMSPE) of the predictor, average width of the interval, and empirical coverage probability of the interval (the proportion of simulation replicates for which the interval contained the simulated response at the prediction location), over 5000 simulations. The results suggest that the difference, if any, in prediction performance using ML or REML estimates of the parameters is small in this scenario, and that the conventional empirical prediction intervals corresponding to both estimators perform very well. Similar conclusions are reached in another scenario where there are many more parameters in the mean structure; see Exercise 9.19.

The second modification to kriging, as it has been presented in this chapter, is motivated by concerns about computations for large datasets. The computations may be burdensome because the number of data sites is large, the number of prediction sites is large, or both. Generally, the issue is more pressing for geostatistical data than for areal data, but we will consider it for both data types, beginning with the geostatistical case. If the number of data sites is large, any of the methods described in Section 8.7 (e.g., covariance tapering, fixed rank kriging, and data partitioning) may help to reduce computations required for parameter estimation. Then, in what amounts to an informal application of similar ideas, the empirical prediction of $Y(\mathbf{s}_0)$ is often based not on the entire data vector $\mathbf{y}$ but on only those observations that lie in a specified neighborhood around $\mathbf{s}_0$. The mean structure, the nugget-to-sill ratio, the range, and the spatial configuration of data locations are important factors in choosing this neighborhood (Chiles & Delfiner, 2009). Generally speaking, near-optimal prediction under models with more complex mean structures and larger nugget-to-sill ratios requires larger neighborhoods. However, there is no simple relationship between the range and the required neighborhood size since, for example, a process with no finite range exists for which the kriging predictor is based on just the two nearest neighbors (see Exercise 9.3) and, conversely, there are processes with finite ranges for which observations at or beyond the range play a nontrivial role in the kriging predictor (see Exercise 9.16 and also Stein & Handcock (1989)). Consequently, a purely geometric criterion for choosing the kriging neighborhood seems ill-advised. Some non-geometric criteria, which focus more on the increase in prediction error variance as observations are removed from the predictor, were proposed by Haslett (1989) and Emery (2009). In any case, whenever a spatial neighborhood is used, the formulas for the empirical ordinary and universal kriging predictors and the corresponding estimates of the prediction error variances are of the same form as (9.13) and (9.14), but with $\tilde{\mathbf{r}}_0$ and $\mathbf{y}$ replaced by subvectors, and $\tilde{\mathbf{R}}$ and $\mathbf{X}$ ($\mathbf{1}$ for the OK case) replaced by submatrices, corresponding to the neighborhood.

Turning to the areal case, recall from Section 5.6 that the inverse of the covariance matrix of observations that follow a SAR or CAR model may be obtained without actually inverting a matrix whose dimensions are the number of observations. In fact, if the model is specified merely in the context of the observation sites, only some scalar reciprocations and matrix multiplication and addition are needed to obtain the inverse. If the model is specified in the context of observation and prediction sites, some additional calculation is necessary, which includes inversion of a matrix whose dimensions are the number of additional sites in which the observation sites are embedded. For prediction purposes, the additional sites are the prediction sites (which may include sites where it was originally intended to take observations). Thus, as described in Section 5.6, the inverse of the covariance matrix of observations, which is needed for both estimation and prediction, may be obtained by inverting a matrix whose dimensions are those of the prediction sites. Suppose that there are multiple prediction sites. From expressions (9.13) and (9.14) for the empirical UK predictor and the conventional estimate of its prediction error covariance matrix, it appears that to obtain those quantities we also need the matrix of covariances between the observations and the predictands ($\mathbf{\Sigma}_{12}$) and the covariance matrix of the predictands ($\mathbf{\Sigma}_{22}$). These could be obtained, of course, by inverting $\mathbf{\Sigma}^{-1}$ and then extracting the upper right and lower right blocks from the result, but the inversion could be computationally burdensome and is exactly the operation we are trying to avoid. It turns out that is possible to obtain the BLUP and its prediction error covariance matrix more efficiently using ideas similar to those used in Section 5.6 to obtain the inverse of the observations' covariance matrix. By Theorem

A.6d in Appendix A,

$$\mathbf{\Sigma}^{21} = -(\mathbf{\Sigma}_{22} - \mathbf{\Sigma}_{21}\mathbf{\Sigma}_{11}^{-1}\mathbf{\Sigma}_{12})^{-1}\mathbf{\Sigma}_{21}\mathbf{\Sigma}_{11}^{-1},$$

whence

$$\mathbf{\Sigma}_{21}\mathbf{\Sigma}_{11}^{-1} = -(\mathbf{\Sigma}^{22})^{-1}\mathbf{\Sigma}^{21}.$$

Then the vector of UK predictors and its prediction error covariance matrix may be expressed as

$$\hat{\mathbf{y}}_2 = \mathbf{X}_2\hat{\boldsymbol{\beta}}_{GLS} - (\mathbf{\Sigma}^{22})^{-1}\mathbf{\Sigma}^{21}(\mathbf{y}_1 - \mathbf{X}_1\hat{\boldsymbol{\beta}}_{GLS}),$$

and

$$\mathrm{Var}(\hat{\mathbf{y}}_2 - \mathbf{y}_2) = (\mathbf{\Sigma}^{22})^{-1} + [\mathbf{X}_2 + (\mathbf{\Sigma}^{22})^{-1}\mathbf{\Sigma}^{21}\mathbf{X}_1][\mathbf{X}_1^T\mathbf{\Sigma}_{11}^{-1}\mathbf{X}_1]^{-1}[\mathbf{X}_2 + (\mathbf{\Sigma}^{22})^{-1}\mathbf{\Sigma}^{21}\mathbf{X}_1]^T.$$

These forms are more computationally efficient than (9.10) and (9.11), as they eliminate the need to obtain $\mathbf{\Sigma}_{12}$ and $\mathbf{\Sigma}_{22}$ and they require the inversion of only the lower right block $\mathbf{\Sigma}^{22}$ of $\mathbf{\Sigma}^{-1}$. Thus, as long as the number of prediction sites is not large, modeling and prediction under SAR and CAR models may be performed efficiently for very large data sets.

In practice, one has to choose a covariance function before one can estimate it and use the estimated function for empirical spatial prediction. Practitioners should therefore be concerned not only with the effects of estimating a correctly specified covariance function on spatial prediction, but also with the effects of incorrectly specifying the covariance function in the first place. It turns out that the impact on the kriging predictor of misspecifying the covariance function is asymptotically negligible, under a fixed-domain asymptotic regime, if the correct and misspecified covariance functions are **compatible** with each other (Stein, 1988; Stein & Handcock, 1989). Compatibility is a technical condition that is beyond our scope to define precisely here, but a necessary condition for it to hold is that the two covariance functions behave similarly at the origin, i.e., they have the same degree of smoothness there. Unfortunately, similar behavior at the origin is not sufficient for compatibility; for example, the spherical and exponential covariance functions both have linear behavior at the origin but are not compatible within a sufficiently large three-dimensional region (Stein & Handcock, 1989). Nevertheless, the necessity of similar behavior at the origin for compatibility and the robustness it provides to the choice of covariance function for kriging supports modeling the covariance using a family of functions that allows for a wide range of smoothnesses, such as the Matérn family or generalized Wendland family.

Compatibility of covariance functions may also explain the superior performance of prediction intervals based on empirical kriging demonstrated in Table 9.6 compared to the mixed performance of likelihood-based estimation of individual covariance parameters described previously in Section 8.2. Zhang (2004) showed that two covariance functions within the isotropic Matérn class

$$C(r; \sigma^2, \alpha, \nu) = \frac{\sigma^2(r/\alpha)^\nu}{\Gamma(\nu)2^{\nu-1}}K_\nu(r/\alpha)$$

are compatible if and only if they have the same value of $\sigma^2/\alpha^{2\nu}$, and that for processes with two compatible functions within the class, predictions are asymptotically equivalent (under a fixed-domain regime). He showed further that for known ν, $\sigma^2/\alpha^{2\nu}$, but neither of the individual parameters σ^2 and α, is consistently estimable. Thus, provided that the smoothness parameter is known or estimated well, empirical kriging benefits from the good performance of the estimation of $\sigma^2/\alpha^{2\nu}$ (shown empirically in Chapter 8 for the exponential covariance function, for which $\nu = 1/2$) and is not adversely affected by the relatively poor

TABLE 9.7

Empirical average length ("PI length") and empirical coverage probability ("PI coverage"), of nominal 90% prediction intervals for ten randomly deleted observations in the scenario described in Section 9.9. The simulated data have three mean structures and three covariance models, which are estimated using REML-EGLS, acting as though each of the covariance models is the true model, and using OLS (fitted model "nocorr"), prior to prediction. Columns under PI length and PI coverage correspond to the three true models: spherical ("sph"), exponential ("exp"), and Gaussian ("Gau"). The fitted models have the same designations, apart from "nocorr."

Mean effects	Fitted model	PI length			PI coverage		
		sph	exp	Gau	sph	exp	Gau
constant	sph	2.159	2.264	2.165	0.894	0.870	0.919
	exp	2.324	2.367	2.344	0.909	0.896	0.921
	Gau	2.312	2.576	1.679	0.815	0.833	0.877
	nocorr	3.246	3.172	3.255	0.893	0.892	0.899
row	sph	2.216	2.335	2.211	0.893	0.868	0.920
	exp	2.375	2.433	2.383	0.909	0.895	0.922
	Gau	2.372	2.667	1.700	0.817	0.845	0.873
	nocorr	3.245	3.147	3.273	0.890	0.893	0.901
row-column	sph	2.278	2.417	2.249	0.892	0.875	0.921
	exp	2.422	2.506	2.412	0.908	0.896	0.920
	Gau	2.431	2.774	1.733	0.824	0.864	0.877
	nocorr	3.222	3.082	3.258	0.893	0.894	0.900

estimation of σ^2 and α individually. One implication of this is that setting either σ^2 or α (but not both) equal to an arbitrary value and estimating the other may produce reasonable results insofar as spatial prediction is concerned.

Results of a simulation study that more directly investigates the robustness of kriging to the choice of covariance function in a specific situation are presented in Table 9.7. For this study, the spatial domain and model were identical to those used in Section 8.6 to investigate the robustness of mean estimation to the choice of covariance function. Specifically, data were simulated from a Gaussian process at the nodes of a 12×12 square, unit-spaced grid in $\mathbb{R}^2$ from models with constant, row-effects, and additive row-column-effects mean structures and isotropic nuggetless spherical, exponential, and Gaussian covariance functions. The covariance functions had variance 1.0, with range parameters set to values that yield a correlation of 0.5 at unit distance. One thousand such datasets were simulated. For each dataset, 10 of the observations were randomly selected and set aside. Using the remaining 134 observations from each dataset, REML-EGLS estimates of model parameters were obtained under the covariance model that generated it, under the other two covariance models, and under a no-correlation model. Quality of prediction was measured by the length and coverage of the (nominal) 90% conventional prediction interval, averaged over the 10 predictions and all 1000 simulations, which are listed in Table 9.7. Several points may be made from these results. First, regardless of the true spatial covariance model, the OLS-based prediction intervals are much wider than those based upon a spatial model, but their coverage probabilities are very close to the nominal 90% level. Second, when the true model is spherical or exponential, prediction intervals obtained by fitting one of those models are no wider, and usually narrower, than those obtained by fitting the Gaussian model, and coverage probabilities obtained by fitting one of those models are much closer to the nominal level. Coverage probabilities of the intervals obtained by fitting a Gaussian model are considerably lower. Third, when the true model is Gaussian, the narrowest intervals are obtained by fitting that model, but their coverage probabilities are slightly too low. The intervals obtained by fitting a spherical or exponential model are wider, and they tend to overcover slightly. Taken as a whole, these results reinforce the notion that spatial prediction is robust to misspecification of the covariance function, provided that the behavior of the covariance at short distances under the misspecified model is similar to what it is under the true model.

A final practical issue associated with kriging is **positional uncertainty**. All of the methodology presented in this chapter is based on an assumption that the data locations and prediction locations are recorded perfectly, whereas in reality they may be measured with some error. If the errors are small, they may be ignored with negligible effect on the results. If they are sufficiently large, however, kriging and the parameter estimation that precedes it should be adjusted to properly account for them. Methods for making these adjustments were proposed by Gabrosek & Cressie (2002), Cressie & Kornak (2003), and Fanshawe & Diggle (2011) (the latter is Bayesian). The main distinguishing feature of the positional-error adjustment is the replacement of point evaluations of the covariance function (for point-to-point kriging) with integrals of the covariance function taken with respect to a location-error density function. In this respect, the positional-error adjustment has some commonalities with ATP and ATA kriging. A difficulty with implementing the adjustment is the requirement that the location-error density is either known or well-estimated, with the latter requiring considerable extra effort. Fortunately, the widespread availability of inexpensive GPS receivers nowadays has made the ascertainment of data locations highly accurate in most environmental studies, so that an adjustment for positional uncertainty is usually not necessary.

9.10   Cross-validation

Chapter 8 included a description of hypothesis testing and penalized likelihood methods for comparing and evaluating spatial general linear models. Those methods used the maximized likelihood function, but not any predictions, for model assessment. Now that we have described how to perform spatial prediction, however, we may consider a model selection method known as **cross-validation**, which utilizes predictions based on fitted models. We consider two main types: **leave-one-out cross-validation** (LOOCV) and n-**fold cross-validation**.

LOOCV eliminates one datum at a time, using the rest of the data to predict the one that was removed. There are slow and fast ways to do this. The slow way is to remove each datum, in turn, and re-estimate all of the parameters (using ML or REML) after each removal. However, removal of just a single datum usually changes the parameter estimates very little. A faster way to carry out LOOCV is based on holding all parameters at their values as estimated using all the data, and then exploiting some known results on inverting partitioned matrices so that we only have to compute the inverse of the covariance matrix once (which can be done at the ML or REML estimation stage and then saved, so that in fact no additional matrix inversion is required). Recall from Section 5.6 that, if the covariance matrix of $\mathbf{y}$ and its inverse are partitioned as

$$\mathbf{\Sigma} = \begin{bmatrix} \mathbf{\Sigma}_{11} & \mathbf{\Sigma}_{12} \\ \mathbf{\Sigma}_{21} & \mathbf{\Sigma}_{22} \end{bmatrix} \quad \text{and} \quad \mathbf{\Sigma}^{-1} = \begin{bmatrix} \mathbf{\Sigma}^{11} & \mathbf{\Sigma}^{12} \\ \mathbf{\Sigma}^{21} & \mathbf{\Sigma}^{22} \end{bmatrix},$$

respectively, then

$$\mathbf{\Sigma}_{11}^{-1} = \mathbf{\Sigma}^{11} - \mathbf{\Sigma}^{12}(\mathbf{\Sigma}^{22})^{-1}\mathbf{\Sigma}^{21}.$$

Moreover, let us order the data such that the datum to be removed is last, so that $\mathbf{\Sigma}^{22}$ is a scalar. Then the inversion of $\mathbf{\Sigma}^{22}$ is trivial and $\mathbf{\Sigma}_{11}^{-1}$ can be computed rapidly. The main computational expense of kriging is associated with the inversion of the covariance matrix of the observed data, but for LOOCV that is given by $\mathbf{\Sigma}_{11}^{-1}$, which may be computed rapidly without any further matrix inverses if we already have $\mathbf{\Sigma}^{-1}$. The only other quantity from $\mathbf{\Sigma}$ needed for prediction is the vector $\mathbf{\Sigma}_{12}$. Operationally, we just re-order the data, one at a time, putting the one to be removed last in the covariance matrices above, and that allows the predictions to be computed quickly.

Let $\hat{\hat{y}}_i$ be the ith predicted value (obtained by empirical ordinary kriging if the model has constant mean and by empirical universal kriging otherwise) using LOOCV where the ith datum has been removed, and let $\sqrt{\hat{\hat{v}}_i} = \sqrt{\widehat{\mathrm{Var}}(\hat{\hat{y}}_i - y_i)}$ be the ith prediction standard error (the square root of the empirical prediction error variance) of that prediction. Then consider two metrics to assess model performance. One is the root-mean-squared prediction error (RMSPE), computed as

$$\sqrt{\frac{1}{n}\sum_{i=1}^{n}(\hat{\hat{y}}_i - y_i)^2}.$$

Models with lower RMSPE have better predictive performance. The other metric is the 90% prediction interval coverage, PIC90, which we compute as

$$\frac{1}{n}\sum_{i=1}^{n}I\left(\hat{\hat{y}}_i - z_{\epsilon/2}\sqrt{\hat{\hat{v}}_i} \le y_i \le \hat{\hat{y}}_i + z_{\epsilon/2}\sqrt{\hat{\hat{v}}_i}\right),$$

where $I(\cdot)$ is an indicator function, equal to 1 if its argument is true, and 0 otherwise, and $z_{\epsilon/2}$ is the $100(1-\epsilon)$th percentile of the standard normal distribution. We chose $\epsilon = 0.1$ for PIC90, resulting in the familiar $z_{0.05} = 1.645$.

One problem with LOOCV is that it may be overly optimistic in assessing actual prediction. A simple thought experiment reveals why. Suppose that 99 data sites were clustered very closely together, and another was separated from the cluster. Then, for LOOCV, as we removed each datum from the cluster, we would have many nearby locations and get very precise predictions with small prediction standard errors, and they would swamp RMSPE and PIC90 if our overall goal was to predict in a region that was substantially larger than that enclosing the cluster of locations. The same is essentially true if all locations were pairs of locations that were very close to each other, but the pairs were scattered. Still, under typical sampling scenarios, it can be a good way to evaluate models as they are all operating under the same sampling scheme.

A second way to use cross-validation is called n-fold cross-validation. Here we divide the data into n groups, and remove one whole group to be predicted – this is often called the **test dataset**. The remaining data are called the **training dataset**, and are used to fit a model and make predictions at the locations of the test dataset. Then, the predictions at the locations for the test dataset can be compared to the actual values that were removed. In fact, RMSPE and PIC90 can be computed for n-fold cross-validation in exactly the same way as for LOOCV. To distinguish them notationally, we use $\mathrm{RMSPE_{LO}}$ and $\mathrm{PIC90_{LO}}$ for LOOCV, and $\mathrm{RMSPE_{Nf}}$ and $\mathrm{PIC90_{Nf}}$ for n-fold cross-validation. With n-fold cross-validation we do not have to fit the model as many times, so we completely refit the model for each group that is removed. Groups are often created randomly, and we will do it this way too. If we want to get the best feel for how well a model will interpolate, or even extrapolate a bit at the edges, it is desirable to have just a few groups. This may result in a more pessimistic assessment of a model's performance than that obtained using the complete data because we are decreasing the sample size substantially. To examine both extremes, in what follows we complement LOOCV with 3-fold cross-validation. Because we create three groups randomly for this purpose, we would like to ensure that our results do not depend too much on any particular randomized grouping. Hence, we do 3-fold cross-validation 10 times and then average the results for RMSPE (by first averaging the mean-squared prediction error, and then taking the square root) and PIC90.

9.11   Conditional simulation

Conditional simulation is the practice of using a fitted geostatistical model to repeatedly simulate whole potential surfaces. Of course, this is considerably more difficult than just predicting, with standard errors, at each point. Conditional simulation requires that we make full use of the joint distribution of predictions at all locations so that we simulate the entire surface while accounting for the correlation among the predictions. Why would we want to do this? The following hypothetical example suggests a possible answer.

Let us suppose that the data plotted in Figure 9.3 represent water depths, on the log scale, at ten sites along a linear transect across a body of water. Our job is to lay a cable across that body of water, and to do so, we must order the cable to be manufactured. We do not want to have a cable that is too short, as that would require a whole new cable to be built. On the other hand, we do not want the cable to be too long (we are assuming it can be easily shortened after laying it), as that would cost more than necessary. We might consider the length of the OK or UK surface in Figure 9.3, which in this one-dimensional setting is really a segmented but continuous curve, to be our best estimate. However, this would be a mistake because the kriged curve is "smoother" than the actual data. That is, the depth at any additional location along the transect, if we were able to observe it, would likely be above or below the kriged surface, and in order to take account of all "ups and downs" at unobserved locations, the cable would need to be longer than the kriged curve. Moreover, we would rather have the cable be a little longer than necessary rather than too short. A conditional simulation attempts to create a surface (values of the response at all sites in the spatial domain of interest) that could have been observed, often called a "realization" of the surface. In our hypothetical example, this surface is a curve, and the conditional simulation of such a curve is much rougher, hence longer, than the kriged curve. Now, if we create 1000 such curves, each a different and equally probable realization, then we can compute the length of cable needed for each realization. From these 1000 lengths, we can compute an empirical distribution of lengths, and from these, choose a value that matches our relative risk for ordering a cable that is too short versus too long. For example, if we wanted to be 95% certain that our cable was long enough, and were able to tolerate that extra expense, we would choose the 950th ordered value from the 1000 lengths computed from the realizations.

The example above is one where a nonlinear mathematical operation is computed on a whole surface. Another example often used in environmental applications is computing the area above a threshold. For example, there might be some contaminant in the environment, and after sampling, we want to estimate the total area impacted above a regulatory threshold on that contaminant. Again, the kriging surface will be too smooth, and because the prediction standard errors are point-wise measures of uncertainty, it is impossible to obtain valid confidence intervals. So instead, we may create multiple realizations of surfaces, compute the area above the threshold for each surface, and then obtain the distribution of those areas to make our inference.

Conditional simulation was recognized from the very beginnings of geostatistics as an important companion methodology to kriging, and one of the earliest methods for conditional simulation was the turning-bands method (Matheron, 1973). An introduction to turning-bands was given by Mantoglou & Wilson (1982). Here, we present a hierarchical formulation that makes conditional simulation more transparent.

In Bayesian terminology and methods, conditional simulation is the use of the posterior predictive distribution within Markov chain Monte Carlo (MCMC) methods. As part of the MCMC chain, hierarchically the algorithm consists of 1) drawing covariance parameters from their posterior distributions; 2) conditional on the covariance parameters, drawing fixed effects from their posterior distributions; and 3) conditional on covariance parameters and fixed effects, drawing predictions from their conditional distributions. We describe a similar algorithm that, rather than being part of an MCMC chain, starts with independent samples of marginal estimates of covariance parameters, and proceeds conditionally.

We describe conditional simulation starting with a REML estimate; the procedure is very similar when starting with the MLE.

- Preliminary item: If $\tilde{\boldsymbol{\theta}}$ is the REMLE of the covariance parameters $\boldsymbol{\theta}$, then compute $\mathcal{I}_R(\tilde{\boldsymbol{\theta}})$ as given by (8.7).

- Preliminary item: Set $k = 0$, and set K equal to the desired total number of simulations.

- Step 1: Set $k = k + 1$.

- Step 2: Sample $\boldsymbol{\theta}_k^*$ randomly from $\mathrm{N}(\tilde{\boldsymbol{\theta}}, [\mathcal{I}_R(\tilde{\boldsymbol{\theta}})]^{-1})$. If any values of $\boldsymbol{\theta}_k^*$ are outside of their parameter space (e.g., a negative variance value), resample until they are not.

- Step 3: Sample $\boldsymbol{\beta}_k^*$ from $\mathrm{N}(\tilde{\boldsymbol{\beta}}_k, [\mathbf{X}^T[\boldsymbol{\Sigma}(\boldsymbol{\theta}_k^*)]^{-1}\mathbf{X}]^{-1})$, where $\tilde{\boldsymbol{\beta}}_k = \{\mathbf{X}^T[\boldsymbol{\Sigma}(\boldsymbol{\theta}_k^*)]^{-1}\mathbf{X}\}^{-1}\mathbf{X}^T[\boldsymbol{\Sigma}(\boldsymbol{\theta}_k^*)]^{-1}\mathbf{y}$.

- Step 4: Create $\mathbf{R}_k^*$, $\mathbf{R}_{\mathbf{uu},k}^*$, and $\mathbf{R}_{\mathbf{yu},k}^*$ by using $\boldsymbol{\theta}_k^*$ for the joint spatial correlation matrix of $\mathbf{y}$ and $\mathbf{u}$. Let $\hat{\mathbf{u}}_k^* = \mathbf{X}_{\mathbf{u}}^T\boldsymbol{\beta}_k^* + \mathbf{R}_{\mathbf{uu},k}^{*T}\mathbf{R}_k^{*-1}(\mathbf{y} - \mathbf{X}\boldsymbol{\beta}_k^*)$ and let $\tilde{\mathbf{C}}_k^*$ be the prediction error covariance matrix $\mathrm{Var}(\hat{\mathbf{u}} - \mathbf{u})$ given by (9.5) where, in its construction, all covariance parameters are replaced with $\boldsymbol{\theta}_k^*$. Then draw the kth conditional simulation $\mathbf{u}_k^*$ from $\mathrm{N}(\hat{\mathbf{u}}_k^*, \tilde{\mathbf{C}}_k^*)$.

- If $k < K$, go to Step 1. Otherwise, stop.

Formally, suppose that f is a nonlinear function computed on a conditional simulation, and let $c_k = f(\mathbf{u}_k^*)$. Then, for K conditional simulations, we obtain the set $\{c_1, c_2, \ldots, c_K\}$ which can be used for inferences; i.e., the mean, median, or mode, and valid confidence intervals, can be computed on this set. This is the primary attraction of conditional simulation. If f is the identity function, then the mean of the set $\{c_1, c_2, \ldots, c_K\}$ will converge to the kriging predictions, and the pointwise standard errors will converge to the kriging standard errors.

9.12   Examples

9.12.1   Prediction of wet sulfate deposition

We continue with our analysis from Section 8.8.1 of the clean wet sulfate deposition data in the continental U.S. Our ultimate goal is to create a map, with prediction intervals, on a

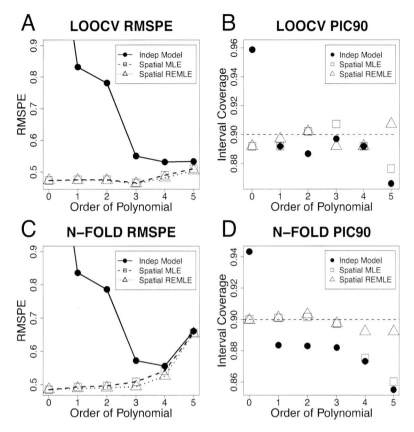

FIGURE 9.6
Root-mean-squared prediction error (RMSPE) and 90% prediction intervals coverage (PIC90) using leave-one-out cross-validation (LOOCV) and 3-fold cross-validation for polynomial surfaces up to fifth-order, fitted to the clean wet sulfate deposition data. The spatial models were fitted with an exponential covariance model using both MLE and REMLE. A) LOOCV RMSPE; B) LOOCV PIC90; C) 3-fold RMSPE; D) 3-fold PIC90.

grid of prediction locations across that domain that will help us assess the spatial patterns in wet sulfate deposition and our confidence in those predictions. We left some model selection choices open in Section 8.8.1 because the predictive ability of the models will help in our selection process. We complete the analysis here, beginning with LOOCV and 3-fold cross-validation for polynomials of order up to 5, again using the exponential covariance function for the spatial models. RMSPE using LOOCV for the independence models generally decreases as the order of the polynomial surface increases (Figure 9.6A), suggesting, like *AIC* and *BIC* did, that among this set of models, a fourth- or fifth-order polynomial is best. However, as regards predictive performance, the spatial models are much superior, having average deviations from true values of around 0.47 for lower order polynomials, versus approximately 0.53 for the higher order polynomials of the independence models. This translates to prediction intervals that are more than 11% shorter. The problem of over-fitting polynomials is revealed more clearly by 3-fold cross-validation (Figure 9.6C), where the RMSPEs of the independence models increase rapidly as the order of the polynomial increases from four to five, and increase over all orders of the spatial models. It also appears that for the spatial models in Figures 9.6A,C, REMLE is just slightly better than MLE.

TABLE 9.8

Performance metrics for constant-mean models with various covariance functions fitted to the clean wet sulfate deposition data by REML. Subscripts "LO" and "3f" correspond to LOOCV and 3-fold cross validation, respectively.

Model	m2LL	*AIC*	*BIC*	RMSPE$_{LO}$	RMSPE$_{3f}$	PIC90$_{LO}$	PIC90$_{3f}$
exponential	302.4	308.4	318.2	0.473	0.473	0.892	0.907
spherical	298.8	304.8	314.6	0.469	0.470	0.887	0.902
gaussian	308.5	314.5	324.3	0.487	0.490	0.871	0.907
circular	302.6	308.6	318.4	0.474	0.480	0.892	0.897
pentaspherical	299.5	305.5	315.3	0.470	0.470	0.887	0.902
wave	323.4	329.4	339.2	0.510	0.515	0.902	0.892
jbessel	328.9	334.9	344.7	0.530	0.527	0.902	0.892
gravity	301.2	307.2	317.0	0.477	0.479	0.887	0.892
rquad	302.1	308.1	317.9	0.478	0.480	0.881	0.897
magnetic	303.0	309.0	318.8	0.479	0.481	0.876	0.897
matern	303.9	311.9	324.9	0.481	0.483	0.871	0.897
cauchy	300.4	308.4	321.5	0.475	0.481	0.887	0.892
pexponential	298.3	306.3	319.4	0.473	0.471	0.887	0.902

We want our prediction surfaces to be as precise as possible, but we also need to evaluate whether the estimated prediction error variances are valid. By valid, we mean that a 90% prediction interval should contain the true value 90% of the time. For the independence models, that is approximately true when the order of the polynomial surfaces is one through four, but the constant mean model is overly conservative, as the prediction intervals constructed under it contain the true values almost 96% (Figure 9.6B) or 94% (Figure 9.6D) of the time. The fifth-order polynomial is overly optimistic (the prediction intervals are too short and contain the true value only about 86% of the time, Figure 9.6B,D). Coverages of the REML-based intervals are very close to the nominal value for all orders of the polynomial surface, while the ML-based intervals start to become too optimistic at order four using 3-fold cross-validation (Figure 9.6D) and order five using LOOCV (Figure 9.6B).

For the spatial models, based on *BIC*, LOOCV, and 3-fold cross-validation, there appears to be little reason to go beyond ordinary kriging. However, we still need to choose a covariance function. Table 9.8 summarizes the fits of many of the models in Tables 6.1 and 6.2 using `spmodel`, based on the model selection metrics discussed so far. With the exceptions of the wave and J-Bessel models (which were markedly inferior), all models performed similarly. Also note that for models with a constant mean, there was very little difference between LOOCV and 3-fold cross-validation. The spherical model is just slightly better than most other covariance models based on *AIC*, *BIC*, and RMSPE. It also appears to have appropriate prediction interval coverage, so we will choose the spherical model for further analyses.

Now we are ready to make maps of wet sulfate deposition predictions across the U.S. We created an evenly spaced grid of points and clipped them to the boundaries of the continental U.S., resulting in 3663 prediction sites. We used five different models to illustrate various features of the resulting maps. Figure 9.7A shows predictions for a fourth-order polynomial assuming independent errors. The prediction error standard errors (Figure 9.7B) show the typical pattern for regression models, where the standard errors are smallest near the mean of the explanatory variables (which are the spatial coordinates in this case). Hence, the prediction error standard errors are smallest near the center of the country, somewhat weighted by the fact that there are more samples in the northeast. For this fourth-order

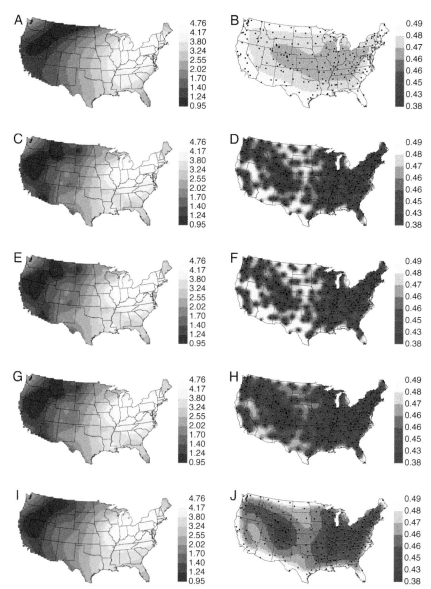

FIGURE 9.7

Wet sulfate deposition prediction maps (left column) and prediction error standard errors (right column) for a variety of models. A-B) Fourth-order polynomial assuming independent errors; C-D) constant mean spherical model; E-F) constant mean exponential model; G-H) Third-order polynomial spherical model; I-J) constant mean Gaussian model.

polynomial model, standard errors are marginally reliable, as shown by PIC90 from the LOOCV and 3-fold cross-validations.

The spatial model with a constant mean and a spherical covariance function is a more flexible surface than the polynomial surface (Figure 9.7C). For example, there is a small area of higher sulfate deposition along the middle southern border of the U.S. near New Orleans, in the state of Louisiana, which is shaped like a boot. The higher concentrations are apparent in the "toe" of the boot. A model with spatial correlation is better able to accommodate smaller fluctuations like this. The corresponding map of prediction error standard errors (Figure 9.7D) exhibits a "bull's eye" pattern around locations with observed data, which is characteristic of many of the covariance models and very different than the smooth surface in Figure 9.7B. Overall, the estimated prediction error standard errors in Figure 9.7D are significantly lower than those of Figure 9.7B, which we believe to be valid based on our cross-validation analysis of these data.

We also show that, while we chose the spherical model, many other covariance functions would give very similar results (e.g., the exponential model in Figures 9.7E,F). Recall that one of the simulation studies in Section 9.9 suggested that the choice of covariance function is not that important for spatial prediction, at least not in comparison to the important choice of a spatial model versus a model that assumes independent errors. For these data, there is even some robustness to the choice of mean structure, for when the spherical model with a constant mean (Figures 9.7C,D) is compared to a spherical model with a third-order polynomial mean function (Figures 9.7G,H), the prediction error standard errors are somewhat more diffuse, but the prediction maps are virtually identical.

However, we suggest that for a practical analysis several models should be tried to see if and where the results differ. One model that often yields very different predictions from the others is the Gaussian covariance model. It creates very smooth surfaces, and Figure 9.7I shows that the area of higher concentration in the toe of Louisiana (Figure 9.7C,E,G) is not apparent when using this model, and the prediction error standard errors change more gradually (Figure 9.7J), rather than having the bull's eye effect. Generally, covariance functions that are flat near the origin and have a sigmoid shape will lead to behavior like that seen in Figures 9.7I,J, while those that drop rapidly and linearly near the origin will result in behavior more like what is seen in Figures 9.7C,D.

9.12.2   Prediction of trends in harbor seal abundance

We continue with our analysis from Section 8.8.2 of the trends in harbor seal abundance in southeast Alaska. There, we made inferences on covariance parameters and fixed effects. Here, our inferential goal will be to make predictions, with prediction intervals, for sites with missing data, and also to investigate leave-one-out cross-validation (LOOCV) as a way to smooth CAR and SAR models.

Based on our evaluation of (penalized) likelihoods in Section 8.8.2, CAR models with row standardization fit much better than all SAR models and CAR models without row standardization. Furthermore, the model with an extra variance parameter for islands (isolated sites) fit better than any model for which all sites were forced to be connected. Do these better fits translate to better predictive performance? We investigated using LOOCV. All models discussed in the following have the stock-effects mean structure unless otherwise noted, and are estimated by REML. The CAR row-standardized model with fourth-order neighbors and provision for islands (three covariance parameters) was still best based on LOOCV RMSPE, with a value of 0.01739. This was followed by the same model without row standardization, with RMSPE = 0.01751. A CAR covariance model without provision for islands, based on a first-order neighbor structure, was next with RMSPE = 0.01765, but the same model with a fourth-order neighbor structure did very poorly, with RMSPE = 0.02775, even though this was the best model in Figure 8.8B according to *AIC*. This reinforces the idea that a variety of model checks should be performed before settling on any one model, and that it can be difficult to come to a decision when different criteria identify different "best" models. A model without any spatial correlation had RMSPE = 0.01777. Finally, we tried a model with constant mean, and a CAR row-standardized covariance with fourth-order neighbors and provision for islands, that resulted in RMSPE = 0.01800.

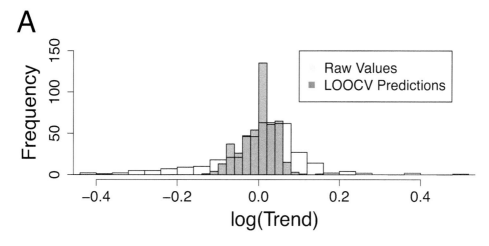

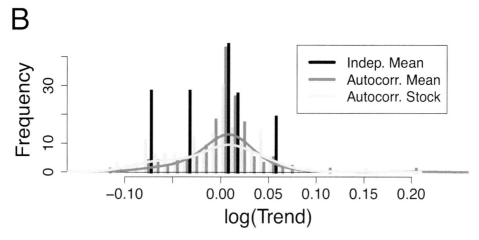

FIGURE 9.8

Histograms of raw harbor seal trends data and predictions: A) Darker shaded histogram of predictions is laid over the histogram of the raw data in a lighter shade; B) Histograms of predictions for three different models. The darkest shade corresponds to a stock-means model with independent errors. The middle shade corresponds to a constant-mean model with spatially correlated errors. The lightest shade corresponds to a stock-means model with spatially correlated errors.

Based on these results, we feel confident in proceeding with the CAR row-standardized model with fourth-order neighbors, provision for islands, and a stock-effects mean structure. An important inference goal for these data was to characterize the effect of genetic stock on trend, which was explored in Section 8.8.2, but now we would also like to predict values for missing data and make smoothed maps of the existing data. To help explain smoothing, Figure 9.8A shows a histogram of the observed values and the LOOCV values. It is possible for predictions to be more extreme than observed values, but here, with distinct stock means and fairly weak spatial correlation, predictions "shrink" away from extremes. LOOCV predictions are a combination of the estimated stock mean and a weighted average of local residuals, which causes the predictions to be much less variable than the original data. Predictions of the missing data are presented in Figure 9.8B. For the independence model, the predictions can take on only one of five possible values, which are the five estimated stock means, shown by the darkest black bars. Predictions for the constant-mean model are shown, as well as those for the stock-effects model. When spatial correlation is included in the model with stock effects, the predictions can take on values that deviate from the stock means because they include weights from the residuals of neighboring sites in the same way that they did for LOOCV. Broadly speaking, the spreads of the predictions are roughly similar for the constant-mean and stock-means models, and they are also very similar to the spread of the LOOCV predictions. LOOCV, combined with prediction of missing data, is one way to smooth maps, but there are many others. Here, we will use LOOCV.

Figure 9.9A shows predictions for all 159 sites with missing responses. These are shown as colored circles and are superimposed on the colored polygons of raw values. Note that the range of predicted values, given by the legend in Figure 9.9, is roughly the same as the range of the raw values in Figure 9.8. The prediction error standard errors (Figure 9.9B) tend to be largest near the edges of the spatial domain, where sites have fewer neighbors, as expected. Occasionally, large standard errors also occur at interior sites that have relatively few neighbors. The predictions of missing responses, along with LOOCV values for locations with raw data, are given in Figure 9.9C. Note (from the legend) that the range of values is much narrower than it is in Figure 9.9A, so this qualifies as a smoothed map. The effect of stock mean is fairly evident in Figure 9.9C. As in Figure 9.9B, the prediction error standard errors (Figure 9.9D) tend to be largest near the edges, where sites have fewer neighbors. Figure 9.9E shows predictions of missing responses, along with LOOCV, for the constant-mean model with the same covariance as previously. This is also a smoothed map, with slightly more range in values than the stock-effects model, and the effect of stock is not as evident. The standard error map (Figure 9.9F) looks very similar to the other standard error maps.

9.12.3   Conditional simulation for the moss heavy metals data

In Section 8.8.4, we used a spatial linear model for the natural logarithm of lead (Pb) concentration in moss tissue samples that included explanatory variables for 1) year of sample (2001 or 2006), 2) natural log of distance-from-road, and 3) side-of-road (north or south). From the fit of the model, we found that the dominant effect was distance-from-road. There also appeared to be some lowering of lead concentration in 2006, possibly due to better coverings on trucks that transported ore on the haul road, although the P-value of 0.118 would be judged non-significant if our tests were of size 0.05 or 0.10.

For the analyses of the wet sulfate deposition data and harbor seal trends data presented earlier in this chapter, we used kriging to make maps of pointwise or polygon-wise predictions, together with estimated standard errors of the corresponding prediction errors. In this section, we focus on conditional simulation for the moss heavy metals data. We have 5 prediction data sets, and they are shown as stratum 1 through stratum 5 in Figure 1.8B. Recall that the data are modeled after transforming them to the log scale, but we wish to make inferences back on the original scale of the data. This nonlinear back-transformation, coupled with any further computations on a whole prediction surface, complicates the construction of the covariance matrix of prediction errors. So instead, we simulate a whole surface on the transformed scale, then back-transform that whole surface, and then perform further computations. Because we simulate many equiprobable surfaces, we perform our inference *after* computing on individual surfaces by using the variation among the computed values on individual surfaces.

As an example, consider whether or not there is a difference between 2001 and 2006 in each stratum for the moss heavy metals data. First, we simulate a surface using conditional simulation, with fixed effects set at 2001 and 2006 (so we essentially simulate the surface

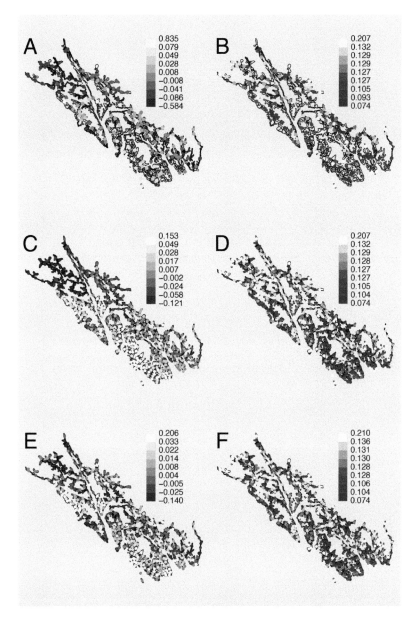

FIGURE 9.9
Harbor seal trend prediction maps (left column) and prediction error standard errors (right column) for three cases. A–B) Model with stock-means and a row-standardized CAR covariance structure with fourth-order neighbors and provision for islands; C–D) LOOCV, rather than raw values, for the same model; E–F) LOOCV and predictions for the same covariance model as above but with constant mean.

twice, but with the same set of fixed effects and covariance parameters). Then, we exponentiate the values comprising the surface, average each of the 2001 and 2006 surfaces over space, and then subtract the averaged 2006 surface from the averaged 2001 surface. Trying to obtain the variance of this difference analytically would be quite difficult based on our

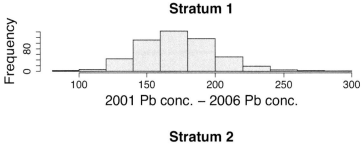

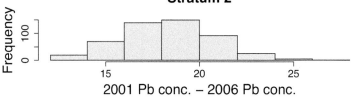

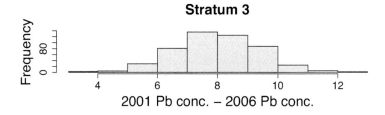

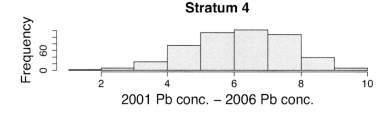

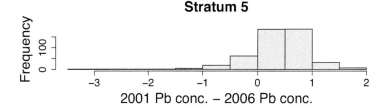

FIGURE 9.10
Histograms of the difference between 2001 and 2006 average lead (Pb) concentrations per stratum.

original model on the log scale. However, using conditional simulation, we simply simulate another surface, and repeat all computations.

For our example, we did this 500 times, and a histogram of the differences between 2001 and 2006 is shown in Figure 9.10. Note that for strata 1 through 4, there is very strong evidence that lead concentrations decreased from 2001 to 2006, as computations from all 500 conditional simulations yielded positive values. To estimate P-values, we can use quantiles from the simulations. Here, because all 500 simulated differences are greater than 0, we can state that $P < 0.002$. The histograms of these differences have approximately a Gaussian shape, so we would also be justified in using a normal distribution approximation by taking the mean of differences divided by the standard deviation of the differences. Doing so estimates that $P < 10^{-5}$ for all strata. Stratum 5 is a control, being far to the south from the road, and here we see a non-significant difference between 2001 and 2006. If we order the differences, then 67 differences are less than 0 and 433 are greater than 0, so a two-tailed P-value is $2(1 - 67/500) = 0.268$. Using the normal approximation, we obtain $P = 0.403$.

This conditional simulation approach indicates that the evidence is very strong that lead concentrations were lower in 2006 than 2001, and because the control (stratum 5) did not change significantly, it is reasonable to conclude that the coverings on the trucks decreased the contamination of lead near the haul road. Recall from Section 8.8.4 that the fitted intercepts were somewhat different between 2001 and 2006, but the P-value was greater than 0.10, so the difference was not conclusive. Why is the evidence here, using conditional simulation, so much stronger? The reason is that previously we were making an inference on a mean parameter in a model, while here we are making an inference on predictions of what was actually realized "in the field." This is an important distinction. Do we want to make an inference on the mean of some statistical process that produced the observed data, where the data may actually be quite far from that mean, or do we want to make an inference on those realized data? More simply, do we want to make an inference on what would happen on average if we let time start over again and our model was correct, or do want to make an inference on what actually happened? There is no right answer, and it depends on how broadly one wants their inference to apply. Notice that this may seem somewhat confusing, as we are using conditional simulation to create equiprobable surfaces, but the key word here is "conditional," as those equiprobable surfaces are constrained by the observed data.

We can also use conditional simulation to make pointwise inferences. In particular, at any spatial location the average of the conditional simulations will be very similar to (essentially approximating) the kriging predictor, and the variance of the conditional simulations, pointwise, will be very similar to the kriging variance. However, conditional simulation can be useful when our pointwise inferences are nonlinear. In our moss heavy metals example, we modeled the data on the log scale. In the original paper by Neitlich *et al.* (2017), the authors were interested in maps that showed where the changes from 2001 to 2006 were greatest, and what the size of those changes were. The data needed to be transformed back to their original scale, and the proportional (or percent) change computed. Thus, if $\hat{u}_{year}$ is the prediction for the uth location for $year \in \{2001, 2006\}$ on the log scale, they wanted to predict $[\exp(\hat{u}_{2006}) - \exp(\hat{u}_{2001})]/\exp(\hat{u}_{2001})$. Creating unbiased predictions of this quantity, along with valid prediction intervals, would be complicated. Alternatively, the delta method (Ver Hoef, 2012b) could be used, but it may not be accurate. Conditional simulation provides a straightforward way to estimate the proportional change, along with prediction intervals. Let

$$\ddot{u}^{[k]} = \frac{\exp(\hat{u}_{2006}^{[k]}) - \exp(\hat{u}_{2001}^{[k]})}{\exp(\hat{u}_{2001}^{[k]})},$$

where $\hat{u}_{year}^{[k]}$ is the kth conditional simulation for $year \in \{2001, 2006\}$. Then an estimate of the proportional change at the uth location is the mean of $\ddot{u}^{[k]}$, and prediction intervals can be obtained from the quantiles of $\{\ddot{u}^{[k]}; k = 1, 2, \ldots, K\}$ for K conditional simulations.

Maps of proportional change obtained via conditional simulation are displayed in Figure 9.11. In general, the proportional changes were often a decrease of about 50%, and sometimes as much as 100%. However, Figure 9.11 also reveals that the area to the northeast, especially near the road, had higher concentrations of lead. This may be due to the

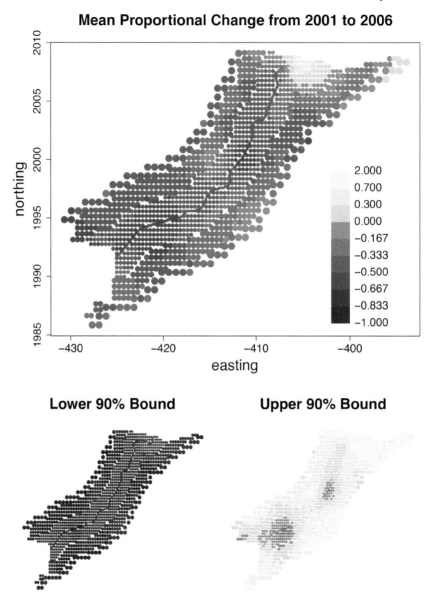

FIGURE 9.11
Predicted proportional change, using conditional simulation, from 2001 to 2006 for each
prediction location.

effects of proximity to the mine site, rather than proximity to the road itself. Maps such
as these help uncover possible additional mitigation strategies. The pointwise 90% prediction bounds, taken as the 5% and 95% quantiles from $\{\ddot{u}^{[k]}; k = 1, 2, \ldots, K\}$, show that a
proportional decrease in lead concentration occurred at sites near the road.

When working with environmental data, a common goal is to estimate the area above
a threshold. For example, environmental law, or some other decision process, may have
determined that action is required if pollutants or other measured variables exceed a critical
threshold. Generally, such action is costly, so the cost of the action needs to be evaluated.

TABLE 9.9
Estimated proportion of strata 1 through 3 above 55 mg/kg concentration of lead (with confidence limits) in 2001 and 2006, using conditional simulation.

	2001			2006		
Stratum	Lower95	Median	Upper95	Lower95	Median	Upper95
1	0.963	0.984	0.996	0.735	0.821	0.884
2	0.169	0.213	0.267	0.030	0.050	0.080
3	0.007	0.017	0.036	0.000	0.000	0.010

The cost may be proportional to the area above a threshold, so an estimate of that area is sought, along with a measure of its uncertainty, as the minimum and maximum costs are important in any decision. Conditional simulation is ideal for estimating these quantities, as they are nonlinear functions of whole surfaces. If τ is the stated threshold, and $\hat{\mathbf{u}}^{[k]}$ is the kth conditional simulation of all prediction locations, then the proportion-above-threshold may be estimated by

$$\hat{p}^{[k]} = \frac{\mathbf{1}^T \mathbb{I}[\hat{\mathbf{u}}^{[k]} > (\tau \mathbf{1})]}{\mathbf{1}^T \mathbf{1}},$$

where $\mathbb{I}(\cdot)$ is a vector indicator function, defined elementwise as equal to 1 if its argument is true and equal to 0 otherwise. Inference on the proportion-above-threshold is then based on $\{\hat{p}^{[k]}; k = 1, 2, \ldots, K\}$, where the mean, median, quantiles, standard deviation, etc., may be computed.

As an example using the moss heavy metals data, we arbitrarily set a threshold of 55 mg/kg for lead concentration, and then estimate the proportion of locations at which the predicted value exceeds this threshold. To compute the area above a threshold, we simply multiply the proportion above the threshold by the total area, so here we only present the proportion above the threshold. Table 9.9 shows that in 2001 the estimated proportion of lead concentration above 55 mg/kg was 0.984 in stratum 1, with 95% confidence limits of 0.963 and 0.996. The proportion was clearly and significantly lower in 2006. Stratum 1 is the area very close to the road, so let us concentrate on strata 2 and 3, as these are the areas of greatest concern. Looking at stratum 2, the area above the threshold decreased by about 75% from 2001 to 2006, in both the estimates and bounds. This reduction in contaminated area can be judged against the cost of adding better, more expensive coverings to trucks hauling the ore on the road. Moreover, looking at stratum 3, the coverings can be judged as a complete success between 2001 and 2006, and again the cost can be evaluated against that success.

With these examples, we have shown that conditional simulation is an important prediction method, in addition to kriging, when working with environmental data. Many problems require functions that are summaries of whole surfaces, and it is important to assess the uncertainty of those summaries. Conditional simulation provides predictions, and prediction bounds, for nonlinear functions of predictions and whole surfaces.

9.13   Exercises

1. Verify that the BLUP coefficients vector $\boldsymbol{\lambda}$ defined implicitly in (9.2) satisfies the unbiasedness constraint given by (9.1).

2. (Adapted from Cressie (1988)). Consider a scenario in $\mathbb{R}$ in which a stationary process with a triangular covariance function is observed at three point sites: $-a/2$, $a/2$, and $a/2 + af$, where a is the range of the covariance function and $0 < f < 1$. Suppose that we wish to predict the response at the origin. Determine the OK weights for the three data sites, and the OK variance, as functions of f. Comment on the signs and magnitudes of the weights as f ranges from 0 to 1.

3. Consider a scenario in $\mathbb{R}$ in which a process with constant (but unknown) mean and nonstationary covariance function $C(s, s') = \sigma^2 \min(s, s')$ is observed at $n \geq 2$ sites at arbitrary distinct locations $0 < s_1 < s_2 < \cdots < s_n$. Obtain the OK weights and OK variance for prediction at a site s_0 lying anywhere between s_i and s_{i+1} $(i = 1, \ldots, n-1)$. You may find the following matrix lemma helpful.

   Lemma. The inverse of an $n \times n$ symmetric positive definite matrix $\mathbf{A}$ of form

   $$\mathbf{A} = \begin{pmatrix} a_1 & a_1 & a_1 & a_1 & \cdots & a_1 \\ a_1 & a_2 & a_2 & a_2 & \cdots & a_2 \\ a_1 & a_2 & a_3 & a_3 & \cdots & a_3 \\ a_1 & a_2 & a_3 & a_4 & \cdots & a_4 \\ \vdots & \vdots & \vdots & \vdots & \ddots & \vdots \\ a_1 & a_2 & a_3 & a_4 & \cdots & a_n \end{pmatrix},$$

   where $0 < a_1 < a_2 < \cdots < a_n$, is of symmetric tridiagonal form

   $$\begin{pmatrix} \frac{a_2}{a_1(a_2-a_1)} & -\frac{1}{a_2-a_1} & 0 & \cdots & 0 & 0 \\ -\frac{1}{a_2-a_1} & \frac{a_3-a_1}{(a_2-a_1)(a_3-a_2)} & -\frac{1}{a_3-a_2} & \cdots & 0 & 0 \\ 0 & -\frac{1}{a_3-a_2} & \frac{a_4-a_2}{(a_3-a_2)(a_4-a_3)} & \cdots & 0 & 0 \\ \vdots & \vdots & \vdots & \ddots & \vdots & \vdots \\ 0 & 0 & 0 & \cdots & \frac{a_n-a_{n-2}}{(a_{n-1}-a_{n-2})(a_n-a_{n-1})} & -\frac{1}{a_n-a_{n-1}} \\ 0 & 0 & 0 & \cdots & -\frac{1}{a_n-a_{n-1}} & \frac{1}{a_n-a_{n-1}} \end{pmatrix}.$$

4. Consider a scenario in which a two-dimensional stationary geostatistical process is observed at two sites $(0, 0)$ and $(1, 0)$, and suppose that the covariance function is $C(u, v) = \sigma^2 \rho_1^{|u|} \rho_2^{|v|}$ (which is separable) where $0 < \rho_1 < 1$ and $0 < \rho_2 < 1$. Obtain the OK weights for predicting the process at $(0, 1)$. Is there anything surprising about these weights?

5. Consider the spatial configuration in Figure 9.1. Obtain the OK weights and OK variance corresponding to Matérn covariance functions with variance 1.0, practical range 4.0, and smoothness parameters 1.5 and 2.5. From these results and those for the exponential and Gaussian covariance functions given in Table 9.1, summarize how the OK weights and OK variance appear to change as the smoothness parameter increases.

6. Again consider the spatial configuration in Figure 9.1. Obtain the OK weights and OK variance corresponding to powered exponential covariance functions with variance 1.0, practical range 4.0, and exponents 0.5 and 1.5. From these results and those for the exponential and Gaussian covariance functions given in Table 9.1, summarize how the OK weights and OK variance appear to change as the exponent increases.

7. Suppose that a Gaussian process with constant mean and nuggetless isotropic spherical covariance function

   $$C(r) = \begin{cases} 1 - 1.5(r/16) + 0.5(r/16)^3 & \text{if } r < 16, \\ 0 & \text{otherwise} \end{cases}$$

is observed at 40 sites taken along a one-dimensional transect at locations $s_1 = 1, s_2 = 2, \ldots, s_{20} = 20, s_{21} = 22, s_{22} = 23, \ldots, s_{40} = 41$. Suppose further that we wish to predict the unobserved value of this process at $s_0 = 21$.

(a) Do the weights decay monotonically as distance from the prediction site increases? Which weights exceed 0.01 in absolute value?

(b) Repeat part (a) for a prediction problem the same as that described above except that the covariance function is $[C(r)]^2$.

8. Observations are taken at the following four sites in $\mathbb{R}^2$: $(0,3), (2,0), (0,-3),$ $(-2,0)$. Spatial prediction is desired at site $\mathbf{s}_0 = (0,0)$. The observations are assumed to follow a spatial linear model with constant mean and geometrically anisotropic covariance function

$$C(\mathbf{h}) = 3\exp(-\mathbf{h}^T \mathbf{B}\mathbf{h}/8), \quad \mathbf{h} \in \mathbb{R}^2,$$

where $\mathbf{B}$ is a positive definite 2×2 matrix whose upper left element is 1.0, but whose remaining elements are to be determined. If the OK predictor of the variable at $\mathbf{s}_0$ is the average of the four observations, what are the remaining elements of $\mathbf{B}$?

9. Repeat the OK analysis for the SAR example in Section 9.2, but with $\rho_{\text{SAR}} = 0.2$, and comment on any differences between the results and those when $\rho_{\text{SAR}} = 0.3$.

10. Repeat the OK analysis for the SAR example in Section 9.2, but for a single multiplicative-parameter CAR model with $\rho_{\text{CAR}} = 0.3$. Comment on any differences between these results and those for the SAR example.

11. Using the following lemma, show that expressions (9.8) and (9.9) coincide with (9.6) and (9.7) when $Y(\cdot)$ is second-order stationary.

Lemma. If $\mathbf{A}$ is an $n \times n$ nonsingular matrix and $\mathbf{a}$ is an arbitrary n-vector, then $\mathbf{A} + \mathbf{a}\mathbf{a}^T$ is nonsingular and

$$(\mathbf{A} + \mathbf{a}\mathbf{a}^T)^{-1} = \mathbf{A}^{-1} - \left(\frac{1}{1 + \mathbf{a}^T \mathbf{A}^{-1}\mathbf{a}}\right) \mathbf{A}^{-1}\mathbf{a}\mathbf{a}^T \mathbf{A}^{-1}.$$

12. Verify that the UK weights in Table 9.3 satisfy the unbiasedness constraints, two of which are given by (9.12).

13. For the same spatial configuration of areal sites and the same three prediction sites considered in the example in Section 9.2 of ordinary kriging under a spatial-weights linear model, obtain the UK weights and UK variances under a SAR model with an additive row-column effects mean structure but otherwise identical to the one considered in that example.

14. Repeat the calculations for the scenario depicting the screening effect in Figure 9.4 and Table 9.5, but this time taking the covariance functions to be Matérn with smoothness parameters $\nu = \frac{3}{2}$ and $\frac{5}{2}$, with practical range equal to 4 in both cases. Summarize these results and those corresponding to the exponential and Gaussian covariance functions and comment on the apparent influence of the smoothness parameter on the screening effect in this scenario.

15. Repeat the calculations for the scenario depicting the negative-weights rain-shadow in Figure 9.5, but this time take the covariance function to be Matérn with smoothness parameters $\nu = \frac{1}{2}, \frac{3}{2},$ and $\frac{5}{2}$, with practical range equal to 3 in all 3 cases. Summarize these results and those corresponding to the Gaussian covariance function and comment on the apparent effect of the smoothness parameter on the negative-weights rainshadow in this scenario.

16. (Adapted from Stein & Handcock (1989)). Consider a scenario in which a one-dimensional stationary geostatistical process is observed at sites $\frac{1}{n-1}, \frac{2}{n-1}, \ldots, 1, \frac{n}{n-1}$ for a given integer $n > 1$. Suppose that the process has unknown constant mean and triangular covariance function with variance 1.0 and range 1.0. The covariance matrix of the observations is then

$$
\Sigma = \frac{1}{n-1}
\begin{pmatrix}
n-1 & n-2 & n-3 & \cdots & 0 \\
n-2 & n-1 & n-2 & \cdots & 1 \\
n-3 & n-2 & n-1 & \cdots & 2 \\
\vdots & \vdots & \vdots & \ddots & \vdots \\
0 & 1 & 2 & \cdots & n-1
\end{pmatrix},
$$

which is positive definite and has an inverse of the form

$$
\Sigma^{-1} = (n-1)
\begin{pmatrix}
\frac{n}{2n-2} & -\frac{1}{2} & 0 & 0 & \cdots & 0 & \frac{1}{2n-2} \\
-\frac{1}{2} & 1 & -\frac{1}{2} & 0 & \cdots & 0 & 0 \\
0 & -\frac{1}{2} & 1 & -\frac{1}{2} & \cdots & 0 & 0 \\
0 & 0 & -\frac{1}{2} & 1 & \cdots & 0 & 0 \\
\vdots & \vdots & \vdots & \vdots & \ddots & \vdots & \vdots \\
0 & 0 & 0 & 0 & \cdots & 1 & -\frac{1}{2} \\
\frac{1}{2n-2} & 0 & 0 & 0 & \cdots & -\frac{1}{2} & \frac{n}{2n-2}
\end{pmatrix}.
$$

(a) Determine the OK weights for prediction at 0. Do you find anything noteworthy about the weights, in particular their behavior as $n \to \infty$?

(b) Determine the OK variance for prediction at 0, and compare it to the prediction error variance of the alternative unbiased predictor given by the observation at $\frac{1}{n-1}$. How much does the inclusion of the observations at the two most distant observations in the OK predictor improve the prediction error variance, relative to that of the aforementioned alternative unbiased predictor?

17. For the soil pH data introduced in Exercise 3.14, obtain empirical OK predictors and OK variances for the response at the following three sites:

(a) halfway between columns 1 and 2, and halfway between rows 1 and 2;

(b) halfway between columns 5 and 6, and halfway between rows 5 and 6;

(c) within row 8, one-fourth of the way from column 3 to column 4.

(Columns are numbered from left to right and rows from top to bottom.) For this purpose, assume that the data are second-order stationary and isotropic, with an exponential covariance function with nugget effect.

18. Repeat the previous problem, except this time obtain UK predictors and UK variances based on the following two mean structures:

(a) planar;

(b) complete second-order polynomial.

19. Conduct a simulation study of the effect of ML versus REML estimation of covariance parameters on kriging, to complement the study presented in Table 9.6, as follows. Simulate observations on a 12×12 square with unit spacing as in Section 8.2, from Gaussian processes with nuggetless isotropic exponential covariance function having correlation 0.5 at unit distance, and three mean structures: constant, row effects only, and row and column effects. Set 10 randomly chosen

observations aside, fit the model by ML and REML, and use the fitted model to perform empirical kriging at the 10 sites without observations. Compute the average bias, RMSPE, average 90% prediction interval length, and average 90% prediction interval coverage over the 10 predictions. Repeat this for 999 more simulated realizations of the process, and average the four metrics over all 1000 simulations. Verify that there is little difference in prediction quality between ML and REML estimation in this scenario, and that the empirical coverage of the nominal 90% prediction interval is very close to 90%.

10

Spatial Sampling Design

Environmental data occur in space (and in time, but we consider only the spatial aspect in this chapter), and precisely where they occur may affect the quality of inferences made from the data. For example, if precipitation is measured at gauges widely separated in space, average precipitation amounts over large regions may be estimated quite well but little may be inferred about the local or small-scale spatial variability of those amounts. Consequently, considerable effort has been put into the development of **spatial sampling designs**, i.e., selections of spatial locations that maximize or otherwise enhance the quality of inferences that can be made from sampling the environment. This chapter reviews such designs.

Because this is a book on spatial linear *models*, the designs featured most prominently here are **model-based**. That is, the sites included in the sample are chosen to optimize a criterion that measures the quality of the design for some type of inference based on the model. However, we will also describe, albeit more briefly, designs that are **probability-based**, i.e., that involve the selection of samples according to a random mechanism that can be described by a probability distribution. Still more types of designs included in this chapter are neither model-based nor probability-based but meet certain well-defined geometric objectives, prime examples being **space-filling designs**.

10.1   Spatial sampling design framework

We consider spatial sampling design for environmental data within the following framework. Assume that the environmental variable of interest, Y, may be adequately described over a region $\mathcal{A}$ by a Gaussian spatial process $\{Y(\mathbf{s}) : \mathbf{s} \in \mathcal{A}\}$. Let $\mathcal{D}$ denote the **design space**, i.e., the set of sites where it is possible to take observations on Y, and suppose that the sample size n, i.e., the number of sites at which Y will be observed, is predetermined. The sampling design problem is then to select a design $D = \{\mathbf{s}_1, \mathbf{s}_2, \ldots, \mathbf{s}_n\}$ of n sites from $\mathcal{D}$ in such a way as to best meet some objective or combination of objectives. A wide variety of objectives are possible, including such things as estimating a spatial average, detecting noncompliance with regulatory limits, determining the environmental impact of an event, measuring the effects of environmental mitigation, and detecting extreme levels (e.g., floods or smog) so that public alerts can be issued. The sampling should be principled, which precludes **haphazard sampling**, in which samples are chosen purely out of convenience, and **judgment sampling**, in which samples are chosen with a goal of being "representative" in some vague sense.

Although $\mathcal{D}$ could in principle be a continuum coinciding with $\mathcal{A}$, administrative, practical, and economic considerations often restrict $\mathcal{D}$ to a finite subset of points in $\mathcal{A}$. Even in cases where there are no such reasons for restricting $\mathcal{D}$, for some approaches to sampling design it may be necessary to discretize $\mathcal{D}$ in order to simplify the search for the best, or at least a good, design.

DOI: 10.1201/9780429060878-10

It is possible for the costs of sampling to vary across $\mathcal{D}$; for example, some sites may be more expensive to travel to than others. However, variable costs are often difficult to incorporate formally into the design criteria to be reviewed here, and we will not attempt to do so. In practice, a design of one of the types described here, which is obtained via statistical, probabilistic, or geometric considerations, may need to be modified to meet budgetary or other constraints.

In some situations where a good sampling design is sought, data may have previously been collected at some sites, perhaps using haphazard or judgment sampling, and the job of the designer is to select sites where additional observations should be taken (augmentation) or sites that, perhaps due to reduced budgets, must be deleted from the sampling network (contraction). In other situations, however, the designer has the opportunity to create the design *de novo*. Designs of the latter type may be constructed all at once or adaptively where, in the adaptive case, parameter estimates or other information from previous stages may be used to guide the selection of sites for subsequent stages.

Some of the sampling design approaches to be described require the use of a design criterion function and a computational algorithm to optimize that function. A variety of algorithms have been proposed. For relatively small augmentation and contraction problems, enumeration of all possible designs may be feasible. In problems too large for complete enumeration, greedy algorithms, exchange algorithms, branch-and-bound algorithms, simulated annealing, and gradient-based methods may be used. We will not attempt to describe any of these. Some type of stopping rule must be supplied to these algorithms, and there is no guarantee that the design obtained when the stopping rule is met will actually be the optimal design, though it is likely to be at least near-optimal in the sense that it has a criterion value close to the global optimum.

10.2 Model-based sampling design

Model-based sampling design under a spatial linear model is concerned with choosing a design to optimize a criterion that measures how suitable the design is for making precise inferences about either the parameters of the model or predictions of unobserved values of Y obtained using the model. The design criterion may depend on the object(s) of primary inferential interest, how the chosen object is estimated/predicted, and how to measure the quality of the chosen estimator/predictor.

We describe model-based design with respect to criteria focused respectively on mean parameter estimation, spatial prediction, covariance parameter estimation, and entropy. Designs well-suited for the first two objectives depend on the nature and strength of the spatial covariance function of the underlying process, so we consider those objectives separately under scenarios in which the covariance parameters are known or unknown. Several quantities in the spatial linear model play an important role in our description. For a given design $D = \{\mathbf{s}_1, \ldots, \mathbf{s}_n\}$, the model matrix $\mathbf{X}$ is the matrix whose ith row is $[\mathbf{x}(\mathbf{s}_i)]^T$, the covariance matrix $\mathbf{\Sigma}$ is the matrix whose (i, j)th element is $C(\mathbf{s}_i, \mathbf{s}_j; \boldsymbol{\theta})$, and $\mathbf{y}$ is the vector with ith element $Y(\mathbf{s}_i)$. Each of $\mathbf{X}$, $\mathbf{\Sigma}$, and $\mathbf{y}$ depend on the design D, but this is not explicitly indicated by the notation. It is assumed that both $\mathbf{\Sigma}$ and $\mathbf{X}^T \mathbf{\Sigma}^{-1} \mathbf{X}$ are invertible, which may slightly restrict the set of candidate designs.

One practical difficulty in choosing a single best design with respect to a given model-based criterion is that the optimal design may depend on the unknown parameter vector $\boldsymbol{\theta}$. In other words, a "globally optimal" design may not exist, and we may have to settle for a **locally optimal design** (a design that is optimal for a specific value of $\boldsymbol{\theta}$ and presumably for values in a neighborhood of it). There are various ways to circumvent this difficulty. One is to use a two-stage adaptive sampling approach, wherein some sites are sampled without regard to their optimality in order to provide an initial estimate of $\boldsymbol{\theta}$, and then subsequent sites are chosen to yield the locally optimal design corresponding to this estimate. Alternatively, a maximin approach chooses the design that maximizes the minimum value of the criterion across all values of $\boldsymbol{\theta}$. Still another approach is Bayesian, in which the uncertainty about $\boldsymbol{\theta}$ in the design criterion is incorporated into the model via a prior distribution, about which we will say more later.

10.2.1   Design for mean estimation when covariance parameters are known

Suppose that the investigator's greatest interest lies in estimating the model's mean parameters, $\boldsymbol{\beta}$, as well as possible. Although a well-developed classical theory exists for exact optimization of a design for the estimation of mean parameters (regression coefficients) in linear models (Federov, 1972; Silvey, 1980), among the assumptions on which it is based are 1) independence of the observations, and 2) the ability to sample repeatedly at any combination of the regressors. These assumptions are not satisfied in typical spatial sampling design problems in which the regressors are merely the spatial coordinates, so attempts to apply the theory to these problems have faced significant challenges. Here we describe instead an approximate approach to optimal design for mean parameter estimation, which was studied by Müller (2005) for use when the covariance parameters are known. An extension of it to the case of unknown covariance parameters is described in a later subsection.

As noted in Chapter 5, if $\boldsymbol{\theta}$ is known then $\boldsymbol{\Sigma}$ is known and the best (minimum variance) linear unbiased estimator (BLUE) of $\boldsymbol{\beta}$ is the generalized least squares estimator $\hat{\boldsymbol{\beta}}_{GLS} = (\mathbf{X}^T \boldsymbol{\Sigma}^{-1} \mathbf{X})^{-1} \mathbf{X}^T \boldsymbol{\Sigma}^{-1} \mathbf{y}$. A standard statistical measure of a design's quality for estimating $\boldsymbol{\beta}$, known as the D-optimality criterion, is the determinant of the Fisher information matrix for $\boldsymbol{\beta}$ or equivalently (assuming a Gaussian process) the determinant of the inverse of the covariance matrix of $\hat{\boldsymbol{\beta}}_{GLS}$, i.e.,

$$\phi_{ME}(D; \boldsymbol{\theta}) = |\mathbf{X}^T [\boldsymbol{\Sigma}(\boldsymbol{\theta})]^{-1} \mathbf{X}|,$$

where "ME" denotes "mean estimation." A design that maximizes $\phi_{ME}(D; \boldsymbol{\theta})$ for a given $\boldsymbol{\theta}$ may be called a (locally) **mean estimation-optimal design**. If the mean function is constant, then, provided that the model errors are stationary, such a design tends to consist of regularly spaced sites that have good coverage of the entire study region. This is intuitively reasonable, since such a configuration maximally reduces the dependence among pairs of observations, which in turn increases the reliability of the GLS estimator of the mean. On the other hand, if the mean function is a planar trend, i.e., $\mathrm{E}[Y(\mathbf{s})] = [\mathbf{x}(\mathbf{s})]^T \boldsymbol{\beta}$, then the optimal design has a preponderance of sites dispersed on or near the boundary of the study region. This is due to the higher leverage that boundary sites have when the mean function is planar rather than constant, with the result that estimation is improved by sampling more points near the boundary despite the stronger correlation among observations that results from taking them in higher density there.

The features of a mean estimation-optimal design just described are illustrated by the toy example design displayed in Figure 10.1A. This design maximizes $\phi_{ME}(D; \boldsymbol{\theta})$ among all designs of size five for a scenario in which $\mathcal{D}$ is a 5×5 square grid with unit spacing and the underlying random field has constant mean and an isotropic exponential covariance function with nugget,

$$C(r; \boldsymbol{\theta}) = \begin{cases} \theta_1 + \theta_0 & \text{for } r = 0, \\ \theta_1 \theta_2^r & \text{for } r > 0, \end{cases}$$

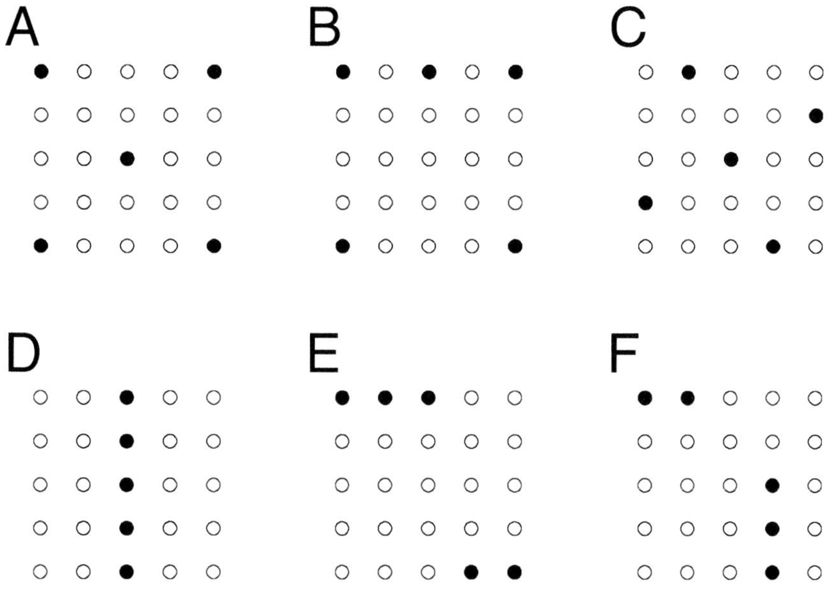

A   B   C

D   E   F

FIGURE 10.1

Locally optimal model-based designs of size 5 for the toy example described in Section 10.2 when $\theta_0 = \theta_1 = \theta_2 = 0.5$ unless noted otherwise and the mean is unknown. A) Optimal design for estimation of constant mean when covariance is known; B) optimal design for estimation of planar mean when covariance is known; C) optimal design for spatial prediction when covariance is known; D) optimal design for covariance parameter estimation when $\theta_2 = 0.5$; E) optimal design for covariance parameter estimation when $\theta_2 = 0.8$; F) optimal design for empirical spatial prediction when $\theta_2 = 0.8$. Candidate design points within each 5×5 grid are represented by open circles, and points belonging to the optimal design are represented by closed circles.

where $\theta_0 = \theta_1 = \theta_2 = 0.5$. This scenario is sufficiently small that $\phi_{ME}(D; \boldsymbol{\theta})$ may be computed for all possible five-point designs so that a truly optimal design can be found. If the scenario is changed so that the mean is a planar function of the spatial coordinates, but everything else is left the same, the locally optimal five-point mean-estimation design merely differs by one point: the center point moves to the center of any of the four edges of the square grid (Figure 10.1B). This comes about as a result of the aforementioned importance of leverage for estimating planar mean functions. It should be noted that the locally optimal mean-estimation design is unique in the constant mean case, but not in the planar case.

10.2.2   Design for spatial prediction when covariance parameters are known

Let $y_0 = Y(\mathbf{s}_0)$ denote the realized but unobserved value of Y at an arbitrary point $\mathbf{s}_0 \in \mathcal{A}$. As was shown in Section 9.3, if the covariance parameter vector $\boldsymbol{\theta}$ is known, then the BLUP (UK predictor) of y_0 is given by

$$\hat{y}_0(\boldsymbol{\theta}, D) = \mathbf{x}_0^T \hat{\boldsymbol{\beta}}_{GLS} + \mathbf{r}_0^T \mathbf{R}^{-1}(\mathbf{y} - \mathbf{X}\hat{\boldsymbol{\beta}}_{GLS}).$$

The universal kriging variance associated with $\hat{y}_0(\boldsymbol{\theta}, D)$ is given by

$$\sigma^2_{UK}(\mathbf{s}_0; \boldsymbol{\theta}, D) = \mathrm{Var}_{\boldsymbol{\theta}}[\hat{y}_0(\boldsymbol{\theta}, D) - y_0]$$
$$= \sigma^2[r_{00} - \mathbf{r}_0^T\mathbf{R}^{-1}\mathbf{r}_0 + (\mathbf{x}_0 - \mathbf{X}^T\mathbf{R}^{-1}\mathbf{r}_0)^T(\mathbf{X}^T\mathbf{R}^{-1}\mathbf{X})^{-1}(\mathbf{x}_0 - \mathbf{X}^T\mathbf{R}^{-1}\mathbf{r}_0)].$$

In both expressions above, we have extended the notation from what it was in Chapter 9 in order to make the dependence on $\boldsymbol{\theta}$ and D more explicit.

The universal kriging variance depends on neither the mean parameters $\boldsymbol{\beta}$ nor the observed data $\mathbf{y}$; it does, however, depend on the prediction site $\mathbf{s}_0$, the covariance parameter $\boldsymbol{\theta}$, and the design D. An inverse measure of the global performance of a design for prediction under a model with known covariance parameters is the maximum prediction error variance over a set $\mathcal{P}$ where prediction is desired, i.e.,

$$\phi_K(D; \boldsymbol{\theta}) = \max_{\mathbf{s} \in \mathcal{P}} \sigma^2_{UK}(\mathbf{s}; \boldsymbol{\theta}, D),$$

where $\mathcal{P}$ is typically either $\mathcal{D}$ or $\mathcal{A}$. The subscript "K" stands for "kriging" and a design that minimizes $\phi_K(D; \boldsymbol{\theta})$ is called a (locally) **kriging-optimal design**. If the mean function is constant, designs that minimize $\phi_K(D; \boldsymbol{\theta})$ tend to be "regular," i.e., their sites are rather uniformly dispersed over the study region. If, instead, the mean function is planar, the sites again have good spatial coverage but tend to be denser near the periphery of the study region. A design that minimizes $\phi_K(D; \boldsymbol{\theta})$ for the same toy example (with constant mean) described in the previous subsection is shown in Figure 10.1C (this is one of two designs that produce the same set of lags; the other is its mirror image about a vertical line through the center column). Though this design has good spatial coverage, it is slightly less dispersed than the (locally) mean estimation-optimal design. The slight contraction improves prediction by reducing the largest average distance from any unoccupied gridpoint to the design locations. Kriging-optimal designs for larger examples (not shown) mimic these features.

10.2.3   Design for covariance parameter estimation

Sometimes understanding the covariance structure of the data is the primary inferential objective of a study. Even when it is not, and estimation of mean parameters or spatial prediction of unobserved values of $Y(\cdot)$ is the primary inferential goal instead, accomplishing this other goal successfully may depend, in part, on how well the covariance parameters are estimated. So in either case it is important to give due attention to the quality of the design for estimating covariance or semivariogram parameters. Relevant early work on this topic considered design criteria for moment-based estimation of the semivariogram. For example, Russo (1984) considered a design to be optimal if it minimized the dispersion of intersite distances within bins used for the classical method-of-moments estimator of the semivariogram, so that the blurring effect described in Section 3.6.1 would be minimized. Müller & Zimmerman (1999) took a more direct approach to optimizing semivariogram estimation quality by choosing a design to maximize the determinant of the variance-covariance matrix of semivariogram parameter estimators obtained by parametric smoothing methods, such as the weighted least squares smoothing described in Section 3.6.2.

Criteria that are more relevant to estimation quality in the context of a spatial linear model were introduced by Zhu & Stein (2005) and Zimmerman (2006). Zhu and Stein's and Zimmerman's approaches are similar to those of Müller & Zimmerman (1999), except that it is the determinant of the Fisher information matrix of variance-covariance parameters (corresponding to the likelihood or residual likelihood function) that is maximized. For example, for ML estimation, the criterion that is maximized is

$$\phi_{CPE}(D; \boldsymbol{\theta}) = |\mathcal{I}(\boldsymbol{\theta}, D)|, \tag{10.1}$$

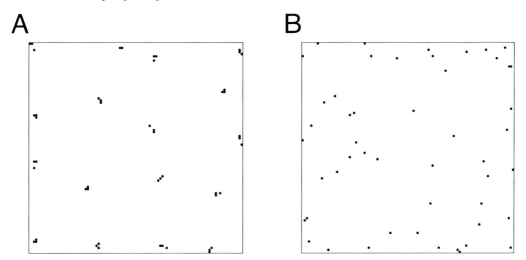

FIGURE 10.2
Model-based sampling designs for the larger of the two examples described in Section 10.2 when $\theta_1 = \theta_2 = \theta = 0.5$ and the mean is constant (but unknown). The designs were obtained by simulated annealing algorithms, seeking to optimize for A) estimation of covariance parameters or B) empirical kriging.

where the subscripts "CPE" denote "covariance parameter estimation" and the (i, j)th element of the Fisher information matrix $\mathcal{I}(\boldsymbol{\theta}, D)$ is (as given originally in Section 8.3) $(1/2)\mathrm{tr}[\boldsymbol{\Sigma}^{-1}(\partial\boldsymbol{\Sigma}/\partial\theta_i)\boldsymbol{\Sigma}^{-1}(\partial\boldsymbol{\Sigma}/\partial\theta_j)]$. Designs that maximize this criterion and others like it may be called (locally) **covariance estimation-optimal designs** and generally consist of several small clusters or linear strands of sites. There is a tendency for many of the clusters to lie along the periphery of the study region, with the remainder spaced more or less evenly across it. Such a design produces relatively many small lags and large lags, and fewer intermediate lags, than most other designs, which facilitates more precise estimation of covariance parameters. Figure 10.1D,E display designs of size five that maximize $\phi_{CPE}(D; \boldsymbol{\theta})$ for the same toy example of the previous two subsections, first when $\theta_2 = 0.5$ and then when $\theta_2 = 0.8$. It should be noted that the locally covariance estimation-optimal designs in these two cases are not unique; rotations and reflections (and translations in the first case) yield lags with the same set of intersite distances and thus the same value of the design criterion.

A toy example of size five may be rather too small to reveal the true nature of covariance estimation-optimal designs for situations more likely to occur in practice, especially since the design criterion given by (10.1) is based on an asymptotic result. With that in mind, Figure 10.2A displays a (nearly) covariance estimation-optimal design for a much larger example in which $\mathcal{D}$ is a 100×100 square grid and $n = 50$. The covariance function for this example is also exponential, with correlation equal to 0.5 for sites 20 grid spacings apart so that the nature and strength of spatial correlation relative to the size of the study area are similar to what they were for the toy example. It is not feasible to enumerate all possible designs in this case, so the design in Figure 10.2A was obtained by a simulated annealing algorithm. It turns out that this design, like its toy example counterpart, consists entirely of clusters. Most of the clusters are small (2–3 points), and they are rather regularly spaced throughout the entire study region, with the majority lying very close to its periphery.

10.2.4   Design for mean estimation or prediction when covariance parameters are unknown

Although designs that minimize $\phi_{ME}(D; \boldsymbol{\theta})$ or $\phi_K(D; \boldsymbol{\theta})$ are of interest in their own right, they do not necessarily perform well for the mean estimation problem or prediction problem of greatest practical importance, which is to estimate $\boldsymbol{\beta}$ or predict unobserved values of Y when the spatial covariance parameters are unknown. We deal with the case of prediction first. Recall from Section 9.9 that the standard predictor in this situation, known as the empirical BLUP (E-BLUP), is given by an expression identical to the known-$\boldsymbol{\theta}$ BLUP but with the covariance function evaluated at a likelihood-based estimate $\tilde{\boldsymbol{\theta}}$, rather than at the hitherto-assumed-known $\boldsymbol{\theta}$. That is, the E-BLUP at $\mathbf{s}_0$ is given by $\hat{Y}(\mathbf{s}_0; \tilde{\boldsymbol{\theta}}, D)$. Unfortunately, an exact expression for the E-BLUP's prediction error variance is unknown (except in very special cases) and simulation-based approaches to approximate it at more than a few sites are computationally prohibitive. Harville & Jeske (1992) suggested the approximation

$$\sigma^2_{EK}(\mathbf{s}_0; \boldsymbol{\theta}, D) \doteq \sigma^2_{UK}(\mathbf{s}_0; \boldsymbol{\theta}, D) + \mathrm{tr}\{\mathbf{A}(\mathbf{s}_0, \boldsymbol{\theta}, D)[\mathcal{I}(\boldsymbol{\theta}, D)]^{-1}\},$$

where "EK" denotes "empirical kriging" and $\mathbf{A}(\mathbf{s}_0; \boldsymbol{\theta}, D) = \mathrm{var}_{\boldsymbol{\theta}}[\partial \hat{Y}(\mathbf{s}_0; \boldsymbol{\theta}, D)/\partial \boldsymbol{\theta}]$. Accordingly, for optimal design for prediction with estimated covariance parameters, Zimmerman (2006) proposed to minimize the criterion

$$\phi_{EK}(D; \boldsymbol{\theta}) = \max_{\mathbf{s} \in \mathcal{P}} \sigma^2_{EK}(\mathbf{s}; \boldsymbol{\theta}, D),$$

which combines a measure of design quality for prediction under a model having known covariance parameters (the first term in Harville and Jeske's approximation) with a measure of quality for covariance parameter estimation (the second term). Not surprisingly, then, such **empirical kriging-optimal designs** represent a compromise between designs that are optimal for its two component objectives. That is, they often have reasonably good spatial coverage while also having a few small clusters. A design that minimizes $\phi_{EK}(D; \boldsymbol{\theta})$ for the aforementioned toy example when $\theta_2 = 0.8$ is displayed in Figure 10.1F. (The design that minimizes $\phi_{EK}(D; \boldsymbol{\theta})$ when $\theta_2 = 0.5$ coincides with the design that minimizes $\phi_{CPE}(D; \boldsymbol{\theta})$.) How much better the spatial coverage is, compared to what it is for the covariance estimation-optimal design, is debatable in this toy example. The large-example counterpart shown in Figure 10.2B does indeed combine decent spatial coverage with a few very compact clusters.

The prediction-based design approach based on $\phi_{EK}(\cdot)$ is focused on good empirical point prediction. Zhu & Stein (2006) carried this approach further by constructing a criterion that gives some attention to the utility of the design for interval prediction in addition to point prediction. The spatial configuration of sites in designs that are good with respect to this extended criterion do not appear to be markedly different from designs that are good for empirical point prediction, however.

Arguably, a more natural way to account for covariance parameter uncertainty when constructing efficient designs for spatial prediction is a Bayesian approach, as expounded by Diggle & Lophaven (2006). They put a prior distribution on the unknown parameters and then minimized an estimate of the spatially averaged prediction variance obtained by Monte Carlo sampling from the posterior distribution for $\boldsymbol{\theta}$ and hence from the predictive distribution of $\{Y(\mathbf{s}) : \mathbf{s} \in \mathcal{A}\}$. The designs obtained in this fashion for the examples they considered were qualitatively very similar to those obtained by the empirical prediction-based approaches of Zimmerman (2006) and Zhu & Stein (2006).

Finally, consider optimal design for mean parameter estimation when the covariance parameters are unknown. The natural estimator of $\boldsymbol{\beta}$ in this case is the empirical generalized least squares estimator $\hat{\boldsymbol{\beta}}_{EGLS} = \{\mathbf{X}^T[\boldsymbol{\Sigma}(\tilde{\boldsymbol{\theta}})]^{-1}\mathbf{X}\}^{-1}\mathbf{X}^T[\boldsymbol{\Sigma}(\tilde{\boldsymbol{\theta}})]^{-1}\mathbf{y}$, which is also the

MLE, $\tilde{\boldsymbol{\beta}}$. This estimator is unbiased under very unrestrictive conditions, but unfortunately an exact expression for its error covariance matrix is unknown, except in very special cases. An approximation for the covariance matrix of $\tilde{\boldsymbol{\beta}}$ akin to the Harville-Jeske approximation for $\sigma_{UK}^2(\mathbf{s}_0; \boldsymbol{\theta}, D)$ may be given, or one may again use a Bayesian approach to incorporate uncertainty about $\boldsymbol{\theta}$. In the former approach, the determinant of the inverse of the approximate covariance matrix of $\tilde{\boldsymbol{\beta}}$ may serve as a design criterion. Designs that maximize this criterion for a given mean function usually do not differ greatly from those that maximize the D-optimality criterion for the case of known $\boldsymbol{\theta}$.

10.2.5   Multi-objective and entropy-based criteria

Rather than optimizing a criterion measuring the quality of a design with respect to a single, narrowly focused objective, we may optimize a criterion that combines several objectives. In fact, the empirical kriging criterion introduced in the previous subsection combines criteria for covariance parameter estimation and prediction, and the criterion of Zhu & Stein (2006) combines criteria for point and interval prediction. Two general strategies for combining criteria yield what are called **compound designs** (Müller & Stehlík, 2010) and **Pareto-optimal designs** (Lu *et al.*, 2011; Müller *et al.*, 2015). A compound design criterion is a weighted linear combination of two or more component design criteria, while Pareto-optimal designs (also called the Pareto frontier) are multi-component designs for which one of the component design criteria cannot be improved without worsening the design with respect to another component criterion. Naturally, both approaches tend to yield compromise optimal designs having some features of designs that are optimal with respect to each of the component criteria.

In some situations, it may be difficult to elicit precise design objectives from the user(s) of the data, and it is also possible that the objectives may change over time. In fact, future uses of the data, and therefore future design objectives, may not necessarily be foreseen at the outset of a long-term environmental monitoring program. For these situations, perhaps the best thing to do, while still basing the choice of design on the assumed model, is to try to minimize the uncertainty of the actual responses rather than the uncertainty of model parameters or predictions based upon them. A sensible way to measure this uncertainty is by the relative entropy of the joint predictive distribution of responses at "ungauged" sites (i.e., unobserved $Y(\mathbf{s})$) given the responses at "gauged" sites (i.e., the observations) (Le & Zidek, 2006). We will not describe this approach further, except to note that it tends to add sites to an existing design that correspond to responses that are the most unpredictable, either because of their relative lack of correlation with responses at gauged sites or because of their greater intrinsic variability.

10.3   Probability-based sampling design

In **probability-based sampling design**, no model is assumed for the spatial process $Y(\cdot)$; in fact, for fixed $\mathbf{s}$, $Y(\mathbf{s})$, though it is unknown prior to sampling (and even after sampling unless $\mathbf{s}$ is included in the sample), is regarded as nonrandom. Instead, randomness is introduced into the system by the investigator through a random process for selecting the sites in the design from $\mathcal{D}$. The locations of sampled sites are determined entirely by this random process; no design criterion is optimized, hence no numerical optimization algorithm is needed. The random process is such that the probability of a site being selected for the design, called the **inclusion probability** if $\mathcal{D}$ is finite or the **inclusion density**

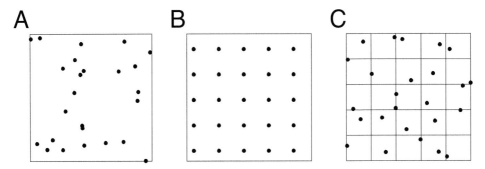

FIGURE 10.3
Probability-based spatial sampling designs, for 25 sites in a square study region: A) Simple random sample; B) systematic sample; C) stratified random sample using strata resulting from a 5×5 square grid partition of the study region.

function if $\mathcal{D} = \mathcal{A}$, is known for every site. Inclusion probabilities and densities need not be constant across sites, and different joint inclusion probabilities/densities correspond to different designs. These quantities provide a mathematical foundation for an inferential paradigm known as design-based inference, which is much different than model-based inference. The objects of interest in design-based inference are usually the spatial average or the spatial cumulative distribution function of Y. Predicting unobserved values of $Y(\mathbf{s})$ or estimating the parameters of a model for $Y(\cdot)$ are not purposes for which probability-based sampling was developed, though one can, of course, still perform those activities for such samples, which is why we include a brief summary of the topic here.

10.3.1   Simple random sampling

In **simple random sampling**, the n sites are selected according to a uniform inclusion density on $\mathcal{D}$, without replacement. All possible samples of n sites, therefore, are equally likely. One such design is displayed in Figure 10.3A. The advantages of this type of sampling include that it is seen as "fair" or unbiased; it requires no knowledge about Y or how it is distributed in space; and it yields a relatively simple theoretical basis for (design-based) inference. The main disadvantages are that it could, by chance, result in a design in which some sample sites are quite close to one another (clusters), yielding redundant information if Y is positively spatially correlated, and that some subregions of appreciable size may not be sampled at all, resulting in poor spatial prediction in those subregions. That the locations of these clusters and empty regions would likely change from one simple random sample to another implies that estimators of population quantities of interest, though they may be unbiased, could vary a lot between samples. As a result, simple random sampling is seldom used in environmental studies.

10.3.2   Systematic random sampling

In **systematic random sampling**, a single initial site is chosen according to simple random sampling and the remaining $n - 1$ sites are located nonrandomly, according to some regular pattern (to the extent possible within $\mathcal{D}$). The pattern could be a square, equilateral triangular, or hexagonal grid, for example. Figure 10.3B shows a systematic square grid design, oriented with the axes of the square study region. It would also be possible to choose the grid's angle of orientation randomly, according to a uniform distribution on $(0°, 90°)$.

The characteristics of a systematic design differ in important ways from those of the simple random design described previously and the stratified random design to be described next. First and foremost, neither large unsampled subregions nor clusters of sites can occur, so the design has a property known as **spatial balance**. Spatial balance may be desirable for several reasons. For example, in air or groundwater pollution studies, it may be politically unacceptable to not sample at all from large areas where significant numbers of citizens reside. Also, from a model-based perspective, spatial balance ensures that no two sites are so close together that they provide essentially redundant information and that no large convex subregions are left unsampled. On the other hand, as noted previously, some clusters in the design may be desirable for the purpose of good estimation of the covariance parameters of a model. Furthermore, from a pure probability-based perspective, a disadvantage of a systematic design is that it carries with it a risk of producing poor estimates if there is a periodicity in $Y(\cdot)$ that coincides with the orientation and spacing of the grid.

Among the various regular grids, which is best? Although site determination in the field is generally easiest for a square grid, triangular and hexagonal grids facilitate estimation of the covariance function in three directions, which is useful for checking for isotropy. Furthermore, the triangular grid appears to be slightly more efficient than other grids for estimating the spatial average of Y (Matérn, 1960) as well as for purposes of covariance parameter estimation and spatial prediction (Olea, 1984; Yfantis *et al.*, 1987).

10.3.3 Stratified random sampling

A sampling scheme that overcomes the weaknesses of simple random sampling without incurring possible problems due to periodicity is **stratified random sampling**. In stratified random sampling, $\mathcal{A}$ is partitioned into strata (subregions), and then simple random samples, typically of size one or two, are taken within each stratum, independently across strata; see Figure 10.3C for an example. Although this can still result in some small clusters, if the strata are chosen appropriately it will make it impossible for large convex subregions to be unsampled, so that the design has some degree of spatial balance. Consequently, a stratified random design can be more efficient than a simple random design.

How should the strata be chosen? Efficiency gains are largest when the variability within strata is as small as possible, relative to the variability between strata. Thus, for example, for sampling a pollutant around a point source for which there is no favored diffusion direction, annular strata centered at the point source would be a sensible choice. For the moss heavy metals study, it would make sense to define strata in terms of distance from the road, and indeed this is how the sampling was performed in that study, with denser sampling near the road. For spatial sampling design more generally, it makes sense to choose strata that are geographically compact since $Y(\cdot)$ is usually positively spatially correlated, hence more homogeneous, within such strata. In the absence of any additional information, approximately square, triangular, or hexagonal partitionings are reasonable possibilities. A particular stratified random design that has received considerable attention is the **randomized-tessellation stratified design** (Overton & Stehman, 1993). The strata for this design are hexagons formed by the tessellation of a triangular grid. The regularity of the hexagonal strata confers good spatial balance upon the design, while the randomization within strata greatly diminishes any phase-correspondence with a periodic surface over $\mathcal{A}$. Consequently, the randomized-tessellation stratified design is more efficient than a systematic design of the same size for estimating the spatial average of Y; moreover, it yields a better estimate of the variance of this estimator (Overton & Stehman, 1993). If $\mathcal{A}$ is so irregularly shaped that it is not possible to form strata that are approximate hexagons, a k-means clustering algorithm may be applied to a fine discretization of the study region to construct geographically compact strata (Walvoort *et al.*, 2010).

In some sampling situations, information may be available on an auxiliary variable (e.g., elevation) that is correlated with Y. In those situations, it would often be desirable to choose strata to minimize within-strata variation of the auxiliary variable, subject to some geographical compactness constraints. An algorithm that accomplishes this for square strata of varying sizes was proposed by Minasny *et al.* (2007).

10.3.4   Spatially balanced designs

Spatial balance, as it was described in Section 10.3.2, is actually just one of two types of spatial balance: spatial balance with respect to geography (the type described previously) and spatial balance with respect to the population. Spatial balance with respect to the population refers to how closely the spatial configuration of the sample resembles the spatial configuration of the population. In settings where the population is rather evenly dispersed in space, these two types are not very different, but in settings where the population is finite and clustered, they differ significantly. Spatially balanced sampling is a probabilistic sampling method in which the selected sites are spatially balanced with respect to the population.

One of the earliest, and still one of the most important, types of spatially balanced designs is the **generalized random-tessellation stratified (GRTS) design** (Stevens & Olsen, 2004). Suppose that $\mathcal{D}$ consists of a finite number of population sites distributed over a two-dimensional study region. A GRTS sample results from superimposing a randomly placed 2×2 square grid over the study region, then repeatedly subdividing each grid cell into another 2×2 set of grid cells until the sum of inclusion probabilities in each of the smallest cells is no larger than one. The grid cells are hierarchically numbered using a base-4 numbering system. Then, the numbered grid cells are permuted using a procedure called hierarchical randomization and mapped, in that permuted order, to the real line. This procedure preserves two-dimensional spatial relationships within a single dimension to a large degree. Finally, the randomly ordered linear structure is systematically sampled, with a random start, and the selected points are mapped back to their respective locations in the study region.

Another, more recently developed type of spatially balanced design is the **balanced acceptance sampling (BAS) design** (Robertson *et al.*, 2013). For this design, a bounding box is placed around the study region, and the region is rescaled so that the dimensions of the box are those of a unit square. Then, sample points are selected as the numbers in a special type of quasi-random number sequence known as the (two-dimensional) Halton sequence, provided that they lie inside the study region. If a sample of size n is desired, the first n points in a random-start Halton sequence that fall into the study region constitute the sample. The method can be extended easily to three or more dimensions, which is very useful when balance is desirable not only in geographic space but also across additional features, such as measures related to habitat suitability or species vulnerability (Brown *et al.*, 2015). Furthermore, it can produce designs with slightly better spatial balance than GRTS designs (Robertson *et al.*, 2013).

GRTS designs have been used with considerable success in numerous environmental monitoring studies carried out within the United States Environmental Protection Agency's Environmental Monitoring and Assessment Program (EMAP). The R package spsurvey Dumelle *et al.* (2023b) is available for implementing GRTS. Stevens & Olsen (2003) proposed the so-called local neighborhood variance estimator for GRTS designs, which is included in spsurvey. This design-based variance estimator was compared to the geostatistical model-based variance estimator decribed in Section 9.7 by Dumelle *et al.* (2022), who showed that using the model-based variance estimator often results in a slight improvement, though it relies on more assumptions. BAS designs and several other spatially balanced designs may be generated using the `SDraw` package (McDonald, 2020).

One major advantage of the spatially balanced designs described in this subsection, compared to systematic or stratified random sample designs, is their ability to maintain spatial balance even when replacement sites must be added due to nonresponse, or when changes to the sample size are needed to deal with budget-induced modifications to the overall sampling effort. For details on how this may be accomplished, see Dumelle *et al.* (2023b) and Robertson *et al.* (2018).

10.3.5 Variable probability sampling

More complex probabilistic sampling designs are possible, including some with unequal marginal inclusion probabilities resulting in what is called **variable probability sampling**. Such a design might be appropriate for sampling lakes for levels of mercury in fish, for example, if one wants the probability that a lake is selected to be proportional to its size because larger lakes tend to hold more fish. The key to making unbiased inferences from a variable probability sampling design is the Horvitz-Thompson theorem (Horvitz & Thompson, 1952) and its analog for sampling in a continuum (Cordy, 1993). By this theorem, an unbiased estimator, known as the Horvitz-Thompson estimator, of the population total is given by the weighted sum of all the sampled values, where the weights are the reciprocals of the inclusion probabilities. Furthermore, the variance of the Horvitz-Thompson estimator is given by an expression involving the marginal and pairwise inclusion probabilities and can be estimated unbiasedly by a function of the data and those same probabilities; details are given by Stevens (1997).

Stevens (1997) also extended the randomized-tessellation stratified design to allow the inclusion density function to vary across strata while retaining good spatial balance. These more general designs are called the **multiple-density, nested, random-tessellation stratified (MD-NRTS)** designs. GRTS and BAS designs are also easily extended to allow for unequal inclusion probabilities (Stevens & Olsen, 2004; Robertson *et al.*, 2017).

10.4 Space-filling and augmented space-filling designs

Space-filling designs are, as their name suggests, designs which leave no large (relative to the sampling intensity) convex subregion of the study region devoid of sampling points. The concept is essentially the same as that of geographic spatial balance described previously for probability-based designs. In contrast to a probability-based design, however, the sample locations of a space-filling design are determined not by randomization but by optimizing a mathematical criterion. Furthermore, optimization of a space-filling design criterion does not require any prior knowledge of, or reference to, a model governing the observations, and is usually less computationally intensive than optimization of a model-based criterion. Space-filling designs have been developed primarily for use in computer experiments (which is probably why there is very little overlap between their literature and that of spatially balanced designs), but they are also relevant to environmental studies. They may be particularly useful when the investigator cannot define a suitable model-based criterion or when there are multiple competing design objectives. Five types of space-filling designs have received the most attention: maximin distance designs, minimax distance designs, Latin hypercube designs, spatial coverage designs, and regular grids.

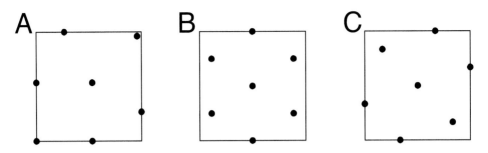

FIGURE 10.4
Space-filling designs, for 7 sites in a square study region: A) Maximin design; B) minimax design; C) maximin (and minimax) Latin hypercube design.

An n-point **maximin distance design** (Johnson *et al.*, 1990) is an n-point design for which the shortest Euclidean distance between sampled sites is maximized, i.e., a design for which

$$\phi_{Mm}(D) = \min_{\mathbf{s}_i,\mathbf{s}_j \in D, i \neq j} \|\mathbf{s}_i - \mathbf{s}_j\|$$

is maximized over all possible n-point designs D. Not surprisingly, such a design necessarily includes points on the boundary of the study region. Maximin distance designs are solutions to the problem of packing circles (in two dimensions) or spheres (in three dimensions) inside a region, about which much is known when the region is square, rectangular, or circular; see, e.g., http://www.packomania.com/. An example of a 7-point maximin distance design in the unit square is displayed in Figure 10.4A.

An n-point **minimax distance design** (Johnson *et al.*, 1990), in contrast to a maximin distance design, is a design for which the maximum distance from any point in the study region to the closest point in the design is minimized, i.e., a design for which

$$\phi_{mM}(D) = \max_{\mathbf{s} \in \mathcal{A}} \min_{\mathbf{s}_i \in D} \|\mathbf{s} - \mathbf{s}_i\|$$

is minimized over all possible n-point designs D. It may be more appropriate to call such a design a "minimaximin" distance design in light of the form of $\phi_{mM}(D)$, but the shorter name has prevailed. For fixed n, minimax distance designs often have fewer points on the boundary than maximin distance designs. Minimax designs are related to so-called covering problems in geometry. An example of a 7-point minimax distance design in the unit square is displayed in Figure 10.4B.

A possibly undesirable feature of maximin and minimax distance designs is that they are not necessarily space-filling when projected to their lower-dimensional subspaces. This can be visualized in the maximin and minimax designs of Figure 10.4, where the projections to either the bottom or the left side of the square would result in some coincident points and relatively large gaps between points. An n-point **Latin hypercube design** (McKay *et al.*, 1979; Pistone & Vicario, 2010) addresses this shortcoming by having the property that all of its one-dimensional projections are n-point maximin designs; this latter property is equivalent to each one-dimensional projection comprising the sequence $\{0, 1/(n-1), 2/(n-1), \ldots, 1\}$ (when $\mathcal{A}$ is a unit square or cube). For any n there are many Latin hypercube designs, some of which may be very poor with respect to global (d-dimensional) space-filling. It is common, therefore, to seek a Latin hypercube design that is globally space-filling within that class. Figure 10.4C shows a 7-point Latin hypercube design that is both globally maximin and globally minimax within the class of 7-point Latin hypercube designs.

Royle & Nychka (1998) defined an n-point **spatial coverage design** as a design that minimizes

$$\phi_{C,r,q}(D) = \left[\sum_{\mathbf{s} \in \mathcal{A}} \left\{ \left(\sum_{\mathbf{s}_i \in D} \|\mathbf{s} - \mathbf{s}_i\|^r \right)^{q/r} \right\} \right]^{1/q}$$

over all possible n-point designs D, where r and q are negative and positive integers, respectively. Note that $\phi_{C,r,q}(D)$ is a function of the same quantities involved in the minimax distance criterion $\phi_{mM}(D)$; more specifically, $\phi_{C,r,q}(D)$ is an L_q-average of generalized distances between candidate points and the design. Since $r < 0$, the generalized distance $\left(\sum_{\mathbf{s}_i \in D} \|\mathbf{s}_i - \mathbf{s}\|^r \right)^{1/r}$ converges to 0 as $\mathbf{s}$ converges to any site in D. The design that minimizes $\phi_{C,r,q}(D)$ depends on r and q, but appears to be insensitive to these values if r is very negative. In the limit as $r \to -\infty$ and $q \to \infty$, $\phi_{C,r,q}(\cdot)$ converges to the minimax distance criterion. Royle & Nychka (1998) featured spatial coverage designs with $r = -5$ and $q = 1$.

The last category of space-filling designs, **regular grids**, has already been discussed in the context of probability-based design. The only difference between regular grids of the probability-based variety and those of the space-filling variety is that the locations of the gridpoints of the latter are not determined randomly but are selected either to maximize the minimum distance between them (a maximin grid design) or to minimize the maximum distance between a point in $\mathcal{A}$ and the nearest gridpoint (a minimax grid design). If $\mathcal{A}$ is rectangular, some points of a maximin grid design will necessarily lie on the boundary of $\mathcal{A}$, whereas points of a minimax grid design will lie entirely within $\mathcal{A}$.

In practice, irregularity of $\mathcal{A}$, constraints on the candidate sampling sites, and/or the sheer size of the design problem may preclude an investigator from finding an optimal maximin, minimax, or Latin hypercube design within the catalog of known ones, in which case the design must be found using a numerical optimization algorithm. Many of the same algorithms used to optimize model-based design criteria may be used successfully to optimize space-filling criteria (Pronzato and Müller, 2012), with exchange algorithms seeming to be especially popular.

How do space-filling designs compare to model-based designs? A comparison of Figure 10.4 with Figure 10.1 makes it plain that the site configuration of a globally space-filling design is similar to that of a design that optimizes a model-based prediction-error variance criterion like $\phi_K(D; \boldsymbol{\theta})$, but very different from that of a design that optimizes some other model-based criterion. Consequently, space-filling designs perform very well for spatial prediction with known covariance parameters; in fact, Zimmerman & Li (2013) and Li & Zimmerman (2015) showed, for cases with $n = 4$ or 5 and $\mathcal{D}$ a 4×4 or 5×5 grid, that the ϕ_K-optimal design is a maximin or minimax Latin hypercube design, and Johnson *et al.* (1990) established a theoretical connection between maximin and minimax distance designs and designs that optimize certain other prediction-oriented criteria. However, for objectives other than spatial prediction, especially for estimation of covariance parameters, space-filling designs can perform very poorly.

In light of this, attempts have been made to augment space-filling designs with some clusters of sites in hopes of improving them for purposes of covariance estimation and empirical kriging. Diggle & Lophaven (2006) introduced two augmented space-filling designs. Their **lattice plus close pairs design** consists of a $k \times k$ square lattice (grid) of $n_s = k^2$ sites with spacing Δ augmented by n_a additional sites, each distributed uniformly within a disk of radius $\alpha\Delta$ centered at a randomly chosen (without replacement) lattice site. Their **lattice plus infill design** starts with the same square lattice of n_s sites but is augmented by sites in a grid with smaller spacing $\alpha\Delta$ within m randomly selected grid cells. Examples of such designs are displayed in Figure 10.5A,B. Diggle & Lophaven (2006) found that both designs performed best for empirical kriging when the augmented sites comprise about

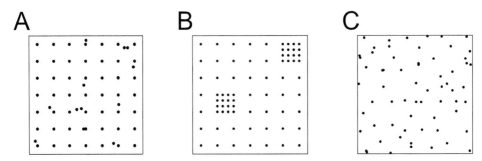

FIGURE 10.5
Augmented space-filling designs in a square study region: A) Lattice plus close pairs design with $(n_s, \Delta, n_a, \alpha) = (49, \frac{1}{7}, 15, \frac{1}{2})$; B) lattice plus infill design with $(n_s, \Delta, m, \alpha) = (49, \frac{1}{7}, 2, \frac{1}{21})$; C) inhibitory plus close pairs design with $(n_s, \Delta, n_a, \alpha) = (49, \frac{1}{9}, 15, \frac{1}{2})$.

20–30% of the total number of sites in the design, but that the close-pairs version was much superior to the infill version. Building off of Diggle and Lophaven's work, Chipeta *et al.* (2017) proposed a class of **inhibitory plus close pairs designs** in which the square lattice of the previous two types of design is replaced by n_s sites obtained by sequentially sampling at random from the study region subject to the constraint that each additional site is no closer than a distance Δ from any previously sampled site. Augmentation with n_a additional sites proceeds as in the lattice plus close pairs design. One such design is displayed in Figure 10.5C. For such designs, it was found again that they were best when about 10–30% of the sample was devoted to augmentation. Leung *et al.* (2021) proposed a somewhat similar class of designs obtained by simulating a realization of a generalized Neyman-Scott point process. Both types of designs appear to be competitive with those that optimize the empirical kriging criterion of Zimmerman (2006) and are more computationally efficient to obtain because they replace the high-dimensional optimization problem of selecting sampling sites with a low-dimensional problem of selecting optimal parameter sets from a specified family of spatial point processes.

10.5   Preferential sampling design

Preferential sampling refers to choosing a sampling design in a manner that is not independent of the responses. Preferential sampling designs usually arise when sampling locations are deliberately concentrated in subregions of the study area where the response is thought to be larger (or smaller) than average. Formally, consider the special case of a geostatistical linear model given by

$$y_i = \mu + W_i + \delta_i \quad (i = 1, \ldots, n),$$

where μ is an unknown constant, the W_i's are unobserved values of a zero-mean, mean-square continuous process $W(\cdot) \equiv \{W(\mathbf{s}) : \mathbf{s} \in \mathcal{A}\}$ at arbitrary sites $\mathbf{s}_1, \ldots, \mathbf{s}_n$, and the δ_i's are measurement (or microscale) errors with common variance σ^2. Allowing the design $D = (\mathbf{s}_1, \ldots, \mathbf{s}_n)$ to be chosen stochastically, the joint distribution of $W(\cdot)$, D, and $\mathbf{y}$ may generally be written as

$$[W(\cdot), D, \mathbf{y}] = [W(\cdot)][D|W(\cdot)][\mathbf{y}|W(\cdot), D],$$

where $[\cdot]$ denotes an arbitrary distribution. If samping is non-preferential, then D and $W(\cdot)$ are independent, so that $[D|W(\cdot)] = [D]$. The sampling is preferential if and only $[W(\cdot), D] \neq [W(\cdot)][D]$.

Under preferential sampling, inference generally is biased. Diggle *et al.* (2010) showed specifically, via simulation, that if a preferentially sampled design D is selected in such a way that the probability of a site's inclusion in the design, conditional on $W(\cdot)$, is proportional to $\exp[\alpha + \beta W(\mathbf{s})]$ where $\beta > 0$, then the sample semivariogram can be substantially negatively biased, and the ordinary kriging predictor is also biased in subregions of the study area where few or no observations were taken. The same authors described a Monte Carlo MLE method for the parameters, which attempts to account for the preferential sampling. Schliep *et al.* (2023) proposed a weighted composite likelihood estimation approach with the same objective. Gelfand *et al.* (2012) also considered the effects of preferential sampling on kriging. They found that not only is kriging biased under preferential sampling, but the kriging variance is less well estimated than it is under simple random sampling. Furthermore, they found that while adjustments that account for preferential sampling can improve prediction performance, kriging with those adjustments is still inferior to classical kriging under simple random sampling.

10.6 Exercises

1. Consider two four-point spatial sampling designs on the unit square in $\mathbb{R}^2$. In Design I, the data sites are located at the corners of the square, and in Design II the sites are located at the midpoints of the sides of the square. Suppose further that the mean function for observations at these sites is constant, and the covariance function is isotropic spherical, with range α.

   (a) Determine which of the two designs maximizes $|\mathbf{X}^T \mathbf{\Sigma}^{-1} \mathbf{X}|$ for each of the following values of α: 0.8, 1.0, 1.5, 2.0.

   (b) Determine which of the two designs minimizes the maximum ordinary kriging variance over the unit square for each of the values of α listed in part (a). (The maximum ordinary kriging variance occurs at the center of the unit square for Design I, and at any of the corner sites for Design II.)

2. Repeat the activities performed in the previous exercise, but for a situation identical in every way except that the mean function is planar rather than constant. For the criterion in part (b), replace the ordinary kriging variance at the two specified prediction sites with the universal kriging variance at those sites.

3. Suppose that $\{Y(\mathbf{s}) : \mathbf{s} \in \mathcal{A}\}$ is a stationary Gaussian process with constant mean μ and covariance function $C(\mathbf{h}; \boldsymbol{\theta})$, which is monotone decreasing in every direction, but possibly anisotropic. Consider the following two designs. Design I is a line transect design, i.e., 25 equally spaced points that lie on a straight line, with unit spacing between points. Design II is a square grid design, i.e., 25 points lying at the nodes of a 5×5 square grid, with unit spacing between adjacent grid points.

   (a) Describe the advantages, if any, of each design relative to the other, insofar as the maximum likelihood estimation of μ is concerned.

   (b) Describe the advantages, if any, of each design relative to the other, insofar as the maximum likelihood estimation of $\boldsymbol{\theta}$ is concerned.

11

Analysis and Design of Spatial Experiments

In this chapter, we shift our focus from design for spatial sampling of the environment to the analysis and design of a **spatial experiment**. A spatial experiment is an experiment in which two or more treatments are applied to experimental units (EUs) that occupy fixed, known positions in one-, two-, or three-dimensional space. Such an experiment is typically conducted in order to compare the effects of the treatments on a variable measured on each experimental unit. Classical analysis-of-variance methods were first developed by Sir Ronald Fisher, Frank Yates, and others at the Rothamsted Experiment Station in the first half of the twentieth century for agricultural field-plot experiments, which were, in fact, spatial experiments, though that term apparently was not introduced into the lexicon until much later (Zimmerman and Harville, 1991; Federov, 1996). The caribou forage experiment introduced in Section 1.1 is an example of a spatial experiment; the main objective of that experiment, as noted previously, was to compare the effects of tarps (a proxy for shade) and water on the nutritional value of caribou forage.

The presence of trends and/or correlation in fertility, moisture, or other attributes among proximate field plots was widely recognized by the early developers of experimental designs. Indeed, **blocking** and **randomization**, two of the pillars of classical experimental design, were introduced, in part, to eliminate or at least neutralize such spatial effects. For example, if an environmental gradient in one direction is known to exist in the study area in which the experiment is to be conducted, it is wise to carry out a randomized block design experiment of one type or another, with blocks oriented perpendicular to the gradient, as doing so (and including block effects in the model used for analysis) will likely reduce the mean squared error, resulting in more powerful inferences for treatment effects and comparisons. The connections between randomization, the classical linear model, and the spatial linear model are important and often confusing, and in this chapter we clarify their relationships, and by extension, how to use them appropriately. We also describe optimal design for spatial experiments under the spatial linear model.

11.1 The randomization-derived linear model for designed experiments

11.1.1 Notation and model form

We begin by re-introducing the role of randomization in designed experiments. Many practitioners of experiments employ randomization in their design but then for purposes of analysis adopt the classical linear model, (1.1), in which the errors are independent. From

DOI: 10.1201/9780429060878-11

a design-based inferential perspective, this is not strictly correct because the distribution implied by the classical linear model is actually an approximation to the distribution induced by the randomization — the so-called **randomization distribution**. To see why, we first need to imagine that responses from all of the EUs have fixed values that occurred, or would have occured, had no treatments been added to them. The experiment is the act of applying treatments to the EUs, thus altering what we measure or observe from the EUs. To keep the focus on the most important concepts, let us consider a one-factor experiment with a balanced design, with t treatments and r replications of that treatment. The total number of EUs, then, is $N = rt$. Our development follows Hinkelmann & Kempthorne (2007, Section 6.3).

Define design variables $\{\delta_{ij}^k\}$ ($i = 1, \ldots, t$; $j = 1, \ldots, r$; $k = 1, \ldots, N$):

$$\delta_{ij}^k = \begin{cases} 1 & \text{if EU } k \text{ gets the } j\text{th replicate of the } i\text{th treatment,} \\ 0 & \text{otherwise.} \end{cases}$$

These are Bernoulli random variables that are identically distributed, but not independent because of restrictions $\sum_k \delta_{ij}^k = 1$ and $\sum_{ij} \delta_{ij}^k = 1$, and they are similar to inclusion probabilities (Section 10.3) used in design-based sampling theory. The restrictions induce negative correlations among the variables, and also among the observed treatment means (which are quantified below). Let T_{ik} be the value that would be observed if treatment i were applied to the kth EU. Then the jth observation on the ith treatment is

$$y_{ij} = \sum_{k=1}^{N} \delta_{ij}^k T_{ik}, \tag{11.1}$$

for some realization of the random variables $\{\delta_{ij}^k\}$ and fixed values T_{ik}.

An important assumption is that treatments are *additive*, so that we can assume that $T_{ik} = T_i + U_k$, where U_k is the value that would be observed on the kth EU if no treatment were added, and T_i is the "treatment effect," which is the (additive) deviation of T_{ik} from U_k. Let $\overline{T}$ be the average treatment effect (mean of the t T_i's), and let $\overline{U}$ be the average value among all EUs (the mean of the N U_k's). Then

$$T_{ik} = (\overline{T} + \overline{U}) + (T_i - \overline{T}) + (U_k - \overline{U}),$$

and letting $\mu_r = \overline{T} + \overline{U}$, $\tau_i = T_i - \overline{T}$, and $u_k = U_k - \overline{U}$, we can write

$$T_{ik} = \mu_r + \tau_i + u_k,$$

where $\sum_i \tau_i = 0$ and $\sum_k u_k = 0$. Then (11.1) can be written as

$$y_{ij} = \sum_k \delta_{ij}^k (\mu_r + \tau_i + u_k) = \mu_r + \tau_i + \omega_{ij},$$

where $\omega_{ij} = \sum_k \delta_{ij}^k u_k$, or alternatively, in vector and matrix notation, as

$$\mathbf{y} = \mathbf{X}\boldsymbol{\beta} + \boldsymbol{\omega}$$

for a suitable $\mathbf{X}$, where $\boldsymbol{\beta} = (\mu_r, \tau_1, \tau_2, \ldots, \tau_t)^T$ and $\boldsymbol{\omega} = (\omega_{11}, \omega_{12}, \ldots, \omega_{1r}, \omega_{21}, \ldots, \omega_{tr})^T$. We call this model the **randomization-derived linear model**.

11.1.2 Distributional properties of the randomization-derived linear model

Note the following properties of $\{\delta_{ij}^k\}$:

$$P(\delta_{ij}^k = 1) = \frac{1}{N}, \tag{11.2}$$

$$P(\delta_{ij}^k = 1, \delta_{i'j'}^{k'} = 1) = \begin{cases} \frac{1}{N(N-1)} & \text{for } k \neq k' \text{ and } (i,j) \neq (i',j'), \\ 0 & \text{for } k = k' \text{ or } (i,j) = (i',j'). \end{cases} \tag{11.3}$$

Thus, using the results from Chapter 2 on expectation, variance, and covariance, we obtain

$$\mathrm{E}(\delta_{ij}^k) = \frac{1}{N}, \tag{11.4}$$

$$\mathrm{Var}(\delta_{ij}^k) = \frac{1}{N}\left(1 - \frac{1}{N}\right), \tag{11.5}$$

$$\mathrm{Cov}(\delta_{ij}^k, \delta_{i'j'}^{k'}) = \begin{cases} \frac{1}{N^2(N-1)} & \text{for } k \neq k' \text{ and } (i,j) \neq (i',j'), \\ -\frac{1}{N^2} & \text{for } k = k' \text{ or } (i,j) = (i',j'). \end{cases} \tag{11.6}$$

Using these results, and letting $\sigma_u^2 = \sum_k u_k^2/(N-1)$, we have

$$\mathrm{E}(\omega_{ij}) = 0, \tag{11.7}$$

$$\mathrm{Var}(\omega_{ij}) = \left(1 - \frac{1}{N}\right)\sigma_u^2, \tag{11.8}$$

$$\mathrm{Cov}(\omega_{ij}, \omega_{i'j'}) = -\frac{1}{N}\sigma_u^2 \quad \text{for } (i,j) \neq (i',j'). \tag{11.9}$$

Finally, letting $\bar{y}_i = (1/r)\sum_{j=1}^r y_{ij}$, we have

$$\mathrm{Var}(y_{ij}) = \left(1 - \frac{1}{N}\right)\sigma_u^2, \tag{11.10}$$

$$\mathrm{Var}(\bar{y}_i) = \frac{1}{r}\left(1 - \frac{r}{N}\right)\sigma_u^2, \tag{11.11}$$

$$\mathrm{Cov}(\bar{y}_i, \bar{y}_{i'}) = -\frac{1}{N}\sigma_u^2 \quad \text{for } i \neq i'. \tag{11.12}$$

Verifying all of these properties is left as an exercise (Exercise 11.1).

11.2 The classical linear model for designed experiments

In this book, we have been focused on linear models (1.1) of the form

$$\mathbf{y} = \mathbf{X}\boldsymbol{\beta} + \mathbf{e},$$

and if $\mathbf{y}$ arises from a designed experiment, we have made the classical assumption that it follows this linear model. A consequence of this model, when used in an experimental design setting where treatments are classification (binary) variables, is that the treatment effects are additive, thus agreeing with the additivity assumption of the randomization-derived linear model. Additional assumptions that constrain the form of the model to make it appropriate for designed experiments are as follows:

1) the random errors (the elements of **e**) are homoscedastic (homogeneous or equal variances);

2) the random errors are stochastically independent;

3) the random errors follow a zero-mean normal (Gaussian) distribution.

Previously we have called this model the (Gaussian) Gauss-Markov model, but in this chapter we will refer to it as the **classical linear model** because of its long history as the model used to analyze data from designed experiments. Recall that ordinary least squares (OLS) estimation yields the best linear unbiased estimators of estimable linear functions of $\boldsymbol{\beta}$ under this model.

It is useful to compare the assumptions just listed for the classical linear model to the properties of the randomization-derived linear model noted in Section 11.1.2. If randomization occurs, (11.10) and (11.11) show that homoscedasticity holds for the observations and (assuming equal treatment replication) for the treatment means under both models. Regarding the second assumption, (11.9) shows that the observations are *not* independent under the randomization-derived linear model, which differs from the situation under the classical linear model. However, for large N, the correlations among the errors of the former model are very close to zero, so the difference in the dependence assumptions between the two models is not large in practice. Finally, although the errors of the randomization-derived linear model are not normally distributed, the standard confidence intervals, t tests, and F tests described in Chapter 4 for inference under the classical linear model are nonetheless approximately valid under the randomization-derived model because they are computed on means, and the central limit theorem drives the distributions of those means toward normality as sample size increases, regardless of the underlying distribution.

For the reasons just described, many experimenters who use the classical linear model to analyze their data claim, with some justification, that the aforementioned inference procedures are robust to mild departures from the assumptions of that model if randomization is used. On the flip side, those who randomize and assume the randomization-derived linear model often (though not always) use the classical linear model as an approximation. This close relationship between the classical and randomization-derived linear models has blurred their distinction and has important consequences for inference that we will discuss in the next section.

11.3   The spatial linear model for designed experiments

The spatial linear model for designed experiments is identical to the classical linear model in every respect except for the latter's assumption of independent errors. Under the spatial linear model, the errors are assumed to be produced by a Gaussian process, i.e., **e** has a multivariate normal distribution with possibly nonzero correlations. The correlations are modeled with just a few parameters, using spatial information such as distances (as in Chapter 6) or neighbor relationships (as in Chapter 7) among the experimental units. Estimation of parameters in the model's mean structure and covariance structure is carried out by likelihood-based methods (ML or REML). For the mean structure parameters, this amounts to performing generalized least squares as if the likelihood-based estimates of the covariance parameters were their true values, yielding what we called (in Chapter 8) the ML-EGLS or REML-EGLS estimators.

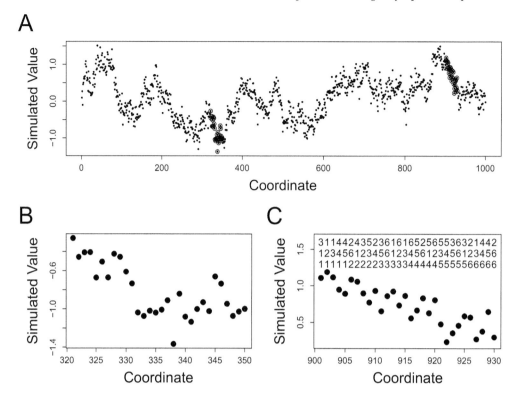

FIGURE 11.1

A) Scatterplot of 1000 observations simulated from a spatial linear model with zero mean and exponential covariance function, versus their (one-dimensional) spatial locations; B) a close-up of the observations at locations 321 to 350; C) a close-up of the observations at locations 901 to 930. The numbers above the points displayed in C are the labels of treatments assigned to the corresponding locations; the top row is a completely randomized design, the middle row is Design 1, and the bottom row is Design 2.

11.3.1   A simulated spatial experiment

We illustrate the use of REML-EGLS estimation of treatment effects and contrasts for a simulated spatial experiment. To connect the simulation to a real example, we simulated observations from the same model that was fitted to the caribou forage data, but to make the visualization easier we simulated for EUs lying in one spatial dimension rather than two. Thus, we simulated zero-mean normal random variables having an exponential covariance function with a range of 49.2, a nugget effect of 0.0222, and a partial sill of 0.269, as estimated in Section 8.8.3 for the caribou forage data. The likelihood surface and fitted autocorrelation for the caribou forage data were shown in Figure 8.12C,D, respectively. A scatterplot of the simulated data (observed response versus spatial coordinate) is shown in Figure 11.1A, where the values of the spatial coordinate are the integers from 1 to 1000. Open circles are drawn around simulated values between coordinates 321 to 350, and between coordinates 901 to 930, which are shown in more detail in Figures 11.1B,C. It may be noted that in neither of Figures 11.1B,C does the range of the data contain the true process mean (which was zero). Furthermore, a downward trend is visible in both scatterplots. In Figure 11.1B, the trend is away from the mean, and in Figure 11.1C, the

TABLE 11.1

Estimates of the first treatment effect and all treatment contrasts and corresponding estimated standard errors, from fits of two models to data simulated as described in Section 11.3.1. CLM is the classical linear model (with independent errors), while SLM is a spatial linear model with an exponential covariance function. The CLM was fitted by OLS. Two REML-EGLS fits of the SLM are included. For one fit, all of the covariance parameters were estimated, but for the other, the range parameter (ρ) was held fixed at 200 and only the nugget effect and partial sill parameters were optimized.

| | Estimate | | | Standard Error | | |
| | **CLM** | **SLM** | **SLM** | **CLM** | **SLM** | **SLM** |
Effect	$\rho = 0$	$\rho = 4394$	$\rho = 200$	$\rho = 0$	$\rho = 4394$	$\rho = 200$
$\mu + \tau_1$	0.872	0.786	0.787	0.124	11.339	0.767
$\tau_2 - \tau_1$	-0.130	0.009	0.008	0.175	0.091	0.091
$\tau_3 - \tau_1$	-0.152	-0.107	-0.107	0.175	0.089	0.089
$\tau_4 - \tau_1$	-0.089	-0.008	-0.008	0.175	0.093	0.093
$\tau_5 - \tau_1$	-0.318	-0.207	-0.208	0.175	0.098	0.099
$\tau_6 - \tau_1$	-0.190	-0.083	-0.083	0.175	0.089	0.089

trend is toward the mean. This example illustrates that autocorrelated data can "wander" away from the mean for long stretches. Indeed, it can be demonstrated that the stronger the (positive) spatial correlation, the larger and longer the excursions from the mean tend to be. Hence, when a model is fit to data that are highly autocorrelated, there will be a lot of uncertainty in estimating the model's mean. Looking at Figures 11.1B,C, without knowledge that the true mean is zero, how can one tell if the data are straddling the mean, wandering away from the mean, or wandering back toward the mean? In fact one cannot, and this has consequences for the estimation of treatment effects and contrasts for the randomization-derived linear model, the classical linear model, and a spatial linear model, on which we now elaborate.

We took the data in Figure 11.1C, specifically, to be realizations of the random errors, $\mathbf{e}$, of a spatial linear model for data from a randomized experiment with a single factor that had 6 treatments. We took the mean of an observation assigned to the ith treatment to be $\mu + \tau_i$, where μ is an overall mean effect and τ_i is the additive effect of the ith treatment. In a completely randomized design, treatments are applied to experimental units at random, subject to equal replication. One such randomization is shown in Figure 11.1C, where the treatment labels assigned to EUs are the top row of numbers over the scatterplot (the numbers in the second and third rows will be described later). We set $\mu = 0$ and also set all 6 treatment effects to zero ($\tau_1 = \tau_2 = \cdots = \tau_6 = 0$), and then we estimated treatment effects and contrasts. Because we know the values of all parameters under the model that created the data, we can evaluate the performance of inference procedures corresponding to the randomization-derived linear model, the classical linear model, and the spatial linear model.

Table 11.1 shows results from fitting the classical and spatial linear models to the simulated data. Here, the classical linear model could be regarded as the model for the process that generated the data, or as an approximation to the randomization-derived linear model. The fitted spatial model was taken to be the same model that generated the data, but with unknown parameters. For this model, REML estimates of the range, partial sill, and nugget effect were 4394, 128.6, and 0.0139, respectively. Although we do not display the fitted REML log-likelihood, it had a strong ridge very similar to the one shown for a completely different dataset in Figure 8.5. Let us examine the effect of a different parameter set, as long as we stay on that ridge. When the range parameter was set to 200, the estimated partial sill was 0.623 and the estimated nugget effect was 0.0137. The estimated mean structure for both fitted spatial models is shown in Table 11.1.

The parameterization used in Table 11.1 is the default parameterization used by R, for which the parameter it labels "intercept" is the mean plus the first treatment effect. Similar to the situation described in Section 4.5.3, a model with an overall mean effect μ and all treatment effects τ_i $(i = 1, \ldots, t)$ is overparameterized, rendering the mean effect and any treatment effect nonestimable. However, the sum of the mean and any treatment effect, $\mu + \tau_i$, is estimable, and so is any treatment contrast. Because the parameter labeled "intercept" in R is $\mu + \tau_1$, the rest of the parameters, which are labeled using the treatment labels in R, actually represent contrasts $\tau_2 - \tau_1$, $\tau_3 - \tau_1$, etc. To obtain the OLS estimate of, say, $\mu + \tau_2$, we observe that $\mu + \tau_2 = (\mu + \tau_1) + (\tau_2 - \tau_1)$, so that its OLS estimate is $\boldsymbol{\ell}^T \hat{\boldsymbol{\beta}}_{OLS}$, where $\boldsymbol{\ell} = (1, 1, 0, 0, 0, 0)$ and $\hat{\boldsymbol{\beta}}_{OLS}$ is the OLS estimate of $\boldsymbol{\beta}$ in the parameterization used by R. Similarly, the REML-EGLS estimate of $\mu + \tau_2$ is $\boldsymbol{\ell}^T \tilde{\boldsymbol{\beta}}$ where $\tilde{\boldsymbol{\beta}}$ is the REML-EGLS estimate of $\boldsymbol{\beta}$. The estimated standard errors of these two estimators, obtained under their respective models, are $[\hat{\sigma}_{OLS}^2 \boldsymbol{\ell}^T (\mathbf{X}^T \mathbf{X})^{-1} \boldsymbol{\ell}]^{1/2}$ and $(\boldsymbol{\ell}^T \tilde{\mathbf{C}}_{\boldsymbol{\beta}} \boldsymbol{\ell})^{1/2}$, respectively, where $\tilde{\mathbf{C}}_{\boldsymbol{\beta}} = (\mathbf{X}^T \tilde{\boldsymbol{\Sigma}}^{-1} \mathbf{X})^{-1}$.

11.3.2 Parameter estimates and standard errors

Let us first consider the estimates of $\mu + \tau_1$. Recall that the data were simulated under a model with true parameters $\mu = 0$ and $\tau_1 = 0$, so that the true value of $\mu + \tau_1$ is 0. The OLS estimate of $\mu + \tau_1$ (obtained by fitting the classical linear model) is 0.872 (Table 11.1), which is very far from the true value. However, if we take the point of view that the classical linear model is an approximation to the randomization-derived linear model, then that OLS estimate is more appropriately regarded as an estimate of $\mu_r + \tau_1 = \overline{T} + \overline{U} + \tau_1$ because the randomization-derived linear model regards the realized values from the simulation as fixed. The computed mean of the 30 of those values displayed in Figure 11.1C is 0.726. With $\tau_1 = 0$, this is the quantity that the randomization-derived linear model should be regarded as estimating, and the OLS estimate is indeed much closer to this value than to 0.

Note here that there can be confusion between "mean" as a parameter, and "mean" as a computed average of some observations. If we assume a classical linear model or a spatial linear model, then the mean we wish to estimate is a parameter of a generating process. If, instead, we assume the randomization-derived linear model, then we should estimate the average of the realized values. When random variables are independent with common mean and variance, the difference between the mean parameter and the computed average of the realized values converges to zero very rapidly as the sample size increases. The consequence of this is that there is little distinction between regarding the classical linear model as the generating process or as an approximation to the randomization-derived linear model for large samples of independent observations. However, for positively correlated data (the usual case in practice), the convergence can require much larger sample sizes, depending on the strength of the spatial correlation, as is evident when considering all 1000 simulated observations in Figure 11.1A. Indeed, some subsets of the observations, such as the subset in Figure 11.1C, may have realized means that are very far from the true mean of the generating process.

Let us now look at the parameter estimates from the fit of the spatial linear model (Table 11.1). The REML-EGLS estimates of $\mu + \tau_1$ for the spatial linear model with the two sets of covariance parameter estimates are very similar to each other, even though the covariance parameter estimates themselves are quite different. In fact, the two REML-EGLS estimates differ substantially more from the OLS estimate obtained under the classical linear model than they do from each other. The reasons for this were already discussed in Sections 5.2 and 5.3, and the same principles apply here. They also differ greatly from the true value of 0, being similar to the OLS estimate in this respect.

Table 11.1 shows large differences in the standard errors for the estimates of $\mu + \tau_1$. The standard error obtained using the classical linear model is the smallest, and in light of the previous discussion we should ask the question, "Does this standard error actually correspond to estimating $\mu + \tau_1$ or to estimating $\mu_r + \tau_1$ (recall that μ_r is the realized mean)?" The 95% confidence interval is $0.872 \pm (1.96 \times 0.124) = (0.629, 1.115)$, which easily covers $\mu_r + \tau_1$, but is not close to covering $\mu + \tau_1$. Of course, this is just one simulated data set and one randomization. We will explore this issue more thoroughly with some further simulations below. Notice the large standard error, 11.339, for the spatial linear model with range 4394 (Table 11.1). The corresponding 95% confidence interval is $0.786 \pm (1.96 \times 11.339)$, which easily covers both $\mu_r + \tau_1$ and $\mu + \tau_1$. Also, while the REML-EGLS estimates of $\mu + \tau_1$ corresponding to the two fits of the spatial linear model are quite similar, there is a large difference in the corresponding standard errors, as the model with range 200 had a standard error of 0.767. The 95% confidence interval for $\mu + \tau_1$ under that model was $0.787 \pm (1.96 \times 0.767) = (-0.716, 2.290)$, which easily covers both $\mu_r + \tau_1$ and $\mu + \tau_1$.

Now let us turn our attention to the estimation of contrasts. Neither mean, μ nor μ_r, is involved in estimating a treatment contrast, so which of the two means is being estimated is not an issue in this context. OLS estimates of contrasts obtained under the classical linear model range from -0.318 to -0.089 (Table 11.1). As was the case for the estimates of treatment effects, there is very little difference in REML-EGLS estimates of contrasts for the two fits of the spatial linear model. However, in comparison to the OLS estimates, they are all closer to 0, the true value. Moreover, their standard errors are nearly identical, in stark comparison to the standard errors for the estimated treatment effect. Additionally, the standard errors of the REML-EGLS estimates are much smaller than those of the OLS estimates, again in stark comparison to the situation when estimating a treatment effect. Confidence intervals for the contrasts in Table 11.1 corresponding to the three model fits covered the true values in all cases except for $\tau_5 - \tau_1$ obtained under the two fits of the spatial linear model, for which coverage just barely failed. So, what is going on here? Are the differences listed above just peculiar results for this particular dataset and randomized design, or are there some general principles worth knowing? In order to answer this question, we will conduct a more extensive simulation study.

11.3.3 A larger simulation study

For this larger simulation study, we use the entire set of simulated values displayed in Figure 11.1A. Starting at positions 1 through 30, we randomly apply 6 treatments with true values of 0, just as we did before. Then we move to positions 2 through 31, and randomly apply 6 treatments again. We continue in this fashion, applying a fresh randomization each time, and stop when we get to positions 971 through 1000. For each randomized assignment of treatments (971 of them), we fit the classical and spatial linear models, where in this case for the spatial linear model we fix the covariance parameters at their true values so that we can better understand how the estimation of treatment effects and contrasts performs. For each simulation, we estimate the treatment effect $\mu + \tau_1$ and the contrast $\tau_2 - \tau_1$. As performance measures, for both the OLS estimates (corresponding to the classical linear model) and the GLS estimates (corresponding to the spatial linear model with known covariance parameters) of these parameters, we compute the root-mean-squared errors and the empirical Type I error rates for size-0.05 tests of null hypotheses that these parameters are equal to zero (i.e., the proportion of times that the confidence interval does not cover 0).

TABLE 11.2
Root-mean-squared error (RMSE), and empirical Type I error rate for size-0.05 tests, of
OLS estimates (corresponding to the classical linear model, labeled as CLM) and GLS
estimates (corresponding to the spatial linear model with known covariance parameters,
labeled as SLM), of three quantities for the simulated data in Figure 11.1A.

Effect	RMSE		Type I Error	
	CLM	**SLM**	**CLM**	**SLM**
$\mu_r + \tau_1$	0.108	0.125	0.037	0.000
$\mu + \tau_1$	0.532	0.518	0.757	0.060
$\tau_2 - \tau_1$	0.168	0.112	0.051	0.048

The results of the simulation experiment are given in Table 11.2. Note that when estimating the treatment effect that is added to the realized mean ($\mu_r + \tau_1$), the RMSE of the
OLS estimator under the clasical linear model is smaller than that of the GLS estimator
under the spatial linear model. Moreover, the Type I error rate corresponding to the classical linear model, 0.037, is reasonably close to the nominal value of 0.05, suggesting that
the standard errors obtained under this model are appropriate, while the Type I error rate
corresponding to the spatial linear model is 0, indicating that the standard errors obtained
under this model are much too large for estimating $\mu_r + \tau_1$. The situation changes when
estimating the treatment effect that is added to the true process mean ($\mu + \tau_1$): the RMSE
corresponding to the spatial linear model is smaller than that corresponding to the classical
linear model, and the Type I error rate for the latter model is near 0.05, while that for
the former model is greater than 0.75, indicating that its standard error is much too small.
Turning to the contrast estimate, the RMSE corresponding to the spatial linear model is
considerably smaller than that corresponding to the classical linear model, but both models have Type I error rates near the nominal level of 0.05, suggesting that they both have
appropriate standard errors.

The simulation results in Table 11.2 suggest that when treatments are randomized, inference based on the classical linear model is appropriate as an approximation to inference for
the randomization-derived linear model, even when the observations are spatially correlated
(it has valid confidence intervals for $\mu_r + \tau_i$ and $\tau_i - \tau_j$, but not for $\mu + \tau_i$.). As mentioned
previously, early developers such as Fisher and Yates recognized spatial patterning in the
environment, and devised blocking, along with randomization, to help increase precision
by accounting for variation among blocks through a mean structure that included block
effects. However, not all spatial patterning can be so easily eliminated by adding simple row
and/or column effects to the model, and blocking does not come without a cost, as it uses
up degrees of freedom.

On the other hand, the spatial linear model *is* appropriate for estimating the parameters
of a spatially correlated process, with some unknown mean parameter, that generated the
observations. The observations may wander far from that mean, and it is therefore quite
appropriate that standard errors of estimates of functions that involve the unknown mean
are so high. But why is the standard error of the GLS estimate of the contrast (obtained
using the spatial linear model) lower than the standard error of the OLS estimate (obtained
under the classical linear model)?

This notion of a difference in model performance for estimating treatment effects versus
contrasts can be clearly understood with the aid of a very simple example. Consider two
random variables, X_1 and X_2, with mean zero, unit variance, and correlation ρ, where
$0 \le \rho \le 1$. If we add the same treatment effect to both X_1 and X_2, and use the average

of the two variables to estimate that treatment effect, then the variance of that average will be $(2 + 2\rho)/2^2 = (1 + \rho)/2$. If $\rho = 0$, then we obtain the familiar variance of a sample mean assuming independence, σ^2/n, which in this case is $1/2$. If, instead, $\rho > 0$, then the variance of the sample mean is larger than $1/2$. The worst-case scenario occurs when $\rho = 1$, in which case the variance of the sample mean is 1. It is easy to extend this idea to more than two random variables, where the variance of the sample mean under the zero correlation (classical linear) model is $1/n$ for n random variables, while the variance of that quantity under a model with positive correlations among all pairs of random variables is larger than $1/n$. Thus, as sample size increases, the classical linear model is a better model under which to estimate a sample mean than a spatial linear model with all positive correlations. However, now consider adding the effect of treatment 1 to X_1 and adding the effect of treatment 2 to X_2, and then estimating their difference by the difference of the two random variables. In this case, the variance of the estimated difference will be $2 - 2\rho$. When the correlation is zero, the variance is 2, but as $\rho \to 1$, the variance goes to zero. Thus, if there is strong correlation among the errors, there is essentially less variation among them, which better reveals any differences due to the treatment effects.

In summary, there is a disconnect between the classical linear model and the spatial linear model when used for designed experiments, and this disconnect goes beyond the mere fact that their covariance matrices differ. As shown above, for spatially-patterned data (which is the usual case in environmental applications), the standard errors obtained under the classical linear model are only appropriate for testing a treatment effect if the classical linear model is being used as an approximation to the randomization-derived linear model. However, the spatial linear model generalizes the classical linear model, and so the standard errors for the spatial linear model, as generally presented, are only appropriate when inference is made on a treatment effect that is added to the process mean rather than the realized mean. In other words, the appropriate estimands for the two models are different. The investigator may be interested in estimating a treatment effect added to the process mean, but what if they instead want to estimate a treatment effect added to the realized mean under the spatial linear model? We turn to that idea next.

11.4 Spatial prediction of functions involving the realized mean

11.4.1 Methodology

Quantities of inferential interest under the randomization-derived linear model may be expressed as elements of the vector

$$\mathbf{L}^T \boldsymbol{\beta} + \mathbf{A}^T \mathbf{e}, \tag{11.13}$$

where the columns of $\mathbf{L}$ contain the weights for any number of estimable functions, such as the process mean-referenced treatment means $\mu + \tau_i$ ($i = 1, \ldots, t$), and the columns of $\mathbf{A}$ contain weights for the errors, where here we are particularly interested in columns of $\mathbf{A}$ for which every element is equal to $1/N$. That is, we are trying to make inferences about treatment effects added to the realized mean of a spatial stochastic process, so for a general formulation, we redefine $\mu_r = \mu + \bar{e}$ and we add treatment effects to it to yield the quantities of interest, $\mu_r + \tau_i = \mu + \tau_i + \bar{e}$ ($i = 1, \ldots, t$), rather than adding treatment effects to the mean of the process that generated the data. Because the quantities in (11.13) involve random errors in addition to fixed effects, the appropriate form of inference is prediction rather than estimation, and a best linear unbiased prediction approach for such quantities can be developed, which we call BLUPTE for "best linear unbiased prediction for treatment effects." This moniker emphasizes both the treatment effects and the prediction of the realized mean.

Expressions for the BLUPTE and the covariance matrix of its prediction errors may be obtained by specializing results for best linear unbiased prediction given in Chapter 9. Here we merely give the expressions; we ask for justifications of them in Exercise 11.4. Specializing (9.4) to this setting yields the BLUPTE as

$$\mathbf{L}^T \hat{\beta}_{GLS} + \mathbf{A}^T (\mathbf{y} - \mathbf{X}\hat{\beta}_{GLS}). \tag{11.14}$$

Its form makes intuitive sense in light of the form of the predictand in (11.13), as it adds the average of the fitted GLS residuals, as predictors of the errors, to the best linear unbiased estimators of the treatment effects. Specialization of (9.5) yields the BLUPTE's covariance matrix of prediction errors as

$$(\mathbf{L}^T - \mathbf{A}^T \mathbf{X})(\mathbf{X}^T \mathbf{\Sigma}^{-1} \mathbf{X})^{-1}(\mathbf{L}^T - \mathbf{A}^T \mathbf{X})^T, \tag{11.15}$$

and this can be compared to the covariance matrix of the standard GLS estimator of treatment effects, $\mathbf{L}^T \hat{\beta}_{GLS}$, which is $\mathbf{L}^T (\mathbf{X}^T \mathbf{\Sigma}^{-1} \mathbf{X})^{-1} \mathbf{L}$ (under the spatial linear model). In practice, of course, $\mathbf{\Sigma}$ in the expressions above must be replaced by its ML or REML estimate $\tilde{\mathbf{\Sigma}}$.

11.4.2   BLUPTE for the simulated data and the simulation study

How does the BLUPTE compare to the REML-EGLS estimator for spatial experiments? As an illustrative example, we carried out BLUPTE on the same dataset of 30 observations used previously to obtain Table 11.1. For the fitted spatial linear model with a range of 4394, the BLUPTE of $\mu_r + \tau_1$ was 0.792, which is slightly different than the REML-EGLS estimate, which was 0.786. Recall that BLUPTE was motivated as a predictor of the realized mean, which in this case is 0.726, whereas the REML-EGLS is an estimator of μ, which in this case is 0. For the spatial linear model with a range of 200, the BLUPTE was 0.792, the same (within rounding errors to the third decimal) as it is when the range was 4394. The BLUPTE's standard error (square root of prediction error variance) under the spatial linear model with range 4394 was 0.0589, which is orders of magnitude less than 11.339 for the REML-EGLS. Moreover, the BLUPTE confidence interval, $0.792 \pm 1.96 \times 0.0589 = (0.676, 0.907)$, covers the true value of 0.726 that it is trying to predict. The BLUPTE's standard error under the spatial linear model with range 200 was 0.0590, which is very similar to the standard error under the model with range 4394, and this contrasts with the REML-EGLS standard errors under these two models (11.339 versus 0.767), which were very different. It is easy to verify that the BLUPTE confidence interval under the model with range 200 also covers the true value.

This single example indicates that BLUPTE might be more appropriate than REML-EGLS for inference under the randomization-derived linear model, so we added BLUPTE to the simulation study where the results for the randomization-derived linear model and GLS are given in Table 11.2. The RMSE for BLUPTE, as a predictor of $\mu_r + \tau_1$, was 0.0721, which is much better than the RMSE under the classical linear model as an approximation to the randomization-derived linear model, and the RMSE under the spatial linear model. Also, the Type I error rate of BLUPTE for $\mu_r + \tau_1$ was 0.0412, close to the nominal value of 5%.

11.4.3 Practical advice on the use of the spatial linear model for designed experiments

When the spatial linear model is used for contrasts, it can be more powerful than the randomization-based linear model, even when treatments are randomized. The spatial linear model uses spatial information such that, when two different treatments are in close proximity, positive spatial correlation tends to reduce the difference in the underlying errors, and thus more easily reveal the difference in those two treatments' effects. When the spatial linear model is used for estimating treatment effects, the investigator needs to carefully consider whether they are interested in inference for the treatment effect added to the mean of the process that generated the data, which is a more global type of inference, in which case they would use the REML-EGLS estimator and its standard error, or whether they are interested in inference for the treatment effect added to the realized mean of the errors contained within just that experiment, in which case it would be better to use BLUPTE and its standard error.

This distinction also occurs for the linear model that assumes the errors are independently generated by a random process (the classical linear model). However, the mean of that process will converge rapidly to the realized mean, so the distinction is not very important if, indeed, the random errors are independent and the sample sizes are moderately large. However, it is a mistake to think that randomization magically creates independent random errors in the underlying process that generated the data; rather, we re-emphasize that randomization creates the randomization-derived linear model on the realized errors, where the near-zero correlations allow independence to be reasonably assumed.

The use of BLUPTE appears to be a novel idea, and it is not implemented in any software. However, it is easy to obtain residuals and coefficients from most software, so (11.14) is easy to compute from any existing fitted model. Equation (11.15) requires $\mathbf{X}$, which in R can be obtained by using the `model.matrix` function, and $(\mathbf{X}^T \mathbf{\Sigma} \mathbf{X})^{-1}$, which can be obtained using the `vcov` function in, for example, the `spmodel` package, among others.

11.5 Optimal spatial design of experiments

So far, we have only investigated the completely randomized design. The spatial linear model relies on modeling a spatial stochastic process, regardless of whether we are interested in inference for $\mu + \tau_i$ or $\mu_r + \tau_i$ (via REML-EGLS or BLUPTE, respectively). If we are relying on the probabilistic structure of that spatial process for inference, then we can abandon completely randomized designs for something that may be more efficient.

11.5.1 Simulated data

To illustrate the impact of design, consider the second row of numbers over the scatterplot in Figure 11.1C, which we call Design 1. It is a systematic design, where we apply Treatments 1 through 6, in order from left to right, and repeat 5 times until all treatments have been applied to the 30 simulated values. Also consider the third row of numbers over the scatterplot, which we call Design 2. It is a "trend-confounded" design, where we apply Treatment 1 to the first 5 samples, Treatment 2 to the second 5 samples, etc., until all treatments have been applied to the 30 simulated values. In Design 2, the treatments are confounded with the (apparent) trend because Treatment 1 was applied to the values on

TABLE 11.3

Estimates and standard errors under several model-design combinations for the simulated data in Figure 11.1C. CLM is the classical linear model, and SLM is a spatial linear model with an exponential covariance function, the covariance parameters of which are fixed at the values used to simulate the data. The treatment assignments for the completely randomized design (CRD), Design 1, and Design 2 respectively are given in the first, second, and third row of numbers above the scatterplot in Figure 11.1C.

| Effect | Estimate | | | | Standard Error | | | |
| | CRD | | Design 1 | Design 2 | CRD | | Design 1 | Design 2 |
	CLM	SLM	SLM	SLM	CLM	SLM	SLM	SLM
$\mu_r + \tau_1$	0.872	0.793	0.809	0.993	0.124	0.076	0.071	0.207
$\mu + \tau_1$	0.872	0.789	0.818	1.063	0.124	0.466	0.464	0.496
$\tau_2 - \tau_1$	−0.130	0.006	−0.001	0.055	0.175	0.117	0.104	0.179

the far left of Figure 11.1C, and Treatment 6 was applied to the values on the far right of Figure 11.1C, so it will be difficult to distinguish the difference of treatment effects, $\tau_6 - \tau_1$, from the trend, and the same goes, albeit to a lesser degree, for all pairwise differences of treatment effects.

For fitting the spatial linear models, we took the covariance parameters that were used to simulate the data (exponential covariance function with a range of 49.2, a nugget effect of 0.0222, and a partial sill of 0.269) as known, so as to eliminate the issue of estimating those parameters when comparing results for the designs. We included the original completely randomized design as a baseline, which can be fit with either the classical linear model as an approximation to the randomization-derived linear model, or with the spatial linear model. Furthermore, under the latter model, we can choose to make inference for either $\mu + \tau_i$ or $\mu_r + \tau_i$ (using GLS or BLUPTE). The results are given in Table 11.3.

First consider the standard errors. When estimating $\mu_r + \tau_1$ in conjunction with the completely randomized design, the standard error obtained using BLUPTE under the spatial linear model is smaller than the standard error obtained using OLS under the classical linear model. However, the systematic design (Design 1) yields an even smaller standard error, suggesting that this design is better than a completely randomized design for estimating $\mu_r + \tau_1$. On the other hand, the standard error using the trend-confounded design (Design 2) is worse (larger) than the standard errors for the completely randomized and systematic designs.

For estimating $\mu + \tau_1$, the standard error obtained using OLS under the classical linear model is identical to that for estimating $\mu_r + \tau_1$, and, as we learned earlier, the standard errors obtained using GLS under the spatial linear model are (appropriately) larger for all designs. The lack of precision in estimating $\mu + \tau_1$ is reflected in the larger standard errors, as compared to estimating $\mu_r + \tau_1$. The completely randomized and systematic designs are somewhat better than the trend-confounded design for estimating this parameter.

Similar patterns occur for estimation of the contrast $\tau_2 - \tau_1$, where the systematic and trend-confounded designs have the smallest and largest standard errors, respectively. For all parametric functions in the table, confidence intervals can be formed from the estimates and their standard errors, and in all cases they cover the true values, or are close, except when OLS under the classical linear model is used to estimate $\mu + \tau_1$.

From the information in Table 11.3, we see that we would make a Type I error for testing the null hypothesis that $\mu + \tau_1$ is equal to 0 if we were to base this test on the spatial linear model and the trend-confounded design, since $1.063/0.496 = 2.143 > 1.96$, although

TABLE 11.4

Root-mean-squared errors (RMSE), and empirical Type I error rates for size-0.05 tests, of OLS estimates (corresponding to the classical linear model, labeled as CLM) and GLS estimates (corresponding to the spatial linear model with known covariance parameters, labeled as SLM), of three quantities for three treatment assignments (designs) applied to the simulated data in Figure 11.1A. The treatment assignments for the completely randomized design (CRD), Design 1, and Design 2 are given in the first, second, and third row of numbers above the scatterplot in Figure 11.1C.

| | RMSE | | | | Type I Error | | | |
| | CRD | | Design 1 | Design 2 | CRD | | Design 1 | Design 2 |
Effect	CLM	SLM	SLM	SLM	CLM	SLM	SLM	SLM
$\mu_r + \tau_3$	0.109	0.073	0.071	0.125	0.028	0.034	0.057	0.044
$\mu + \tau_3$	0.535	0.516	0.513	0.537	0.756	0.058	0.056	0.063
$\tau_2 - \tau_3$	0.170	0.113	0.106	0.173	0.054	0.037	0.061	0.049

it is close. Note that the quality of inferences would likely worsen if we were to estimate the covariance parameters (which we must do with real data, as we do not know the true generating process in practice), rather than fixing them at their true values. Indeed, when we estimate the covariance parameters of an exponential covariance model by REML, the estimated range is 15.6, which is smaller than the true range 49.2, and the estimated partial sill is 0.0335, which is much smaller than the true partial sill 0.269. The spatial confounding associated with the trend-confounded design essentially transfers spatial pattern from the errors to the mean structure (see Section 5.5), so that the estimated variance and correlation of those errors decreases; this, in turn, leads to a much larger Type I error. Specifically, using the estimated covariance parameters, the estimate of $\mu + \tau_1$ was 1.007 with a standard error of 0.121, yielding a z-value of $1.007/0.121 = 8.302$, which is highly significant and much larger than 1.96. Plainly, it is important to avoid trend-confounded designs, for the sake of both reliable estimation of covariance parameters and efficient estimation of mean structure parameters.

11.5.2 Another simulation study

Table 11.3 considered just one simulated experiment, so the results given therein are merely anecdotal. To gain a more general understanding of the role the design plays in the quality of inferences, we conducted another simulation study. We conducted this study in the same manner as the one described in Section 11.3.5, moving along positions in Figure 11.1A, starting with positions 1 through 30, then 2 through 31, etc., and always applying the three designs shown in Figure 11.1C. Each completely randomized design used a new randomized treatment assignment, but the systematic design (Design 1) and trend-confounded design (Design 2) used the same arrangement of treatments for each successive group of 30 positions that were used for the original set of 30 positions in Figure 11.1C. For each data set, we computed the same estimators and performance measures that we computed for the single experiment in the previous subsection. The results are presented in Table 11.4. Note that here we estimate the effects $\mu_r + \tau_3$, $\mu + \tau_3$, and $\tau_2 - \tau_3$ to eliminate possible edge effects, since τ_1 was always in the first position for the systematic and trend-confounded designs.

The RMSE for the completely randomized design is smaller under the spatial linear model than under the classical linear model (Table 11.4), consistent with earlier results. The RMSE is slightly smaller for the systematic design (0.071 vs. 0.073), but much larger for the trend-confounded design. These patterns are true for $\mu_r + \tau_3$ (estimated using BLUPTE) and for $\mu + \tau_3$ and $\tau_2 - \tau_3$ (estimated using GLS). In looking at the Type I error rates, it seems that the standard errors are properly estimated except when estimating $\mu + \tau_3$ under the classical linear model, which is again consistent with earlier results. The main takeaway from Table 11.4 is that, when using a spatial linear model, the design can have a large impact on precision when estimating treatment effects or contrasts.

11.5.3   More on designs

Spatial interspersion of treatments among experimental units, though not foundational to classical experimental design, has nonetheless been recognized as a feature of good spatial experimental design when plots near one another tend to be more alike than those farther apart (Cochran, 1976; Hurlbert, 1984). Indeed, prior to Fisher's work, many agricultural researchers used systematic designs, with no randomization, to achieve adequate spatial interspersion. The problem with this was that without randomization (or an explicit model for the underlying fertility) there was no valid method for comparing treatments. Fisher's advocacy for randomized rather than systematic designs was based on the fact that randomization allows for a correct design-based statistical analysis. However, in many spatial experiments conducted in the natural environment, the number of replicates per treatment is not large, with the consequence that a completely randomized process of assigning treatments to experimental units has a good chance of producing assignments in which some treatments are concentrated in relatively small subregions of the experiment's study area. This leaves the experiment vulnerable to either unknown gradients existing in the field prior to conducting the experiment, or to chance events during its conduct that impact only a small portion of the study area. Thus, blocking to achieve some level of spatial interspersion of treatments and some degree of local control may prove useful even when a pre-existing environmental gradient cannot be identified.

In the case of no trend but appreciable spatial correlation among observations on neighboring experimental units, classical randomized designs, including randomized block designs, yield inferences that are statistically valid but not as precise as is possible if the design, not merely the analysis, were to take more explicit account of the spatial positions of the experimental units. The following toy example illustrates the point. Consider a complete block spatial experiment in three treatments and three blocks, for a total of nine observations. Suppose that the three experimental units (plots) within each block are rectangular, have the same size, and form a contiguous "strip" with two end plots and a middle plot. The design in Figure 11.2A depicts the layout. Suppose further that blocks, contrary to how they are portrayed in the figure, are sufficiently far apart that responses from different blocks may be regarded as uncorrelated, but that the covariance between responses on plots j and k within block i is given by the stationary first-order autoregressive covariance model $C(y_{ij}, y_{ik}) = \sigma^2 \rho^{|j-k|}$ $(i,j,k=1,2,3)$ where $\sigma^2 > 0$ and $\rho = 0.5$. (Recall that this covariance model and the exponential covariance model are equivalent.) Then the covariance matrix of the entire observational vector $\mathbf{y} = (y_{11}, y_{12}, y_{13}, y_{21}, \ldots, y_{33})^T$ is given by

$$\boldsymbol{\Sigma} = \sigma^2 \mathbf{I}_3 \otimes \begin{pmatrix} 1 & 0.5 & 0.25 \\ 0.5 & 1 & 0.5 \\ 0.25 & 0.5 & 1 \end{pmatrix}.$$

Finally, suppose that

$$\mathrm{E}(y_{ij}) = \mu + \beta_i + \tau_{[ij]} \quad (i,j=1,2,3),$$

where μ is an overall effect, β_i is a fixed block effect, and $\tau_{[ij]}$ is a fixed effect of the treatment assigned to plot j within block i. Our goal is to minimize *AVTD*, the average variance of

A

1	2	3
3	1	2
1	3	2

B

1	2	3	4	5
2	4	1	5	3
3	1	5	2	4
4	5	2	3	1
5	3	4	1	2

C

5	4	1	3	2
2	5	4	1	3
3	2	5	4	1
1	3	2	5	4
4	1	3	2	5
5	1	2	4	3
3	5	1	2	4
4	3	5	1	2
2	4	3	5	1
1	2	4	3	5

FIGURE 11.2
Three nearest-neighbor balanced complete block designs. Each row within a given design represents a block. A) The toy example in 3 blocks of size 3 described in the text; B) a NNBBD in blocks of size 5; C) another NNBBD in blocks of size 5.

the generalized least squares estimators of the three pairwise treatment differences $\tau_1 - \tau_2$, $\tau_1 - \tau_3$, and $\tau_2 - \tau_3$ under this model. How should the treatments be assigned to plots within blocks to accomplish this goal? There are $3! = 6$ possible distinct treatment assignments within each block, for a total of $6^3 = 216$ possible assignments (designs) overall. It turns out, however, that these assignments yield only three distinct values of $AVTD$, according to the number of blocks in which the same treatment is assigned to the middle plot. Specifically,

$$AVTD = \begin{cases} 0.3571\sigma^2 & \text{if each treatment occurs exactly once in the middle plot,} \\ 0.3671\sigma^2 & \text{if exactly two treatments occur in the middle plot,} \\ 0.3889\sigma^2 & \text{if the same treatment occurs in every middle plot.} \end{cases} \tag{11.16}$$

(see Exercise 11.5). Thus, the 216 distinct designs fall into three equivalence classes, and any design in the class for which each treatment occurs exactly once in the middle plot (as is the case for the particular design depicted in Figure 11.2A) is optimal for minimizing $AVTD$. The gain in efficiency for the equivalence class of optimal designs compared to the class of worst designs is rather modest, about 8%, but would be larger for larger ρ; for example, when $\rho = 0.8$ it is 18%. The average gain in efficiency over the entire population of randomized complete block designs is slightly less (e.g., 6% when $\rho = 0.5$), but all of these gains are rather small because of the small number of blocks and treatments involved in the toy example. For larger designs, gains in efficiency for designs that explicitly account for spatial location can be more substantial, as seen below.

An equivalent and, as it happens, more relevant characterization of the equivalence class of optimal designs for the toy example just described is the subclass of complete block designs for which each unordered pair of distinct treatments occurs on adjacent plots the same number of times. This characterization, unlike the "middle-plot characterization" given in (11.16), also applies to block designs with arbitrary block sizes and defines the class of **nearest-neighbor balanced block designs** (NNBBDs). Given a number of blocks, b, a number of treatments, t, and a block size, k, a NNBBD may or may not exist. When t is even, a complete-block NNBBD exists with $b = t/2$ (but no smaller), whereas when t is odd, $b = t$ (or larger) is required. Two examples of NNBBDs for 5 treatments and complete blocks of size 5 are displayed in Figure 11.2B,C.

Gill & Shukla (1985a) and Kunert (1987) showed, under a (one-dimensional) first-order autoregressive model of any positive correlation strength, that those NNBBDs for which each pair of treatments occurs equally often on edge plots, when they exist for a given block size and given numbers of blocks and treatments, are **universally optimal** (Kiefer, 1975). Universal optimality is a rather general type of optimality that includes minimization of the average variance of estimated pairwise treatment differences. The 10-block design in Figure 11.2C, but not the 5-block design in Figure 11.2B, is universally optimal under the aforementioned autoregressive model. NNBBDs, whether or not they belong to the universally optimal class, also perform better than randomized block designs under other spatial correlation models. For the more general case of a (one-dimensional) second-order autoregressive model, Grondona & Cressie (1993) characterized the universally optimal designs, for which two necessary (but not sufficient) properties are that the design is a NNBBD and that each pair of treatments occurs on plots separated by a single plot the same number of times. Such designs are called **second-order NNBBDs**. It can be verified that the 10-block design in Figure 11.2C is second-order nearest-neighbor balanced. Second-order NNBBDs may demand a large number of blocks to exist; a necessary condition for existence is that the number of blocks be an integer multiple of $t(t-1)/2$ (Grondona & Cressie, 1993). However, it turns out that first-order NNBBDs, which generally require fewer blocks to exist, remain highly efficient under second-order autoregressive models.

Similar types of designs have been developed for other spatial situations. Williams (1952) considered NNBBDs for settings in which blocks are oriented linearly and contiguously end-to-end (so that responses on units from nearby blocks may be correlated), such as could occur for an experiment conducted along a river, road, or shoreline. Two-dimensional **nearest-neighbor balanced designs** (NNBDs), for which there is blocking by rows and columns, are such that each unordered pair of distinct treatments are adjacent equally often over the rows and columns. (It is not necessary that the numbers of such adjacencies in rows are the same as in columns.) One class of NNBDs is the quasi-complete Latin squares, for which $b = t = k$, each treatment occurs exactly once within each row and once within each column, and the number of adjacencies for each pair of treatments is two within rows and two within columns. Quasi-complete Latin squares exist for every $t > 3$. If the blocks of the five-block NNBBD displayed in Figure 11.2B were laid contiguously side by side, just as they are portrayed there, then they would form a quasi-complete Latin square. Gill & Shukla (1985b) showed that the class of quasi-complete Latin squares is universally optimal when the model is such that the expectation of each observation is equal to an overall mean plus the fixed effect of the treatment applied to the corresponding plot and the covariance matrix among observations is that of a CAR model with the rook's definition of contiguity. As an illustration, the gain in efficiency of the quasi-complete Latin square in Figure 11.2B relative to a completely randomized design under a single-parameter CAR model with $\rho = 0.25$ is approximately 18%. If the covariance matrix is instead that of a CAR model with the queen's definition of contiguity, then for universal optimality nearest-neighbor balance is required not only in rows and columns but also along diagonals. In general, satisfying this last condition requires a design comprising more than one Latin square.

Methods for actually constructing or finding universally optimal spatial experimental designs, NNBBDs, or NNBDs for those combinations of t, b, and k for which they exist include direct combinatorial techniques that are beyond the scope of this brief review, and algorithmic methods. Algorithmic searches for these designs or for near-optimal or nearly neighbor-balanced designs may utilize some of the same types of algorithms that are used to search for good spatial sampling designs (e.g., exchange or simulated annealing algorithms); see, for example, Russell & Eccleston (1987); Chan & Eccleston (2003); Butler & Eccleston (2008). To illustrate, let $\mathbf{L}^T$ be a matrix, each row of which contains weights for treatment effects and contrasts, and let $\mathbf{L}^T \hat{\boldsymbol{\beta}}_{GLS}$ be the vector of GLS estimators of those effects or contrasts. Then Ver Hoef (2012a) defined the function

$$h(\mathbf{X}; \boldsymbol{\Sigma}, \mathbf{L}) = \text{tr}[\mathbf{L}^T (\mathbf{X}^T \boldsymbol{\Sigma}^{-1} \mathbf{X})^{-1} \mathbf{L}],$$

where $\text{tr}(\cdot)$ is the matrix trace, and defined an optimal design as that for which the model matrix $\mathbf{X}_B$ satisfies

$$\mathbf{X}_B = \arg \min_i h(\mathbf{X}_i; \boldsymbol{\Sigma}, \mathbf{L}).$$

This is a discrete optimization problem, and it may be difficult to find a sequence $\{\mathbf{X}_i\}$ that converges toward $\mathbf{X}_B$, but Ver Hoef (2012a) showed that some greedy algorithms like simulated annealing and genetic algorithms work well in practice. The optimization of $h(\mathbf{X}_i; \boldsymbol{\Sigma}, \mathbf{L})$ requires knowledge of $\boldsymbol{\Sigma}$, but Ver Hoef (2012a) showed that the optimal design under a given value for a spatial correlation parameter was nearly optimal for many different values of that parameter. In other words, it was not very sensitive to the actual specification of $\boldsymbol{\Sigma}$, and any reasonable estimate should yield better results than a completely randomized design (up to 30% improvement when there is a high degree of spatial correlation).

A very simple idea that would, nonetheless, find a "good" design is to create $\{\mathbf{X}_i\}$ randomly subject to any design constraints (number of treatments, replicates, blocks, etc.), evaluate $h(\mathbf{X}_i; \boldsymbol{\Sigma}, \mathbf{L})$ for each design, and select the "best" among potentially millions of randomly generated designs. This would be better than relying on a single randomized design, which, by chance, might be a bad design. Two examples implementing this idea are considered in Exercises 11.6 and 11.7. Functions other than the trace may be chosen for $h(\mathbf{X}_i; \boldsymbol{\Sigma}, \mathbf{L})$, which could include weighting among the treatment effects and contrasts, depending on their importance.

The main approach taken by Ver Hoef (2012a) was based on a genetic algorithm, where the design matrix $\mathbf{X}$ is envisioned as a chromosome. Improvements are continually made by keeping the best $\mathbf{X}_i$, then creating $\mathbf{X}_{i+1}^*$ from $\mathbf{X}_i$ by swapping rows within $\mathbf{X}_i$ under any design constraints, such as blocking, then evaluating $h(\mathbf{X}_{i+1}^*; \boldsymbol{\Sigma}, \mathbf{L})$ and updating $\mathbf{X}_i$ to $\mathbf{X}_{i+1}$ by either keeping $\mathbf{X}_i$ or moving to $\mathbf{X}_{i+1}^*$, whichever one is better. Because we are swapping rows within $\mathbf{X}_i$, the number of treatments remains fixed. We might also consider randomly changing a row of $\mathbf{X}_i$ from one treatment to another. This will allow the number of replications for treatments to vary, which may be beneficial if some treatments are used more often than others in the contrasts contained within $\mathbf{L}$. For example, suppose that we are interested in various treatments contrasted with a control, but not so interested in contrasting non-control treatments among themselves. Then it is sensible to replicate the control more often than any of the other treatments. How much more? The attraction of using a numerical optimizer is to find a good design and answer that question.

Relative to nearest-neighbor balanced designs, classical designs (randomized complete and incomplete block designs, Latin squares, etc.) have the advantages of being more familiar, more likely to exist for a given number of treatments and block size, and easier to construct. We suspect that many practitioners would take the view that pairing a classical design with a spatial analysis (i.e., accounting for the spatial nature of the experiment just once) is a reasonable compromise between pairing an optimal (or near-optimal) nearest-neighbor balanced design with a spatial analysis (twice accounting) and pairing a classical design with a non-spatial ordinary least squares analysis (no accounting). This, however, was not the perspective taken by the investigators who conducted the caribou forage experiment. Those investigators optimized the design as described above, and then performed a spatial analysis on the data. The next section describes this in detail.

TABLE 11.5

Mean structure effects (first column) and their OLS or REML-EGLS estimates (columns 2–4) and corresponding estimated standard errors (columns 5–7) for three models fitted to the caribou forage data. CLM is the classical linear model (with estimated variance $\hat{\sigma}_{OLS}^2 = 0.0372$), SLM-SAR is a row-standardized SAR model with first-order neighbors (with REML covariance parameter estimates $\tilde{\sigma}_R^2 = 0.031$ and $\tilde{\rho}_R = 0.424$), and SLM-EXP is a spatial linear model with an exponential covariance function (with REML covariance parameter estimates $\tilde{\sigma}_R^2 = 0.269$, nugget effect $= 0.0222$, and range parameter $= 49.2$).

Effect	Estimate			Standard Error		
	CLM	**SLM-SAR**	**SLM-EXP**	**CLM**	**SLM-SAR**	**SLM-EXP**
$\mu_r + \alpha_1 + \gamma_1 + \tau_{11}$	1.906	1.913	1.896	0.086	0.066	0.069
$\mu + \alpha_1 + \gamma_1 + \tau_{11}$	1.906	1.917	1.992	0.086	0.086	0.506
$\alpha_2 - \alpha_1$	0.074	0.057	0.048	0.122	0.104	0.104
$\gamma_2 - \gamma_1$	0.112	0.143	0.173	0.122	0.099	0.107
$\gamma_3 - \gamma_1$	0.352	0.361	0.396	0.122	0.098	0.105
$\tau_{22} - \tau_{11}$	−0.124	−0.175	−0.185	0.173	0.136	0.148
$\tau_{23} - \tau_{11}$	−0.203	−0.226	−0.218	0.173	0.139	0.145

11.6 Caribou forage example

We now use the caribou forage data to illustrate the ideas in this chapter. In Section 8.8.3, we fitted several models to these data and obtained estimates of covariance parameters. Here, we use those fitted models to make inferences on treatment effects and contrasts. Recall from Section 4.5.3 that our model for the data is

$$y_{ijk} = \mu + \alpha_i + \gamma_j + \tau_{ij} + e_{ijk} \quad (i = 1, 2; \; j = 1, 2, 3; \; k = 1, 2, 3, 4, 5),$$

with identifying constraints given by, say, either (4.9) or (4.10). Here, α_i is an effect for watering, γ_j is an effect for tarps, and τ_{ij} is a water×tarp interaction effect. The e_{ijk}'s are random variables assumed to be normally distributed with zero means, and with one of several possible variance-covariance structures.

11.6.1 Estimating the mean structure

REML-EGLS estimates of mean structure parameters for three models are given in Table 11.5. The three models are the classical linear model (CLM), the spatial linear model with the row-standardized SAR covariance structure fitted in Section 8.8.3, and the spatial linear model with the pseudo-geostatistical isotropic exponential covariance function also fitted in Section 8.8.3. The difference in standard errors for the estimate of the watering=1 and tarp=1 effects (and their interaction) added to μ_r (BLUPTE) versus μ (REML-EGLS) in the first two rows of the table is noteworthy. For the former, the standard errors obtained under the SLM-SAR and SLM-EXP models are similar and smaller than their counterpart obtained under the CLM model. For the latter, the standard errors obtained under the CLM and SLM-SAR models are very similar and much smaller than their counterpart obtained under the SLM-EXP model. For all of the contrasts in the remaining rows of Table 11.5, the standard errors obtained under the two spatial models are smaller than their counterpart obtained under the CLM model, with those for SLM-SAR slightly smaller than those for

TABLE 11.6
Spatial ANOVA table corresponding to the REML-estimated SLM-EXP model for the caribou forage data.

Source	df	Sum of Squares	Mean Square	F	P-value
Water	1	0.498	0.498	1.710	0.2033
Tarp	2	4.607	2.304	7.906	0.0023
Water × Tarp	2	0.756	0.378	1.297	0.2919
Error	24	6.993	0.291		
Total	29	12.855			

SLM-EXP. Recall that the likelihood-based model selection procedure presented in Section 8.8.3 indicated that the SLM-SAR model was preferred, and here we see that it also gives results with the best estimated precision. Based on our simulations in the previous sections, the estimated standard errors in Table 11.5 should be reliable. All of these results are sensible in light of our simulations and discussions in the preceding sections.

11.6.2 ANOVA results

A larger question is, are the effects listed in Table 11.5 statistically and practically significant? Might we be able to remove some effects from the model, such as the interaction effect? Consider the spatial sequential ANOVA corresponding to the REML-estimated SLM-EXP model. Let $\mathbf{D}$ be the distance matrix between all pairwise sites, and let σ_1^2, σ_0^2, and ρ be the partial sill, nugget effect, and range parameter, respectively. Write $\mathbf{\Sigma} = \sigma^2 \mathbf{R} = \sigma^2[\xi \exp(-\mathbf{D}/\rho) + (1 - \xi)\mathbf{I}]$ where $\sigma^2 = \sigma_1^2 + \sigma_0^2$ and $\xi = \sigma_1^2/\sigma^2$, and let $\tilde{\mathbf{R}}_R$ represent $\mathbf{R}$ with REML estimates $\tilde{\sigma}_{1,R}^2 = 0.269$, $\tilde{\sigma}_{0,R}^2 = 0.0222$, and $\tilde{\rho}_R = 49.2$ substituted for their true unknown values. Then we created $\mathbf{y}_* = \tilde{\mathbf{R}}_R^{-\frac{1}{2}}\mathbf{y}$ and $\mathbf{X}_* = \tilde{\mathbf{R}}_R^{-\frac{1}{2}}\mathbf{X}$ and used them as described in Section 5.1 to obtain the spatial ANOVA table shown in Table 11.6. Note that Table 11.6 provides Type I tests of hypotheses, which are sequential and test an effect after accounting for every effect above it in the table, but not any effects below it. Here, we see that the interaction is not significant in a model that has both main effects, but the tarp effect is significant (at the 0.05 level) in a model that includes the water effect, and that the water effect, on its own, is not significant. Thus, it appears that the interaction effect may be removed from the model. Note also that the sums of squares in Table 11.6 are quite different — roughly an order of magnitude larger — than those in the non-spatial sequential ANOVA presented in Table 4.5. The pre-multiplication of $\mathbf{y}$ and $\mathbf{X}$ by $\tilde{\mathbf{R}}_R^{-1/2}$ rescales the data, conferring more variation to it than was originally present.

Notice that the REML estimates of the two variance components, $\tilde{\sigma}_{1,R}^2$ and $\tilde{\sigma}_{0,R}^2$, sum to 0.291, which is also the mean-squared error (MSE) in Table 11.6. Although it may not be obvious, the MSE from the generalized overall ANOVA table (constructed using $\mathbf{y}_*$ and $\mathbf{X}_*$ rather than $\mathbf{y}$ and $\mathbf{X}$), is exactly the estimator $\tilde{\sigma}_R^2 = \mathbf{y}^T[\mathbf{Q}(\tilde{\boldsymbol{\theta}}_{-1,R})]\mathbf{y}/(n-p)$ first described in Section 8.2, where the overall variance was "profiled" out of the likelihood to save one parameter during optimization, and then this parameter was estimated with the analytical solution $\tilde{\sigma}_R^2$ given all of the other covariance parameters. Also, from Table 11.6, we obtain the spatial version of the coefficient of determination, $R^2 = (0.498 + 4.607 + 0.756)/12.855 = 0.456$. This is slightly larger than the non-spatial R^2, which, from performing the same type of computation on the sums of squares from Table 4.5, is 0.317.

Recall that because this design is balanced, i.e., all treatment combinations are equally replicated, the Type I and Type III sums of squares for Water and Tarp were equal in the non-spatial sequential ANOVA table (Table 4.5), so that it made no difference which type is used for testing those effects. However, the pre-multiplication of $\mathbf{X}$ by $\tilde{\mathbf{R}}_R^{-1/2}$ alters the situation, so for a spatial ANOVA the type of sum of squares used to perform a test does matter. In order to test Type III hypotheses, it would be possible to re-order the rows of Table 11.6, putting each term in the last position. However, an alternative was given in (4.3). A test of no difference among all p levels of a treatment is equivalent to a test that $p - 1$ linearly independent contrasts among those levels are all equal to zero. One such possibility (there are many, all leading to the same value of the test statistic) is to consider the difference of each level of the treatment with the first one. This is exactly the way that design matrices in R are coded, so in Table 11.5, the single contrast to test the water effect is $\alpha_2 - \alpha_1$. Hence, a test statistic for this effect is

$$t = \frac{\boldsymbol{\ell}^T \tilde{\boldsymbol{\beta}} - 0}{\sqrt{\text{MSE}}\sqrt{\boldsymbol{\ell}^T (\mathbf{X}_*^T \mathbf{X}_*)^{-1} \boldsymbol{\ell}}} = 0.457, \tag{11.17}$$

where $\boldsymbol{\ell}^T = (0, 1, 0, 0, 0, 0)$. The probability of obtaining $|t| = 0.457$ for a t-distribution with 24 degrees of freedom, or a value more extreme, under the null hypothesis that the effect is zero, is 0.652. Thus, we do not reject the null hypothesis that water has no effect. Note that this is a test of a main effect, Water, in a model with an interaction, Water $\times$ Tarp, that includes the main effect being tested. In general, interactions are removed from models before main effects (although there are exceptions). In any case, we have computed this test mainly as an illustration.

The R package `spmodel` takes a slightly different testing approach. The t-value above would be valid if we truly knew $\mathbf{R}$ for the Aitken equations, but in practice it must be estimated from the data. Thus, all distributional results are approximate. The t-value above is simply the estimate divided by its standard error, which is often compared to a standard normal distribution. That approximately standard normal variate can also be squared, which is then a chi-squared random variable with 1 degree of freedom. For example, $0.457^2 = 0.209$, and the probability that a chi-squared variable with one degree of freedom equals or exceeds 0.209 is 0.647. It is easier to generalize to the case with multiple levels for an effect using the chi-squared distribution, which we show next.

In Table 11.5, two suitable contrasts for testing the hypothesis that all tarp effects are equal to zero are $\gamma_2 - \gamma_1$ and $\gamma_3 - \gamma_1$, which can be used in

$$C = \frac{(\mathbf{L}^T \tilde{\boldsymbol{\beta}} - \mathbf{0})^T [\mathbf{L}^T (\mathbf{X}_*^T \mathbf{X}_*)^{-1} \mathbf{L}]^{-1} (\mathbf{L}^T \tilde{\boldsymbol{\beta}} - \mathbf{0})}{\text{MSE}} = 14.451,$$

where

$$\mathbf{L}^T = \begin{pmatrix} 0 & 0 & 1 & 0 & 0 & 0 \\ 0 & 0 & 0 & 1 & 0 & 0 \end{pmatrix}.$$

Then, C may be compared to quantiles of a chi-squared distribution with q degrees of freedom, as in `spmodel`, where q is the number of rows in $\mathbf{L}^T$, or C/q can be compared to an F-distribution, as in (4.3). In the former case, the probability of obtaining 14.451 or larger under the null hypothesis is 0.00073. In order to test the interaction of water and tarp, we simply change $\mathbf{L}^T$ to

$$\mathbf{L}^T = \begin{pmatrix} 0 & 0 & 0 & 0 & 1 & 0 \\ 0 & 0 & 0 & 0 & 0 & 1 \end{pmatrix},$$

and obtain $C = 2.594$ with a P-value of 0.273. Note that the F test of the interaction term in Table 11.6 is a Type III test of hypothesis because the interaction sum of squares is in the last position, and thus the test's P-value, 0.292, is directly comparable to 0.273. Use of C and the chi-square distribution to obtain the P-value essentially assumes infinite denominator degrees for freedom, compared to the use of the F-distribution with 24 denominator degrees of freedom.

11.6.3 Estimating treatment effects and contrasts

After determining which effects in the mean structure are significant, we could exclude the remainder from the model, or we could leave them in if we are interested in specific treatment effects or contrasts that involve those effects. In fact, prior to the caribou forage experiment, Ver Hoef (2012a) identified 16 different contrasts of interest, some of which were interactions, so we leave all main effects and interactions in the model. Several contrasts of interest were comparisons of specific levels of the tarps. The value of C for the test of a clear tarp versus no tarp (control) contrast ($\gamma_2 - \gamma_1$ in Table 11.5) is $(0.173/0.107)^2 = 2.614$, which is less than 3.84, the critical value for a size-0.05 test, so we do not reject that null hypothesis. In that case, we want to compare the shade tarp to the average of the clear tarp and control, or $(\gamma_3 - \gamma_1) - 0.5(\gamma_2 - \gamma_1) = \gamma_3 - 0.5(\gamma_1 + \gamma_2)$. If we let $\boldsymbol{\ell}^T = (0, 0, -0.5, 1, 0, 0)$ in (11.17), then $t^2 = 12.021$ with a P-value of 0.000526, so we conclude that the shade tarp increases nitrogen content in plants, relative to the average nitrogen content obtained using a control or clear tarp.

Moreover, we might want to estimate the nitrogen content under shade conditions and low moisture as input into a nutritional model for caribou. First, let us assume that we are interested in current nitrogen content in the vicinity of the experimental site. Hence, we will estimate the "shade" tarp treatment effect and the "no water added" water treatment effect, and their interaction, added to the realized mean of the spatial process. In terms of the mean parameters listed in the first column of Table 11.5, we will estimate $\mu_r + \alpha_1 + \gamma_3 + \tau_{13}$, so let $\boldsymbol{\ell}^T = (1, 0, 0, 1, 0, 0)$. Note that in constructing $\boldsymbol{\ell}$, we are using the parameterization in Table 11.5 (where α_1, γ_1, and τ_{11} are absorbed into the overall mean μ_r) and the constraint $\tau_{13} = 0$. (If either $i = 1$ or $j = 1$, then $\tau_{ij} = 0$, i.e., only τ_{22} and τ_{23} are nonzero). Then, using (11.14) and (11.15), we obtain an estimate of 2.292 with a standard error of 0.0676. Now instead let us suppose that the experimental site represents just a small fraction of a much larger area of interest, over which a spatial process operates that we believe may vary more widely than it does over our experimental site. This is analogous to conducting the experiment in a small portion of Figure 11.1A, but wanting to make inference on nitrogen content throughout a much larger area, encompassing all of the variation of the spatial process (even over multiple realizations, which might include another spatial process in future years). In terms of the mean parameters listed in the first column of Table 11.5, we are interested in $\mu + \alpha_1 + \gamma_3 + \tau_{13}$, so again let $\boldsymbol{\ell}^T = (1, 0, 0, 1, 0, 0)$, but with the first element of $\boldsymbol{\beta}$ taken to be μ rather than μ_r. In this case, we obtain an estimate of 2.388 with a standard error of 0.507 using empirical GLS, as in Chapter 8. This illustrates, once again, that inference on treatment effects in terms of the realized mean is much more precise than inference on treatment effects in terms of the overall process mean.

11.6.4 Optimal design for the caribou forage example

Here we focus on the use of planned contrasts, as in Ver Hoef (2012a), where there were 16 such contrasts contained in the matrix $\mathbf{L}^T$, which would eventually be estimated by $\mathbf{L}^T \tilde{\boldsymbol{\beta}}$. Using some pre-treatment data, Ver Hoef (2012a) obtained an initial estimate of $\boldsymbol{\Sigma}$ and then used a genetic algorithm, as described in Section 11.5.3, to obtain a "near-optimal" design for the estimation of $\mathbf{L}^T \boldsymbol{\beta}$. Because the design was obtained via a discrete numerical optimization routine rather than by complete enumeration, we can never be sure that we have found the absolute optimal design, but we should get very close. The design obtained by the genetic algorithm was used for the experiment, and can be seen in Figure 1.5. Notice that there is a certain symmetry in the design. If we flip the design from top to bottom, then the shade treatments do not change, but all of the water treatments change from Y to N, or N to Y. Experiments in the environment are often expensive, requiring extensive logistics, and it is important to get as much information as we can from our experiment by using the best design possible. Optimizing a design prior to actual experimentation can be a good investment of time and money, even when the data are to be analyzed by a spatial linear model.

11.7 Exercises

1. Verify (11.2)–(11.12).

2. Consider the linear model $y_i = \mu + \tau_i + e_i$ ($i = 1, 2$), where μ is an overall mean, τ_1 and τ_2 are treatment effect parameters, and e_1 and e_2 are random errors with zero means and covariance matrix

$$\Sigma = \begin{pmatrix} 1 & \rho \\ \rho & 1 \end{pmatrix}.$$

Let $\mathbf{X}$ represent a full-rank model matrix corresponding to the model with τ_1 set to zero and the second parameter set equal to the difference between the first and econd treatment effects. (This is the default behavior of R which uses option `contr.treatment` for contrasts). This can also be viewed as having τ_1 absorbed into the overall mean. What are the variances of the GLS estimators of the two estimable functions $\mu + \tau_1$ and $\tau_2 - \tau_1$ for any $-1 < \rho < 1$? What is their covariance? Make a plot of $\text{Var}(\widehat{\mu + \tau_1})_{GLS}$ and $\text{Var}(\widehat{\tau_2 - \tau_1})_{GLS}$ as a function of ρ. For what value(s) of ρ are the variances equal?

3. Consider a linear model with the same overall mean and treatment parameters as the model considered in the previous exercise, $y_{ij} = \mu + \tau_i + e_j$, but for which there are 10 locations along a line at integers $j = 1, 2, \ldots, 10$ with random errors e_j. The errors have mean zero and covariance matrix $\mathbf{\Sigma} = (\text{Cov}(e_j, e_k)) = (\rho^{|j-k|})$ for $-1 < \rho < 1$. Consider two designs, one where the first treatment is assigned to sites at integers 1 through 5 and the second treatment is assigned to sites at integers 6 through 10. Denote the model matrix corresponding to this design as $\mathbf{X}_1$. For the second design, the first treatment is assigned to sites at the odd integers and the second treatment is assigned to sites at the even integers; denote the corresponding model matrix as $\mathbf{X}_2$. As in the previous exercise, let the model matrices be full rank and such that τ_1 is set to zero. Write computer code to plot the variances of the GLS estimators of the two estimable functions $\mu + \tau_1$ and $\tau_2 - \tau_1$ for all $-1 < \rho < 1$, for both designs, as a function of ρ. Can you explain the patterns that you see? It may help to plot some realizations of e_j for extreme values of ρ at ± 0.99.

4. Show the details of obtaining (11.14) and (11.15) by appropriately specializing the general expressions for the BLUP of a vector of predictands given by (9.4) and its associated covariance matrix of prediction errors given by (9.5).

5. For the toy example described in Section 11.5.3, write a computer program to verify expression (11.16) for three designs, one for each case of the expression. Repeat for the case $\rho = 0.8$.

6. Consider a designed experiment with 4 treatments, on a 4×4 square grid with unit spacing, that is balanced (equally replicated). Take the model to be the pure treatment-effects model $y_{ij} = \tau_i + e_{ij}$, where $i = 1, 2, 3, 4$ and j indexes the locations on the spatial grid to which treatment i is assigned, and the e_{ij}'s have mean zero and covariance matrix

$$\text{Var}(\mathbf{e}) = \boldsymbol{\Sigma} = \exp(-\mathbf{D}/\alpha).$$

Here $\alpha > 0$ is the range parameter of the exponential covariance function and $\mathbf{D}$ is the distance matrix among the 16 locations. Consider the matrix of all possible pairwise contrasts,

$$\mathbf{L}_a = \begin{pmatrix} 1 & -1 & 0 & 0 \\ 1 & 0 & -1 & 0 \\ 1 & 0 & 0 & -1 \\ 0 & 1 & -1 & 0 \\ 0 & 1 & 0 & -1 \\ 0 & 0 & 1 & -1 \end{pmatrix}.$$

(Here the subscript a denotes "all.") For $\alpha = 1$, write R code to generate 100,000 balanced designs at random and compute the average variance among all pairwise contrasts, $\text{mean}\{\text{diag}[\mathbf{L}_a(\mathbf{X}^T\boldsymbol{\Sigma}^{-1}\mathbf{X})^{-1}\mathbf{L}_a^T]\}$, for each of the 100,000 randomly generated designs.

(a) Among those 100,000 designs, which design has the lowest average variance among all pairwise contrasts?

(b) Which design has the highest average variance among all pairwise contrasts?

(c) How much better is the best design (lowest average contrast variance) compared to the design with the mean average contrast variance over all 100,000 designs?

(d) After looking at the best design among all 100,000 randomly generated designs, can you guess at an even better design?

(e) Repeat the process for $\alpha = 6$, and answer all four of the questions above again.

(f) How much better is the best design found for $\alpha = 6$ than the best design found for $\alpha = 1$ when, in fact, $\alpha = 6$? Use this finding to speculate on how robust the optimal design is to the value of α.

7. Consider a scenario very similar to that considered in the previous exercise, but where the contrast matrix corresponds to contrasting only treatment 1 with the other treatments, as might happen if treatment 1 were a control, i.e.,

$$\mathbf{L}_1 = \begin{pmatrix} 1 & -1 & 0 & 0 \\ 1 & 0 & -1 & 0 \\ 1 & 0 & 0 & -1 \end{pmatrix}.$$

Again generate 100,000 balanced designs at random, taking $\alpha = 4$, and compute the average contrast variance (among the $\mathbf{L}_1$-contrasts). Then, consider randomized unbalanced designs where treatment 1 is replicated 5 times at the expense of treatment 4, which is only replicated 3 times. Generate 100,000 such designs at random, and compute the average contrast variance for each design. Similarly, consider randomized unbalanced designs where treatment 1 is replicated 6 times at the expense of treatments 3 and 4, which are only replicated 3 times, generate 100,000 such designs at random, and compute the average contrast variance of each design. Finally, consider randomized unbalanced designs where treatment 1 is replicated 7 times and treatments 2, 3, and 4 are only replicated 3 times, generate 100,000 such designs at random, and compute the average contrast variance of each design.

(a) Find the best design for each case where treatment 1 is replicated 4, 5, 6, and 7 times. What patterns do you see?

(b) How many times should treatment 1 be replicated to get the best overall design, and how much better is it than a balanced design?

12

Extensions

Throughout this book, we have presented spatial linear models, and statistical methodology associated with those models, for continuous-scale univariate spatial data in Euclidean domains. In this final chapter, we describe extensions of the models and methodology to five additional types of data: global data (data extending over much of the surface of the earth), data on stream networks, space-time data, multivariate spatial data, and discrete (binary or count) spatial data. Objectives for the analysis of data of these additional types are very much like those for univariate spatial data in Euclidean domains: to characterize the spatial (and spatio-temporal) dependence of the variable(s) of interest, to estimate the effects of various explanatory variables on the variable(s), and to predict the variable(s) at unsampled locations (and times). The models and methodologies used to address these objectives, though they are similar to those used for univariate Euclidean data, have some unique and interesting features. However, our treatment of these topics is rather superficial compared to the rest of the book, and no examples or exercises are provided. Many of the topics we cover are active areas of research and will likely remain so in the foreseeable future.

12.1   Global data

Some important environmental, geophysical, and climate science datasets are global in extent, or nearly so. Examples include global land surface temperatures measured by thermometers at weather stations, global sea surface temperatures measured remotely by satellite on a grid of latitudes and longitudes, and "data" produced as output of various global climate models. Such data occur at spatial locations on the Earth's surface, whose coordinates may be represented in a spherical coordinate system by (R, L, ℓ)=(radius, latitude, longitude), where the origin is at the Earth's center, R is the average radius of the Earth (≈ 6371 km), $0 \leq L \leq \pi$, and $-\pi < \ell \leq \pi$. Although the coordinates may be projected in various ways to a planar region, any such projection distorts at least some of the distances between data locations and disrupts their longitudinal periodicity by moving some sites that are very close to one another on the sphere to opposite sides of the region. These distortions and disruptions may induce spurious anisotropy (since the distances corresponding to one degree of latitude and one degree of longitude are unequal except at the equator) and spurious nonstationarity (since the distance corresponding to one degree of longitude depends strongly on latitude, and some sites that are very close to each other on the sphere may be quite distant in the planar region). Thus, satisfactory spatial modeling of global data generally requires the use of a geostatistical (or areal) process indexed by points (or regions) on a sphere rather than in Euclidean space, with a well-defined distance metric on the sphere (in the geostatistical case).

DOI: 10.1201/9780429060878-12

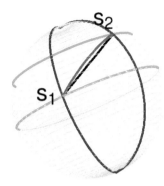

FIGURE 12.1
Two points on the surface of a sphere and two distances between them. The length of the solid red curve connecting the points is the great circle distance, and the length of the solid black line segment between them is the chordal distance.

Figure 12.1 displays a sphere representing the Earth, where $\mathbf{s} = (L, \ell)$ and $\mathbf{s}' = (L', \ell')$ are two points on the surface of the Earth and we have assumed that the Earth's average radius has been rescaled to equal 1.0. The shortest distance between $\mathbf{s}$ and $\mathbf{s}'$ along the surface of the sphere is called the **great-circle distance** (or spherical or geodesic distance) and is given by

$$gcd(\mathbf{s}, \mathbf{s}') = \arccos(\sin L \sin L' + \cos L \cos L' \cos(\ell' - \ell)).$$

A different distance metric, the **chordal distance**, is the Euclidean distance (in $\mathbb{R}^3$) between $\mathbf{s}$ and $\mathbf{s}'$ and is related to the great-circle distance via $chd(\mathbf{s}, \mathbf{s}') = 2\sin(gcd(\mathbf{s}, \mathbf{s}')/2)$. Both distances are depicted in Figure 12.1. Note that two points that are the same great-circle distance apart are also the same chordal distance apart, and vice versa.

12.1.1   Spatially structured dependence

Let $\mathcal{S}^2$ denote the surface of the unit sphere representing the Earth, and let $\{Y(\mathbf{s}) : \mathbf{s} \in \mathcal{S}^2\}$ be a Gaussian process on $\mathcal{S}^2$ with mean function $\mu(\mathbf{s})$ and covariance function $C(\mathbf{s}, \mathbf{s}')$. Models for the mean and covariance structures of data taken on the sphere, like those for data taken on a Euclidean domain, must have some type of spatial structure if they are to be successfully estimated. The analog of Euclidean stationarity for a sphere is **homogeneity** (sometimes also called spherical stationarity). A process on $\mathcal{S}^2$ is said to be **homogeneous** if its mean and covariance functions are invariant to all rotations of the sphere (Jones, 1963). Thus, $Y(\cdot)$ is homogeneous if and only if its mean is constant and its covariance function is (geodesically) **isotropic**, meaning that

$$C(\mathbf{s}, \mathbf{s}') = \psi(gcd(\mathbf{s}, \mathbf{s}')) \quad \text{for all } \mathbf{s}, \mathbf{s}' \in \mathcal{S}^2$$

for some function ψ mapping $[0, \pi]$ to $\mathbb{R}$. The requirement that the covariance function depends only on great-circle distance may be replaced with the equivalent requirement that the covariance function depends only on chordal distance.

An assumption of homogeneity is unrealistic for most global data. In fact, the mean and covariance functions of many global datasets appear to depend strongly on latitude. A more realistic assumption is that of axial symmetry. A (Gaussian) process on the sphere is

axially symmetric if its mean and covariance function are invariant to rotations about the sphere's rotational axis, i.e., if the mean is a function of latitude only and

$$C(\mathbf{s}, \mathbf{s}') = K(L, L', \ell - \ell')$$

for some function K mapping $[0, \pi)^2 \times [-2\pi, 2\pi)$ to $\mathbb{R}$ (Jones, 1963), where the subtraction of longitudes is performed modulo 2π. Furthermore, an axially symmetric process is **longitudinally reversible** if its covariance function depends on the longitudinal lag $\ell - \ell'$ through only its magnitude, not its sign (Stein, 2007). This property is somewhat analogous to quadrant symmetry for a covariance function in Euclidean space (as it was defined in Section 2.2.2).

Two additional properties of spatial dependence on the sphere are completely analogous to properties of Euclidean spatial dependence. These are **separability** and **additivity** of latitude and longitude, defined by the requirements that

$$C(L, L', \ell, \ell') = C_1(L, L')C_2(\ell, \ell')$$

and

$$C(L, L', \ell, \ell') = C_1(L, L') + C_2(\ell, \ell'),$$

respectively, for some functions $C_1(\cdot)$ mapping $[0, \pi]^2$ to $\mathbb{R}$ and $C_2(\cdot)$ mapping $(-\pi, \pi]^2$ to $\mathbb{R}$.

As in Euclidean geostatistics, some of the aforementioned types of spatial structure may provide opportunities to reduce the computational burden of model-fitting and other inference procedures for global data. For example, Jun & Stein (2008) showed that axial symmetry, in conjunction with gridded data on the sphere, confers a block-circulant structure upon the covariance matrix that can be exploited using the Discrete Fourier Transform to greatly speed up computations associated with likelihood-based estimation. Also as in Euclidean geostatistics, parallel definitions of spatially structured dependence may be given in terms of the semivariogram of the process rather than the covariance function; see Huang *et al.* (2011).

12.1.2 Exploratory spatial data analysis

The same tools used for ESDA of Euclidean data may be applied to global data, but with some differences due to the different topology of a sphere. For example, choropleth maps and bubble plots covering the entire Earth cannot be shown in a static two-dimensional plot. They must either be shown separately, in static plots for each of two hemispheres (e.g., eastern and western, or northern and southern), or displayed dynamically as the sphere rotates about its axis. A number of plots may be constructed for the specific purpose of investigating whether axial symmetry or isotropy are plausible assumptions. Side-by-side boxplots of the response, with one boxplot for each of many latitude or longitude bands, can reveal systematic latitudinal or longitudinal trends in the response or its variance. Stein (2007) constructed such a plot for global Total Ozone Mapping Spectrometer (TOMS) data, which showed a strong increase in ozone with increasing latitude in the northern hemisphere. The same type of plot was also constructed for residuals of those data from a fitted mean function, which showed that the fitted mean effectively removed the trend. For a different TOMS dataset, Jun & Stein (2008) plotted site-specific standard deviations (computed in this case from observations taken at multiple times) versus longitude, and also versus latitude, and found that they depended strongly on latitude but not so much on longitude.

Because the first two moments of global data are so often affected by latitude, a single sample semivariogram based on all of the data is of little use. Instead, several sample semivariograms should be computed, one for each of several latitudinal bands, with lags defined in terms of great circle distance. Stein (2007) displayed latitude-specific directional sample semivariograms for his TOMS data at five different latitudes, which revealed substantial latitudinal variation in the nature and strength of the spatial dependence and the implausibility of longitudinal reversibility. For those semivariograms, the first site in a site-pair was located within one of five one-degree latitude bands, while the second was within a spherically rectangular subregion of dimensions one degree latitude by one degree longitude, lagged no more than 9 degrees latitude and 20 degrees longitude from the first site. The sample semivariances were smoothed and plotted as a contour plot in two dimensions after projecting the lags onto a plane. In a similar vein, Jun & Stein (2008) computed latitude-specific lag-one semivariances in several directions. The semivariances showed that the spatial dependence was neither homogeneous nor longitudinally reversible, but possibly axially symmetric.

A formal diagnostic test for isotropy on the sphere was proposed by Sahoo *et al.* (2019); however, it requires spatiotemporal data rather than mere spatial data. Tests for axial symmetry or other spatially structured properties of the covariance structure of data on the sphere do not yet appear to be available.

12.1.3 Parametric models for the mean structure

Owing to the periodic nature of longitude, a polynomial mean function is generally not appropriate for data on $\mathcal{S}^2$. An exception occurs when all of the data locations lie within a thin longitudinal band, in which case the longitudes may perhaps be ignored and a polynomial in latitude only may be useful. The natural family of models for the mean function on a sphere is based on **spherical harmonics**. These are functions $\{Y_k^m(L, \ell)\}$ that are orthogonal (when integrated over $\mathcal{S}^2$) and complete (so that any continuous function on $\mathcal{S}^2$ may be approximated arbitrarily well by a linear combination of sufficiently many of them). Specifically,

$$Y_k^m(L, \ell) \propto \begin{cases} P_k^m(\cos L) \cos(m\ell) & \text{for } m = 0, 1, \ldots, k, \\ P_k^{-m}(\cos L) \sin(m\ell) & \text{for } m = -k, -k+1, \ldots, -1, \end{cases}$$

where $P_k^m(\cdot)$ is the mth associated Legendre function of degree k (Kupper, 1972). (See, e.g., http://en.citizendium.org/wiki/Spherical_harmonics/Catalogs for a list of the spherical harmonics up to order 4, and see http://en.citizendium.org/wiki/Spherical_harmonics for a picture of some of them.) The mean function in the spatial linear model is then

$$\mu(L, \ell) = \sum_{k=0}^{K} \sum_{m=-k}^{k} \beta_{km} Y_k^m(L, \ell),$$

which has $(K + 1)^2$ regressors. Practitioners using this mean function have usually taken K to be quite large, resulting in dozens of regressors: Jeong & Jun (2015) took $K = 6$, Stein (2007) used 78 of the regressors available when $K = 12$, and Jun & Stein (2008) and Jeong *et al.* (2017) used all available regressors when $K = 12$. To actually fit this mean function by ordinary least squares using R, the `sphericalHarmonics` function within the `rcosmo` package can be used to obtain the spherical harmonics at specified points from their three-dimensional Cartesian coordinates, after which those regressors may be supplied in standard fashion to the `lm` function.

12.1.4 Parametric models for the covariance structure

Recall that for a function to be a covariance function of a geostatistical process, it must be positive definite on the spatial domain of the process. It turns out that a function that is

TABLE 12.1

Some parametric, geodesically isotropic correlation models on the sphere. Here $(x)_+$ equals x if $x > 0$, and equals 0 otherwise.

Name	Functional form	Parameter space
Powered exponential	$\rho(s; \alpha, \xi) = \exp[-(\frac{s}{\alpha})^\xi]$	$\alpha > 0, 0 < \xi \leq 1$
Matérn	$\rho(s; \alpha, \nu) = \frac{2^{1-\nu}}{\Gamma(\nu)}(\frac{s}{\alpha})^\nu K_\nu(\frac{s}{\alpha})$	$\alpha > 0, 0 < \nu \leq \frac{1}{2}$
Generalized Cauchy	$\rho(s; \alpha, \tau, \xi) = (1 + (\frac{s}{\alpha})^\xi)^{-\tau/\xi}$	$\alpha > 0, \tau > 0, 0 < \xi \leq 1$
Dagum	$\rho(s; \alpha, \tau, \xi) = 1 - [(\frac{s}{\alpha})^\xi / (1 + \frac{s}{\alpha})^\xi]^{\tau/\xi}$	$\alpha > 0, 0 < \xi \leq 1,$ $0 < \tau < \xi$
Poisson	$\rho(s; \lambda) = \exp[\lambda(\cos s - 1)]$	$\lambda > 0$
Negative binomial	$\rho(s; \delta, \tau) = \left(\frac{1-\delta}{1-\delta\cos s}\right)^\tau$	$0 < \delta < 1, \tau > 0$
Multiquadric	$\rho(s; p, \tau) = \left(\frac{(1-p)^2}{1+p^2-2p\cos s}\right)^\tau$	$0 < p < 1, \tau > 0$
Sine power	$\rho(s; \xi) = 1 - 2^{-\xi}(1 - \cos s)^{\xi/2}$	$0 < \xi \leq 2$
Spherical	$\rho(s; \alpha) = (1 + \frac{s}{2\alpha})(1 - \frac{s}{\alpha})_+^2$	$\alpha > 0$
Askey	$\rho(s; \alpha, \tau) = (1 - \frac{s}{\alpha})_+^\tau$	$\alpha > 0, \tau \geq 2$
C^2-Wendland	$\rho(s; \alpha, \tau) = (1 + \tau\frac{s}{\alpha})(1 - \frac{s}{\alpha})_+^\tau$	$0 < \alpha \leq \pi, \tau \geq 4$
C^4-Wendland	$\rho(s; \alpha, \tau) = (1 + \tau\frac{s}{\alpha} + \frac{\tau^2-1}{3}\frac{s^2}{\alpha^2})(1 - \frac{s}{\alpha})_+^\tau$	$0 < \alpha \leq \pi, \tau \geq 6$
Legendre-Matérn	$\rho(s; \alpha, \nu) = \sum_{k=0}^\infty \frac{1}{(\alpha^2+k^2)^{\nu+1/2}}P_k(\cos s)$	$\alpha > 0, \nu > 0$
Circular Matérn	$\rho(s; \alpha, \nu) = \frac{1}{2\pi}\sum_{k=-\infty}^\infty \frac{\cos s}{(\alpha^2+k^2)^{\nu+1/2}}$	$\alpha > 0, \nu > 0$

positive definite in $\mathbb{R}^3$ (using Euclidean distance) is not necessarily positive definite on $\mathcal{S}^2$ (using great circle distance). A real continuous function $C(s)$ is a valid isotropic covariance function on $\mathcal{S}^2$ if and only if it is of the form

$$C(s) = \sum_{k=0}^\infty b_k P_k(\cos s) \quad \text{for } s \in [0, \pi),$$

where $b_k \geq 0$, $\sum_{k=0}^\infty b_k < \infty$, and $P_k(\cdot)$ are the Legendre polynomials (Schoenberg, 1942; Gneiting, 2013). Using this characterization, it can be shown that some isotropic covariance functions in $\mathbb{R}^3$ are not valid on the sphere for any values of their parameters. For example, the wave, Gaussian, and (generalized) Cauchy models listed in Table 6.2 are not valid on $\mathcal{S}^2$ (Huang *et al.*, 2011). Some other isotropic covariance functions on $\mathbb{R}^3$ are valid on $\mathcal{S}^2$ only over a portion of the parameter space for which they are valid in $\mathbb{R}^3$. Examples of such functions include the powered exponential and Matérn covariance functions. However, some isotropic covariance functions in $\mathbb{R}^3$ *are* valid on $\mathcal{S}^2$ over their entire parameter space. For example, every compactly supported covariance function on $\mathbb{R}^3$ is valid on $\mathcal{S}^2$ (Gneiting, 2013). Thus, the spherical, cubic, and penta models from Table 6.1 are valid on the sphere. Table 12.1 lists several parametric correlation functions on $\mathcal{S}^2$ and the parameter spaces over which they are valid, which are compiled from Huang *et al.* (2011) and Gneiting (2013), among other sources.

Recall that the Matérn covariance function has the flexibility to model data in a Euclidean space with a wide range of smoothnesses and is therefore very useful in that context. The nonpositive definiteness of the Matérn covariance function on the sphere for all values of the smoothness parameter greater than $\frac{1}{2}$, however, severely limits its usefulness for modeling global data. In fact, none of the models described so far in this section allow the underlying process to be mean-square differentiable. Numerous attempts have been made to construct covariance functions in $\mathcal{S}^2$ that are "Matérn-like" in the sense that they allow for a wide range of smoothnesses. Jeong & Jun (2015) obtained such a covariance function by integrating a process having Matérn covariance function with smoothness parameter less than or equal to $\frac{1}{2}$ over small neighborhoods on the surface of the sphere. Guinness & Fuentes (2016) introduced several Matérn-like covariance functions, of which two, the Legendre-Matérn and circular Matérn, are listed in Table 12.1. Alegría *et al.* (2021) introduced another family of Matérn-like functions they called the $\mathcal{F}$-family. In general, these last three types of functions, like the Matérn function itself, do not have closed form expressions, so in practice the infinite sums must be truncated, resulting in functions that are no longer guaranteed to be positive definite. However, similar to what occurs with the Matérn function in a Euclidean domain, when the smoothness parameters of the circular Matérn function and $\mathcal{F}$-family are half-integers, closed-form expressions for the functions exist.

A completely different approach to the modeling of smooth global data, suggested first by Yadrenko (1983), is to regard the data locations as points in $\mathbb{R}^3$ and use a Matérn covariance function with chordal (Euclidean) distance, rather than great circle distance, as the function's argument. Modeling the covariance using chordal distance may not be as reasonable scientifically as using great circle distance and may result in physically unrealistic distortions when the great circle distance is large, but it has the advantage of providing practitioners with a panoply of familiar covariance functions (not merely the Matérn) for which software for implementation is widely available.

This subsection has focused entirely on covariances of geostatistical, not areal, models for data on a sphere. This is because the response variable in most, if not all, global datasets to which a spatial linear model has been fitted is observed on very small areas that can effectively be regarded as points. However, if each datum was observed on a comparatively large subregion of the globe, the neighbor relationships among those subregions could be used to formulate spatial weights matrices for use within SAR, CAR, and SMA models in exactly the same fashion as in Euclidean domains. Interestingly, if the subregions comprised a complete partition of the sphere's surface, there would be no edge effects, with the result that the marginal variances under those models might be more homogeneous than they are within a Euclidean context.

12.1.5 Inference and design

Once a suitable model has been formulated, parameter estimation and kriging may proceed on the sphere exactly as they do in a Euclidean space. The R packages `GpGp` and `RandomFields` have functions for accomplishing this; the latter package, in particular, can estimate an impressive variety of covariance functions on the sphere.

As for design, the design criteria described previously for Euclidean geostatistics may be used within the global context as well, and lead to qualitatively similar designs: space-filling designs for good prediction with known covariance parameters, clustered designs for good estimation of covariance parameters, and hybrid designs for good prediction with unknown covariance parameters.

12.2 Stream network data

Rivers and streams are an important environmental resource. Accordingly, some environmental investigations involve the collection of data in or along a stream. The data could consist, for example, of water quality measurements (nitrate level, biological oxygen demand, etc.) taken at discrete points (essentially) on the stream, or counts of fish over stream reaches. If the data locations are sufficient in geographic extent, they may lie along not merely a single main stream channel (which, though curved on the landscape, could be mapped to a one-dimensional spatial domain), but along a main channel and several of its tributaries (which cannot be so mapped). That is, the data locations lie on a stream network. Although the network may be conceptualized as lying within two-dimensional space, it may not be sensible to use Euclidean two-dimensional spatial models for many stream variables, since such variables, whether they flow passively (nitrates) or swim under their own power (fish), can only move *within* the network. In fact, two locations very close together in space could be quite far apart within the network. It seems that spatial linear modeling in this situation calls for a completely different class of models tailored specifically for data on a stream network.

A stream network may be viewed mathematically as a particular type of **linear network** (or graph) with **Euclidean edges** (Anderes *et al.*, 2020), for which nodes are defined by stream junctions and edges by stream segments between junctions. Furthermore, in this case the graph is **directed**, with directions of the edges defined by the direction of streamflow (Ver Hoef *et al.*, 2006). If we assume that the stream segments in the portion of the network of interest split only as one moves upstream, not downstream (e.g., the stream is not braided and there is no delta), then there will be a single most downstream point, the "outlet," whose spatial coordinate may be set to 0. Any point in the network can be connected by a continuous curve through the network to the outlet, and the length of that curve is called the "upstream distance" of that point. To uniquely define individual points and keep track of their upstream distances, each point may be denoted by $s_i \equiv s_{R_i}$, where R_i $(i = 1, \ldots, m)$ represents the ith segment (according to an arbitrary labeling system) on which the point lies, and s is its upstream distance. Note that two points on the network may be distinct, yet have the same upstream distance. The **stream distance** between points s_i and t_j is the shortest distance between them through the network, and is denoted by $d(s_i, t_j)$.

For each s_i, let U_i denote the set of stream segments that lie upstream of s_i, including the ith segment. Points s_i and t_j are said to be **flow-unconnected** (which we sometimes abbreviate as FU) if $U_i \cap U_j = \emptyset$, and are **flow-connected** (which we sometimes abbreviate as FC) otherwise. If locations s_i and t_j are flow-connected, then $d(s_i, t_j) = |s_i - t_j|$; see Figure 12.2A. If, alternatively, s_i and t_j are flow-unconnected, then $d(s_i, t_j) = (s_i - q_{ij}) + (t_j - q_{ij})$, where q_{ij} is the upstream distance of the "common junction" of segments i and j, i.e., the junction where flows from segments i and j first combine; see Figure 12.2B.

12.2.1   Spatially structured dependence

Due to the hierarchical branching structure and directional flow within a stream network, somewhat different types of structured spatial dependence than those described previously for data on Euclidean domains are relevant. Within a given stream segment, the network is one-dimensional and the concept of stationarity, i.e., constancy of variance and correlations that depend only on stream distance, makes sense. For observations on different segments, however, it is plausible that the correlations could depend on whether the segments are flow-connected or flow-unconnected. In fact, for such variables as point-source pollutants, which flow passively downstream, it is quite possible for observations on flow-connected segments to be strongly correlated while observations on flow-unconnected segments are completely uncorrelated. This type of structured dependence is called **tail-up dependence**, for reasons that will become clear in Section 12.2.4. If, in addition to the absence of correlation between flow-unconnected observations, the correlation between flow-connected observations is a function of stream distance only and the variance is constant throughout the network, we say that the tail-up dependence is **quasi-stationary**. If, alternatively, the correlation between observations at sites on flow-unconnected segments is a function of the two stream

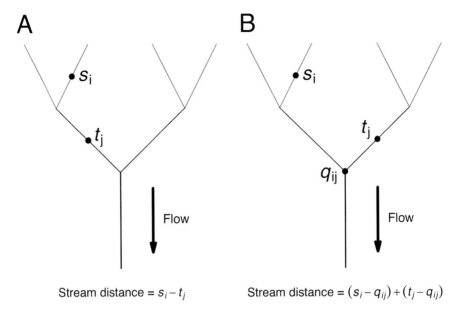

A) Flow-connected sites, and B) flow-unconnected sites, on a stream network, and the definition of stream distance between sites of each type.

FIGURE 12.2

distances from those sites to their common junction, while that between flow-connected observations is a function of stream distance only, and again the variance is constant, we say that the dependence is **tail-down**. This type of dependence might be more realistic for such variables as counts of fish, since fish can move both upstream and downstream. A special case of tail-down dependence is **isotropy**, for which the dependence is a function of stream distance only, regardless of flow-connectedness.

12.2.2 Exploratory spatial data analysis

Many of the same ESDA tools used for data on Euclidean domains are easily adapted for use with stream network data. For example, a common tool for visualizing the data's large-scale variation is a map of the stream network with color-coded circles superimposed at each data location, the color representing an ordinal scale for the response. Figure 12.3 shows such a map for trout density data from the Salt River network in Wyoming, USA. Because upstream distance often affects the response, a simple two-dimensional scatterplot of the response versus upstream distance may also be helpful. Outliers may be detected using nearest-neighbor scatterplots, but due to the possibility of tail-up dependence, it may be prudent to construct such plots separately for flow-connected and flow-unconnected sites. To investigate whether the residual variance is relatively constant over the network, the residuals from an ordinary least squares regression of the response on upstream distance, and possibly other explanatory variables of interest, may be plotted against upstream distance. A simple diagnostic for assessing the stationarity of the dependence is a series of three nearest-neighbor scatterplots based on data in the headwaters, near the outlet, and in-between. Together, these last two types of plots can determine whether quasi-stationarity is a plausible assumption for the data. Song & Zimmerman (2021) constructed such plots for the Salt River network trout density data and found substantial evidence against

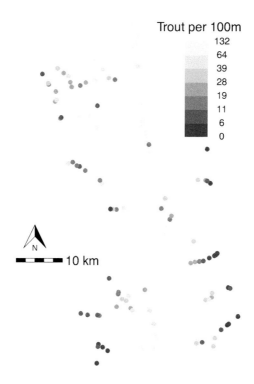

FIGURE 12.3
Map of the Salt River network with color-coded trout density.

quasi-stationarity: residual variance decreased, but spatial correlation increased, as one moved upstream.

As noted in the previous subsection, the correlation among responses on a stream network may depend not only on stream distance but also on flow connectedness and/or distances to a common junction. Thus, in contrast to Euclidean geostatistics, one sample semivariogram is insufficient for exploring spatial dependence on a stream network. Instead, no less than three distinct sample semivariograms, each a modification of the Euclidean sample semivariogram, are needed. Collectively, this set of sample semivariograms is called the **Torgegram** (Zimmerman & Ver Hoef, 2017). The remainder of this subsection describes the three components of the Torgegram. The R package SSN (Ver Hoef *et al.*, 2014) provides functions for computation and plotting of the first and third of these components.

The first component of the Torgegram is the flow-unconnected stream-distance (FUSD) sample semivariogram, which is computed from only those site-pairs that are flow-unconnected and, for such pairs, is a function of stream distance only. It is defined formally as

$$\hat{\gamma}_{FUSD}(h_k) = \frac{1}{2N(\mathcal{U}_k)} \sum_{(s_i, t_j) \in \mathcal{U}_k} \left(Y(s_i) - Y(t_j) \right)^2 \qquad (k = 1, \ldots, K_{\mathcal{U}}),$$

where $\mathcal{U}_k = \{(s_i, t_j) : d(s_i, t_j) \in \mathcal{H}_k, \ U_i \cap U_j = \emptyset\}$, $\{\mathcal{H}_k : k = 1, \ldots, K\}$ is a partition of the stream distances into bins, h_k is a representative distance within $\mathcal{U}_k$, $N(\mathcal{U}_k)$ is the number of distinct site-pairs in $\mathcal{U}_k$, and $K_{\mathcal{U}}$ is the number of stream-distance bins for those site-pairs. If responses at flow-unconnected sites are uncorrelated, then $\hat{\gamma}_{FUSD}(\cdot)$ is unbiased (apart from a "blurring" effect due to binning similar but unequal total stream distances) for the flow-unconnected portion of the semivariogram, which in this case is "flat," i.e., a constant function. Thus, in this case, a superior estimate of each $\gamma(h_k)$ is given by the weighted overall average

$$\overline{\gamma}_{FUSD} = \frac{\sum_{k=1}^{K_{\mathcal{U}}} N(\mathcal{U}_k)\hat{\gamma}_{FUSD}(h_k)}{\sum_{k=1}^{K_{\mathcal{U}}} N(\mathcal{U}_k)}.$$

On the other hand, if responses at flow-unconnected sites are correlated, then the flow-unconnected portion of the semivariogram may be a function not of total stream distance but of the two stream distances from sites within a site-pair to their common junction. In that case $\hat{\gamma}_{FUSD}(\cdot)$ may not be fully relevant, i.e., what it purports to estimate doesn't exist.

The second component of the Torgegram, like the first, is computed using only those site-pairs that are flow-unconnected, but it is not a function of the stream distance between sites. Rather, it is a function of the two stream distances from each site in the pair to their common junction. This component is called the flow-unconnected distances-to-common-junction (FUDJ) sample semivariogram. Let $\{\mathcal{J}_k : k = 1, \ldots, K_{\mathcal{J}}\}$ be the bins of a partition of the stream distances to common junction that occur among the flow-unconnected site-pairs, let $N(\mathcal{J}_k, \mathcal{J}_l)$ be the number of such site-pairs for which one site's distance to common junction lies in $\mathcal{J}_k$ and the other's lies in $\mathcal{J}_l$, and let j_k be a representative distance within $\mathcal{J}_k$. Without loss of generality, assume that $j_k \leq j_l$. Then the FUDJ sample semivariogram is

$$\hat{\gamma}_{FUDJ}(j_k, j_l) = \frac{1}{2N(\mathcal{J}_k, \mathcal{J}_l)} \sum_{s_i \in \mathcal{J}_k, t_j \in \mathcal{J}_l} \left(Y(s_i) - Y(t_j) \right)^2 \qquad (k \leq l = 1, \ldots, K_{\mathcal{J}}).$$

Regardless of whether responses at flow-unconnected sites are correlated or uncorrelated, $\hat{\gamma}_{FUDJ}(\cdot, \cdot)$ is unbiased (apart from blurring) for the flow-unconnected portion of its semivariogram. This is its advantage over $\hat{\gamma}_{FUSD}(\cdot)$, which (as noted previously) is not always fully relevant. However, because $\hat{\gamma}_{FUDJ}(\cdot, \cdot)$ is a function of two distances rather than one, it must be displayed in three dimensions (unless contour or gray-scale plotting is used). Hence it is a bit more cumbersome to plot and examine than $\hat{\gamma}_{FUSD}(\cdot)$. Furthermore, due to a reduction in sample sizes within bins, it has greater uncertainty, as each $\hat{\gamma}_{FUDJ}(j_k, j_l)$ is generally computed from only a subset of the site-pairs used to compute the corresponding $\hat{\gamma}_{FUSD}(h_k)$.

The third component, the flow-connected stream distance (FCSD) sample semivariogram, is based on stream distance only but differs from the FUSD semivariogram by being computed from site-pairs that are flow-connected rather than flow-unconnected. Thus it is defined as

$$\hat{\gamma}_{FCSD}(h_k) = \frac{1}{2N(\mathcal{C}_k)} \sum_{(s_i, t_j) \in \mathcal{C}_k} \left(Y(s_i) - Y(t_j) \right)^2 \qquad (k = 1, \ldots, K_{\mathcal{C}}),$$

where $\mathcal{C}_k = \{(s_i, t_j) : d(s_i, t_j) \in \mathcal{H}_k, \ U_i \cap U_j \neq \emptyset\}$, h_k is a representative distance within $\mathcal{C}_k$, $N(\mathcal{C}_k)$ is the number of distinct site-pairs in $\mathcal{C}_k$, and $K_{\mathcal{C}}$ is the number of stream-distance bins for those site-pairs. If $Y(\cdot)$ is pure tail-down, then $\hat{\gamma}_{FCSD}(\cdot)$ is unbiased (apart from blurring) for the flow-connected portion of its semivariogram. If, however, $Y(\cdot)$ is pure tail-up or a tail-up/tail-down mixture, $\hat{\gamma}_{FCSD}(\cdot)$ is not fully relevant because in those cases the flow-connected semivariogram is a function of not merely stream distance, but of stream distance and the spatial weights. Furthermore, the FCSD semivariogram does not account for differential flow volumes in the network, which causes it to be biased if the correlation depends on volume. The first of these issues may be circumvented by using only those site-pairs that lie on the same stream segment to compute $\hat{\gamma}_{FCSD}(\cdot)$. And for the second issue, if the flow volumes or proxies such as watershed areas are known, it may be possible to modify the sample semivariogram to yield an unbiased estimator of an unweighted flow-connected semivariogram. Details on both of these modifications are provided by Zimmerman & Ver Hoef (2017).

Torgegram semivariances may be used to perform formal tests of the Lu-Zimmerman type for assessing whether the spatial dependence is tail-up, tail-down, isotropic, or none of these, as proposed by Zimmerman *et al.* (2024).

12.2.3 Parametric models for the mean structure

In contrast to the situation for global data, there is not a natural, specialized family of models for the mean structure of spatial network data. Polynomials of upstream distance (and nonspatial explanatory variables, if any are observed) are commonly used. A "piecewise" polynomial model, in which the polynomials are grafted at junctions in such a way that immediately below a junction the mean is equal to the flow-weighted average of the means on the two segments immediately upstream of the junction, would appear to be a more sensible and flexible alternative, but such models do not appear to have ever been used.

12.2.4 Parametric models for the covariance structure

Although the spatial domain in a sufficiently small neighborhood around any point in a stream network except a junction is one-dimensional, valid covariance functions in $\mathbb{R}$ are not necessarily valid on a stream network, as demonstrated by Ver Hoef *et al.* (2006); additional relevant discussion was given by Curriero (2006). Cressie *et al.* (2006), Ver Hoef *et al.* (2006), and Ver Hoef & Peterson (2010) obtained valid geostatistical covariance models for stream network variables by adapting the auto-convolution approach described in Section 6.1 to the unique topology of stream networks. More specifically, they created a Gaussian process $\{\epsilon(s_i) : i = 1, \ldots, m\}$ on a stream network S with m segments via

$$\epsilon(s_i) = \int_S g(x - s_i; \boldsymbol{\theta}) \, dW(x), \qquad (12.1)$$

where s_i and t_j are locations on S, $g(\cdot)$ is a moving average function on S, and $\{W(t_j) : j = 1, \ldots, m\}$ is a Gaussian white noise process on S. They recommended in particular using only **unilateral** moving average functions, i.e. moving-average functions that are positive in only one direction (either upstream or downstream) and identically equal to zero in the other direction. Such functions contrast with the bilateral moving average functions typically used in Euclidean space, such as the one depicted in Figure 6.1. Consider first a model for which the moving-average function is positive only upstream, or **tail-up**. Figure 12.4A,B depict one such model along a stream network, evaluated respectively at flow-connected and flow-unconnected sites. Because of the "upstream" construction, the only positive contribution to the integral analogous to (6.5) occurs over that portion of S that lies upstream of the site that is furthest upstream. Thus, Figure 12.4B reveals that the correlation between observations at flow-unconnected sites is zero because the moving-average functions emanating from such sites do not overlap. In contrast, Figure 12.4A shows that the correlation between observations at flow-connected sites may be positive because the moving average functions emanating from such sites may overlap to some degree. Note that the tail-up moving average function splits at confluences as one moves upstream, and it turns out that the variances of observations at flow-connected sites separated by one

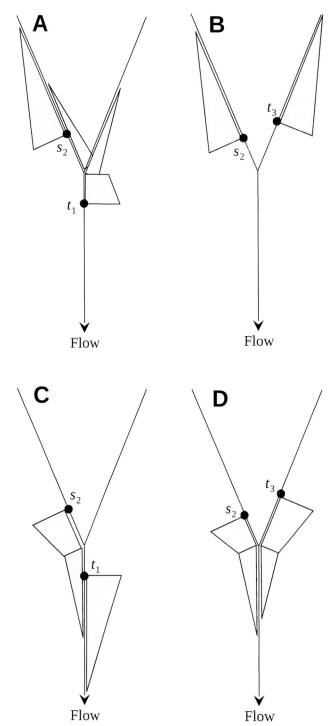

FIGURE 12.4
Tail-up and tail-down models, in combination with flow-connected and flow-unconnected sites on a stream network: A) Tail-up model, flow-connected sites; B) tail-up model, flow-unconnected sites; C) tail-down model, flow-connected sites; D) tail-down model, flow-unconnected sites.

or more confluences will be heterogeneous unless the Gaussian process definition (12.1) is modified in a particular way (see Ver Hoef & Peterson, 2010, Section 2.2). This modification results in downweighting the moving-average function above each junction; common choices for the weights are based upon Shreve's stream order or the proportion of the flow volume contributed by each immediately upstream tributary. Thus, the family of tail-up covariance functions is given by

$$C_{TU}(s_i, t_j; \{\pi_{ij}\}, \sigma^2, \boldsymbol{\theta}) = \begin{cases} \pi_{ij}\sigma^2 \rho_{UW}(|s - t|; \boldsymbol{\theta}) & \text{if } s_i \text{ and } t_j \text{ are FC,} \\ 0 & \text{if } s_i \text{ and } t_j \text{ are FU.} \end{cases} \quad (12.2)$$

Here, the π_{ij}'s are the specified weights, and $\rho_{UW}(\cdot)$, the *unweighted* flow-connected correlation function, is a valid correlation function on $\mathbb{R}$. If, for example, flow volumes are used to construct the weights, then they are given by $\pi_{ij} = \prod_{k \in B_{ij}} \sqrt{\omega_k}$ where B_{ij} is the set of segments that lie between the ith and jth (including the jth but excluding the ith) and ω_k is the proportion of flow volume contributed by the kth segment to the junction at its terminus.

For moving-average functions that are positive only in the downstream direction, i.e., **tail-down** functions, there may be overlap regardless of whether sites are flow-connected or flow-unconnected; see Figure 12.4C,D. Furthermore, because the tail heading downstream does not have to split, there is no need for weighting as there is for tail-up functions. The class of tail-down covariance functions is given by

$$\begin{aligned} C_{TD}(s_i, t_j; \sigma^2, \boldsymbol{\theta}) &= \sigma^2 \rho_{TD}(s_i, t_j; \boldsymbol{\theta}) \\ &= \begin{cases} \sigma^2 \rho_{FC}(|s - t|; \boldsymbol{\theta}) & \text{if } s_i \text{ and } t_j \text{ are FC,} \\ \sigma^2 \rho_{FU}(s - q_{ij}, t - q_{ij}; \boldsymbol{\theta}) & \text{if } s_i \text{ and } t_j \text{ are FU,} \end{cases} \end{aligned}$$

where $\rho_{FC}(\cdot)$ is a valid correlation function on $\mathbb{R}$ and $\rho_{FU}(\cdot)$ is related to $\rho_{FC}(\cdot)$ through its functional dependence on the same moving average function. Thus, the flow-connected portion of a tail-down covariance function is a function of only the stream distance between locations, but the flow-unconnected portion is generally a function of the two stream distances to the common junction.

Table 12.2 lists several tail-up and tail-down correlation models. (It suffices to give correlation, rather than covariance, models because all of the covariance models are variance stationary.) The tail-down exponential model, which is obtained using a unilaterally downstream exponential moving-average function, is noteworthy because it is the only tail-down auto-convolution model that is isotropic (Tang & Zimmerman, 2024). An alternative to the tail-up and tail-down models are two-tailed models, for which the moving average function is bilateral, but not necessarily symmetric about 0. Although they are flexible, two-tailed models are often computationally challenging to fit, so Ver Hoef & Peterson (2010) recommended the use of a mixed linear model with tail-up and tail-down components as a more practical alternative. A model of this last type takes the error at site s_i to be given by

$$\epsilon(s_i) = \epsilon_{tu}(s_i) + \epsilon_{td}(s_i) + \epsilon_0(s_i), \quad (12.3)$$

where $\epsilon_{tu}(\cdot)$, $\epsilon_{td}(\cdot)$, and $\epsilon_0(\cdot)$ are (Gaussian) random processes that are uncorrelated with each other and have pure tail-up, pure tail-down, and pure nugget effect covariance functions, respectively. Observations with errors that follow this model have a covariance matrix given by

$$\boldsymbol{\Sigma}(\boldsymbol{\theta}) = \boldsymbol{\Sigma}(\sigma_{tu}^2, \sigma_{td}^2, \sigma_0^2, \boldsymbol{\rho}_{tu}^T, \boldsymbol{\rho}_{td}^T)^T = \sigma_{tu}^2 \mathbf{R}_{tu}(\boldsymbol{\rho}_{tu}) + \sigma_{td}^2 \mathbf{R}_{td}(\boldsymbol{\rho}_{td}) + \sigma_0^2 \mathbf{I}, \quad (12.4)$$

where $\mathbf{R}_{tu}(\boldsymbol{\rho}_{tu})$ is a matrix of autocorrelations among the tail-up components; $\mathbf{R}_{td}(\boldsymbol{\rho}_{td})$ is a matrix of autocorrelations among the tail-down components; σ_{tu}^2, σ_{td}^2, and σ_0^2 (the nugget

TABLE 12.2
Some parametric, quasi-stationary tail-up and tail-down, and isotropic completely monotone, correlation models on stream networks. For the tail-up models, only the unweighted flow-connected portion is displayed; the complete model is given by substitution in (12.2). Throughout, α is a positive correlation decay parameter, and $s \geq t$ without loss of generality.

Name	Functional form
Tail-up linear-with-sill	$\rho_{UW}(s_i, t_j; \alpha) = \left(1 - \frac{s-t}{\alpha}\right) I\left(s - t \leq \alpha\right)$
Tail-up spherical	$\rho_{UW}(s_i, t_j; \alpha) = \left(1 - \frac{3(s-t)}{2\alpha} + \frac{(s-t)^3}{\alpha^3}\right) I\left(s - t \leq \alpha\right)$
Tail-up exponential	$\rho_{UW}(s_i, t_j; \alpha) = \exp(-(s-t)/\alpha)$
Tail-up Mariah	$\rho_{UW}(s_i, t_j; \alpha) = \begin{cases} 1 & \text{if } s = t, \\ \frac{\log(1+(s-t)/\alpha)}{(s-t)/\alpha} & \text{if } s > t \end{cases}$
Tail-down linear-with-sill	$\rho_{TD}(s_i, t_j; \alpha) = \begin{cases} \left(1 - \frac{s-t}{\alpha}\right) I\left(s - t \leq \alpha\right) & \text{if FC}, \\ \left(1 - \frac{s-q_{ij}}{\alpha}\right) I\left(s - q_{ij} \leq \alpha\right) & \text{if FU} \end{cases}$
Tail-down spherical	$\rho_{TD}(s_i, t_j; \alpha) = \begin{cases} \left(1 - \frac{3(s-t)}{2\alpha} + \frac{(s-t)^3}{2\alpha^3}\right) I\left(s - t \leq \alpha\right) \\ \quad \text{if FC}, \\ \left(1 - \frac{3(t-q_{ij})}{2\alpha} + \frac{s-q_{ij}}{2\alpha}\right) \left(1 - \frac{s-q_{ij}}{\alpha}\right)^2 \\ \quad \times I\left(s - q_{ij} \leq \alpha\right) & \text{if FU} \end{cases}$
Tail-down exponential	$\rho_{TD}(s_i, t_j; \alpha) = \exp(-(s-t)/\alpha)$
Tail-down Mariah	$\rho_{TD}(s_i, t_j; \alpha) = \begin{cases} 1 & \text{if FC and } s = t, \\ \frac{\log(1+(s-t)/\alpha)}{(s-t)/\alpha} & \text{if FC and } s > t, \\ \frac{1}{1+(s-q_{ij})/\alpha} & \text{if FU and } s = t, \\ \frac{\log(1+(s-q_{ij})/\alpha)-\log(1+(t-q_{ij})/\alpha)}{(s-t)/\alpha} \\ \quad \text{if FU and } s > t \end{cases}$
Isotropic Matérn	$\rho_I(s_i, t_j : \alpha, \nu) = \frac{2^{1-\nu}}{\Gamma(\nu)}(d(s_i, t_j)/\alpha)^{\nu} K_{\nu}(d(s_i, t_j)/\alpha),$ $0 < \nu \leq \frac{1}{2}$
Isotropic power exponential	$\rho_I(s_i, t_j; \alpha, \xi) = \exp[-(d(s_i, t_j)/\alpha)^{\xi}],$ $0 < \xi \leq 1$
Isotropic generalized Cauchy	$\rho_I(s_i, t_j; \alpha, \xi, \tau) = [1 + (d(s_i, t_j)/\alpha)^{\xi}]^{-\tau/\xi},$ $0 < \xi \leq 1, \ \tau > 0$
Isotropic Dagum	$\rho_I(s_i, t_j; \alpha, \xi, \tau) = \left[1 - \left(\frac{(d(s_i, t_j)/\alpha)^{\xi}}{1+(d(s_i, t_j)/\alpha)^{\xi}}\right)\right]^{\tau/\xi},$ $0 < \xi \leq 1, \ 0 < \tau \leq 1$

effect) are variance components; and $\boldsymbol{\rho}_{tu}$ and $\boldsymbol{\rho}_{td}$ are vectors of correlation parameters. An additional error component may be added to (12.3) to account for Euclidean spatial dependence, if any, such as might occur if the response is a water chemistry variable that is affected substantially by the underlying bedrock formations through which the stream flows (Ver Hoef *et al.*, 2019). In that case a corresponding additional variance component and autocorrelation matrix would appear in the covariance matrix given by (12.4).

Another class of valid covariance models on stream networks is the isotropic (in stream-distance) completely monotone class, which may be written as

$$C_I(s_i, t_j; \sigma^2, \boldsymbol{\theta}) = \sigma^2 \rho_I(d(s_i, t_j); \boldsymbol{\theta}), \tag{12.5}$$

where $\rho_I(\cdot)$ belongs to a certain subset of stationary correlation functions on $\mathbb{R}$. Anderes, Møller, and Rasmussen (2020) established that many isotropic correlation models used in

Euclidean geostatistics are positive definite on a subclass of linear networks with Euclidean edges (which includes braidless stream networks) after restricting the parameter space to that for which the models are **completely monotone**. (A function f mapping $(0, \infty)$ to $[0, \infty)$ is completely monotone if $(-1)^k f^{(k)}(x) \geq 0$ for $x > 0$ and $k = 1, 2, \ldots$.) For example, a Matérn correlation function with smoothness parameter less than or equal to $\frac{1}{2}$ is completely monotone and thus may be used to construct, via (12.5), an isotropic stream-network covariance model. The Matérn correlation model and three other isotropic completely monotone correlation models are listed in Table 12.2.

Song & Zimmerman (2021) developed two classes of nonstationary covariance models, called elastic models and spatially varying moving-average models, for geostatistical data on a stream network. An elastic model may be obtained from any positive definite quasi-stationary model by replacing upstream distances s, t, and q_{ij} with s^ψ, t^ψ, and q_{ij}^ψ, where $\psi \in \mathbb{R}$ and it is assumed that $s \geq 1$, $t \geq 1$, and $q_{ij} \geq 1$. The "stretch parameter" ψ deforms the network nonlinearly, so that sites close to the outlet are stretched farther apart relative to sites farther upstream if $\psi < 1$, and vice versa if $\psi > 1$. Spatially varying moving-average models are obtained by adapting the auto-convolution approach described in Section 6.4, using a moving-average function whose shape is site-specific.

To date, little attention has been given to the formulation of covariance structures for areal data on stream networks, such as counts of fish over stream reaches. One reference is Liu (2018), who proposed an ICAR model for areal stream network data using precision matrices based on three types of neighbor relationships. Taking a similar approach with a SAR, CAR, or SMA model, it could be assumed that the spatial weights matrix is of form

$$\mathbf{C} = \rho_1 \mathbf{W}_1 + \rho_2 \mathbf{W}_2 + \rho_3 \mathbf{W}_3,$$

where $\mathbf{W}_1$, $\mathbf{W}_2$, and $\mathbf{W}_3$ are binary adjacency matrices for, respectively, reaches on the same stream segment, reaches on distinct flow-connected segments, and reaches on flow-unconnected segments (which are adjacent if and only if they share the same junction). In this formulation, $\rho_3 = 0$ corresponds to a spatial-weights version of pure tail-up dependence, while $\rho_1 = \rho_2 = \rho_3$ is a spatial-weights version of isotropy.

12.2.5 Inference

Likelihood-based parameter estimation, prediction, and smoothing may be performed in essentially the same way on stream networks as they were on Euclidean domains. The R package SSN (Ver Hoef *et al.*, 2014) provides functions for ML and REML estimation of parameters of various tail-up and tail-down models, including those listed in Table 12.2, and for kriging on stream networks.

The approach to spatial modeling on stream networks we have described thus far is based on mean models that take upstream distance, and covariance models that take stream distance and flow connectedness/unconnectedness, respectively, as their arguments. Prediction associated with this modeling approach has been called **network kriging** (Okabe & Sugihara, 2012), in reference to its exclusive use of distances within the network. Two alternative approaches that have also been used for prediction on networks are based on mean and covariance models for a geostatistical process on a two-dimensional region containing the watershed of the network. The first of these takes the geostatistical process on that region to be isotropic, with Euclidean distance between sites or catchment area centroids as the argument of the covariance function. This approach, called **Euclidean kriging**, takes no account of flow direction or flow connectedness within the network. The second alternative conceptualizes stream flow as the integral of the process over catchment areas, with covariances computed by integrating the covariance function of the aforementioned geostatistical process over catchments. This approach accounts for the shape of catchments and their nestedness (hence accounting for flow direction and flow-connectedness). Spatial prediction within this last approach is a stream-network version of block kriging from block-level data, and is known as **top-kriging** (topological kriging). Skoien *et al.* (2006) and Laaha *et al.* (2012) argued that top-kriging will generally outperform Euclidean kriging, as the latter tends to give too much weight according to distance in geographic space and too little weight according to stream topology. Laaha *et al.* (2012) argued further that top-kriging is also preferable to network kriging, which in their view gives too much weight according to stream topology and too little according to geographic distance. To the authors' knowledge, no comprehensive empirical comparisons of these methods have been published.

12.2.6 Sampling design

The same broad categories of sampling design methodologies (model-based, probability-based, and space-filling) described in Chapter 10 for data on Euclidean domains may be applied to the stream network context, but there are some important differences as a result of the unique topology of stream networks (Dobbie *et al.*, 2008) and the distinct characteristics of tail-up and tail-down models for spatial dependence on streams. These differences may affect the relevance and utility of various sampling designs. For example, for stratified random sampling, Liebetrau (1979) proposed taking stream segments as the sampling units and stratifying not on geography but on stream order (number of upstream tributaries), drainage area, or mean annual flow. A space-filling approach was proposed by Dixon *et al.* (1999), with the wrinkle that the maximum subcatchment area (defined as the drainage area of a sampling site minus the drainage area of any other sampling sites farther upstream), rather than maximum intersite distance, was minimized. Som *et al.* (2014) and Falk *et al.* (2014) investigated model-based sampling design for stream networks from frequentist and (pseudo-)Bayesian perspectives, respectively. Som *et al.* (2014) found in particular that: (1) for estimating the covariance parameters of tail-up models, the best designs have clusters of sites on flow-unconnected segments; (2) for estimating the covariance parameters of tail-down models, the best designs tend to have triadic clusters at stream junctions (one site just downstream of the junction and one on each tributary just above the junction); and (3) for all other inference objectives, the best designs usually have a site on the outlet segment and sites on many headwater segments, with the remaining sites evenly dispersed throughout the remainder of the network. Cost considerations may preclude the use of a design with too many headwater segments, however, as those segments are usually the least accessible. The findings of Falk *et al.* (2014) were similar.

12.3 Space-time data

Another domain to which geostatistics and areal data analysis may be extended is the space-time, or spatio-temporal, domain. In fact, more often than not, data collected over space are also collected over time. Examples include wet sulfate deposition totals measured monthly at U.S. National Weather Service stations or ambient carbon monoxide levels measured hourly at air pollution monitors located throughout a large metropolitan area. Such data can be denoted as $Y(S_1, t_1)$, $Y(S_2, t_2)$, ..., $Y(S_n, t_n)$ where $S_1, \ldots, S_n$ are the data locations (points or subregions) in $\mathbb{R}^d$ for $d = 1, 2$, or 3 and $t_1, \ldots, t_n$ are the measurement times. The response variable at some of the data locations may be observed at multiple measurement times, or vice versa. Often, the observations are denser in time than they are in space. An especially convenient case occurs when the measurement times are common across locations. In this case, the data are said to be **rectangular** and may be re-expressed as

$$Y(S_1, t_1),\ Y(S_1, t_2), \ldots, Y(S_1, t_{n_T}),$$
$$Y(S_2, t_1),\ Y(S_2, t_2), \ldots, Y(S_2, t_{n_T}),$$
$$\vdots$$
$$Y(S_{n_S}, t_1),\ Y(S_{n_S}, t_2), \ldots, Y(S_{n_S}, t_{n_T}),$$

where n_S is the number of data locations and n_T is the number of measurement times. Note that the overall sample size, n, is equal to $n_S \times n_T$ in this case. Fitting space-time models is generally easier when the data are rectangular than otherwise. Rectangular data may be regularly spaced in either space or time (or both), but need not be. If they are regular in time, then the data may be viewed as a collection of spatially correlated time series.

12.3.1 Modeling spatio-temporal dependence

For modeling geostatistical spatio-temporal data (observations taken at sites that are essentially points in Euclidean space and time), a Gaussian spatial process may be extended to a Gaussian spatio-temporal process $\{Y(\mathbf{s}, t) : \mathbf{s} \in \mathcal{A} \subset \mathbb{R}^d, t \in \mathcal{T} \subset \mathbb{R}\}$, which is completely characterized by its mean function $\mu(\mathbf{s}, t) = \mathrm{E}(Y(\mathbf{s}, t))$ and covariance function $C((\mathbf{s}, t), (\mathbf{s}', t')) = \mathrm{Cov}(Y(\mathbf{s}, t), Y(\mathbf{s}', t'))$. Just like its spatial counterpart, the covariance function of a spatio-temporal process must be positive definite, i.e.,

$$\sum_{i=1}^{n} \sum_{j=1}^{n} a_i a_j C((\mathbf{s}_i, t_i), (\mathbf{s}_j, t_j)) \geq 0$$

for all n, all real numbers $a_1, \ldots, a_n$, and all space-time locations $(\mathbf{s}_1, t_1), \ldots, (\mathbf{s}_n, t_n) \in \mathbb{R}^d \times \mathbb{R}$. (Similarly, the space-time semivariogram must be conditionally negative definite.) And once again, some assumptions on the geometric structure of the covariance function are often made to facilitate modeling. One of these is **full symmetry**, which specifies that $C((\mathbf{s}, t), (\mathbf{s}', t')) = C((\mathbf{s}, t'), (\mathbf{s}', t))$ for all $\mathbf{s}, \mathbf{s}', t, t'$. Others are versions of stationarity and isotropy appropriate for space-time models. A space-time covariance function $C((\mathbf{s}, t), (\mathbf{s}', t'))$ is said to be **spatially stationary** if $C((\mathbf{s}, t), (\mathbf{s}', t')) = C(\mathbf{s} - \mathbf{s}', t, t')$ for all $\mathbf{s}, \mathbf{s}', t, t'$, **temporally stationary** if $C((\mathbf{s}, t), (\mathbf{s}', t')) = C(\mathbf{s}, \mathbf{s}', t - t')$ for all $\mathbf{s}, \mathbf{s}', t, t'$, or **stationary** if it is both spatially and temporally stationary. Similarly, the function is said to be **spatially isotropic** if $C((\mathbf{s}, t), (\mathbf{s}', t')) = C(\|\mathbf{s} - \mathbf{s}'\|, t, t')$ for all $\mathbf{s}, \mathbf{s}', t, t'$ and **temporally symmetric** if $C((\mathbf{s}, t), (\mathbf{s}', t')) = C(\mathbf{s}, \mathbf{s}', |t - t'|)$ for all $\mathbf{s}, \mathbf{s}', t, t'$. The function may be both spatially isotropic and temporally symmetric, in which case it is also fully symmetric (the converse is not true). "Fully isotropic" covariance functions, which depend on locations only through $\sqrt{\|\mathbf{s} - \mathbf{s}'\|^2 + |t - t'|^2}$ are not meaningful in space-time, owing to the different physical scales of time and space. And finally, a stationary space-time covariance function $C(\mathbf{s} - \mathbf{s}', t - t'))$ is **compactly supported** if $C(\mathbf{s} - \mathbf{s}', t - t') = 0$ whenever $\|\mathbf{s} - \mathbf{s}'\|$ and $|t - t'|$ are sufficiently large.

Clearly, any covariance function that is valid in $d + 1$ spatial dimensions can serve as a spatio-temporal covariance function in $\mathcal{A} \times \mathcal{T}$. However, as noted above, the units describing spatial position and lag differ from those describing temporal position and lag, so that a function that depends on locations only through a distance metric that weights each dimension equally, such as $\sqrt{\|\mathbf{s} - \mathbf{s}'\|^2 + |t - t'|^2}$, is inappropriate. **Metric** spatio-temporal covariance functions deal with this by introducing a parameter that accounts for the different physical scale of time from that of space. Specifically, let $C^*(\mathbf{h}, u)$ be a $(d + 1)$-dimensional stationary covariance function that depends on the elements of $\mathbf{h} = \mathbf{s} - \mathbf{s}'$ and $u = |t - t'|$ via a distance metric that weights each dimension equally. Then, a metric covariance function corresponding to Euclidean distance is

$$C(\mathbf{h}, u; \psi) = C^*(\sqrt{\|\mathbf{h}\|^2 + \psi|u|^2}),$$

where $\psi \geq 0$ (Dimitrakopoulos & Luo, 1994). Another, corresponding to L_1 (city-block) distance, is

$$C(\mathbf{h}, u; \psi) = C^*(|h_1| + |h_2| + \psi|u|),$$

where again $\psi \geq 0$. In the spatially isotropic and temporally symmetric case, ψ can be viewed as the analog of a geometric anisotropy parameter for a spatial covariance function.

Scaling is not the only new issue that arises in modeling spatio-temporal dependence. Owing to the fundamental physical difference between time and space, there is no reason why the dependence in time should be of the same functional form as the dependence in space. Accordingly, several classes of spatio-temporal covariance functions have been developed that, unlike metric functions, are structured to allow for different forms of dependence in these two domains. Two such classes are the **separable** (or multiplicatively separable) and **additive** (or additively separable) functions. A space-time covariance function $C(\mathbf{s}, t), (\mathbf{s}', t'))$ is separable if

$$C((\mathbf{s}, t), (\mathbf{s}', t')) = C_S(\mathbf{s}, \mathbf{s}')C_T(t, t')$$

for all $\mathbf{s}, \mathbf{s}', t, t'$, where $C_S(\cdot)$ and $C_T(\cdot)$ are valid covariance functions in $\mathbb{R}^d$ and $\mathbb{R}$, respectively. Such a function arises as the covariance function of a process $Y(\mathbf{s}, t) = Y_S(\mathbf{s})Y_T(t)$ where $\{Y_S(\cdot)\}$ and $\{Y_T(\cdot)\}$ are independent stationary Gaussian processes on $\mathbb{R}^d$ and $\mathbb{R}$ with positive definite covariance functions $C_S(\mathbf{s})$ and $C_T(t)$, respectively. A separable covariance function, in conjunction with a rectangular data configuration, can lead to considerable computational simplification (see Section 12.3.4). Note that a separable space-time covariance function necessarily is fully symmetric, but not vice versa. The stationary case reduces to $C(\mathbf{s} - \mathbf{s}', t - t') = C_S(\mathbf{s} - \mathbf{s}')C_T(t - t')$. A space-time covariance function $C((\mathbf{s}, t), (\mathbf{s}', t'))$ is additive if

$$C((\mathbf{s}, t), (\mathbf{s}', t')) = C_S(\mathbf{s}, \mathbf{s}') + C_T(t, t'),$$

where again $C_S(\mathbf{s})$ and $C_T(t)$ are valid covariance functions in $\mathbb{R}^d$ and $\mathbb{R}$, respectively. This type arises as the covariance function of a process $Y(\mathbf{s}, t) = Y_S(\mathbf{s}) + Y_T(t)$ where $\{Y_S(\cdot)\}$ and $\{Y_T(\cdot)\}$ are defined as they were for the separable process above. The stationary case reduces to $C(\mathbf{s} - \mathbf{s}', t - t') = C_S(\mathbf{s} - \mathbf{s}') + C_T(t - t')$. A shortcoming of an additive function is that it may not be strictly positive definite; hence, certain space-time configurations can result in a singular covariance matrix (Rouhani & Myers, 1990). Furthermore, an additive function does not lead to any particular computational simplifications. For these reasons, separable models are employed much more often than additive models.

Both separable and additive covariance functions are highly restrictive, however, as they do not allow for space-time interaction. Consequently, various types of nonseparable (and nonadditive) spatio-temporal covariance functions have been developed that allow for such an interaction while also still allowing for different forms of dependence in space and time. One relatively simple class of such functions is the **product-sum** class (De Cesare *et al.*, 2001), for which

$$C((\mathbf{s}, t), (\mathbf{s}', t')) = \kappa_{ST} C_S((\mathbf{s}, \mathbf{s}')C_T(t, t') + \kappa_S C_S(\mathbf{s}, \mathbf{s}') + \kappa_T C_T(t, t'),$$

where $\kappa_{ST}, \kappa_S, \kappa_T$ are nonnegative scalars (or, if one wishes to ensure strict positive definiteness, κ_{ST} is positive). Clearly, this class includes the separable and additive classes as special cases.

TABLE 12.3
Some completely monotone functions $\phi(x), x \geq 0$, and positively-valued functions $\psi(u), u \geq 0$, for constructing space-time covariance functions within the Gneiting class. In every function in which they appear, α and τ are positive parameters. Adapted from Tables 1 and 2 of Gneiting (2002b).

$\phi(x)$	Parameter space
$\exp(-x^\gamma/\alpha)$	$0 < \gamma < 1$
$\{2^{\nu-1}\Gamma(\nu)\}^{-1}((x/\alpha)^{1/2})^\nu K_\nu((x/\alpha)^{1/2})$	$\nu > 0$
$(1 + (x/\alpha)^\gamma)^{-\nu}$	$\nu > 0, 0 < \gamma \leq 1$
$2^\nu \{\exp((x/\alpha)^{1/2}) + \exp(-(x/\alpha)^{1/2})\}^{-\nu}$	$\nu > 0$

$\psi(u)$	Parameter space
$(\tau u^\eta + 1)^\beta$	$0 < \eta \leq 1, 0 \leq \beta \leq 1$
$\log(\tau u^\eta + \kappa)/\log(\kappa)$	$\kappa > 1, 0 < \eta \leq 1$
$(\tau u^\eta + \kappa)/\{\kappa(\tau u^\eta + 1)\}$	$0 < \kappa \leq 1, 0 < \eta \leq 1$

Another class of nonseparable functions, which includes stationary product-sum covariance functions as a discrete special case, is constructed by mixing separable stationary functions as follows (De Iaco *et al.*, 2002; Ma, 2002; Fonseca & Steel, 2011):

$$C(\mathbf{h}, u) = \int C_S(\mathbf{h}; w)C_T(u; z) \, d\mu(w, z),$$

where $\mu(w, z)$ is a bivariate cumulative distribution function. A different but similarly-motivated class of functions is the **Rodrigues-Diggle** class, given by

$$C(\mathbf{h}, u) = C_{S1}(\mathbf{h})C_{T1}(u) + C_{S2}(\mathbf{h})C_{T2}(u)$$

(Rodrigues & Diggle, 2010).

Cressie & Huang (1999) constructed nonseparable stationary covariance functions through Fourier inversion. Their functions, which comprise the **Cressie-Huang** class, are of the form

$$C(\mathbf{h}, u) = \int \exp(i\mathbf{h}^T\mathbf{x})\eta(\mathbf{x}, u)F(d\mathbf{x}),$$

where, for each $\mathbf{x} \in \mathbb{R}^d$, $\eta(\mathbf{x}, \cdot)$ is a continuous, integrable correlation function in $\mathbb{R}$ and $F(d\mathbf{x})$ is a positive finite measure. Another important class of nonseparable stationary functions is the **Gneiting class** (Gneiting, 2002b),

$$C(\mathbf{h}, u) = \frac{1}{\psi(u^2)^{d/2}}\phi\left(\frac{\|\mathbf{h}\|^2}{\psi(u^2)}\right),$$

where $\phi(\cdot)$ is a completely monotone function on $\mathbb{R}$ satisfying $\phi(0) = 1$ and $\psi(\cdot)$ is a positively-valued function on $\mathbb{R}$ with a completely monotone derivative satisfying $\psi(0) = 1$. This class is somewhat richer than the Cressie-Huang class, as it is not limited to functions obtainable as closed-form Fourier inversions of a spectral density, and it includes all the valid examples given by Cressie & Huang (1999). Table 12.3 lists four and three examples, respectively, of suitable functions $\phi(x)$ and $\psi(u)$ as given by Gneiting (2002b), from which 12 members of the class can be constructed. Gneiting-class functions built using the first $\psi(u)$ listed in the table are particularly useful because the parameter β can be interpreted as a space-time interaction parameter: $\beta = 0$ corresponds to a separable model, and the space-time interaction increases as β increases.

Covariance functions belonging to any of the aforementioned nonseparable, nonadditive classes are necessarily fully symmetric, and may therefore be inappropriate when, as with many atmospheric processes, there is a dominant spatial flow direction over time. Also, for functions belonging to all but the Cressie-Huang class, there is a potential problem with lack of smoothness away from the origin. More specifically, if either the spatial or temporal component function of a separable, additive, product-sum, or Rodrigues-Diggle function is nondifferentiable at the origin, and not merely because of a nugget effect, then the space-time covariance function is nondifferentiable away from the origin as well, with some undesirable consequences for inference (Stein, 2005b). For spatially isotropic cases within these classes, the function is nondifferentiable on a "ridge" that extends along the time axis if $C_S(\|\mathbf{h}\|)$ is nondifferentiable and along the spatial distance axis if $C_T(u)$ is nondifferentiable. A similar statement can be made about members of the Gneiting class. Stein (2005b) developed classes of stationary, nonseparable (and nonadditive) functions that do not have these shortcomings. A disadvantage of Stein's method for constructing functions that are smooth away from the origin is that explicit expressions for the covariance functions are available only for some very limited special cases. Stein's method of constructing asymmetric covariance functions, however, has the advantage that it can be used to model processes that have different smoothness across space than across time.

One feature that may distinguish one nonseparable model from another is the "sign" of its nonseparability. Consider stationary, nonnegative covariance functions only, and note that an equivalent characterization of separability for such a function is that $C(\mathbf{h}, u)/\{C(\mathbf{h}, 0)C(\mathbf{0}, u)\} = 1$ for all $\mathbf{h}$ and u for which $C(\mathbf{h}, u)$, $C(\mathbf{h}, 0)$, and $C(\mathbf{0}, u)$ are strictly positive, and is equal to zero otherwise. The covariance function $C(\mathbf{h}, u)$ is said to be (uniformly) **positively nonseparable** if $C(\mathbf{h}, u)/\{C(\mathbf{h}, 0)C(\mathbf{0}, u)\} > 1$ for all $\mathbf{h}$ and u satisfying the aforementioned positivity requirements, and (uniformly) **negatively nonseparable** if $C(\mathbf{h}, u)/\{C(\mathbf{h}, 0)C(\mathbf{0}, u)\} < 1$ for the same such $\mathbf{h}$ and u. Rodrigues & Diggle (2010) showed that the class of product-sum covariance models cannot accommodate positive nonseparability and the Gneiting class cannot accommodate negative nonseparability, but the Rodrigues-Diggle class can accommodate both positive and negative (and zero) nonseparability.

We have described several available classes of spatio-temporal covariance functions and methods for their construction, but there are many more that we have left unmentioned. Some additional classes and methods of construction may be found in the articles of Ma (2003a,b), Porcu *et al.* (2006, 2007a, 2008), Fuentes *et al.* (2008), De Iaco & Posa (2013), Shand & Li (2017), Ip & Li (2017), the book of Montero *et al.* (2015), and recent reviews by Porcu *et al.* (2020a) and Chen *et al.* (2021).

Models for areal space-time data have also received much attention. Natural extensions of the spatial SAR, CAR, and SMA models described in Chapter 7 can be constructed using the same model equations used to construct those models, but with "nearby" site-times designated as neighbors. For example, a spatio-temporal extension of the SAR model may be defined as

$$Y(A_i, t) = \mathbf{x}_{it}^T \boldsymbol{\beta} + \sum_{j \in \mathcal{N}_i, t' \in \mathcal{P}_t} b_{it,jt'}(Y(A_j, t') - \mathbf{x}_{jt'}^T \boldsymbol{\beta}) + d_{it} \quad (i = 1, \ldots, n_S; \, t = 1, \ldots, n_T),$$

where (A_i, t) is an arbitrary site-time, $\mathbf{x}_{it}$ is a vector of explanatory variables associated with that site-time, $\mathcal{N}_i$ comprises (as before) the spatial neighbors of A_i, $\mathcal{P}_t$ comprises the recent temporal predecessors of time t (for example, $t-1$ and $t-2$), $b_{it,jt'}$ is a measure of the spatio-temporal proximity of (A_i, t) and (A_j, t'), and the joint distribution of the d_{it}'s is multivariate normal with mean vector zero and positive definite diagonal covariance matrix. (Here, and throughout this discussion, we assume that spatial neighbors are time-invariant and that times are equally spaced.) As in the purely spatial case, the model is highly overparameterized, and simplification of the model will be required to reduce the number of parameters. Here, rather than the single multiplicative-parameter models used for spatial data, a model with several multiplicative parameters, one for each of the distinct types of neighboring site-times, might make more sense. For example, one parameter might be used for the contemporaneous site-times of (A_i, t), defined as $\{(A_j, t) : j \in \mathcal{N}_i\}$, while another might be used for its site-specific recent predecessors, i.e., $\{(A_i, t') : t' \in \mathcal{P}_t\}$, and still another for the recent predecessors at the neighbors of A_i. Furthermore, in keeping with longstanding practice in modeling time series data, a different parameter might be associated with each lag $1, 2, \ldots, p$ of recent temporal predecessor, yielding the following spatio-temporal SAR model:

$$
\begin{aligned}
Y(A_i, t) &= \mathbf{x}_{it}^T \boldsymbol{\beta} + \sum_{j \in \mathcal{N}_i} \rho_{\text{SAR},0} w_{ij}(Y(A_j, t) - \mathbf{x}_{jt}^T \boldsymbol{\beta}) + \sum_{l=1}^{p} \rho_{\text{SAR},l}(Y(A_i, t-l) - \mathbf{x}_{i,t-l}^T \boldsymbol{\beta}) \\
&\quad + \sum_{l=1}^{p} \sum_{j \in \mathcal{N}_i} \rho_{\text{SAR},p+l} w_{ij}(Y(A_j, t-l) - \mathbf{x}_{j,t-l}^T \boldsymbol{\beta}) + d_{it}, \\
&\quad (i = 1, \ldots, n_S; \ t = 1, \ldots, n_T).
\end{aligned}
$$

Here, as in Chapter 7, w_{ij} is the (i, j)th element of the spatial proximity matrix $\mathbf{W}$.

Spatio-temporal CAR and MA models may be defined analogously. In fact, the CAR alternative to the SAR model we have just described is essentially the same as the STCAR (spatio-temporal conditional auto-regressive) model of Mariella & Tarantino (2010). If dependence on contemporaneous site-times is removed, the model can be viewed as a case of a well-studied multivariate time series model called the pth-order vector autoregressive, or VAR(p), model. (The observations taken over the n_S sites at time t form the vector in this setting.) A general version of the VAR(p) model for spatio-temporal data is given by

$$
\mathbf{Y}_t = \mathbf{X}_t \boldsymbol{\beta} + \sum_{l=1}^{p} \mathbf{B}_l (\mathbf{Y}_{t-l} - \mathbf{X}_l \boldsymbol{\beta}) + \mathbf{d}_t,
$$

where

$$
\mathbf{Y}_t = \begin{pmatrix} Y(A_1, t) \\ \vdots \\ Y(A_{n_S}, t) \end{pmatrix}, \quad \mathbf{X}_t = \begin{pmatrix} \mathbf{x}_{1t}^T \\ \vdots \\ \mathbf{x}_{n_S t}^T \end{pmatrix}, \quad \mathbf{d}_t = \begin{pmatrix} d_{1t} \\ \vdots \\ d_{n_S, t} \end{pmatrix},
$$

and $\mathbf{B}$ is a matrix of n_S^2 (in general) autoregressive coefficients, which again may be given a parsimonious parameterization. For various special cases and extensions of this model, we refer the reader to Sections 6.4 and 7.2 of Cressie & Wikle (2011).

Most models that have actually been fitted to spatiotemporal areal data are hierarchical, with the spatio-temporal dependence accounted for at the second stage by a spatial CAR (often improper) model that evolves over time. The data are most often count data (usually disease counts), for which the likelihood function is either Poisson or binomial, and inference for the models has been almost exclusively Bayesian. These models and methodologies lie outside the scope of this book, and because the literature associated with them is rather voluminous, we will not attempt to list any of it here.

12.3.2 Exploratory spatio-temporal data analysis

Due to the additional dimension, exploratory data analyses, especially those involving graphics, are more challenging for spatio-temporal data than they are for spatial data. Unless the spatial locations lie on a one-dimensional transect, the domain of the data is at least three-dimensional, leaving no dimension available for plotting the response. **Marginal plots**, i.e., plots of the spatial data at each of several discrete snapshots of time, or time series plots at each of several spatial locations, can help the data analyst to visualize the data and explore large-scale variation. The former is particularly informative when the plots are dynamic, i.e., animated continuously over time. None of these plots are very useful, however, when the data are severely non-rectangular, unless some type of interpolation is used to create pseudo-data at the "missing nodes" of a rectangular configuration of locations and times.

Methods described in Chapter 3 for detecting spatial outliers and exploring spatial variability in the variance may be extended to spatio-temporal data in natural ways. A **spatio-temporal outlier** is an observation that is unusual in comparison to its spatial neighbors, its temporal neighbors (i.e. the preceding and subsequent observations at the same site), or a combination of both. Nearest-neighbor scatterplots at each time point and time series plots at each location will identify the first two types of outliers; for identifying an outlier of the last type, nearest-neighbor scatterplots of the *difference* in responses at each pair of consecutive observation times, rather than of the responses themselves, is a useful graphical diagnostic.

For exploring spatio-temporal correlation, if the data are rectangular one may construct the **lag-u spatial sample autocovariance matrices**

$$\hat{\mathbf{C}}_u = \frac{1}{n_T - u} \sum_{t=u+1}^{n_T} (\mathbf{y}_t - \hat{\boldsymbol{\mu}})(\mathbf{y}_{t-u} - \hat{\boldsymbol{\mu}})^T \quad (u = 0, 1, \ldots, n_T - 1),$$

where $\mathbf{y}_t$ is the vector of n_S observations taken over the spatial domain at time t and

$$\hat{\boldsymbol{\mu}} = \frac{1}{n_T} \sum_{t=1}^{n_T} \mathbf{y}_t.$$

The lag-u spatial sample correlation matrix for each u can be obtained from $\hat{\mathbf{C}}_u$ by standard manipulations. The matrices may be plotted in space as an image plot to enhance interpretation. Note that these matrices do not impose spatial or temporal stationarity, but they do average over the observation times. Analogously defined lag-**s** temporal sample autocovariance matrices, which likewise do not impose stationarity but average over the observation locations, may also be examined.

If spatio-temporal stationarity is assumed, then the **space-time sample semivariogram** and **space-time sample autocovariance function** may be defined, as follows:

$$\hat{\gamma}(\mathbf{h}_k, u_l) = \frac{1}{2N(\mathbf{h}_k, u_l)} \sum_{\mathbf{s}_i - \mathbf{s}_j \in H_k, t_i - t_j \in U_l} (y_i - y_j)^2,$$

$$\hat{C}(\mathbf{h}_k, u_l) = \frac{1}{N(\mathbf{h}_k, u_l)} \sum_{\mathbf{s}_i - \mathbf{s}_j \in H_k, t_i - t_j \in U_l} (y_i - \bar{y})(y_j - \bar{y}) \quad (k = 1, \ldots, K; l = 1, \ldots, L),$$

where y_i is the observed value of $Y(\mathbf{s}_i, t_i)$ and the meaning of the remaining notation can be surmised from the definitions of the purely spatial sample semivariogram and covariance function given in Chapter 3. The data need not be rectangular. If, in addition, spatial isotropy is assumed, then this function can be plotted as an image plot or contour plot in two dimensions, with spatial lag $\|\mathbf{h}\|$ on one axis and temporal lag u on the other. A lack of parallelism in $\{\hat{\gamma}(\mathbf{h}_k, u_l) : k = 1, \ldots, K\}$ and $\{\hat{\gamma}(\mathbf{h}_k, u_{l'}) : k = 1, \ldots, K\}$ (or its sample autocovariance counterpart) for any fixed pair (l, l'), or in $\{\hat{\gamma}(\mathbf{h}_k, u_l) : l = 1, \ldots, L\}$ and $\{\hat{\gamma}(\mathbf{h}_{k'}, u_l) : l = 1, \ldots, L\}$ (or its sample autocovariance counterpart) for any fixed pair (k, k') may be taken as an indication of nonadditivity. Similarly, a lack of proportionality in those quantities may be taken as evidence of nonseparability. Also, a consistent positive or negative sign of $\hat{C}(\mathbf{h}, u)/\{\hat{C}(\mathbf{h}, 0)\hat{C}(\mathbf{0}, u)\}$ among those lags whose sample autocovariances are not close to zero indicates that the covariance function is positively or negatively nonseparable. Additional attention may be given to the marginal cases of these functions, i.e., $\hat{C}(\mathbf{h}, 0)$ and $\hat{C}(\mathbf{0}, u)$. Their behavior at the origin (e.g., their continuity or lack thereof) and toward infinity (e.g., whether they vanish), their convexity or concavity, and other features can guide the analyst to select an appropriate, physically meaningful class of spatio-temporal covariance functions to fit to the data (De Iaco, 2010; De Iaco *et al.*, 2013, 2016). Furthermore, any obvious disagreements between the marginal sample autocovariance functions and those estimated by a likelihood-based approach may indicate model misspecification (Stein, 2005b) or problems with the implementation of the likelihood-based approach. Space-time sample semivariograms and covariance functions may be computed and plotted using `gstat` in concert with `spacetime` (Pebesma, 2012; Pebesma & Gräler, 2023). Some additional graphical tools for visualizing and assessing geometric properties of the covariance function were introduced by Huang & Sun (2019).

Tests of Lu-Zimmerman type for testing for geometric and other structural characteristics of space-time covariance functions (e.g., full symmetry and separability) were proposed by Li *et al.* (2007), and have been modified for testing for positive or negative nonseparability by Capello *et al.* (2018). Further variations for testing for certain classes of covariance functions (e.g., product-sum and Gneiting) were also introduced by Capello *et al.* (2018). The `covatest` package in R (Capello *et al.*, 2020) implements these tests.

For exploring spatio-temporal data with areal spatial support, extensions of Moran's I and LISA statistics are available; see Hardisty & Klippel (2010).

12.3.3 Modeling the mean function

For the mean function, polynomial functions of the spatial and temporal coordinates could be used. Such functions could be **additive** with respect to space and time, i.e.,

$$\mu(\mathbf{s}, t) = \mu_S(\mathbf{s}) + \mu_T(t),$$

or not. As an example of the latter, a full (non-additive) second-order space-time polynomial mean function (in two-dimensional space, with $\mathbf{s} = (u, v)^T$) is

$$
\begin{aligned}
\mu(\mathbf{s}, t) = {} & \beta_{000} + \beta_{100}u + \beta_{010}v + \beta_{001}t + \beta_{200}u^2 + \beta_{020}v^2 \\
& + \beta_{002}t^2 + \beta_{110}uv + \beta_{101}ut + \beta_{011}vt.
\end{aligned}
$$

An additive second-order polynomial mean function would exclude the last two terms; the remaining two terms involving t characterize the temporal trend and would likely be of great interest in many applications.

Another feature to consider for the spatio-temporal mean function is **temporal periodicity** (e.g. seasonality). Usually this can be dealt with effectively by using a mean model that includes sine and cosine terms, akin to that described for modeling spatial periodicity in Section 4.4.3. For example, an additive case of such a model, with a second-order polynomial spatial trend and first-order polynomial and periodic temporal trend, is

$$
\begin{aligned}
\mu(\mathbf{s}, t) = {} & \beta_{00} + \beta_{10}u + \beta_{01}v + \beta_{20}u^2 + \beta_{11}uv + \beta_{02}v^2 \\
& + \gamma_0 t + \gamma_1 \cos(2\pi\gamma_3 t) + \gamma_2 \sin(2\pi\gamma_3 t).
\end{aligned}
$$

Here γ_1 and γ_2 are related to the amplitude and phase angle of the sine-cosine type of periodicity, and γ_3 is the reciprocal of the period. In some settings, the period is known (e.g., 24 hours or 12 months), in which case the mean structure is linear in its parameters and can be estimated by (generalized) least squares. If γ_3 is unknown, however, this model is nonlinear in the parameters and the fitting method must be extended to nonlinear (generalized) least squares.

For additional models for spatio-temporal mean structure, including some that allow for interaction between spatial trend and temporal periodicity, see Dimitrakopoulos & Luo (1997).

12.3.4   Inference

Likelihood-based estimation and prediction for spatio-temporal models may be carried out in a very similar manner as for spatial models, with no additional conceptual difficulty. One difference is that spatio-temporal prediction is usually (but not always) interpolation in space and (forward) extrapolation in time. That is, the predictand of interest is usually $Y(S_0, t_0)$, where S_0 lies within the convex hull of data locations (in fact, it is often a data location) but $t_0 > t_{n_T}$. Three consequences of this are: (1) the prediction error variance tends to be larger than it is for spatial interpolation at one of the observation times; (2) it is very important to appropriately model temporal trend; (3) for areal data, spatio-temporal prediction is much more relevant than purely spatial prediction. Another difference is that there is an additional computational burden with spatio-temporal prediction, due to required inversion of the much larger space-time covariance matrix. A substantial reduction in computation is possible in the case of a separable covariance function fitted to rectangular data because in this case the covariance matrix has Kronecker product form, for which (it may be recalled from Section 5.6) the inverse may be obtained very efficiently. But this case is quite restrictive. Some efficiencies are also possible for the more general case of a product-sum covariance function, even when the data are nonrectangular (Dumelle *et al.*, 2021). To make inference feasible when the model is not of this form and the dataset is large, some of the same strategies described in Section 8.7 (data partitioning, composite likelihood, covariance tapering, etc.) may be employed.

Regarding the choice of an asymptotic regime for parameter estimation and prediction within spatio-temporal models, there does not appear to be a consensus. By treating time as an additional dimension, it is clear that the conditions specified by Mardia & Marshall (1984), which guarantee that the ML estimator of the parameters of a spatial process is consistent and asymptotically normal under increasing-domain asymptotics, can be extended without difficulty to the ML estimator of the parameters of a spatio-temporal process under an asymptotic regime in which both the spatial and temporal domains are increasing. Bevilacqua *et al.* (2012) appealed to this type of asymptotic regime for (composite) likelihood-based parameter estimation of several spatio-temporal models. In contrast, Ip & Li (2017) adopted an asymptotic regime of fixed-domain type in both space and time to study the consistent estimability of the parameters of a general Matérn class of spatio-temporal covariance functions. Perhaps the most appropriate regime would be fixed-domain in space and increasing-domain in time, but curiously this appears to have not yet been employed. Though it is to be expected that all parameters will be consistently estimable (subject to appropriate mixing conditions) in this case, different asymptotic distributions may be expected in comparison to the increasing-spatial-domain case.

Several R packages support inference for spatio-temporal models for geostatistical data. Package `CompRandFld` can fit a wide variety of spatio-temporal covariance structures, but requires the data to be rectangular and the mean to be constant; furthermore, it estimates parameters only by weighted least squares and composite likelihood methods. Package `gstat`, in concert with package `spacetime`, does not require rectangularity and supports the estimation of a linear mean structure, but the spatio-temporal covariance structures it can fit are limited to the metric, separable, and product-sum models (and a few minor variations thereof), and parameter estimation of those structures is by weighted least squares only. Perhaps the most comprehensive package for implementing inference for spatio-temporal models is `RandomFields`. Like `gstat` it allows the data to be nonrectangular and to have a linear mean structure, but like `CompRandFld` it can estimate many spatio-temporal covariance structures. Furthermore, it can estimate those structures by weighted least squares and maximum likelihood methods. A detailed review of these packages was given by Network *et al.* (2017).

R packages with functionality for implementing frequentist inference for spatio-temporal models for areal data apparently do not exist.

12.3.5 Spatio-temporal sampling design

Environmental monitoring studies typically occur in space *and* time, and often have as one of their aims the detection of change or trend in Y over time. Consequently, sampling design for such studies may have a temporal component in addition to a spatial component. That said, for some studies temporal design need not be considered. For example, if the monitoring instruments measure more or less continuously in time (e.g., water temperature loggers in streams) or aggregate over time (air filters), or if the nature of Y or the use to which it will be put demands that it be sampled at specific times (e.g. annual April 1 snow water equivalent measurements in the western U.S. mountain snowpack used to predict summer water supplies), then temporal design is not relevant. Temporal design is potentially useful only when measurements represent values of Y on a sparsely sampled portion of the time domain $\mathcal{T}$ to which one desires to make inferences.

Designs for which the temporal component is given consideration may be classified as either **static** or **dynamic**. In a static design, the sites do not change over time; that is, the data from a static design are rectangular. A static design may be appropriate when measuring stations are expensive to set up and cannot be moved easily and the cost of sampling and measurement at those stations is so low that it makes little sense to not sample at all of them whenever one is sampled. Furthermore, because it yields rectangular data (barring any missing data), a static design facilitates several exploratory and diagnostic procedures, as noted previously in Section 12.3.2. Although a static design generally will be suboptimal, optimizing a design criterion over the class of static designs will usually result in little loss of efficiency (Li & Zimmerman, 2015). Analogous to the purely spatial sampling design approaches reviewed previously in Chapter 10, model-based, probability-based, and "space-time filling" approaches may be taken to choose the sampling sites and times of a static design. Of these, model-based designs seem to be much less common. Perhaps the most common choice of times is simply regular time points $1, 2, \ldots, n_T$ (in appropriate time-scale units), due partly to its time-filling properties and partly to the sheer convenience of a constant sampling interval.

If a spatio-temporal design is not static, it is dynamic. A class of dynamic designs for environmental monitoring that has received much attention is the class of **panel** designs (Urquhart *et al.*, 1993), also called revisit designs (McDonald, 2003). Here, a "panel" refers to a set of sites that are sampled contemporaneously; an extreme case is a panel consisting of a single site. Two important types of panel designs are **rotating panel designs** and **serially alternating designs**. For concreteness, suppose that the basic sampling interval is a year. Then, a rotating panel design prescribes that each panel will be sampled in each of several consecutive years and subsequently removed permanently from future consideration; as each panel is removed, another is started. Table 12.4 shows a rotating panel design in which three panels are sampled in each of six years. In contrast, a serially alternating design prescribes that each panel is sampled in every rth year, where r is an integer. Table 12.5

TABLE 12.4

An example of an augmented rotating panel design. An asterisk in a column indicates that the panel is sampled in the year given by the column label.

Panel	Year					
	1	**2**	**3**	**4**	**5**	**6**
1	*					
2	*	*				
3	*	*	*			
4		*	*	*		
5			*	*	*	
6				*	*	*
7					*	*
8						*
Augmented	*	*	*	*	*	*

displays a serially alternating design in which there are four panels, each of which is sampled every fourth year, for a total sampling period of 12 years.

Both types of panel designs just described may be augmented with a panel that is sampled every year. The designs in Tables 12.4 and 12.5, with the bottom portions included, are an augmented rotating panel design and an augmented serially alternating design, respectively. Augmentation improves the designs for certain purposes; for example, without augmentation, the estimation of a change in Y from one year to the next is nonestimable in the serially alternating design of Table 12.5. Urquhart *et al.* (1993) recommended allocating 10-20% of the sites to the augmented panel for best performance of either panel design. Urquhart & Kincaid (1999) showed that an augmented serially alternating design has greater power to detect a trend than an augmented rotating panel design with the same total number of sites, but that the latter estimates current "status" (the mean of Y over space at a particular time) more precisely than the former.

How should the panels of sites be chosen? In light of results from Chapter 10, it should come as no surprise that sites within a panel should be spatially balanced for good estimation of status and trend. Wang & Zhu (2019) proposed a design algorithm that results in a augmented panel design that is spatially balanced in each panel and also in each pair of consecutive panels, and they showed that such panels result in a much more efficient design than panels obtained by simple random sampling.

TABLE 12.5

An example of an augmented serially alternating panel design. An asterisk in a column indicates that the panel is sampled in the year given by the column label.

Panel	Year											
	1	**2**	**3**	**4**	**5**	**6**	**7**	**8**	**9**	**10**	**11**	**12**
1	*				*				*			
2		*				*				*		
3			*				*				*	
4				*				*				*
Augmented	*	*	*	*	*	*	*	*	*	*	*	*

Rotating and serially alternating panel designs require that every site be sampled in multiple years. In contrast, a fully dynamic design allows each site to be sampled only once. Wikle & Royle (1999) considered fully dynamic model-based design for monitoring spatiotemporal processes. For a relatively simple Gaussian separable model with first-order Markovian temporal dependence, they found that the degree to which the average prediction error variance at time t is reduced by a dynamic design, from what it is for a static design, is largest when the temporal dependence is strong. Furthermore, in that circumstance, the spatial locations of the optimal dynamic design for spatiotemporal prediction at time t are typically quite far removed from those at time $t - 1$. This is intuitively reasonable, since $Y(\mathbf{s}, t-1)$ is strongly informative of $Y(\mathbf{s}^*, t)$ when $\mathbf{s}^*$ is at or near $\mathbf{s}$ and the temporal dependence is strong. Not surprisingly, when the spatiotemporal process has more complicated dynamics or is nonseparable or non-Gaussian, the dynamic prediction-optimal design can be more complicated, even nonintuitive, but is still more efficient than the optimal static design (Wikle & Royle, 1999, 2005).

12.4 Multivariate spatial data

In many environmental investigations, monitors distributed over space collect information simultaneously (more or less) on not one but multiple response variables. For example, temperature, solar radiation, wind speed, and precipitation may be measured at weather stations, or concentrations of various pollutants (e.g. ozone, sulfur dioxide, ammonium) may be measured at air monitors. The methodology described in this book may be extended to build spatial linear models for multivariate spatial data of this type, and to make inferences about those models from the data. The notation for multivariate spatial data is an extension of that defined previously for univariate spatial data. Let q represent the number of response variables, let y_{li} represent the ith observation $(i = 1, \ldots, n_l)$ on the lth response variable $(l = 1, \ldots, q)$, and let S_{li} represent the site where that observation was taken. The site may be alternatively represented as $\mathbf{s}_{li}$ or A_{li}, according to whether the data are geostatistical or areal. Explanatory variables measured at that site, if any, are collected into a vector denoted by $\mathbf{x}_{li}$; these may include, but are not limited to, the spatial coordinates or functions thereof.

Three noteworthy situations regarding the variables and sites have special names. First, if all q response variables are expressed in the same units, the variables are said to be **commensurate**. In practice, commensurate variables tend to have similar variabilities, whereas incommensurate variables may need to be rescaled (to different units) or rendered unitless to make their variabilities comparable. Furthermore, only commensurate variables may be added or subtracted from one another unambiguously to create new variables that have the same units as the original variables. Second, if all q response variables are observed at each data location, we say that the data are **collocated**. Collocation is entirely analogous to rectangularity for space-time data, with multiplicity of variables playing the role of repeated observations over time. Collocation is often a practically and economically advantageous sampling strategy relative to taking the same total number of observations with some variables observed only at some sites and other variables only at other sites. The extreme opposite of collocation occurs when only one variable is observed at each site. Third, **balance** refers to a situation in which the number of sites at which one variable is observed is equal to the number of sites at which each of the other variables is observed. Departures from balance may be practically necessary when one variable is much more expensive to collect or measure than another. Collocation implies, but is not required for, balance. When the data are balanced, $n_1 = n_2 = \cdots = n_q \equiv n$, say. When in addition the data are collocated, the sites can be represented using a single subscript, as for univariate data, by $S_1, \ldots, S_n$.

12.4.1 Multivariate spatial dependence

Spatial linear models for multivariate geostatistical data are based on multivariate Gaussian spatial processes. A **q-dimensional Gaussian spatial process** $\{\mathbf{Y}(\mathbf{s}) : \mathbf{s} \in \mathcal{A}\}$ is a collection of q-dimensional random vectors indexed by points in a region $\mathcal{A} \subset \mathbb{R}^d$ such that the joint distribution of any finite collection of n of the vectors is qn-variate normal. Such a process is completely characterized by its vector-valued mean function $\boldsymbol{\mu}(\mathbf{s}) \equiv \mathrm{E}(\mathbf{Y}(\mathbf{s}))$ and its matrix-valued multivariate spatial covariance function $\mathbf{C}(\mathbf{s}, \mathbf{s}') \equiv \mathrm{Cov}(\mathbf{Y}(\mathbf{s}), \mathbf{Y}(\mathbf{s}'))$. For multivariate spatial *linear* models, each functional element of $\boldsymbol{\mu}(\mathbf{s})$ is taken to be a linear function of unknown parameters; more details on this will be given subsequently. The multivariate spatial covariance function, which is composed of real-valued marginal covariance functions $C_{ll}(\mathbf{s}, \mathbf{s}') \equiv \mathrm{Cov}(Y_l(\mathbf{s}), Y_l(\mathbf{s}'))$ $(l = 1, \ldots, q)$ and real-valued cross-covariance functions $C_{lm}(\mathbf{s}, \mathbf{s}') \equiv \mathrm{Cov}(Y_l(\mathbf{s}), Y_m(\mathbf{s}'))$ $(l \neq m = 1, \ldots, q)$, is necessarily positive definite, i.e.,

$$
\mathrm{Var} \begin{pmatrix} \mathbf{Y}(\mathbf{s}_1) \\ \mathbf{Y}(\mathbf{s}_2) \\ \vdots \\ \mathbf{Y}(\mathbf{s}_n) \end{pmatrix} = \begin{pmatrix} \mathbf{C}(\mathbf{s}_1, \mathbf{s}_1) & \mathbf{C}(\mathbf{s}_1, \mathbf{s}_2) & \cdots & \mathbf{C}(\mathbf{s}_1, \mathbf{s}_n) \\ \mathbf{C}(\mathbf{s}_2, \mathbf{s}_1) & \mathbf{C}(\mathbf{s}_2, \mathbf{s}_2) & \cdots & \mathbf{C}(\mathbf{s}_2, \mathbf{s}_n) \\ \vdots & \vdots & \ddots & \vdots \\ \mathbf{C}(\mathbf{s}_n, \mathbf{s}_1) & \mathbf{C}(\mathbf{s}_n, \mathbf{s}_2) & \cdots & \mathbf{C}(\mathbf{s}_n, \mathbf{s}_n) \end{pmatrix}
$$

must be a nonnegative definite matrix for all positive integers n and all locations $\mathbf{s}_1, \ldots, \mathbf{s}_n \in \mathcal{A}$. This matrix must also be symmetric, i.e., $C_{lm}(\mathbf{s}_i, \mathbf{s}_j) = C_{ml}(\mathbf{s}_j, \mathbf{s}_i)$ for all $l, m = 1, \ldots, q$, all positive integer n, and all $\mathbf{s}_1, \ldots, \mathbf{s}_n \in \mathcal{A}$, but this does not imply that $C_{lm}(\mathbf{s}, \mathbf{s}') = C_{lm}(\mathbf{s}', \mathbf{s})$ for any $l \neq m$ and $\mathbf{s} \neq \mathbf{s}'$. However, if this last equality does hold for all $\mathbf{s}, \mathbf{s}' \in \mathcal{A}$ and all l and m, then $\mathbf{C}(\mathbf{s}, \mathbf{s}')$ is said to be **symmetric**. An equivalent definition of symmetry is that $C_{lm}(\mathbf{s}, \mathbf{s}') = C_{ml}(\mathbf{s}, \mathbf{s}')$ for all l, m and all $\mathbf{s}, \mathbf{s}' \in \mathcal{A}$. Stationarity and isotropy of the multivariate covariance function are defined just as would be expected, i.e., as $\mathbf{C}(\mathbf{s}, \mathbf{s}') = \mathbf{C}(\mathbf{s} - \mathbf{s}')$ and $\mathbf{C}(\mathbf{s}, \mathbf{s}') = \mathbf{C}(\|\mathbf{s} - \mathbf{s}'\|)$, respectively.

Satisfying the multivariate covariance function's positive definiteness requirement is the primary challenge of building models for multivariate geostatistical data. In particular, despite the plethora of symmetric positive definite functions available to serve as marginal spatial covariance functions for the individual variables, many of which were described in Chapter 6, it is not obvious where to find functions that can serve as valid models for the between-variable spatial dependence, nor is it clear how those functions may be combined with each other and with the marginal covariance functions to yield a valid multivariate spatial model for $\mathbf{Y}$. A large number of strategies have been proposed for meeting this challenge, and we give only a brief overview of some of them here. More thorough reviews may be found in Genton & Kleiber (2015) and Gelfand (2021).

One very simple approach to constructing a multivariate covariance function is to suppose that

$$
\mathbf{C}(\mathbf{s}, \mathbf{s}') = \rho(\mathbf{s}, \mathbf{s}') \cdot \mathbf{A} \tag{12.6}
$$

for some univariate spatial correlation function $\rho(\cdot, \cdot)$ and some $q \times q$ symmetric nonnegative definite matrix $\mathbf{A} = (a_{lm})$. This type of multivariate covariance structure, which is known as **intrinsic coregionalization** (Matheron, 1982), separates the between-variable covariance structure from the spatial correlation structure. As one consequence, $\mathbf{C}(\mathbf{s}, \mathbf{s}')$ is symmetric. An explicit model for $\mathbf{Y}(\mathbf{s})$ that yields (12.6) is

$$
\mathbf{Y}(\mathbf{s}) = \boldsymbol{\mu}(\mathbf{s}) + \mathbf{A}^{1/2}\mathbf{w}(\mathbf{s}), \tag{12.7}
$$

where $\mathbf{A}^{1/2}$ is the Cholesky square root of $\mathbf{A}$ and $\mathbf{w}(\mathbf{s})$ is a q-variate Gaussian process with independent and identically distributed univariate component processes $w_l(\mathbf{s})$ having zero means, unit variances, and common spatial correlation function $\rho(\mathbf{s}, \mathbf{s}')$.

Intrinsic coregionalization has advantages and disadvantages. In addition to its conceptual simplicity, it is convenient computationally, as will be shown subsequently. However, it is very restrictive with respect to the properties of the variables' spatial dependence. Because intrinsic coregionalization specifies the joint correlation structure of all variables in terms of only one correlation function, it imposes equality on the correlation ranges (or practical ranges) of all variables, and also on their cross-correlation ranges. Similarly, it imposes equality of smoothnesses across variables. As it stands in (12.6), intrinsic coregionalization specifies a common nugget effect for all variables, but it may be extended to allow for variable-specific nugget effect variances by supposing that

$$\mathbf{C}(\mathbf{s}, \mathbf{s}') = \rho(\mathbf{s}, \mathbf{s}') \cdot \mathbf{A} + \mathrm{diag}(\nu_1^2, \ldots, \nu_q^2).$$

A less restrictive form of the multivariate covariance function arises by assuming a model for $\mathbf{Y}(\mathbf{s})$ that has the same model equation as (12.7), where the component processes $w_l(\mathbf{s})$ likewise have zero means and unit variances but have variable-specific correlation functions $\rho_1(\mathbf{s}, \mathbf{s}'), \ldots, \rho_q(\mathbf{s}, \mathbf{s}')$. Under such a model,

$$\mathbf{C}(\mathbf{s}, \mathbf{s}') = \sum_{l=1}^{q} \rho_l(\mathbf{s}, \mathbf{s}') \mathbf{A}_l,$$

where $\mathbf{A}_l = \mathbf{a}_l \mathbf{a}_l^T$ with $\mathbf{a}_l$ the lth column of $\mathbf{A}^{1/2}$. This form is known as the **linear model of coregionalization** (Wackernagel, 1998). If the variable-specific correlation functions are stationary and/or isotropic, then so is the multivariate covariance function. One of this form's nice features in the isotropic case is that it allows each variable to have its own correlation range and smoothness, though the range and smoothness parameters of the lth variable are not necessarily those of $\rho_l(\cdot, \cdot)$. The cost of this greater flexibility is a loss of computational efficiency. Variable-specific nugget effect variances can be incorporated in the same fashion as they are for intrinsic coregionalization.

Two other approaches to constructing stationary multivariate spatial covariance functions from univariate covariance functions use convolution techniques. The moving average approach (Ver Hoef & Barry, 1998; Ver Hoef *et al.*, 2004) is an extension of the auto-convolution method described in Section 6.1. This approach begins with the creation of a white noise process from two other white noise processes,

$$W_l(\mathbf{u}) = \sqrt{1 - \rho_l^2} B_l(\mathbf{u}) + \rho_l B_0(\mathbf{u} - \boldsymbol{\Delta}_l),$$

where $\mathrm{E}[B_l(\mathbf{u})] = \mathbf{0}$, $\mathrm{Var}[\int_A B_l(\mathbf{u}) d\mathbf{u}] = |A|$, $\mathrm{Cov}[\int_A B_l(\mathbf{u}) d\mathbf{u}, \int_{A'} B_l(\mathbf{u}) d\mathbf{u}] = \mathbf{0}$ if $A \cap A' = \varnothing$, and $B_l(\mathbf{u})$ is independent of $B_m(\mathbf{u})$ if $l \neq m$. The parameter $\boldsymbol{\Delta}$ controls a shift in the $B_0(\mathbf{u})$ white noise process. Under this construction, $W_l(\mathbf{u})$ and $W_m(\mathbf{u})$ are cross-correlated because they share the $B_0(\mathbf{u})$ white noise process. Next, a random variable is defined as in Section 6.1:

$$e_l(\mathbf{s}) = \int_{\mathbb{R}^d} g_l(\mathbf{u} - \mathbf{s}) W_l(\mathbf{u}) d\mathbf{u},$$

where the $g_l(\cdot)$'s are square-integrable real-valued functions on $\mathbb{R}^d$. Then

$$C_{lm}(\mathbf{h}) = \mathrm{Cov}[e_l(\mathbf{s}), e_m(\mathbf{s} + \mathbf{h})] = \rho_l \rho_m \int_{\mathbb{R}^d} g_l(\mathbf{u}) g_m(\mathbf{u} - \mathbf{h} + \boldsymbol{\Delta}_l - \boldsymbol{\Delta}_m) d\mathbf{u}, \qquad (12.8)$$

and (12.8) reduces to (6.5) when $l = m$. Note that, in contrast to the coregionalization model, the functional form of $C_{lm}(\mathbf{h})$ is different from $C_{ll}(\mathbf{h})$ and $C_{mm}(\mathbf{h})$ and is not a linear combination of them. Additionally, $C_{lm}(\mathbf{h})$ can accommodate a spatial shift in cross-correlation through the $\mathbf{\Delta}$ parameters. In practice, one $\mathbf{\Delta}$ is set to $\mathbf{0}$, and then all shifts are relative to it. Also, part of the cross-correlation is controlled by ρ_l and ρ_m, which is how much of the $B_0(\mathbf{u})$ white noise process is shared between any two variables. If there are only two variables, only the product $\rho_1 \rho_2$ is identifiable. A distinct nugget effect for each variable may be added to the marginal covariance functions. Finally, $C_{lm}(\mathbf{h})$ is generally a covariance model for random errors, and mean structures may be added to each variable. A special case of (12.8), which occurs when $g_l(\cdot)$ and $g_m(\cdot)$ are covariance functions, is called the covariance convolution approach (Majumdar & Gelfand, 2007). These multivariate convolution methods generally require numerical integration, and one fast approach was given by Ver Hoef *et al.* (2004).

In the construction above, the shift $\mathbf{\Delta}$ creates an **asymmetric cross-correlation function**. In fact, an asymmetric version of any stationary multivariate covariance function $\{C_{lm}(\cdot)\}$ can be obtained by modifying the argument of those functions from $\mathbf{h}$ to $\mathbf{h} + \mathbf{\Delta}_l - \mathbf{\Delta}_m$, where $\sum_{i=1}^q \mathbf{\Delta}_i = \mathbf{0}$ (Li & Zhang, 2011).

Because of its popularity for modeling univariate spatial dependence, which is attributable to its flexibility with respect to modeling smoothness, the Matérn correlation function has received the most attention as a parametric model from which a multivariate correlation function may be built. Gneiting *et al.* (2010) introduced the multivariate Matérn model (and special cases thereof), for which each marginal correlation function is the usual Matérn function and the cross-correlation functions are Matérn correlation functions multiplied by a quantity specific to the variable pair, i.e.,

$$\rho_{ij}(\mathbf{h}; \boldsymbol{\theta}) = \begin{cases} M(\mathbf{h}; \alpha_i, \nu_i) & \text{if } i = j, \\ \gamma_{ij} M(\mathbf{h}; \alpha_{ij}, \nu_{ij}) & \text{if } i \neq j, \end{cases}$$

where $M(\mathbf{h}; \alpha, \nu)$ is the Matérn correlation function as defined in Chapter 6. This construction does not yield a valid correlation function in general, but it does so when certain restrictions are imposed upon the parameters. One valid special case is the **parsimonious Matérn** model, for which the correlation range parameters, both marginal and cross-, are all equal ($\alpha_i = \alpha_{ij} = \alpha$, say, for all i, j) and the cross-smoothnesses are equal to the average of the marginal smoothnesses ($\nu_{ij} = (\nu_i + \nu_j)/2$ for all i, j). While this case allows each component of the multivariate process to have its own smoothness, it is highly restrictive with respect to the correlation ranges. Another valid special case is the **full bivariate Matérn** model, which is simply the two-variable case of the multivariate Matérn model. In this model, the processes may have variable-specific marginal and pairwise cross-correlation ranges and smoothnesses. The **flexible Matérn** model, due to Apanasovich *et al.* (2012), overcomes the limitations of the parsimonious and bivariate Matérn models by allowing each of $q > 2$ variables to have its own range parameter and smoothness.

Several families of compactly supported stationary cross-covariance functions were provided by Porcu *et al.* (2013), Du & Ma (2013), and Daley *et al.* (2015). Many of these are stationary models derived from the construction

$$C_{lm}(\mathbf{h}) = \int \left(1 - \frac{\|\mathbf{h}\|}{x} \right)_+^\nu g_{lm}(x)\, dx,$$

where $\mathbf{h} \in \mathbb{R}^d$, $\nu \geq (d+1)/2$, and $\{g_{lm}\}_{l,m=1}^q$ comprise a valid multivariate covariance function. Others are obtained by the aforementioned auto-convolution approach with $g(\cdot)$ taken to be the indicator function of a compact set in $\mathbb{R}^d$.

Three more methods for constructing multivariate spatial covariance functions are based on either the creation of latent dimensions (Apanasovich & Genton, 2010), the use of basis functions at multiple scales of resolution (Kleiber *et al.*, 2019), or modeling joint dependence through conditioning (Cressie & Zammit-Mangion, 2016). The interested reader may consult the references just provided for descriptions of those methods.

The previous discussion concerning the characterization of multivariate spatial dependence for geostatistical data has focused on the multivariate covariance function. An alternate characterization may be given in terms of a multivariate version of the semivariogram, consisting of marginal and **cross-semivariograms**. Two types of cross-semivariogram have been proposed for measuring spatial dependence between variables. In chronological order of when they were proposed, they are the **covariance-based cross-semivariogram** (Myers, 1982),

$$\gamma_{lm}^{(c)}(\mathbf{s}, \mathbf{s}') = \frac{1}{2}\text{Cov}(Y_l(\mathbf{s}) - Y_l(\mathbf{s}'), Y_m(\mathbf{s}) - Y_m(\mathbf{s}')),$$

and the **variance-based cross-semivariogram** (Clark *et al.*, 1989; Myers, 1991),

$$\gamma_{lm}^{(v)}(\mathbf{s}, \mathbf{s}') = \frac{1}{2}\text{Var}(Y_l(\mathbf{s}) - Y_m(\mathbf{s}')).$$

Both cross-semivariograms reduce to the ordinary semivariogram when $l = m$. The covariance-based cross-semivariogram is an even function and may have any sign; its variance-based counterpart is generally not even but is necessarily nonnegative. Furthermore, $\gamma_{lm}^{(c)}$ may be expressed in terms of $\gamma_{lm}^{(v)}$, but not vice versa (Papritz *et al.*, 1993). Relationships of both to the cross-covariance function were elucidated by Ver Hoef & Cressie (1993). Under intrinsic stationarity, both cross-semivariograms may be written as functions of $\mathbf{s} - \mathbf{s}'$, and under isotropy they may be written as functions of $\|\mathbf{s} - \mathbf{s}'\|$. Because intrinsic stationarity is more general than second-order stationarity, the marginal and cross-semivariograms exist for some processes for which the multivariate covariance function does not exist. A still more general form of stationarity for multivariate processes is possessed by multivariate intrinsic random functions of order k, for which the spatial dependence is characterized by generalized covariance and generalized cross-covariance functions (Künsch *et al.*, 1997; Huang *et al.*, 2009).

It appears that the cross-covariance function is used more often than either cross-semivariogram for modeling between-variable spatial dependence; however, the cross-semivariograms may have some advantages for exploratory data analysis, as described in the next subsection. The variance-based cross-semivariogram has sometimes been called the "pseudo-cross semivariogram" but, as noted by Cressie & Wikle (1998), this is a misnomer. In fact, there are good reasons for preferring the variance-based cross-semivariogram over its covariance-based counterpart (Ver Hoef & Cressie, 1993; Cressie & Wikle, 1998), which are also described subsequently.

For multivariate areal data, the univariate SAR, CAR, and SMA models may be extended in various ways. Typically, and in contrast to the situation for spatio-temporal data, there is not a meaningful ordering among the variables, so it makes little sense to contemplate "variable weights" as analogues of spatial weights. However, it is reasonable to imagine that, as for multivariate geostatistical data, the value of a given response variable at a given site is related to the values of not only that response variable but other response variables at neighboring sites. For ease of exposition, assume that the data are collocated and likewise that all explanatory variables are measured at every data site. Then, for example, a sensible extension of the univariate areal SAR model is as follows:

$$\mathbf{y}_i = \overline{\boldsymbol{\beta}}\mathbf{x}_i + \sum_{j \in \mathcal{N}_i} \mathbf{B}_{ij}(\mathbf{y}_j - \overline{\boldsymbol{\beta}}\mathbf{x}_j) + \mathbf{d}_i \quad (i = 1, \ldots, n). \tag{12.9}$$

Here $\mathbf{y}_i$ and $\mathbf{x}_i$ are respectively the q-vector of response variables and p-vector of explanatory variables observed at site i, $\overline{\boldsymbol{\beta}}$ is a $q \times p$ matrix of regression coefficients whose lth row relates the mean of the lth response variable at site i to the explanatory variables at site i, $\mathbf{B}_{ij}$ is a $q \times q$ matrix of autoregressive coefficients whose lth row relates the lth response variable at site i to the q response variables at site j (with $\mathbf{B}_{ii} = \mathbf{0}$ for all i), and $\mathbf{d}_i$ is a vector of model residuals assumed to have a q-variate normal distribution with mean vector $\mathbf{0}$ and symmetric positive definite covariance matrix $\mathbf{K}_{ii}$, with $\mathbf{d}_i$ and $\mathbf{d}_k$ independent for $i \neq k$. Upon stacking the ith model equation in (12.9) on top of the $(i+1)$th model equation $(i = 1, \dots, n-1)$ and rearranging analogously to the univariate case, the model can be written as

$$\mathbf{Y} = (\mathbf{I}_n \otimes \overline{\boldsymbol{\beta}})\mathbf{x} + (\mathbf{I}_{nq} - \overline{\mathbf{B}})^{-1}\mathbf{D},$$

provided that $(\mathbf{I}_{nq} - \overline{\mathbf{B}})$ is nonsingular, where $\mathbf{Y}$ is the nq-vctor of stacked $\mathbf{y}_i$'s, $\mathbf{x}$ is the np-vector of stacked $\mathbf{x}_i$'s, and $\overline{\mathbf{B}}$ is the $nq \times nq$ block matrix whose (i,j)th block is $\mathbf{B}_{ij}$. The nq-vector $\mathbf{D}$ of stacked $\mathbf{d}_i$'s has mean vector $\mathbf{0}$ and covariance matrix $\mathbf{K}_{\text{MSAR}} = \text{blockdiag}(\mathbf{K}_{11}, \dots, \mathbf{K}_{nn})$. The joint distribution of $\mathbf{Y}$ implied by this model is multivariate normal with mean $(\mathbf{I}_n \otimes \overline{\boldsymbol{\beta}})\mathbf{x}$ and covariance matrix $(\mathbf{I}_{nq} - \overline{\mathbf{B}})^{-1}\mathbf{K}_{\text{MSAR}}(\mathbf{I}_{nq} - \overline{\mathbf{B}}^T)^{-1}$. In similar fashion, and with notation that is self-evident, the univariate CAR and SMA models may be extended to multivariate areal data, yielding joint distributions

$$\mathbf{Y} \sim \text{N}((\mathbf{I}_n \otimes \overline{\boldsymbol{\beta}})\mathbf{x}, (\mathbf{I}_{nq} - \overline{\mathbf{C}})^{-1}\mathbf{K}_{\text{MCAR}})$$

and

$$\mathbf{Y} \sim \text{N}((\mathbf{I}_n \otimes \overline{\boldsymbol{\beta}})\mathbf{x}, (\mathbf{I}_{nq} + \overline{\mathbf{M}})\mathbf{K}_{\text{MSMA}}(\mathbf{I}_{nq} + \overline{\mathbf{M}}^T)),$$

provided that $\mathbf{K}_{\text{MCAR}}^{-1}(\mathbf{I}_{nq} - \overline{\mathbf{C}})$ is symmetric, $\mathbf{I} - \overline{\mathbf{C}}$ is positive definite, $\mathbf{I}_{nq} + \overline{\mathbf{M}}$ is nonsingular, and the blocks $\{\mathbf{K}_{ii}\}$ of $\mathbf{K}_{\text{MCAR}}$ and $\mathbf{K}_{\text{MSMA}}$ are symmetric and positive definite.

As in the univariate situation, effective inference is possible only for parsimonious cases of the multivariate SAR, CAR, and SMA models. Since spatial cross-dependence can be incorporated via the spatial weights (the elements of $\overline{\mathbf{B}}$, $\overline{\mathbf{C}}$, and $\overline{\mathbf{M}}$), it may be reasonable to take the nonzero blocks of $\mathbf{K}_{\text{MSAR}}$, $\mathbf{K}_{\text{MCAR}}$, and $\mathbf{K}_{MSAM}$ to be diagonal matrices, but to allow for variable-specific variances. Thus one might take $\mathbf{K}_{ii} = \text{diag}(\sigma_1^2, \sigma_2^2, \dots, \sigma_q^2)$ for all i. Sensible parsimonious choices of the weights matrix might include various special cases of the Kronecker product form (illustrated for the case of a multivariate SAR) $\overline{\mathbf{B}} = \mathbf{W} \otimes \mathbf{A}$, where $\mathbf{W}$ is the $n \times n$ neighbor incidence matrix introduced in Chapter 2 and $\mathbf{A}$ is a $q \times q$ matrix whose (l,m)th element a_{lm} describes whether and how the mth variable affects the lth variable. Gelfand & Vounatsou (2003) considered the special case $\mathbf{A} = \rho\mathbf{I}$, which stipulates that the only variable that affects the lth variable on the ith site is the lth variable on the ith site's neighbors, i.e., there is no cross-weighting, and that the strength of that effect (ρ) is common across variables. They also considered the more general case $\mathbf{A} = \text{diag}(\rho_1, \dots, \rho_q)$, which allows the strength of the same-variable effect to vary across variables (though there is still no cross-weighting). Though not considered by Gelfand & Vounatsou (2003), a still more general case that allows for cross-weighting is $\mathbf{A} = \text{diag}(\rho_1, \dots, \rho_q) \odot [(1-\rho)\mathbf{I} + \rho\mathbf{J}]$ where $\odot$ denotes the Hadamard (elementwise) product; here ρ represents a common cross-weighting multiplier. There are many additional possibilities, some of which were considered by Gelfand & Vounatsou (2003) and Sain & Cressie (2007). It is worth emphasizing that the requirement that $\mathbf{K}_{\text{MCAR}}^{-1}(\mathbf{I}_{nq} - \overline{\mathbf{C}})$ be symmetric restricts these possibilities much more for the MCAR model than for the other two models.

12.4.2   Exploratory multivariate spatial data analysis

All of the tools used for ESDA of univariate geostatistical data may be applied to the variables of multivariate geostatistical data individually. For example, the univariate methods described in Sections 3.3 and 3.5 may be used to explore the large-scale variation of each of the q response variables and the spatial variability of their variances. Similarly, spatial outliers and within-variable spatial dependence may be explored using nearest-neighbor scatterplots and sample semivariograms, as detailed in Sections 3.4 and 3.6. And for exploring non-spatial dependence between variables Y_l and Y_m, standard tools of multivariate analysis, such as scatterplots of $Y_l(\mathbf{s}_k)$ versus $Y_m(\mathbf{s}_k)$ and the corresponding numerical summary, the sample correlation coefficient between these two variables, may be helpful, provided that Y_l and Y_m are collocated at a sufficient number of sites. For exploring the spatial dependence between variables, however, new tools are needed. For geostatistical data assumed to be stationary, two such tools are sample-based versions of the cross-covariance function and variance-based cross-semivariogram, defined as follows for $l, m = 1, \ldots, q$:

$$\hat{C}_{lm}(\mathbf{h}_k) = \frac{1}{N(\mathbf{h}_k)} \sum_{\mathbf{s}_{li} - \mathbf{s}_{mj} \in H_k} (y_l(\mathbf{s}_{li}) - \overline{y}_l)(y_m(\mathbf{s}_{mj}) - \overline{y}_m),$$

$$\hat{\gamma}_{lm}^{(v)}(\mathbf{h}_k) = \frac{1}{2N(\mathbf{h}_k)} \sum_{\mathbf{s}_{li} - \mathbf{s}_{mj} \in H_k} [(y_l(\mathbf{s}_{li}) - \overline{y}_l) - (y_m(\mathbf{s}_{mj}) - \overline{y}_m)]^2 \quad (k = 1, \ldots, K).$$

Here, as before, $\mathbf{h}_k$ is a representative lag within lag class H_k. In addition to providing a visualization of the spatial cross-dependence, these functions may be used to informally assess the plausibility of a symmetry assumption for the data, via a comparison of $\hat{C}_{lm}(\mathbf{h}_k)$ (or $\hat{\gamma}_{lm}^{(v)}(\mathbf{h}_k)$) to $\hat{C}_{ml}(\mathbf{h}_k)$ (or $\hat{\gamma}_{ml}^{(v)}(\mathbf{h}_k)$) for $k = 1, \ldots, K$. Furthermore, comparisons of the shapes of these functions and their ranges may permit an assessment of the plausibility of intrinsic coregionalization.

As was the case for the (marginal) sample semivariogram and covariance function, the variance-based sample cross-semivariogram has slightly better bias properties than the sample cross-covariance function and is thereby preferred for exploratory data analysis. Neither function requires any collocation, but the variance-based sample cross-semivariogram would appear to require the variables to be commensurate. However, this can be finessed by dividing the mean-corrected observations of Y_l and Y_m by their sample standard deviations. A possible third tool would be a sample-based version of the covariance-based cross-semivariogram, but it is not recommended for general use because it cannot allow for asymmetry and it can make use of only those observations that are collocated for variables l and m. In fact, if there is no collocation whatsoever for those variables, then a covariance-based sample cross-semivariogram cannot be computed for any lags.

If the multivariate process is stationary, then it is easy to see that the sill of the variance-based cross-semivariogram is equal to the average of the sills of the two marginal semivariograms. Thus, under stationarity we should expect this equality to hold approximately for the sills of their sample-based counterparts $\{\hat{\gamma}_{lm}^{(v)}(\cdot)\}$ also. It is also worth noting that if $C_{lm}(\cdot)$ is negative and increases monotonically to zero with increasing lag, then $\gamma_{lm}^{(v)}(\cdot)$ is largest at lag 0 and *decreases* to its sill, so this type of behavior, which would not be permissible in the marginal sample semivariogram, can occur in $\hat{\gamma}_{lm}^{(v)}(\cdot)$ for $l \neq m$.

As the number of variables increases, the information provided by the $q(q-1)/2$ sample cross-covariance functions (or variance-based cross-semivariograms) can become difficult to summarize and interpret. Parametric smoothing methods similar to those described in Section 3.6.2 for use with univariate spatial data may facilitate interpretation (Gribov *et al.*, 2006). However, some potential properties of the multivariate spatial dependence, especially intrinsic and linear coregionalization, are not easy to assess merely by examining these graphical summaries. More formal and more complete assessments of those properties are possible using nonparametric tests of the Lu-Zimmerman type devised by Li *et al.* (2008).

For collocated multivariate areal data at sites $A_1, \ldots, A_n$, Wartenberg (1985) proposed a multivariate version of a Moran-like spatial autocorrelation statistic, given by

$$I_{lm} = \frac{\sum_{i=1}^{n} \sum_{j=1}^{n} w_{ij} z_l(A_i) z_m(A_j)}{n},$$

where the $z_l(\cdot)$'s and $z_m(\cdot)$'s are the observations on the lth and mth variables standardized to have mean zero and variance one. The statistical significance of each I_{lm} can be assessed by comparing it to its randomization distribution, obtained by fixing observations of one of the two variables at their corresponding sites and randomly permuting the observations of the other variable among the sites. A Bonferroni adjustment may be used to account for the multiplicity of hypothesis tested. Multivariate generalizations of the Moran scatterplot and the Moran-I local indicator of spatial association (LISA) also exist; see Anselin *et al.* (2002) for details.

12.4.3   Parameter estimation

The mean structure of a spatial linear model for multivariate geostatistical data is an amalgamation of the mean structures of the individual variables. Denote the vector of observations on the lth variable by $\mathbf{y}_l \equiv [Y_l(\mathbf{s}_{l1}), \ldots, Y_l(\mathbf{s}_{ln_l})]^T$, and let $\mathbf{x}_{l1}, \ldots, \mathbf{x}_{ln_l}$ denote the corresponding vectors of p_l explanatory variables at those sites. (Note that this implies that the same p_l explanatory variables are measured at all of the locations where the lth response variable is observed.) Let the mean function of the lth response variable be given by $\mathrm{E}[Y_l(\mathbf{s}_{li})] = \mathbf{x}_{li}^T \boldsymbol{\beta}_l$ where $\boldsymbol{\beta}_l$ is a p_l-dimensional vector of unknown parameters, and put $\boldsymbol{\beta} = (\boldsymbol{\beta}_1^T, \boldsymbol{\beta}_2^T, \ldots, \boldsymbol{\beta}_q^T)^T$. Furthermore, define

$$\mathbf{X}_l = \begin{pmatrix} \mathbf{x}_{l1}^T \\ \mathbf{x}_{l2}^T \\ \vdots \\ \mathbf{x}_{ln_l}^T \end{pmatrix},$$

$$\mathbf{X} = \begin{pmatrix} \mathbf{X}_1 & \mathbf{0} & \cdots & \mathbf{0} \\ \mathbf{0} & \mathbf{X}_2 & \cdots & \mathbf{0} \\ \vdots & \vdots & \ddots & \vdots \\ \mathbf{0} & \mathbf{0} & \cdots & \mathbf{X}_q \end{pmatrix},$$

and

$$\boldsymbol{\Sigma} = \begin{pmatrix} \boldsymbol{\Sigma}_{11} & \boldsymbol{\Sigma}_{12} & \cdots & \boldsymbol{\Sigma}_{1q} \\ \boldsymbol{\Sigma}_{21} & \boldsymbol{\Sigma}_{22} & \cdots & \boldsymbol{\Sigma}_{2q} \\ \vdots & \vdots & & \vdots \\ \boldsymbol{\Sigma}_{q1} & \boldsymbol{\Sigma}_{q2} & \cdots & \boldsymbol{\Sigma}_{qq} \end{pmatrix},$$

where $\boldsymbol{\Sigma}_{lm}$ is the $n_l \times n_m$ matrix whose (i,j)th element is $C_{lm}(\mathbf{s}_{li}, \mathbf{s}_{mj}; \boldsymbol{\theta})$. Assume, without loss of generality, that $\mathbf{X}$ has full column rank. Then the model for the data may be written in typical spatial linear model form, i.e., as

$$\mathbf{y} = \mathbf{X}\boldsymbol{\beta} + \mathbf{e},$$

where $\mathrm{E}(\mathbf{e}) = \mathbf{0}$ and $\mathrm{Var}(\mathbf{e}) = \boldsymbol{\Sigma} = \boldsymbol{\Sigma}(\boldsymbol{\theta})$.

Parameter estimation for this model, as for the univariate spatial linear model, may be carried out by likelihood-based methods. For the mean structure, this is equivalent to empirical generalized least squares estimation, yielding the estimator

$$\tilde{\boldsymbol{\beta}} = \{\mathbf{X}^T[\boldsymbol{\Sigma}(\tilde{\boldsymbol{\theta}})]^{-1}\mathbf{X}\}^{-1}\mathbf{X}^T[\boldsymbol{\Sigma}(\tilde{\boldsymbol{\theta}})]^{-1}\mathbf{y},$$

where $\tilde{\boldsymbol{\theta}}$ is either the maximum likelihood or REML estimator of $\boldsymbol{\theta}$.

Considerable computational efficiency in likelihood-based estimation can occur for collocated geostatistical data that satisfy intrinsic coregionalization. In such a data-model situation, $\boldsymbol{\Sigma} = \mathbf{A} \otimes \boldsymbol{\rho}$ where $\boldsymbol{\rho} = (\rho(\mathbf{s}_i, \mathbf{s}_j))$, implying that $\boldsymbol{\Sigma}^{-1} = \mathbf{A}^{-1} \otimes \boldsymbol{\rho}^{-1}$ and $|\boldsymbol{\Sigma}| = |\mathbf{A}|^n |\boldsymbol{\rho}|^q$. Thus, the inversion and calculation of the determinant of the $nq \times nq$ matrix $\boldsymbol{\Sigma}$ can be performed by performing those operations on two much smaller matrices. Furthermore, if in addition the explanatory variables are identical for all response variables, implying that $\mathbf{X}_1 = \mathbf{X}_2 = \cdots = \mathbf{X}_*$, say, then it turns out that

$$\tilde{\boldsymbol{\beta}}_l = (\mathbf{X}_*^T \tilde{\boldsymbol{\rho}}^{-1} \mathbf{X}_*)^{-1} \mathbf{X}_*^T \tilde{\boldsymbol{\rho}}^{-1} \mathbf{y}_l \quad (l = 1, \ldots, q),$$

where $\tilde{\boldsymbol{\rho}}$ is the likelihood-based estimator of $\boldsymbol{\rho}$. That is, in this special case, the between-variable dependence has no impact on the point estimation of $\boldsymbol{\beta}$. However, the between-variable covariance structure does still affect interval estimation and other inferences for the elements of $\boldsymbol{\beta}$. Computational efficiencies can also be realized, as in the univariate case, if the covariance and cross-covariance functions are compactly supported and the resulting sparsity is exploited using sparse matrix routines.

A comprehensive R package implementing estimation of multivariate spatial models does not exist, but several packages have some relevant capabilities. Package `gstat` can fit a linear model of coregionalization to a multivariate sample semivariogram using least squares methods. Package `RandomFields` can obtain maximum likelihood estimates of parameters of several multivariate covariance functions including the bivariate Gneiting model and multivariate Matérn model. For more details, see Schlather *et al.* (2015).

12.4.4 Co-kriging and multivariable spatial prediction

When there are $q \geq 2$ response variables of interest, prediction of the values of one, several, or perhaps even all of those variables at unsampled sites may be desired. For each response variable we wish to predict at a site $\mathbf{s}_0$, we could merely use the universal kriging predictor, which may be computed from the observed values of that variable and the explanatory variables in the manner described in Chapter 9. However, if the other response variables are correlated with that variable, then a better predictor can result from basing the prediction on the observations of all variables. The best linear unbiased predictor of any single response variable based on all these latter quantities is called the **universal co-kriging predictor** of that variable.

To elaborate, let $\mathbf{s}_0 \in \mathcal{A}$ and suppose without loss of generality that the prediction of $Y_1(\mathbf{s}_0)$ is desired (prediction of the corresponding noiseless version would follow along the same lines as described in Section 9.4). Let $\mathbf{x}_0$ denote the p-vector of explanatory variables observed at $\mathbf{s}_0$, and let $\mathbf{c}_{10} = (C_{11}(\mathbf{s}_{11}, \mathbf{s}_0), \cdots, C_{11}(\mathbf{s}_{1n_1}, \mathbf{s}_0), C_{21}(\mathbf{s}_{21}, \mathbf{s}_0), \cdots, C_{21}(\mathbf{s}_{2n_2}, \mathbf{s}_0), \cdots, C_{q1}(\mathbf{s}_{q1}, \mathbf{s}_0), \cdots, C_{q1}(\mathbf{s}_{qn_q}, \mathbf{s}_0))^T$. Also let $\boldsymbol{\Sigma}_{00}$ be the $q \times q$ matrix with (l, m)th element $C_{lm}(\mathbf{s}_0, \mathbf{s}_0; \boldsymbol{\theta})$. Under the assumption that the covariance function's parameter vector $\boldsymbol{\theta}$ is known, the universal co-kriging predictor of $Y_1(\mathbf{s}_0)$ is given by

$$\hat{Y}_1(\mathbf{s}_0; \boldsymbol{\theta}) = \mathbf{x}_0^T \hat{\boldsymbol{\beta}}_{1,GLS} + \mathbf{c}_{10}^T \boldsymbol{\Sigma}^{-1}(\mathbf{y} - \mathbf{X}\hat{\boldsymbol{\beta}}_{GLS}). \tag{12.10}$$

The universal co-kriging predictor's prediction error variance, i.e., $\text{Var}(\hat{Y}_1(\mathbf{s}_0) - Y_1(\mathbf{s}_0))$, also called the co-kriging variance, is given by

$$v_1(\mathbf{s}_0; \boldsymbol{\theta}) \equiv C_{11}(\mathbf{s}_0, \mathbf{s}_0) - \mathbf{c}_{10}^T \boldsymbol{\Sigma}^{-1} \mathbf{c}_{10} + (\mathbf{x}_0 - \mathbf{X}^T \boldsymbol{\Sigma}^{-1} \mathbf{c}_{10})^T (\mathbf{X}^T \boldsymbol{\Sigma}^{-1} \mathbf{X})^{-1}$$
$$\times (\mathbf{x}_0 - \mathbf{X}^T \boldsymbol{\Sigma}^{-1} \mathbf{c}_{10}). \tag{12.11}$$

Since $\boldsymbol{\theta}$ in reality is unknown, the ML or REML estimate $\tilde{\boldsymbol{\theta}}$ may be substituted for $\boldsymbol{\theta}$ in (12.10) and (12.11) to yield the empirical co-kriging predictor $\hat{Y}_1(\mathbf{s}_0; \tilde{\boldsymbol{\theta}})$ and an estimate

$v_1(\mathbf{s}_0; \tilde{\boldsymbol{\theta}})$ of its prediction error variance. An approximate $100(1-\epsilon)\%$ prediction interval for $Y_1(\mathbf{s}_0)$ is then given by

$$\hat{Y}_1(\mathbf{s}_0; \tilde{\boldsymbol{\theta}}) \pm z_{\epsilon/2}\sqrt{v_1(\mathbf{s}_0; \tilde{\boldsymbol{\theta}})}.$$

To show the gain in precision possible by co-kriging rather than kriging, consider a toy example in which $q = 2$ and the observations are collocated at the $n = 3$ sites $s_1 = 0, s_2 = 2, s_3 = 5$ on a linear transect. Suppose further that the mean of each response variable is a variable-specific constant along the transect, and the bivariate covariance function is given by

$$
\begin{aligned}
C_{11}(i,j) &= C_{22}(i,j) = \exp(-|i-j|/2), \\
C_{12}(i,j) &= 0.7\exp(-|i-j-1|/2), \\
C_{21}(i,j) &= 0.7\exp(-|i-j+1|/2).
\end{aligned}
$$

This function is built by starting with a bivariate covariance function that satisfies intrinsic coregionalization with $\mathbf{A} = \begin{pmatrix} 1 & 0.7 \\ 0.7 & 1 \end{pmatrix}$ and exponential correlation function $\rho(d) = \exp(-d/2)$. Then, it is made asymmetric using the aforementioned approach of Li & Zhang (2011), with $\Delta_1 = -1/2$ and $\Delta_2 = 1/2$ (which are scalars in this case). For predicting $Y_1(s_0)$ where in this case $s_0 = 3$ on the transect, the co-kriging weights assigned to the six observations (with the weights for the three observations on the first variable appearing first) are $(0.054, 0.387, 0.189, -0.129, 0.558, -0.058)$, and the co-kriging variance is 0.354. The weights make sense because $s_2 = 2$ is closest to $s_0 = 3$, so the first variables at those sites have strong covariance $C_{11}(2,3) = \exp(-1/2) = 0.606$, conferring a substantial weight of 0.387 upon the observation of the first variable at s_2. However, with the shift of -1, $C_{21}(2,3) = 0.7\exp(0) = 0.7$, so there is even stronger correlation between the second variable at s_2 and the first variable at s_0, leading to an even larger weight of 0.558 for the second variable at s_2. For comparison, if we were to base the prediction of $Y_1(s_0)$ on the observed values of the first response variable only, the kriging weights assigned to those observations would be $(0.071, 0.605, 0.324)$ and the universal kriging variance would be 0.595. Thus, using the observed data on the second response variable in addition to the first to predict the value of the first variable at $\mathbf{s}_0$ results in about a 40% reduction in prediction error variance. Even greater gains in precision from co-kriging are possible when the variable being predicted is less densely sampled than the other variables, perhaps because it is more expensive to measure. However, no gain in precision occurs when the data are collocated and intrinsic coregionalization is satisfied; in that case the universal kriging predictor and co-kriging predictor of each variable coincide (Zhang & Cai, 2015).

A special case of co-kriging, called **complementary co-kriging** (Zimmerman & Holland, 2005), occurs when observations of the same variable are taken on more than one sampling network, possibly using distinct measurement devices operated by different organizations, and possibly with different measurement error means and variances. This methodology can calibrate one measurement system to the other and yield spatial predictions of higher quality than those obtained by kriging, using the data from either network by itself.

When prediction of the entire vector of response variables at $\mathbf{s}_0$, i.e., $\mathbf{Y}(\mathbf{s}_0)$, is desired, the problem is referred to as **multivariable spatial prediction**. For ease of exposition assume that the mean of each element of $\mathbf{Y}(\mathbf{s}_0)$ is a linear function of the *same* set of explanatory variables observed at $\mathbf{s}_0$, denoted by $\mathbf{x}_0$. Then, expressions for the vector of best linear unbiased predictors of $\mathbf{Y}(\mathbf{s}_0)$ and the corresponding covariance matrix of prediction errors may be obtained by replacing $C_{11}(\mathbf{s}_0, \mathbf{s}_0)$, $\mathbf{x}_0$, and $\mathbf{c}_{10}$ in (12.10) and (12.11) with $\boldsymbol{\Sigma}_{00}$,

$\mathbf{I}_q \otimes \mathbf{x}_0$, and

$$
\mathbf{C}_0 = \begin{pmatrix}
C_{11}(\mathbf{s}_{11}, \mathbf{s}_0) & C_{12}(\mathbf{s}_{11}, \mathbf{s}_0) & \cdots & C_{1q}(\mathbf{s}_{11}, \mathbf{s}_0) \\
\vdots & \vdots & & \vdots \\
C_{11}(\mathbf{s}_{1n_1}, \mathbf{s}_0) & C_{12}(\mathbf{s}_{1n_1}, \mathbf{s}_0) & \cdots & C_{1q}(\mathbf{s}_{1n_1}, \mathbf{s}_0) \\
C_{21}(\mathbf{s}_{21}, \mathbf{s}_0) & C_{22}(\mathbf{s}_{21}, \mathbf{s}_0) & \cdots & C_{2q}(\mathbf{s}_{21}, \mathbf{s}_0) \\
\vdots & \vdots & & \vdots \\
C_{21}(\mathbf{s}_{2n_2}, \mathbf{s}_0) & C_{22}(\mathbf{s}_{2n_2}, \mathbf{s}_0) & \cdots & C_{2q}(\mathbf{s}_{2n_2}, \mathbf{s}_0) \\
\vdots & \vdots & & \vdots \\
C_{q1}(\mathbf{s}_{q1}, \mathbf{s}_0) & C_{q2}(\mathbf{s}_{q1}, \mathbf{s}_0) & \cdots & C_{qq}(\mathbf{s}_{q1}, \mathbf{s}_0) \\
\vdots & \vdots & & \vdots \\
C_{q1}(\mathbf{s}_{qn_q}, \mathbf{s}_0) & C_{q2}(\mathbf{s}_{qn_q}, \mathbf{s}_0) & \cdots & C_{qq}(\mathbf{s}_{qn_q}, \mathbf{s}_0)
\end{pmatrix}.
$$

The vector of predictors so obtained coincides, element by element, with the predictors obtained by co-kriging each variable separately at $\mathbf{s}_0$, and the diagonal elements of the corresponding covariance matrix of prediction errors coincide with the co-kriging variances of each variable. Thus, if marginal predictive inferences on the elements of $\mathbf{Y}(\mathbf{s}_0)$ are all that is desired, then multivariable spatial prediction provides nothing beyond that already available from co-kriging each variable separately. However, if joint prediction regions or simultaneous confidence intervals for the elements of $\mathbf{Y}(\mathbf{s}_0)$ are desired, then multivariable spatial prediction is preferable because it properly accounts for the covariances between the elements of the vector of prediction errors (the co-kriging covariances).

Co-kriging and multivariate spatial prediction can also be formulated in terms of variance-based cross-semivariograms, which yield the same optimal unbiased predictors obtained using the formulation in terms of the cross-covariance functions. In contrast, the use of covariance-based cross-semivariograms yields optimal unbiased predictors only when the multivariate covariance function is symmetric (Ver Hoef & Cressie, 1993), and should therefore be avoided.

12.4.5   Multivariate spatial sampling design

Many of the same principles of spatial sampling design presented in Chapter 10 also apply to situations involving two or more response variables. The literature on probability-based and space-filling sampling design generally has not been concerned with the number of variables of interest, but has assumed, either implicitly or explicitly, that if more than one variable is of interest then the data are balanced or, even more strongly, that the observations are collocated. Consequently, most treatments of multivariate sampling design from a model-based perspective (e.g., Le and Zidek, 1994; Bueso et al., 1999; Yeh et al., 2006; Vašát et al., 2010) have also restricted attention to collocated designs. Within the model-based design paradigm, however, neither balance nor collocation is necessarily statistically optimal.

Li and Zimmerman (2015) presented an extensive investigation of multivariate model-based optimal sampling design, allowing for collocation but not requiring it. They found that, generally, the within-variable characteristics of optimal designs with respect to multivariate criteria are similar to those of designs that are optimal with respect to the analogous univariate criteria. For example, designs optimal for covariance and cross-covariance parameter estimation consist of a few clusters or strands, while designs optimal for co-kriging have excellent spatial coverage. Figures 12.5 and 12.6 display examples of such designs, as determined by a simulated annealing algorithm, for a scenario in which there are two variables of interest, the study region is a square, candidate design points lie on a 25×25 square grid

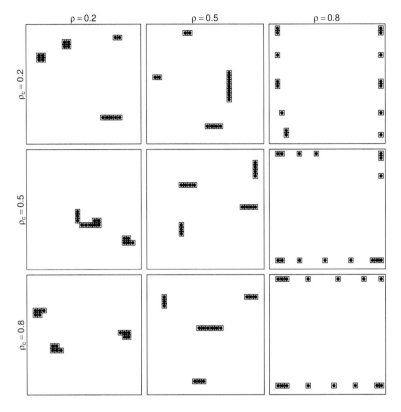

FIGURE 12.5

Optimal designs for covariance and cross-covariance estimation for the scenario described in Section 12.4.5. Closed circles correspond to one variable, and open squares to the other.

within that square, and $n = 15$ for each variable. The observations are assumed to arise from a second-order stationary, isotropic, separable (across variables) bivariate Gaussian process in the square having unknown constant means and a bivariate exponential covariance function with equal correlation decay parameters. The correlation decay for both variables is such that the correlation is equal to ρ at the unit of grid spacing. Nine cases, corresponding to all combinations of three correlation decay parameters ($\rho \in \{0.2, 0.5, 0.8\}$) and three cross-correlation parameters ($\rho_c \in \{0.2, 0.5, 0.8\}$), are displayed in the figure for each design criterion. The design criteria for which the displayed designs are optimal are, in Figure 12.5, the determinant of the information matrix associated with the REML estimator of covariance and cross-covariance parameters (which is maximized), and in Figure 12.6, the maximum determinant of the 2×2 covariance matrix of prediction errors over a fine discretization of the square (which is minimized). What is at least as interesting as the spatial configuration of sites in these optimal designs is the extent of collocation. When the process is separable across variables, the optimal design for covariance and cross-covariance estimation tends to be completely collocated, while the optimal design for co-kriging (with either known or unknown covariance and cross-covariance parameters) often has much less collocation. The extent of collocation in the optimal co-kriging design tends to decrease as either the spatial correlation gets stronger or the cross-correlation gets weaker. Moreover, the optimal design (for any of these objectives) has little to no collocation when the process is nonseparable across variables. On the other hand, in most cases in which collocated designs are not optimal for co-kriging, it turns out that they are reasonably efficient.

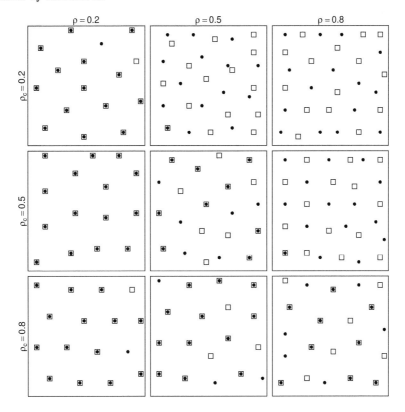

FIGURE 12.6
Optimal designs for co-kriging with known covariance and cross-covariance parameters for
the scenario described in Section 12.4.5. Closed circles correspond to one variable, and open
squares to the other.

Therefore, unless the spatial correlation or degree of nonseparability is very strong, the practical and economic advantages of a balanced, collocated design may outweigh its modest
suboptimality.

12.5   Combinations of extensions

Some work has been done to develop modeling capabilities for data that combine the extensions surveyed separately thus far in this chapter. For space-time covariance functions
for global data, see Jun & Stein (2007), Porcu *et al.* (2016), White & Porcu (2017, 2019),
and the review by Porcu *et al.* (2018). For multivariate covariance functions for global data,
see Jun (2011, 2014), Porcu *et al.* (2016), Bevilacqua *et al.* (2020), or the most up-to-date
summary by Emery *et al.* (2022). Alegría *et al.* (2019) introduced multivariate space-time
covariance functions for global data. Tang & Zimmerman (2024) and Porcu *et al.* (2023) presented some space-time covariance functions for data on stream networks, but multivariate
covariance functions on that domain remain to be developed. For multivariate space-time
covariance functions on Euclidean domains, see De Iaco *et al.* (2005), Bourotte *et al.* (2016),
Salvaña & Genton (2020), and Allard *et al.* (2022).

12.6   Spatial generalized linear models

Many types of data are binary, counts, or positive continuous. Early attempts to model such data relied on transformations to "near normal" so that the methods of classical linear models could be used. For example, a square root transformation was often used for count data. However, Nelder & Wedderburn (1972) introduced a natural extension to linear models using parametric distributions such as the Poisson distribution for counts, the Bernoulli distribution for binary data, etc., called generalized linear models (GLM, McCullagh & Nelder, 1989), which have become very popular and are usually preferred to data transformations. A natural extension of GLMs occurs by introducing latent random effects as a linear mixed model to create a class of generalized linear mixed models (GLMM, Breslow & Clayton, 1993). The latent random effects are usually assumed to be independent and identically distributed normal variables. However, it is also possible for the latent random effects to be spatially correlated, leading to the **spatial generalized linear model** (SGLM, Gotway & Stroup, 1997; Diggle *et al.*, 1998), which we review here.

There are counterparts to the models for continuous areal data described in Chapter 7, for use with discrete data such as binary or count data. These have been termed **auto-models**, including in particular the autologistic, auto-Poisson, autobinomial, and auto negative binomial models, with obvious connections to their nonspatial distributions (Besag, 1974; Cressie, 1993). These models have not been very popular because the conditional specification does not always lead to a recognizable likelihood. For example, for the auto-Poisson, the likelihood may not have a closed form under positive autocorrelation. We will not discuss these models further.

Another approach comes from **spatial quasi-likelihood models** (Breslow & Clayton, 1993; Wolfinger & O'Connell, 1993). These models are an extension of GLMs that use the first and second moments of distributions in the regular exponential family but extend their variance structures to include spatial correlation. These models have been implemented in popular software such as the `glmmPQL` function in the `MASS` package in `R` and the `GLIMMIX` package in `SAS`.

A final class of models is based on a hierarchical constuction, where the mean of any of the distributions in GLMs is allowed to vary by using spatial random effects in the mean structure. Here, there are two broad methods of analysis. The most obvious method is to take a Bayesian approach and compute the posterior distribution of all latent spatial variables and parameters. This has been extremely popular, beginning with disease-mapping (Clayton & Kaldor, 1987) and the introduction of the `WinBUGS` software (Lunn *et al.*, 2000). Less common is a likelihood approach that attempts to estimate covariance parameters and perhaps mean structure parameters simultaneously, while integrating out over all spatial random effects. This can be done using Markov chain Monte Carlo methods (e.g., Christensen, 2004) or more directly using a Laplace approximation (e.g., Evangelou *et al.*, 2011; Bonat & Ribeiro, 2016). Here, we will focus primarily on Bonat & Ribeiro (2016), and improve their methods.

12.6.1   Parametric models for the mean structure

If our linear model is $\boldsymbol{\eta} = \mathbf{X}\boldsymbol{\beta}$, then generalized linear models establish a link function between $\boldsymbol{\mu}$ and $\boldsymbol{\eta}$, denoted as $g(\boldsymbol{\mu}) = \boldsymbol{\eta}$, where $g(\cdot)$ is called the link function. For the Poisson example, $g(\cdot)$ is often the log function. Link functions are monotonic so that $g^{-1}(\cdot)$ is one-to-one with $g(\cdot)$, and $g^{-1}(\cdot)$ is often called the mean function because $\boldsymbol{\mu} = g^{-1}(\boldsymbol{\eta})$. Recall that the mean of a Poisson distribution must be positive, and if $g(\cdot)$ is the log function, then $g^{-1}(\cdot)$ is the exponential function. Hence $\boldsymbol{\mu} = g^{-1}(\boldsymbol{\eta})$ is always positive and $\boldsymbol{\eta}$ is unconstrained on the real line, as is typical for a linear model $\boldsymbol{\eta} = \mathbf{X}\boldsymbol{\beta}$.

Most GLMs are motivated by the exponential family of distributions, where

$$f(y) = \exp\left\{\frac{y\kappa - b(\kappa)}{\delta} + c(y, \delta)\right\}.$$

Many common distributions are special cases of the exponential family, including the normal, gamma, Poisson, binomial, negative binomial, and beta distributions. The first and second moments are $\mu = b'(\kappa)$ and $\text{Var}(y_i) = b''(\kappa)\delta$. The variance can be related to the mean by solving $\text{Var}(y) = a(\mu)\delta$, where $a(\mu)$ is called the variance function. Note that we are now parameterizing the distribution through μ, $a(\mu)$, and δ, rather than $b''(\theta)\delta$. The attraction of this parameterization is that now we can establish the relationship between the mean and the linear model through $\boldsymbol{\mu} = g^{-1}(\boldsymbol{\eta})$.

A fully parametric way to create spatially structured dependence for GLMMs is through a hierarchical construction. We will use the notation $[\mathbf{y}|\boldsymbol{\mu}]$ to denote any probability density function of the vector of random variables $\mathbf{y}$ conditional on a vector of parameters, or other fixed variables, $\boldsymbol{\mu}$. We can have a joint distribution on the left side of the conditional bar, and multiple parameter and fixed value vectors on the right, e.g., $[\mathbf{y}_1, \mathbf{y}_2, \ldots, \mathbf{y}_k|\boldsymbol{\mu}_1, \boldsymbol{\mu}_2, \ldots, \boldsymbol{\mu}_k]$. For example, let $[\mathbf{y}|\boldsymbol{\mu}]$ be the product of independent Poisson distributions with mean parameters contained in the vector $\boldsymbol{\mu}$. Although not strictly necessary, it makes most sense for $\boldsymbol{\mu}$ to be the mean of $\mathbf{y}$, so $\text{E}(\mathbf{y}) = \boldsymbol{\mu}$. The model for the data $\mathbf{y}$ can have more parameters than just the mean, in which case we write it as $[\mathbf{y}|\boldsymbol{\mu}, \boldsymbol{\phi}]$. For an example with extra parameters for $\mathbf{y}$, consider the negative binomial distribution, which can be parameterized with a mean and an extra parameter that allows for overdispersion, which we would write as $[\mathbf{y}|\boldsymbol{\mu}, \phi]$ where ϕ is the overdispersion parameter.

For the hierarchical construction of a generalized linear mixed model, we condition on random effects, and we will change the notation to $\mathbf{w}$ rather than $\boldsymbol{\eta}$ to reflect the fact that $\mathbf{w}$ has a probability distribution. Thus we write $[\mathbf{y}|\mathbf{w}]$, where $\text{E}(\mathbf{y}) = g^{-1}(\mathbf{w})$, and $\mathbf{w}$ is generally considered to have a multivariate normal distribution, which we denote as $[\mathbf{w}|\mathbf{X}, \boldsymbol{\beta}, \boldsymbol{\theta}]$, where $\mathbf{X}$ is the model matrix of fixed explanatory variables, $\boldsymbol{\beta}$ is a vector of fixed effects parameters, $\text{E}(\mathbf{w}) = \mathbf{X}\boldsymbol{\beta}$, and the vector $\boldsymbol{\theta}$ contains covariance parameters. Thus, we use the notation $[\mathbf{w}|\mathbf{X}, \boldsymbol{\beta}, \boldsymbol{\theta}]$ to indicate the probability density function $\mathbf{w} \sim \text{N}(\mathbf{X}\boldsymbol{\beta}, \boldsymbol{\Sigma}(\boldsymbol{\theta}))$. For the hierarchical construction of a SGLM, we simply let $\mathbf{w}$ follow a spatial linear model where spatial information is used to parameterize $\boldsymbol{\Sigma}(\boldsymbol{\theta})$. We turn to this next.

12.6.2 Spatially structured dependence

To develop a spatial covariance matrix for the spatial quasi-likelihood models, which is a moment-based approach of SGLMs, we start with a spatial correlation matrix, which in earlier chapters was denoted as $\mathbf{R}$. For the stationary case and normally-distributed data we obtain the full covariance matrix simply by scaling it by an overall variance parameter, $\boldsymbol{\Sigma} = \sigma^2\mathbf{R}$. A generalization for SGLMs is $\text{Var}(\mathbf{y}) = \mathbf{A}^{1/2}\mathbf{R}\mathbf{A}^{1/2}$, where $\mathbf{A}$ is a diagonal matrix that contains the variance function $a(\mu_i)\delta$ for our chosen model from the exponential family. Note that in the case of normally-distributed data, $a(\mu_i) = 1$ and $\delta = \sigma^2$, so $\boldsymbol{\Sigma} = \sigma^2\mathbf{R}$ is a special case. As another example, consider the Poisson distribution. Here $\delta = 1$ and $a(\mu_i) = \mu_i = \exp(\mathbf{x}_i^T\boldsymbol{\beta})$, where $\mathbf{x}_i^T$ is the ith row of $\mathbf{X}$. In Section 12.6.4, we will develop inference for these models with spatial quasi-likelihood and use iterative fitting algorithms for δ and $\boldsymbol{\beta}$.

For the fully parametric, hierarchical models, let

$$\mathbf{w} = \mathbf{X}\boldsymbol{\beta} + \mathbf{e},$$

where this model for $\mathbf{w}$ is the same one for $\mathbf{y}$ defined in (1.1), with $\mathrm{Var}(\mathbf{e}) = \boldsymbol{\Sigma}(\boldsymbol{\theta})$, and we use the subscript to show the dependence of $\boldsymbol{\Sigma}$ on $\boldsymbol{\theta}$. Then, a very general model can be constructed hierarchically as

$$[\mathbf{y}, \mathbf{w}|\boldsymbol{\phi}, \mathbf{X}, \boldsymbol{\beta}, \boldsymbol{\theta}] = [\mathbf{y}|\mathbf{w}, \boldsymbol{\phi}][\mathbf{w}|\mathbf{X}, \boldsymbol{\beta}, \boldsymbol{\theta}]. \tag{12.12}$$

As a concrete example, suppose that $[\mathbf{y}|\mathbf{w}]$ is Poisson, and $[\mathbf{w}|\mathbf{X}, \boldsymbol{\beta}, \boldsymbol{\theta}]$ is multivariate normal, then the joint likelihood is

$$[\mathbf{y}, \mathbf{w}|\boldsymbol{\phi}, \mathbf{X}, \boldsymbol{\beta}, \boldsymbol{\theta}] = \left(\prod_{i=1}^{n} \frac{\exp(w_i)^{y_i} \exp(-\exp(w_i))}{y_i!} \right) \frac{\exp\{-(\mathbf{w} - \mathbf{X}\boldsymbol{\beta})^T [\boldsymbol{\Sigma}(\boldsymbol{\theta})]^{-1} (\mathbf{w} - \mathbf{X}\boldsymbol{\beta})/2\}}{(2\pi)^{n/2} |\boldsymbol{\Sigma}(\boldsymbol{\theta})|^{1/2}},$$

and note the use of $\mathrm{E}(y_i) = \mu_i = g^{-1}(w_i) = \exp(w_i)$.

The joint distribution given by (12.12) forms the basis for inference, with popular choices given by

- putting prior distributions on $\boldsymbol{\phi}$, $\boldsymbol{\beta}$, and $\boldsymbol{\theta}$ and computing, or sampling from, the joint posterior distribution $[\mathbf{w}, \boldsymbol{\phi}, \boldsymbol{\beta}, \boldsymbol{\theta}|\mathbf{y}, \mathbf{X}]$ using any of a variety of Bayesian methods, or

- integrating over $\mathbf{w}$ using a Laplace approximation, and integrating over $\boldsymbol{\beta}$ as in REML, and then using maximum likelihood to estimate $\boldsymbol{\phi}$ and $\boldsymbol{\theta}$ marginally, followed by GLS estimation of $\boldsymbol{\beta}$ and prediction for $\mathbf{w}$.

Bayesian inference is beyond the scope of this book. Below, we briefly outline spatial quasi-likelihood, and then give more details on the hierarchical model with the Laplace approximation and the marginal maximum likelihood approach.

12.6.3 Parametric models for the covariance structure

In (12.12), we still need parametric models for $\boldsymbol{\Sigma}(\boldsymbol{\theta})$ in $\mathbf{w} \sim \mathrm{N}(\mathbf{X}\boldsymbol{\beta}, \boldsymbol{\Sigma}(\boldsymbol{\theta}))$. There are no theoretical constraints here, and any valid spatial model for $\boldsymbol{\Sigma}(\boldsymbol{\theta})$ is possible. For example, $\boldsymbol{\Sigma}(\boldsymbol{\theta})$ may be constructed from geostatistical models (Chapter 6), where $\boldsymbol{\theta}$ often contains the partial sill, range, and nugget effect, or $\boldsymbol{\Sigma}(\boldsymbol{\theta})$ may be constructed from spatial weights models (Chapter 7), where $\boldsymbol{\theta}$ often contains a spatial dependence parameter and a variance parameter.

12.6.4 Inference using spatial quasi-likelihood

Spatial quasi-likelihood makes assumptions on the first and second moments rather than creating a hierarchical model, but the underlying models are similar. Borrowing from the GLM framework, recall that $\mathrm{E}(\mathbf{y}) = g^{-1}(\mathbf{X}\boldsymbol{\beta})$ and $\mathrm{Var}(\mathbf{y}) = \mathbf{A}^{1/2}\mathbf{R}\mathbf{A}^{1/2}$, where $\mathbf{R}$ is a spatial correlation matrix using one of the models in Chapters 6 or 7 (minus the overall variance parameter), and $\mathbf{A}$ is a diagonal matrix that contains the variance functions of the model.

A multivariate Taylor-series approximation to a nonlinear function $g(\mathbf{x})$ around some value $\mathbf{a}$ is

$$g(\mathbf{x}) \approx g(\mathbf{a}) + g'(\mathbf{a})(\mathbf{x} - \mathbf{a}),$$

so let us expand $g^{-1}(\mathbf{X}\boldsymbol{\beta})$ around some value $\tilde{\boldsymbol{\beta}}$ of $\boldsymbol{\beta}$,

$$g^{-1}(\mathbf{X}\boldsymbol{\beta}) \approx g^{-1}(\mathbf{X}\tilde{\boldsymbol{\beta}}) + \boldsymbol{\Delta}\mathbf{X}(\boldsymbol{\beta} - \tilde{\boldsymbol{\beta}}),$$

where $\boldsymbol{\Delta}$ is a diagonal matrix with diagonal elements

$$\boldsymbol{\Delta} = \frac{\partial g^{-1}(\boldsymbol{\eta})}{\partial \boldsymbol{\eta}},$$

evaluated at $\tilde{\boldsymbol{\beta}}$, and recall that $\boldsymbol{\eta} = \mathbf{X}\boldsymbol{\beta}$. Notice that

$$\mathrm{E}[\mathbf{y} - g^{-1}(\mathbf{X}\tilde{\boldsymbol{\beta}}) - \boldsymbol{\Delta}\mathbf{X}(\boldsymbol{\beta} - \tilde{\boldsymbol{\beta}})] \approx g^{-1}(\mathbf{X}\boldsymbol{\beta}) - g^{-1}(\mathbf{X}\tilde{\boldsymbol{\beta}}) - \boldsymbol{\Delta}\mathbf{X}(\boldsymbol{\beta} - \tilde{\boldsymbol{\beta}})],$$

and, with some rearrangement,

$$\mathbf{X}\boldsymbol{\beta} \approx \boldsymbol{\Delta}^{-1}[g^{-1}(\mathbf{X}\boldsymbol{\beta}) - g^{-1}(\mathbf{X}\tilde{\boldsymbol{\beta}})] + \mathbf{X}\tilde{\boldsymbol{\beta}}.$$

This suggests creating pseudo-data from $\mathbf{y}$ as

$$\tilde{\mathbf{p}} = \boldsymbol{\Delta}^{-1}[\mathbf{y} - g^{-1}(\mathbf{X}\tilde{\boldsymbol{\beta}})] + \mathbf{X}\tilde{\boldsymbol{\beta}}, \tag{12.13}$$

whose approximate expectation is $\mathbf{X}\boldsymbol{\beta}$. From our assumed covariance model for $\mathbf{y}$, we have

$$\mathrm{Var}(\tilde{\mathbf{p}}) = \boldsymbol{\Delta}^{-1}\mathbf{A}^{1/2}\mathbf{R}\mathbf{A}^{1/2}\boldsymbol{\Delta}^{-1} \equiv \mathbf{V}.$$

Treating $\tilde{\mathbf{p}}$ as data and $\mathbf{V}$ as a covariance matrix with unknown parameters, we can use maximum likelihood or residual maximum likelihood methods (Chapter 8) to estimate the covariance parameters of $\mathbf{R}$, and upon plugging them into $\mathbf{V}$ we obtain an estimated covariance matrix that we denote as $\tilde{\mathbf{V}}$. From $\tilde{\mathbf{V}}$, we can get updated values,

$$\tilde{\boldsymbol{\beta}} = (\mathbf{X}^T\tilde{\mathbf{V}}^{-1}\mathbf{X})^{-1}\mathbf{X}^T\tilde{\mathbf{V}}^{-1}\tilde{\mathbf{p}},$$

and from the updated $\tilde{\boldsymbol{\beta}}$, we can update the pseudo-data from (12.13). This iterative procedure continues until convergence in $\tilde{\boldsymbol{\beta}}$ and the covariance parameters in $\mathbf{R}$. The algorithm only needs starting values for $\tilde{\boldsymbol{\beta}}$, which can be obtained by assuming the data are independent and using iteratively reweighted least squares (IRLS), which is the default parameter estimation method for nearly all generalized linear model software. Note that while IRLS converges to the maximum likelihood solution under most common conditions (Green, 1984), the spatial quasi-likelihood approach described in this section does not converge toward any true likelihood, and, in fact, is not guaranteed to converge at all (Boykin *et al.*, 2010; Kleinschmidt *et al.*, 2001; Li *et al.*, 2016).

If the spatial quasi-likelihood approach converges, then all inferences are based on the final values of $\tilde{\mathbf{p}}$ and $\tilde{\mathbf{V}}$. These can be used to estimate fixed effects with generalized least squares, make kriging predictions, etc. Note that all of these inferences are on the link scale, similar to most GLM software for nonspatial models.

12.6.5 Estimating covariance parameters for the hierarchical model

When considering the hierarchical model formulation of the SGLMs, we would like to marginalize the distribution $[\mathbf{w}, \mathbf{y}|\boldsymbol{\phi}, \boldsymbol{\beta}, \boldsymbol{\theta}] = [\mathbf{y}|\mathbf{w}, \boldsymbol{\phi}][\mathbf{w}|\boldsymbol{\beta}, \boldsymbol{\theta}]$ over $\mathbf{w}$ and be free of $\boldsymbol{\beta}$ as well to obtain a distribution of only the data and variance/covariance parameters. First, consider integrating over $\boldsymbol{\beta}$ as well as $\mathbf{w}$:

$$[\mathbf{y}|\boldsymbol{\phi}, \boldsymbol{\theta}] = \int_{\mathbf{w}}\int_{\boldsymbol{\beta}}[\mathbf{w}, \mathbf{y}|\boldsymbol{\phi}, \boldsymbol{\beta}, \boldsymbol{\theta}]\, d\boldsymbol{\beta}\, d\mathbf{w} = \int_{\mathbf{w}}[\mathbf{y}|\mathbf{w}, \boldsymbol{\phi}]\int_{\boldsymbol{\beta}}[\mathbf{w}|\boldsymbol{\beta}, \boldsymbol{\theta}]\, d\boldsymbol{\beta}\, d\mathbf{w}.$$

When $[\mathbf{w}|\boldsymbol{\beta}, \boldsymbol{\theta}]$ is Gaussian, $\int_{\boldsymbol{\beta}}[\mathbf{w}|\boldsymbol{\beta}, \boldsymbol{\theta}]\, d\boldsymbol{\beta}$ is the likelihood for residual maximum likelihood estimation (REML) (this was the topic of Exercise 8.7). Alternatively, consider $[\mathbf{w}|\boldsymbol{\theta}]$ where $\boldsymbol{\beta}$ has been replaced by its conditional (on $\mathbf{w}$) maximum likelihood estimator, $\hat{\boldsymbol{\beta}} = \{\mathbf{X}^T[\boldsymbol{\Sigma}(\boldsymbol{\theta})]^{-1}\mathbf{X}\}^{-1}\mathbf{X}^T[\boldsymbol{\Sigma}(\boldsymbol{\theta})]^{-1}\mathbf{w}$, as in (8.1). Then, both cases can be written as

$$[\mathbf{w}|\boldsymbol{\theta}] = \frac{1}{C_n}\exp\{-(\mathbf{w} - \mathbf{X}\hat{\boldsymbol{\beta}})^T[\boldsymbol{\Sigma}(\boldsymbol{\theta})]^{-1}(\mathbf{w} - \mathbf{X}\hat{\boldsymbol{\beta}})/2\},$$

where, for ML estimation, $C_n = \sqrt{2\pi^n |\boldsymbol{\Sigma}(\boldsymbol{\theta})|}$, and for REML estimation, $C_n = \sqrt{2\pi^{(n-p)} |\boldsymbol{\Sigma}(\boldsymbol{\theta})| |\mathbf{X}^T [\boldsymbol{\Sigma}(\boldsymbol{\theta})]^{-1} \mathbf{X}|}$. Now to get the marginal distribution of the data and covariance parameters, we just need the integral,

$$[\mathbf{y}|\boldsymbol{\phi},\boldsymbol{\theta}] = \int_{\mathbf{w}} [\mathbf{y}|\mathbf{w},\boldsymbol{\phi}][\mathbf{w}|\boldsymbol{\theta}]\, d\mathbf{w}.$$

Let $\ell(\mathbf{w};\mathbf{y},\boldsymbol{\phi},\boldsymbol{\theta}) = \log([\mathbf{y}|\mathbf{w},\boldsymbol{\phi}][\mathbf{w}|\boldsymbol{\theta}])$, and consider $\int e^{\ell(\mathbf{w};\mathbf{y},\boldsymbol{\phi},\boldsymbol{\theta})} d\mathbf{w}$. Let $\mathbf{v}$ be the gradient vector with ith element

$$v_i = \frac{\partial \ell(\mathbf{w};\mathbf{y},\boldsymbol{\phi},\boldsymbol{\theta})}{\partial w_i},$$

and let $\mathbf{H}$ be the Hessian matrix with (i,j)th element,

$$H_{ij} = \frac{\partial^2 \ell(\mathbf{w};\mathbf{y},\boldsymbol{\phi},\boldsymbol{\theta})}{\partial w_i \partial w_j}.$$

Using the multivariate Taylor series expansion around some point $\mathbf{a}$,

$$\int_{\mathbf{w}} e^{\ell(\mathbf{w};\mathbf{y},\boldsymbol{\phi},\boldsymbol{\theta})} d\mathbf{w} \approx \int_{\mathbf{w}} e^{\ell(\mathbf{a};\mathbf{y},\boldsymbol{\phi},\boldsymbol{\theta}) + \mathbf{v}^T(\mathbf{w}-\mathbf{a}) + 1/2(\mathbf{w}-\mathbf{a})^T \mathbf{H}(\mathbf{w}-\mathbf{a})} d\mathbf{w}.$$

Now, if $\mathbf{a}$ is a value for $\ell(\mathbf{a};\mathbf{y},\boldsymbol{\phi},\boldsymbol{\theta})$ such that $\mathbf{v} = \mathbf{0}$, then

$$\int_{\mathbf{w}} e^{\ell(\mathbf{w};\mathbf{y},\boldsymbol{\phi},\boldsymbol{\theta})} d\mathbf{w} \approx e^{\ell(\mathbf{a};\mathbf{y},\boldsymbol{\phi},\boldsymbol{\theta})} \int_{\mathbf{w}} e^{-1/2(\mathbf{w}-\mathbf{a})^T(-\mathbf{H})(\mathbf{w}-\mathbf{a})} d\mathbf{w}.$$

Let $\mathbf{H_a}$ denote $\mathbf{H}$ evaluated at $\mathbf{a}$, and assume that $\mathbf{H_a}$ is negative definite. We know from the normalizing constant of a multivariate Gaussian distribution that

$$\int_{\mathbf{w}} e^{-1/2(\mathbf{w}-\mathbf{a})^T(-\mathbf{H_a})(\mathbf{w}-\mathbf{a})} d\mathbf{w} = (2\pi)^{n/2} |-\mathbf{H_a}^{-1}|^{1/2},$$

so

$$\int_{\mathbf{w}} e^{\ell(\mathbf{w};\mathbf{y},\boldsymbol{\phi},\boldsymbol{\theta})} d\mathbf{w} \approx e^{\ell(\mathbf{a};\mathbf{y},\boldsymbol{\phi},\boldsymbol{\theta})} (2\pi)^{n/2} |-\mathbf{H_a}|^{-1/2} = [\mathbf{y}|\mathbf{a},\boldsymbol{\phi}][\mathbf{a}|\boldsymbol{\theta}](2\pi)^{n/2} |-\mathbf{H_a}|^{-1/2}.$$

Thus, an approximate marginal maximum likelihood estimator for $(\boldsymbol{\phi},\boldsymbol{\theta})$, given $\mathbf{a}$, is

$$\{\hat{\boldsymbol{\phi}}, \hat{\boldsymbol{\theta}}\} = \underset{\boldsymbol{\phi},\boldsymbol{\theta}}{\arg\max} \left\{ \log[\mathbf{y}|\mathbf{a},\boldsymbol{\phi}] + \log[\mathbf{a}|\mathbf{X},\boldsymbol{\Sigma_\theta}] - (1/2)\log(|-\mathbf{H_a}(\boldsymbol{\phi},\boldsymbol{\theta})|) \right\}, \qquad (12.14)$$

where we drop terms that do not contain $\boldsymbol{\phi}$ or $\boldsymbol{\theta}$ and also show the dependence of $\mathbf{H_a}$ on $\boldsymbol{\phi}$ and $\boldsymbol{\theta}$. Note that $\log[\mathbf{a}|\mathbf{X},\boldsymbol{\Sigma}(\boldsymbol{\theta})]$ has exactly the same form of the log-likelihood for ML or REML obtained in Chapter 8, but here it is evaluated at $\mathbf{a}$ and appended by two additional terms. The result (12.14) depends on finding $\mathbf{a}$ so that $\mathbf{v} = \mathbf{0}$. To achieve this, we use Newton-Raphson, conditional on $\boldsymbol{\phi}$ and $\boldsymbol{\theta}$, which we describe next.

Assuming the conditional independence of $\mathbf{y}$ given $\mathbf{w}$,

$$\log([\mathbf{y}|\mathbf{w},\boldsymbol{\phi}][\mathbf{w}|\boldsymbol{\theta}]) = \sum_{i=1}^{n} \log[y_i|w_i,\boldsymbol{\phi}] - \frac{1}{2}(\mathbf{w}-\mathbf{X}\hat{\boldsymbol{\beta}})^T [\boldsymbol{\Sigma}(\boldsymbol{\theta})]^{-1}(\mathbf{w}-\mathbf{X}\hat{\boldsymbol{\beta}}) + C, \qquad (12.15)$$

where C consists of terms that do not contain $\mathbf{w}$. Let $\mathbf{d_\phi}$ be the vector with ith element,

$$d_i \equiv \frac{\partial \log[y_i|w_i,\boldsymbol{\phi}]}{\partial w_i},$$

TABLE 12.6
Distributions, inverse link functions, and first and second partial derivatives with respect to w_i for the data model part of the log-likelihood.

Distribution	$\mu = g^{-1}(\eta)$	d_i	D_{ii}
Binomial	$\mu = \frac{\exp(\eta)}{1+\exp(\eta)}$	$y_i - \frac{n_i \exp(w_i)}{1+\exp(w_i)}$	$-\frac{n_i \exp(w_i)}{(1+\exp(w_i))^2}$
Poisson	$\mu = \exp(\eta)$	$y_i - \exp(w_i)$	$-\exp(w_i)$
Negative Binomial	$\mu = \exp(\eta)$	$\frac{\phi(y_i - e^{w_i})}{\phi + e^{w_i}}$	$-\frac{\phi e^{w_i}(\phi + y_i)}{(\phi + e^{w_i})^2}$

and note that

$$\frac{\partial\{-\frac{1}{2}(\mathbf{w} - \mathbf{X}\hat{\boldsymbol{\beta}})^T[\boldsymbol{\Sigma}(\boldsymbol{\theta})]^{-1}(\mathbf{w} - \mathbf{X}\hat{\boldsymbol{\beta}})\}}{\partial \mathbf{w}} = -[\boldsymbol{\Sigma}(\boldsymbol{\theta})]^{-1}\mathbf{w} + [\boldsymbol{\Sigma}(\boldsymbol{\theta})]^{-1}\mathbf{X}\hat{\boldsymbol{\beta}},$$

so the gradient of (12.15) is

$$\mathbf{v} = \mathbf{d}_\phi - [\boldsymbol{\Sigma}(\boldsymbol{\theta})]^{-1}\mathbf{w} + [\boldsymbol{\Sigma}(\boldsymbol{\theta})]^{-1}\mathbf{X}\hat{\boldsymbol{\beta}} = \mathbf{d}_\phi - \mathbf{P}_{\boldsymbol{\theta}}\mathbf{w},$$

where $\mathbf{P}_{\boldsymbol{\theta}}$ was defined in (8.2). For the Hessian, let $\mathbf{D}_\phi$ be a diagonal matrix with ith diagonal element

$$D_{ii} \equiv \frac{\partial^2 \log[y_i|w_i, \phi]}{\partial w_i^2},$$

where all off-diagonal elements are zero because all second partials are 0 when $i \neq j$ due to conditional independence. A table of d_i and D_{ii} for a few common distributions and link functions is given in Table 12.6. In Table 12.6, the parameterization used for the negative binomial is

$$[y|\mu, \phi] = \frac{\Gamma(y + \phi)}{\Gamma(\phi)y!}\left(\frac{\mu}{\mu + \phi}\right)^y \left(\frac{\phi}{\mu + \phi}\right)^\phi,$$

where $\mathrm{E}(y) = \mu$ and $\mathrm{Var}(y) = \mu + \mu^2/\phi$.

Next, notice that

$$
\begin{aligned}
\frac{\partial^2\{-\frac{1}{2}(\mathbf{w} - \mathbf{X}\hat{\boldsymbol{\beta}})^T[\boldsymbol{\Sigma}(\boldsymbol{\theta})]^{-1}(\mathbf{w} - \mathbf{X}\hat{\boldsymbol{\beta}})\}}{\partial \mathbf{w}\partial \mathbf{w}^T} &= -[\boldsymbol{\Sigma}(\boldsymbol{\theta})]^{-1} \\
&\quad + [\boldsymbol{\Sigma}(\boldsymbol{\theta})]^{-1}\mathbf{X}\{\mathbf{X}^T[\boldsymbol{\Sigma}(\boldsymbol{\theta})]^{-1}\mathbf{X}\}^{-1}\mathbf{X}^T[\boldsymbol{\Sigma}(\boldsymbol{\theta})]^{-1} \\
&= -\mathbf{P}_{\boldsymbol{\theta}},
\end{aligned}
$$

and so the Hessian of (12.15) is

$$\mathbf{H} = \mathbf{D}_\phi - \mathbf{P}_{\boldsymbol{\theta}}. \tag{12.16}$$

Conditional on ϕ and $\boldsymbol{\theta}$, a Newton-Raphson update is,

$$\mathbf{w}^{[k+1]} = \mathbf{w}^{[k]} - \mathbf{H}^{-1}\mathbf{v},$$

and upon convergence we set $\mathbf{a} = \mathbf{w}$ in (12.14) for any evaluation of the likelihood for given ϕ and $\boldsymbol{\theta}$. Notice that this makes the marginal MLE doubly iterative, as $\mathbf{a}$ is obtained while optimizing for ϕ and $\boldsymbol{\theta}$. It is possible to use other maximization routines, such as the EM algorithm, but generally the Newton-Raphson algorithm converges rapidly (often around 10 iterations in our experience). However, on occasion, the stepsize needs to be adjusted so that $\mathbf{v}$ does not diverge. For example, it is easy and fast to check $\mathbf{v}^{[k+1]} = \mathbf{d}_\phi - \mathbf{P}_{\boldsymbol{\theta}}\mathbf{w}^{[k+1]}$, and if $\mathbf{v}^{[k+1]}$ is "larger" than $\mathbf{v}$ by some criterion (e.g., largest or average element of $\mathbf{v}$), then we recommend taking

$$\mathbf{w}^{[k+1]} = \mathbf{w}^{[k]} - \alpha \mathbf{H}^{-1}\mathbf{v},$$

where $0 < \alpha < 1$. In the simulations below, $\mathbf{v}^{[k+1]}$ is checked as described above, and α is set to 0.1 if the largest element of $\mathbf{v}^{[k+1]}$ is larger than the largest element of $\mathbf{v}$. The advantage of using Newton-Raphson is that it provides (an estimate of) $\mathbf{H}$, which is useful for making adjustments to variances when estimating fixed effects and making predictions, as we describe next.

12.6.6   Further inferences for the hierarchical model

In order to estimate ϕ and $\boldsymbol{\theta}$, it was necessary to optimize the likelihood for $\mathbf{w}$, which was called $\mathbf{a}$, using Newton-Raphson, for each evaluation of the likelihood. Thus, upon convergence in estimating ϕ and $\boldsymbol{\theta}$, we also have optimized with respect to $\mathbf{w}$. Let us denote the optimizer as $\hat{\mathbf{w}} = \mathbf{a}$. Also, recall that, in contrast to Bonat & Ribeiro (2016), we integrated over $\boldsymbol{\beta}$, so an estimator of it is needed.

An obvious estimator of $\boldsymbol{\beta}$ results from acting as if $\hat{\mathbf{w}}$ was observed data, and then using the empirical generalized least squares estimator $\hat{\boldsymbol{\beta}}_{EGLS} = \mathbf{B}\hat{\mathbf{w}}$, where $\mathbf{B} = \{\mathbf{X}^T[\boldsymbol{\Sigma}(\hat{\boldsymbol{\theta}})]^{-1}\mathbf{X}\}^{-1}\mathbf{X}^T[\boldsymbol{\Sigma}(\hat{\boldsymbol{\theta}})]^{-1}$. However, $\mathbf{w}$ consists of predictions of unobserved, latent random variables, rather than observed values. In order to estimate the variance of $\hat{\boldsymbol{\beta}}_{EGLS}$, it is convenient to condition on $\mathbf{w}$ as if it was observed, and then use the following well-known result, often called the law of total variance,

$$\text{Var}(\hat{\mathbf{w}}) = \text{E}_\mathbf{w}[\text{Var}(\hat{\mathbf{w}}|\mathbf{w})] + \text{Var}_\mathbf{w}[\text{E}(\hat{\mathbf{w}}|\mathbf{w})]. \tag{12.17}$$

Due to the optimization of $\hat{\mathbf{w}}$ from the likelihood, we will assume that $\hat{\mathbf{w}}|\mathbf{w}$ is approximately distributed as $\text{N}(\mathbf{w}, \mathbf{F}_\mathbf{w}^{-1})$, where $\mathbf{F}_\mathbf{w}$ is the observed Fisher information, or, less strictly, that $\text{E}(\hat{\mathbf{w}}|\mathbf{w}) = \mathbf{w}$ and $\text{Var}(\hat{\mathbf{w}}|\mathbf{w}) = \mathbf{F}_\mathbf{w}^{-1}$, approximately. Thus, the second term in (12.17), $\text{Var}_\mathbf{w}[\text{E}(\hat{\mathbf{w}}|\mathbf{w})]$, may be approximated by $\boldsymbol{\Sigma}(\boldsymbol{\theta})$, which we approximate further by $\boldsymbol{\Sigma}(\hat{\boldsymbol{\theta}})$ after substituting estimated parameters $\hat{\boldsymbol{\theta}}$ for $\boldsymbol{\theta}$. For the first term in (12.17), the observed Fisher information is equivalent to $-\mathbf{H}_\mathbf{w}(\phi, \boldsymbol{\theta})^{-1}$, where we show the dependence on parameters ϕ and $\boldsymbol{\theta}$, and on $\mathbf{w}$ that comes from $\mathbf{D}_\phi$ in (12.16) (see the examples of D_{ii} in Table 12.6). To obtain the Fisher information would require taking the expectation, $\text{E}_\mathbf{w}[\text{Var}(\hat{\mathbf{w}}|\mathbf{w})]$, but this is complicated, and Efron & Hinkley (1978) argued for using the observed Fisher information instead, so we simply replace $\mathbf{w}$ in $-\mathbf{H}_\mathbf{w}(\phi, \boldsymbol{\theta})^{-1}$ with $\hat{\mathbf{w}} = \mathbf{a}$, and we also replace ϕ and $\boldsymbol{\theta}$ by their estimates $\hat{\phi}$ and $\hat{\boldsymbol{\theta}}$, and denote this as $-\mathbf{H}_{\hat{\mathbf{w}}}(\hat{\phi}, \hat{\boldsymbol{\theta}})^{-1}$. Then an estimator of the covariance matrix of estimated fixed effects is

$$\widehat{\text{Var}}(\hat{\boldsymbol{\beta}}_{EGLS}) = \mathbf{B}[\text{Var}(\hat{\mathbf{w}})]\mathbf{B}^T = \mathbf{B}[-\mathbf{H}_{\hat{\mathbf{w}}}(\hat{\phi}, \hat{\boldsymbol{\theta}})^{-1}]\mathbf{B}^T + \mathbf{C}_{\hat{\boldsymbol{\beta}}}, \tag{12.18}$$

where $\mathbf{C}_{\hat{\boldsymbol{\beta}}} = \mathbf{B}\boldsymbol{\Sigma}(\hat{\boldsymbol{\theta}})\mathbf{B}^T$, which simplifies to $\mathbf{C}_{\hat{\boldsymbol{\beta}}} = \{\mathbf{X}^T[\boldsymbol{\Sigma}(\hat{\boldsymbol{\theta}})]^{-1}\mathbf{X}\}^{-1}$, the usual estimated variance-covariance matrix of fixed effects when using generalized least squares if $\mathbf{w}$ were observed.

We proceed in a similar fashion for predictions and prediction error variances. Suppose that we want to predict the latent spatial variable W_k at unobserved locations for $k = 1, \ldots, K$, and we denote this vector of random variables as $\mathbf{u}$. If $\mathbf{w}$ was observed, from (9.4) the best linear unbiased predictor (BLUP) of $\mathbf{u}$ would be $\boldsymbol{\Lambda}\mathbf{w}$, where $\boldsymbol{\Lambda} = \mathbf{X}_\mathbf{u}\mathbf{B} + \boldsymbol{\Sigma}_{\mathbf{wu}}^T[\boldsymbol{\Sigma}(\boldsymbol{\theta})]^{-1} - \boldsymbol{\Sigma}_{\mathbf{wu}}^T[\boldsymbol{\Sigma}(\boldsymbol{\theta})]^{-1}\mathbf{X}\mathbf{B}$. Since, however, $\mathbf{w}$ is unobserved, an alternative predictor of $\mathbf{u}$ may be obtained by substituting $\hat{\mathbf{w}}$ for $\mathbf{w}$ in this expression, yielding $\hat{\mathbf{u}} = \boldsymbol{\Lambda}\hat{\mathbf{w}}$.

To determine the sampling properties of $\hat{\mathbf{u}}$, again we need to make some adjustments that account for the substitution of $\hat{\mathbf{w}}$ for $\mathbf{w}$ in the BLUP. Assuming again that $\hat{\mathbf{w}}$ is unbiased for $\mathbf{w}$, it is easily seen that this alternative predictor is unbiased, i.e., $\mathrm{E}(\boldsymbol{\Lambda}\hat{\mathbf{w}}) = \mathrm{E}(\mathbf{u})$. Now we want an estimator of the covariance matrix of prediction errors associated with this predictor, which is $\mathrm{Var}(\hat{\mathbf{u}} - \mathbf{u}) = \mathrm{Var}(\boldsymbol{\Lambda}\hat{\mathbf{w}} - \mathbf{u})$. Note, from (9.5), that the covariance matrix of the BLUP's prediction error vector is

$$\mathrm{Var}(\boldsymbol{\Lambda}\mathbf{w} - \mathbf{u}) = \boldsymbol{\Sigma}_{\mathbf{uu}} - \boldsymbol{\Sigma}_{\mathbf{wu}}^T [\boldsymbol{\Sigma}(\boldsymbol{\theta})]^{-1} \boldsymbol{\Sigma}_{\mathbf{wu}} + \mathbf{K} C_{\boldsymbol{\beta}} \mathbf{K}^T, \qquad (12.19)$$

where $\mathbf{K} = \mathbf{X_u} - \boldsymbol{\Sigma}_{\mathbf{wu}}^T [\boldsymbol{\Sigma}(\boldsymbol{\theta})]^{-1} \mathbf{X}$. To obtain the prediction error covariance matrix corresponding to our alternative predictor, it is convenient to condition on $\mathbf{w}$ as we did above, and also on $\mathbf{u}$, i.e.,

$$\mathrm{Var}(\boldsymbol{\Lambda}\hat{\mathbf{w}} - \mathbf{u}) = \mathrm{E}_{\mathbf{w},\mathbf{u}}[\mathrm{Var}(\boldsymbol{\Lambda}\hat{\mathbf{w}} - \mathbf{u}|\mathbf{w},\mathbf{u})] + \mathrm{Var}_{\mathbf{w},\mathbf{u}}[\mathrm{E}(\boldsymbol{\Lambda}\hat{\mathbf{w}} - \mathbf{u}|\mathbf{w},\mathbf{u})].$$

Owing to the assumed unbiasedness of $\hat{\mathbf{w}}$ for $\mathbf{w}$, we have $\mathrm{E}(\boldsymbol{\Lambda}\hat{\mathbf{w}} - \mathbf{u}|\mathbf{w},\mathbf{u}) = \boldsymbol{\Lambda}\mathbf{w} - \mathbf{u}$, and the covariance matrix of this vector is given by (12.19). Conditionally, $\mathrm{Var}_{\hat{\mathbf{w}}}(\boldsymbol{\Lambda}\hat{\mathbf{w}} - \mathbf{u})$ does not depend on $\mathbf{u}$, so $\mathrm{E}_{\mathbf{w},\mathbf{u}}[\mathrm{Var}(\boldsymbol{\Lambda}\hat{\mathbf{w}} - \mathbf{u}|\mathbf{w},\mathbf{u})] = \mathrm{E}_{\mathbf{w}}(\boldsymbol{\Lambda}[-\mathbf{H}_{\mathbf{w}}(\boldsymbol{\phi},\boldsymbol{\theta})^{-1}]\boldsymbol{\Lambda}^T)$, and, as we did above, rather than take expectation, we simply use the observed Fisher information and then replace $\mathbf{w}$ in $\mathbf{H}$ with its estimator $\hat{\mathbf{w}} = \mathbf{a}$ and replace $\boldsymbol{\phi}$ and $\boldsymbol{\theta}$ with $\hat{\boldsymbol{\phi}}$ and $\hat{\boldsymbol{\theta}}$. Putting them together, we obtain

$$\widehat{\mathrm{Var}}(\boldsymbol{\Lambda}\hat{\mathbf{w}} - \mathbf{u}) = \boldsymbol{\Lambda}[-\mathbf{H}_{\hat{\mathbf{w}}}(\hat{\boldsymbol{\phi}},\hat{\boldsymbol{\theta}})^{-1}]\boldsymbol{\Lambda}^T + \boldsymbol{\Sigma}_{\mathbf{uu}} - \boldsymbol{\Sigma}_{\mathbf{wu}}^T [\boldsymbol{\Sigma}(\hat{\boldsymbol{\theta}})]^{-1} \boldsymbol{\Sigma}_{\mathbf{wu}} + \mathbf{K} C_{\hat{\boldsymbol{\beta}}} \mathbf{K}^T. \quad (12.20)$$

All covariance matrices depend on $\boldsymbol{\theta}$, although the notation makes this explicit only for the covariance matrix of $\mathbf{w}$. We replace $\boldsymbol{\theta}$ in these matrices by its estimator $\hat{\boldsymbol{\theta}}$, where the fitted covariance function that is used to estimate $\boldsymbol{\Sigma}(\boldsymbol{\theta})$ is also used to estimate $\boldsymbol{\Sigma}_{\mathbf{wu}}$ and $\boldsymbol{\Sigma}_{\mathbf{uu}}$.

How well do all of these approximations work? We will illustrate with some simulations where we know the true values of all parameters and $\mathbf{w}$. The R package `spmodel` is able to simulate and fit all of the models in this section, and several more besides. Consider a square grid of 20×20 data locations on a $(0,1) \times (0,1)$ (unit square) domain. Let $\mathbf{X} = \mathbf{1}$, corresponding to a single overall mean parameter $\beta_0 = 2$. We generated $\mathbf{w}$ on this grid from a spatial linear model with an isotropic exponential covariance function with nugget, $C(r) = \exp(-r) + 0.0001 I(r = 0)$, where $I(\cdot)$ is the indicator function, equal to one if its argument is true, and equal to zero otherwise. The 400 simulated elements in $\mathbf{w}$ are shown in Figure 12.7B. Conditional on $\mathbf{w}$, at each data location we independently simulated a Poisson random variable with mean equal to $\exp(w_i)$, and these are shown in Figure 12.7A.

Using the values in Figure 12.7A, we assumed an unknown mean and isotropic exponential covariance function with nugget, $C(r; \sigma_1^2, \sigma_0^2, \alpha) = \sigma_1^2 \exp(-r/\alpha) + \sigma_0^2 I(r = 0)$, where recall that σ_1^2 is the partial sill, σ_0^2 is the nugget effect, and α is the range parameter. Optimizing the likelihood for (12.14) for $\boldsymbol{\theta} = (\sigma_1^2, \sigma_0^2, \alpha)$ we obtain the values $\hat{\sigma}_1^2 = 0.950$, $\hat{\sigma}_0^2 = 0.046$, and $\hat{\alpha} = 0.894$. The likelihood surface for σ_1^2 and α is shown in Figure 12.7C. A pronounced ridge reveals a positive association in the likelihood between σ_1^2 and α, which we also saw in Chapter 8; e.g., Figure 8.5. The estimation of $\boldsymbol{\theta} = (\sigma_1^2, \sigma_0^2, \alpha)$ also produced $\hat{\mathbf{w}}$, which is shown in Figure 12.7D, and it appears that we are able to recover the spatial patterning within the true simulated $\mathbf{w}$ quite well. A similar example using Bernoulli data is provided in the R code that accompanies this book.

Of course, this is just one simulation. How does this methodology work on average? Are the estimators and predictors unbiased, and are we estimating their variance correctly so that confidence and prediction intervals have proper coverage? In order to answer these questions, we conducted a larger simulation experiment. We simulated data at 200 randomly selected locations in the unit square, from a multivariate normal distribution with exactly the same isotropic exponential covariance model (with nugget) used above, but with mean structure

$$\mathrm{E}(w_i) = \beta_0 + \beta_1 x_i + \beta_2 \tau_i + \beta_3 (x{:}\tau)_i.$$

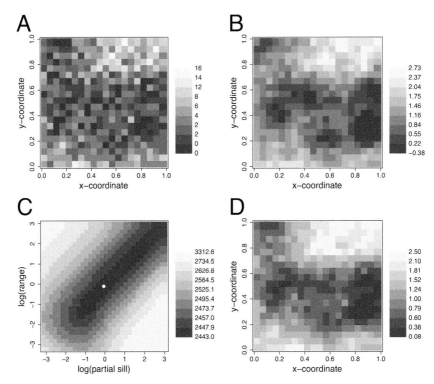

FIGURE 12.7

Estimation for simulated data: A) Simulated count data using the model described in the text; B) the true simulated **w**; C) the likelihood surface of the simulated data, where the white circle shows the maximum likelihood estimators of the partial sill and range, both on the natural log scale; D) the predicted $\hat{\mathbf{w}}$.

Here, x_i is randomly and independently simulated from $N(0, 1)$, τ_i is a randomly and independently simulated Bernoulli variable with probability $p = 0.5$, and $(x{:}\tau)_i$ is the interaction between the normally-distributed and Bernoulli-distributed explanatory variables (i.e., the product of x_i and τ_i). We set $\boldsymbol{\beta} = (0.5, 0.5, -0.5, 0.5)^T$. We also created 100 prediction locations on a 10×10 square grid within the unit square, and also simulated explanatory variables at those locations. In total, 300 w_i values (200 observed, 100 for prediction) were simulated from $N(\mathbf{X}\boldsymbol{\beta}, \boldsymbol{\Sigma}(\boldsymbol{\theta}))$. We then created the observed data as counts from a Poisson distribution conditional on **w**, where at each of the 200 data locations we independently simulated a Poisson random variable with mean equal to $\exp(w_i)$. In this manner, we simulated 2000 data sets to assess bias and confidence/prediction interval coverage.

For each simulated data set, we first estimated the covariance parameters using (12.14), where $\log[\mathbf{a}|\mathbf{X}, \boldsymbol{\Sigma}(\boldsymbol{\theta})]$ is the REML log-likelihood, and again we used an isotropic exponential covariance (with nugget) model. Then, we used the estimated covariance parameters as plug-in values for the covariance model to obtain $\boldsymbol{\Sigma}(\hat{\boldsymbol{\theta}})$, and along with the estimated $\hat{\mathbf{w}}$, we estimated fixed effects as $\hat{\boldsymbol{\beta}}_{EGLS} = \mathbf{B}\hat{\mathbf{w}}$. To estimate bias, we computed the average of $\hat{\boldsymbol{\beta}}_{EGLS} - \boldsymbol{\beta}$ (elementwise) over all 2000 simulated data sets. Standard error was estimated by the square root of the sample variance of the appropriate element of $\hat{\boldsymbol{\beta}}_{EGLS}$. We also formed nominal 90% confidence intervals as $\hat{\beta}_j \pm 1.645\widehat{\text{se}}(\hat{\beta}_{j,EGLS})$ $(j = 0, 1, 2, 3)$, where $\widehat{\text{se}}(\hat{\beta}_{j,EGLS})$ is the square root of the $(j+1)$th diagonal element of the matrix in (12.18). We

TABLE 12.7

Empirical biases, standard errors, and coverage probabilities for estimation of fixed effects $\boldsymbol{\beta}$ and for prediction of $\mathbf{u}$ at unobserved locations. Coverage is for nominal 90% confidence and prediction intervals, where CI90$_c$ uses the corrected versions in (12.18) and (12.20), while CI90$_u$ uses the uncorrected standard error estimator based on $\mathbf{C}_{\boldsymbol{\beta}}$ and the uncorrected prediction error standard error estimator (9.5).

Effect	Bias	Std. error	CI90$_u$	CI90$_c$
β_0	0.069	0.719	0.743	0.762
β_1	−0.005	0.088	0.384	0.899
β_2	0.005	0.161	0.298	0.898
β_3	−0.004	0.144	0.333	0.902
$\hat{\mathbf{u}}$	0.040	0.369	0.701	0.898

computed the proportion of times, over the 2000 simulations, that the confidence interval contained the true value. If the variances of the elements of $\hat{\boldsymbol{\beta}}_{EGLS}$ are estimated well, the coverages should be close to 90%. We also computed the confidence interval coverages based on using the diagonal elements of the naive unadjusted $\mathbf{C}_{\boldsymbol{\beta}}$.

The results are shown in Table 12.7, where it is seen that there is very little bias, relative to the standard errors, in estimating any of the parameters in $\boldsymbol{\beta}$. When using (12.18) to approximate the variance of $\hat{\boldsymbol{\beta}}_{EGLS}$, the confidence interval coverage for β_0 is slightly low, but the confidence interval coverages for β_1, β_2, and β_3 are very close to 90%. When using the naive $\mathbf{C}_{\boldsymbol{\beta}}$, the coverages are less than 90% for all four confidence intervals — much less for the intervals for β_1, β_2, and β_3.

We also used the estimated covariance parameters in $\boldsymbol{\Sigma}(\hat{\boldsymbol{\theta}})$ and the predicted $\hat{\mathbf{w}}$ to make predictions, using $\hat{\mathbf{u}} = \boldsymbol{\Lambda}\hat{\mathbf{w}}$, at all 100 prediction locations. To estimate prediction bias, we computed the average of the elements of $\hat{\mathbf{u}} - \mathbf{u}$ for each simulated data set, where $\mathbf{u}$ contains the 100 simulated values at the prediction locations, and then averaged those across the 2000 simulated data sets. Standard error of prediction was estimated as the square root of the sample variance of the same quantities. We also formed 90% prediction intervals as $\hat{u}_k \pm 1.645\hat{se}(\hat{u}_k)$ $(k = 1, \ldots, 100)$, where $\hat{se}(\hat{u})$ is the square root of the kth diagonal element of the matrix in (12.20). Finally, we computed the proportion of times, over the 100 predictions and 2000 simulations, that the prediction intervals contained the true values, which should be about 90%. Table 12.7 gives the prediction results, which show little indication of bias. Coverage was very close to 90% when using (12.20), but too low when using the naive (12.19).

Appendix A: Some Matrix Results

We give here, as theorems, some matrix results used in the book. For each theorem, a reference is provided where a proof may be found.

Theorem A.1. *Let* $\mathbf{A}$ *represent an* $n \times n$ *real symmetric matrix with (real) eigenvalues* $\lambda_1, \ldots, \lambda_n$. *Then:*

    (a) $\mathbf{A}$ *is nonnegative definite if and only if* $\lambda_i \geq 0$ *for all* $i = 1, \ldots, n$,

    (b) $\mathbf{A}$ *is positive definite if and only if* $\lambda_i > 0$ *for all* $i = 1, \ldots, n$.

Proof. A proof may be found in Schott (2016, p. 117).

Theorem A.2. *Any positive definite matrix is nonsingular.*
Proof. A proof may be found in Harville (1997, p. 213).

Theorem A.3. *Let* $\mathbf{A}$ *and* $\mathbf{P}$ *represent two* $n \times n$ *matrices. If* $\mathbf{A}$ *is positive definite and* $\mathbf{P}$ *is nonsingular, then* $\mathbf{PAP}^T$ *is positive definite.*
Proof. A proof may be found in Harville (1997, p. 213).

Theorem A.4 (Square Root Decomposition Theorem). *Let* $\mathbf{A}$ *represent an* $n \times n$ *real symmetric nonnegative definite matrix. Then a unique* $n \times n$ *symmetric nonnegative definite matrix* $\mathbf{A}^{1/2}$ *exists such that* $\mathbf{A} = \mathbf{A}^{1/2}\mathbf{A}^{1/2}$. *Furthermore, in the special case in which* $\mathbf{A}$ *is positive definite, so is* $\mathbf{A}^{1/2}$.
Proof. A proof may be found in Harville (1997, pp. 544–545).

Theorem A.5 (Spectral Decomposition Theorem). *Let* $\mathbf{A}$ *represent an* $n \times n$ *real symmetric matrix with (real) eigenvalues* $\lambda_1, \ldots, \lambda_n$ *and corresponding orthonormal eigenvectors* $\mathbf{q}_1, \ldots, \mathbf{q}_n$. *Then* $\mathbf{A}$ *may be expressed as*

$$\mathbf{A} = \sum_{i=1}^{n} \lambda_i \mathbf{q}_i \mathbf{q}_i^T,$$

or equivalently as $\mathbf{A} = \mathbf{Q}\boldsymbol{\Lambda}\mathbf{Q}^T$ *where* $\boldsymbol{\Lambda}$ *is a diagonal matrix with ith diagonal element* λ_i *and* $\mathbf{Q} = (\mathbf{q}_1, \ldots, \mathbf{q}_n)$.
Proof. A proof may be found in Harville (1997, pp. 537–538).

Theorem A.6. *Let* $\mathbf{A}$ *represent an* $n \times n$ *nonsingular matrix that is partitioned as*

$$\mathbf{A} = \left(\begin{array}{cc} \mathbf{A}_{11} & \mathbf{A}_{12} \\ \mathbf{A}_{21} & \mathbf{A}_{22} \end{array} \right),$$

where $\mathbf{A}_{11}$ *is* $n_1 \times n_1$, $\mathbf{A}_{12}$ *is* $n_1 \times n_2$, $\mathbf{A}_{21}$ *is* $n_2 \times n_1$, *and* $\mathbf{A}_{22}$ *is* $n_2 \times n_2$. *Let* $\mathbf{B} = \mathbf{A}^{-1}$. *Then* $\mathbf{B}$ *is nonsingular, and* $\mathbf{B}^{-1} = \mathbf{A}$. *Furthermore, let us partition* $\mathbf{B}$ *as*

$$\mathbf{B} = \left(\begin{array}{cc} \mathbf{B}_{11} & \mathbf{B}_{12} \\ \mathbf{B}_{21} & \mathbf{B}_{22} \end{array} \right),$$

DOI: 10.1201/9780429060878-A

where the submatrices of $\mathbf{B}$ *have the same dimensions as the corresponding submatrices of* $\mathbf{A}$. *If* $\mathbf{A}_{11}$ *and* $\mathbf{A}_{22\cdot1} \equiv \mathbf{A}_{22} - \mathbf{A}_{21}\mathbf{A}_{11}^{-1}\mathbf{A}_{12}$ *are nonsingular, then*

(a) $\mathbf{B}_{11} = \mathbf{A}_{11}^{-1} + \mathbf{A}_{11}^{-1}\mathbf{A}_{12}\mathbf{A}_{22\cdot1}^{-1}\mathbf{A}_{21}\mathbf{A}_{11}^{-1}$,

(b) $\mathbf{B}_{22} = \mathbf{A}_{22\cdot1}^{-1}$,

(c) $\mathbf{B}_{12} = -\mathbf{A}_{11}^{-1}\mathbf{A}_{12}\mathbf{A}_{22\cdot1}^{-1}$,

(d) $\mathbf{B}_{21} = -\mathbf{A}_{22\cdot1}^{-1}\mathbf{A}_{21}\mathbf{A}_{11}^{-1}$.

Similarly if $\mathbf{A}_{22}$ *and* $\mathbf{A}_{11\cdot2} \equiv \mathbf{A}_{11} - \mathbf{A}_{12}\mathbf{A}_{22}^{-1}\mathbf{A}_{21}$ *are nonsingular, then*

(e) $\mathbf{B}_{11} = \mathbf{A}_{11\cdot2}^{-1}$,

(f) $\mathbf{B}_{22} = \mathbf{A}_{22}^{-1} + \mathbf{A}_{22}^{-1}\mathbf{A}_{21}\mathbf{A}_{11\cdot2}^{-1}\mathbf{A}_{12}\mathbf{A}_{22}^{-1}$,

(g) $\mathbf{B}_{12} = -\mathbf{A}_{11\cdot2}^{-1}\mathbf{A}_{12}\mathbf{A}_{22}^{-1}$,

(h) $\mathbf{B}_{21} = -\mathbf{A}_{22}^{-1}\mathbf{A}_{21}\mathbf{A}_{11\cdot2}^{-1}$.

Proof. A proof may be found in Schott (2016, pp. 256–257).

Theorem A.7. *Let* $\mathbf{A}$ *represent an* $n \times n$ *matrix with eigenvalues* $\lambda_1, \ldots, \lambda_n$. *Then:*

(a) $\text{tr}(\mathbf{A}) = \sum_{i=1}^{n} \lambda_i$,

(b) $|\mathbf{A}| = \prod_{i=1}^{n} \lambda_i$.

Proof. A proof may be found in Schott (2016, pp. 91–92).

Theorem A.8. *Let* $\mathbf{A}$ *represent an* $n \times n$ *real symmetric matrix with eigenvalues* $\lambda_1, \ldots, \lambda_n$. *Then those eigenvalues are real, and real orthonormal eigenvectors exist that correspond to those eigenvalues.*

Proof. A proof may be found in Schott (2016, pp. 95–97).

Theorem A.9. *Let* $\mathbf{A} = (a_{ij})$ *be an* $n \times n$ *symmetric matrix with positive diagonal elements. If* $\mathbf{A}$ *is (strictly) diagonally dominant, meaning that* $a_{ii} > \sum_{j \neq i} |a_{ij}|$ *for all* $i = 1, \ldots, n$, *then* $\mathbf{A}$ *is positive definite.*

Proof. A proof may be found in Harville (2001, pp. 99–100).

Theorem A.10. *A square matrix* $\mathbf{A}$ *is nonsingular if and only if none of its eigenvalues are equal to 0.*

Proof. A proof may be found in Horn & Johnson (1985, p. 37).

Theorem A.11. *Let* $\mathbf{A}$ *and* $\mathbf{B}$ *be two* $n \times n$ *matrices. If* $\mathbf{B}$ *is nonsingular, then* $\mathbf{A}$ *and* $\mathbf{B}^{-1}\mathbf{A}\mathbf{B}$ *are said to be similar, and the eigenvalues of* $\mathbf{B}^{-1}\mathbf{A}\mathbf{B}$ *and their multiplicities are the same as those of* $\mathbf{A}$.

Proof. A proof may be found in Horn & Johnson (1985, p. 45).

Theorem A.12. *Let* $\mathbf{A}$ *and* $\mathbf{B}$ *be two* $n \times n$ *matrices. If* $\mathbf{A}$ *and* $\mathbf{B}$ *are nonsingular, then so is their matrix product* $\mathbf{A}\mathbf{B}$.

Proof. A proof of a general result, of which this theorem is a special case, may be found in Harville (1997, p. 83).

Bibliography

Abramowitz, M., & Stegun, I.A. 1972. *Handbook of Mathematical Functions with Formulas, Graphs, and Mathematical Tables. National Bureau of Standards Applied Mathematics Series 55. Tenth Printing.* ERIC.

Acosta, J., & Vallejos, R. 2018. Effective sample size for spatial regression models. *Electronic Journal of Statistics*, **12**, 3147–3180.

Adler, R.J. 2010. *The Geometry of Random Fields.* SIAM.

Alegría, A., Porcu, E., Furrer, R., & Mateu, J. 2019. Covariance functions for multivariate Gaussian fields evolving temporally over planet earth. *Stochastic Environmental Research and Risk Assessment*, **33**, 1593–1608.

Alegría, A., Cuevas-Pacheco, F., Diggle, P., & Porcu, E. 2021. The F-family of covariance functions: A Matérn analogue for modeling random fields on spheres. *Spatial Statistics*, **43**, 100512.

Allard, D., Senoussi, R., & Porcu, E. 2016. Anisotropy models for spatial data. *Mathematical Geosciences*, **48**, 305–328.

Allard, D., Clarotto, L., & Emery, X. 2022. Fully nonseparable Gneiting covariance functions for multivariate space–time data. *Spatial Statistics*, **52**, 100706.

Anderes, E., Møller, J., & Rasmussen, J.G. 2020. Isotropic covariance functions on graphs and their edges. *Annals of Statistics*, **48**, 2478–2503.

Anselin, L. 1994. Exploratory spatial data analysis and geographic information systems. *Pages 45–54 of: New Tools for Spatial Analysis.* Eurostat, Luxembourg.

Anselin, L. 1995. Local indicators of spatial association — LISA. *Geographical Analysis*, **27**, 93–115.

Anselin, L. 1996. The Moran scatterplot as an ESDA tool to assess local instability in spatial association. *Pages 111–126 of:* Fischer, M., Scholten, H.J., & Unwin, D. (eds), *Spatial Analytical Perspectives on GIS.* Taylor and Francis, London.

Anselin, L., Syabri, I., & Smirnov, O. 2002. Visualizing multivariate spatial correlation with dynamically linked windows. *In:* Anselin, L., & Rey, S. (eds), *New tools for spatial data analysis: proceedings of the specialist meeting.* University of California, Santa Barbara: Center for Spatially Integrated Social Science (CSISS).

Apanasovich, T.V., & Genton, M.G. 2010. Cross-covariance functions for multivariate random fields based on latent dimensions. *Biometrika*, **97**, 15–30.

Apanasovich, T.V., Genton, M.G., & Sun, Y. 2012. A valid Matérn class of cross-covariance functions for multivariate random fields with any number of components. *Journal of the American Statistical Association*, **107**, 180–193.

Assunção, R., & Krainski, E. 2009. Neighborhood dependence in Bayesian spatial models. *Biometrical Journal*, **51**, 851–869.

Baba, A.M., Midi, H., & Rahman, N.H. Abd. 2022. Spatial outlier accommodation using a spatial variance shift outlier model. *Mathematics*, **10**, 3182.

Banerjee, S., Gelfand, A.E., Finley, A.O., & Sang, H. 2008. Gaussian predictive process models for large spatial data sets. *Journal of the Royal Statistical Society Series B*, **70**, 825–848.

Barendregt, L.G. 1987. The estimation of the generalized covariance when it is a linear combination of two known generalized covariances. *Water Resources Research*, **23**, 583–590.

Barnes, R.J., & Johnson, T.B. 1984. Positive kriging. *Pages 231–244 of:* Verly, G., David, M., G. Journel, A., & Marechal, A. *(eds), Geostatistics for Natural Resources Characterization*. Springer, Dordrecht.

Belsley, D.A., Kuh, E., & Welsch, R.E. 1980. *Regression Diagnostics: Identifying Influential Observations and Sources of Collinearity*. Wiley, New York.

Berg, C., Mateu, J., & Porcu, E. 2008. The Dagum family of isotropic correlation functions. *Bernoulli*, **14**, 1134–1149.

Besag, J. 1974. Spatial interaction and the statistical analysis of lattice systems. *Journal of the Royal Statistical Society, Series B*, **36**, 192–236.

Besag, J. 1975. Statistical analysis of non-lattice data. *The Statistician*, **24**, 179–195.

Besag, J., & Kooperberg, C. 1995. On conditional and intrinsic autoregressions. *Biometrika*, **82**, 733–746.

Bevilacqua, M., Gaetan, C., Mateu, J., & Porcu, E. 2012. Estimating space and space-time covariance functions: A weighted composite likelihood approach. *Journal of the American Statistical Association*, **107**, 268–280.

Bevilacqua, M., Faouzi, T., Furrer, R., & Porcu, E. 2019. Estimation and prediction using generalized Wendland covariance functions under fixed domain asymptotics. *The Annals of Statistics*, **47**, 828–856.

Bevilacqua, M., Diggle, P.J., & Porcu, E. 2020. Families of covariance functions for bivariate random fields on spheres. *Spatial Statistics*, **40**, 100448.

Bevilacqua, M., Caamaño-Carrillo, C., & Porcu, E. 2022. Unifying compactly supported and Matérn covariance functions in spatial statistics. *Journal of Multivariate Analysis*, **189**, 104949.

Bivand, R.S., & Wong, D.W.S. 2018. Comparing implementations of global and local indicators of spatial association. *TEST*, **27**, 716–748.

Bochner, S. 1959. *Lectures on Fourier Integrals*. Princeton University Press, Princeton.

Bonat, W.H., & Ribeiro, P.J. 2016. Practical likelihood analysis for spatial generalized linear mixed models. *Environmetrics*, **27**, 83–89.

Bourotte, M., Allard, D., & Porcu, E. 2016. A flexible class of non-separable cross-covariance functions for multivariate space–time data. *Spatial Statistics*, **18**, 125–146.

Bowman, A.W., & Crujeiras, R.M. 2013. Inference for variograms. *Computational Statistics and Data Analysis*, **66**, 19–31.

Boykin, D., Camp, M.J., Johnson, L., Kramer, M., Meek, D., Palmquist, D., Vinyard, B., & West, M. 2010. Generalized linear mixed model estimation using PROC GLIMMIX: Results from simulations when the data and model match, and when the model is misspecified. *Pages 137–156 of: Conference on Applied Statistics in Agriculture*.

Breslow, N.E., & Clayton, D.G. 1993. Approximate inference in generalized linear mixed models. *Journal of the American Statistical Association*, **88**, 9–25.

Brook, D. 1964. On the distinction between conditional probability and the joint probability approaches in the specification of nearest neighbour systems. *Biometrika*, **51**, 481–483.

Brown, J.A., Robertson, B.L., & McDonald, T. 2015. Spatially balanced sampling: Application to environmental surveys. *Procedia Environmental Sciences*, **27**, 6–9.

Brown, M.B., & Forsythe, A.B. 1974. Robust tests for the equality of variances. *Journal of the American Statistical Association*, **69**, 364–367.

Burnham, K.P, & Anderson, D.R. 2002. *Model Selection and Multimodel Inference: A Practical Information-Theoretic Approach*. Springer-Verlag, New York.

Butler, D.G., & Eccleston, J.A. 2008. On an approximate optimality criterion for the design of field experiments under spatial dependence. *Australian & New Zealand Journal of Statistics*, **50**, 295–307.

Capello, C., Iaco, S. De, & Posa, D. 2018. Testing the type of non-separability and some classes of space-time covariance function models. *Stochastic Environmental Research and Risk Assessment*, **32**, 17–35.

Capello, C., Iaco, S. De, & Posa, D. 2020. covatest: An R package for selecting a class of space-time covariance functions. *Journal of Statistical Software*, **94**, 1–42.

Caragea, P.C., & Smith, R.L. 2007. Asymptotic properties of computationally efficient alternative estimators for a class of multivariate normal models. *Journal of Multivariate Analysis*, **98**, 1417–1440.

Caragea, P.C., & Smith, R.L. 2008. Approximate likelihoods for spatial processes. *Unpublished manuscript*, 169–184.

Carroll, C.W. 1961. The created response surface technique for optimizing nonlinear, restrained systems. *Operations Research*, **9**, 169–184.

Cerioli, A., & Riani, M. 1999. The ordering of spatial data and the detection of multiple outliers. *Journal of Computational and Graphical Statistics*, **8**, 239–258.

Chan, B.S.P., & Eccleston, J.A. 2003. On the construction of nearest-neighbour balanced row and column designs. *Australian & New Zealand Journal of Statistics*, **45**, 97–106.

Chen, D., Liu, C.-T., Kou, Y., & Chen, F. 2008. On detecting spatial outliers. *Geoinformatica*, **12**, 455–475.

Chen, W., Genton, M.G., & Sun, Y. 2021. Space-time covariance structures and models. *Annual Review of Statistics and Its Application*, **8**, 191–215.

Chiles, J.-P., & Delfiner, P. 2009. *Geostatistics: Modeling Spatial Uncertainty.* Vol. 497. John Wiley & Sons, Chichester.

Chipeta, M.G., Terlouw, D.J., Phiri, K.S., & Diggle, P.J. 2017. Inhibitory geostatistical designs for spatial prediction taking account of uncertain covariance structure. *Environmetrics*, **28**, e2425.

Choi, I., Li, B., & Wang, X. 2013. Nonparametric estimation of spatial and space-time covariance functions. *Journal of Agricultural, Biological, and Environmental Statistics*, **18**, 611–630.

Christensen, O.F. 2004. Monte Carlo maximum likelihood in model-based geostatistics. *Journal of Computational and Graphical Statistics*, **13**, 702–718.

Christensen, R., Johnson, W., & Pearson, L.M. 1992. Prediction diagnostics for spatial linear models. *Biometrika*, **79**, 583–591.

Clark, I., Basinger, K.L., & Harper, W.V. 1989. MUCK-a novel approach to co-kriging. *In: Geostatistical, Sensitivity, and Uncertainty Methods for Ground-Water Flow and Radionuclide Transport Modeling. Proceedings.*

Clayton, D., & Kaldor, J. 1987. Empirical Bayes estimates of age-standardized relative risks for use in disease mapping. *Biometrics*, **43**, 671–681.

Clayton, D.G., Bernardinelli, L., & Montomoli, C. 1993. Spatial correlation in ecological analysis. *International Journal of Epidemiology*, **22**, 1193–1202.

Cliff, A.D., & Kelly, F.P. 1977. Regional taxonomy using trend-surface coefficients and invariants. *Environment and Planning A*, **9**, 945–955.

Cliff, A.D., & Ord, J.K. 1973. *Spatial Autocorrelation.* Pion, London.

Cliff, A.D., & Ord, J.K. 1975. The choice of a test for spatial autocorrelation. *Pages 54–77 of:* Davis, J.C., & McCullagh, M.J. (eds), *Display and Analysis of Spatial Data.* Wiley, Chichester, UK.

Cliff, A.D., & Ord, J.K. 1981. *Spatial Processes: Models and Applications.* Pion, London.

Cordy, C. 1993. An extension of the Horvitz-Thompson theorem to point sampling from a continuous universe. *Statistics and Probability Letters*, **18**, 353–362.

Cressie, N. 1985. Fitting variogram models by weighted least squares. *Journal of the International Association for Mathematical Geology*, **17**, 563–586.

Cressie, N. 1986. Kriging nonstationary data. *Journal of the American Statistical Association*, **81**, 625–634.

Cressie, N. 1988. Spatial prediction and ordinary kriging. *Mathematical Geology*, **20**, 405–421.

Cressie, N. 1991. *Statistics for Spatial Data.* Wiley, New York.

Cressie, N., & Hawkins, D.M. 1980. Robust estimation of the variogram. *Journal of the International Association for Mathematical Geology*, **12**, 115–125.

Cressie, N., & Huang, H.C. 1999. Classes of nonseparable, spatio-temporal stationary covariance functions. *Journal of the American Statistical Association*, **94**, 1330–1340.

Cressie, N., & Johannesson, G. 2008. Fixed rank kriging for very large spatial data sets. *Journal of the Royal Statistical Society, Series B*, **70**, 209–226.

Cressie, N., & Kapat, P. 2008. Some diagnostics for Markov random fields. *Journal of Computational and Graphical Statistics*, **17**, 726–749.

Cressie, N., & Kornak, J. 2003. Spatial statistics in the presence of location error with an application to remote sensing of the environment. *Statistical Science*, **18**, 436–456.

Cressie, N., & Lahiri, S.N. 1996. Asymptotics for REML estimation of spatial covariance parameters. *Journal of Statistical Planning and Inference*, **50**, 327–341.

Cressie, N., & Wikle, C.K. 1998. The variance-based cross-variogram: You can add apples and oranges. *Mathematical Geology*, **30**, 789–799.

Cressie, N., & Wikle, C.K. 2011. *Statistics for Spatio-Temporal Data*. Wiley, Hoboken.

Cressie, N., & Zammit-Mangion, A. 2016. Multivariate spatial covariance models: A conditional approach. *Biometrika*, **103**, 915–935.

Cressie, N., Frey, J., Harch, B., & Smith, M. 2006. Spatial prediction on a river network. *Journal of Agricultural, Biological, and Environmental Statistics*, **11**, 127–150.

Cressie, Noel A. C. 1993. *Statistics for Spatial Data, Revised Edition*. John Wiley & Sons, New York.

Curriero, F.C. 2006. On the use of non-Euclidean distance measures in geostatistics. *Mathematical Geology*, **38**, 907–926.

Curriero, F.C., & Lele, S. 1999. A composite likelihood approach to semivariogram estimation. *Journal of Agricultural, Biological, and Environmental Statistics*, **4**, 9–28.

Daley, D.J., Porcu, E., & Bevilacqua, M. 2015. Classes of compactly supported covariance functions for multivariate random fields. *Stochastic Environmental Research and Risk Assessment*, **29**, 1249–1263.

Datta, A., Banerjee, S., Finley, A.O., & Gelfand, A.E. 2016a. Hierarchical nearest-neighbor Gaussian process models for large geostatistical datasets. *Journal of the American Statistical Association*, **111**, 800–812.

Datta, A., Banerjee, S., Finley, A.O., & Gelfand, A.E. 2016b. On nearest-neighbor Gaussian process models for massive spatial data. *Wiley Interdisciplinary Reviews: Computational Statistics*, **8**, 162–171.

De Cesare, L., Myers, D.E., & Posa, D. 2001. Estimating and modeling space-time correlation structures. *Statistics and Probability Letters*, **51**, 9–14.

De Iaco, S. 2010. Space-time correlation analysis: A comparative study. *Journal of Applied Statistics*, **37**, 1027–1041.

De Iaco, S., & Posa, D. 2013. Positive and negative non-separability for space-time covariance models. *Journal of Statistical Planning and Inference*, **143**, 378–391.

De Iaco, S., Myers, D.E., & Posa, D. 2002. Nonseparable space-time covariance models: Some parametric families. *Mathematical Geology*, **34**, 23–42.

De Iaco, S., Palma, M., & Posa, D. 2005. Modeling and prediction of multivariate space–time random fields. *Computational Statistics & Data Analysis*, **48**, 525–547.

De Iaco, S., Posa, D., & Myers, D.E. 2013. Characteristics of some classes of space-time covariance functions. *Journal of Statistical Planning and Inference*, **143**, 2002–2015.

De Iaco, S., Palma, M., & Posa, D. 2016. A general procedure for selecting a class of fully symmetric space-time covariance functions. *Environmetrics*, **27**, 212–224.

De Iaco, S., Posa, D., Cappello, C., & Maggio, S. 2019. Isotropy, symmetry, separability and strict positive definiteness for covariance functions: A critical review. *Spatial Statistics*, **29**, 17–24.

de Jong, P., Sprenger, C., & van Veen, F. 1984. On extreme values of Moran's I and Geary's c. *Geographical Analysis*, **16**, 17–24.

De Oliveira, V., & Han, Z. 2022. On information about covariance parameters in Gaussian Matérn random fields. *Journal of Agricultural, Biological and Environmental Statistics*, **27**, 1–23.

Deutsch, C.V. 1996. Correcting for negative weights in ordinary kriging. *Computers & Geosciences*, **22**, 765–773.

Diblasi, A., & Bowman, A.W. 2001. On the use of the variogram in checking for independence. *Biometrics*, **57**, 211–218.

Dietrich, C.R., & Osborne, M.R. 1991. Estimation of covariance parameters in kriging via restricted maximum likelihood. *Mathematical Geology*, **23**, 119–135.

Diggle, P., & Lophaven, S. 2006. Bayesian geostatistical design. *Scandinavian Journal of Statistics*, **33**, 53–64.

Diggle, P. J., Tawn, J. A., & Moyeed, R. A. 1998. Model-based geostatistics (with discussion). *Journal of the Royal Statistical Society, Series C: Applied Statistics*, **47**, 299–326.

Diggle, P.J., Menezes, R., & Su, T. 2010. Geostatistical inference under preferential sampling. *Journal of the Royal Statistical Society, Series C: Applied Statistics*, **59**, 191–232.

Dimitrakopoulos, R., & Luo, X. 1994. Spatiotemporal modeling: Covariances and ordinary kriging systems. *Pages 88–93 of: Geostatistics for the Next Century*. Springer, Dordrecht.

Dimitrakopoulos, R., & Luo, X. 1997. Joint space-time modelling in the presence of trends. *Pages 138–149 of:* Baafi, E.Y., & Schofield, N.A. (eds), *Geostatistics Wollongong '96, Volume I*. Kluwer Academic Publishers, Dordrecht.

Diniz-Filho, J.A.F., Bini, L.M., & Hawkins, B.A. 2003. Spatial autocorrelation and red herrings in geographical ecology. *Global Ecology and Biogeography*, **12**, 53–64.

Diniz-Filho, J.A.F., Hawkins, B.A., Bini, L.M., Marco, P. De, & Blackburn, T.M. 2007. Are spatial regression methods a panacea or a Pandora's box? A reply to Beale et al. (2007). *Ecography*, **30**, 848–851.

Dixon, W., Smyth, G.K., & Chiswell, B. 1999. Optimized selection of river sampling sites. *Water Resources*, **33**, 971–978.

Dobbie, M.J., Henderson, B.L., & Stevens, D.L. 2008. Sparse sampling: Spatial design for monitoring stream networks. *Statistics Surveys*, **2**, 113–153.

Dormann, C.F. 2007. Effects of incorporating spatial autocorrelation into the analysis of species distribution data. *Global Ecology and Biogeography*, **16**, 129–138.

Dowd, P.A. 1984. The variogram and kriging: robust and resistant estimators. *Pages 91–106 of:* Verly, G., David, M., Journel, A.G., & Marechal, A. (eds), *Geostatistics for Natural Resources Characterization*. D. Reidel, Dordrecht.

Du, J., & Ma, C. 2013. Vector random fields with compactly supported covariance matrix functions. *Journal of Statistical Planning and Inference*, **143**, 457–467.

Dumelle, M., Ver Hoef, J.M., Fuentes, C., & Gitelman, A. 2021. A linear mixed model formulation for spatio-temporal random processes with computational advances for the product, sum, and product–sum covariance functions. *Spatial Statistics*, **43**, 100510.

Dumelle, M., Higham, M., Ver Hoef, J.M., Olsen, A.R., & Madsen, L. 2022. A comparison of design-based and model-based approaches for finite population spatial sampling and inference. *Methods in Ecology and Evolution*, **13**, 2018–2029.

Dumelle, M., Higham, M., & Ver Hoef, J.M. 2023a. spmodel: Spatial statistical modeling and prediction in R. *PLoS ONE*, **18**, e0282524.

Dumelle, M., Kincaid, T., Olsen, A.R., & Weber, M. 2023b. spsurvey: Spatial sampling design and analysis in R. *Journal of Statistical Software*, **105**, 1–29.

Durbin, J., & Watson, G.S. 1950. Testing for serial correlation in least squares regression, I. *Biometrika*, **37**, 409–428.

Durbin, J., & Watson, G.S. 1951. Testing for serial correlation in least squares regression, II. *Biometrika*, **38**, 159–178.

Durbin, J., & Watson, G.S. 1971. Testing for serial correlation in least squares regression, III. *Biometrika*, **58**, 1–19.

Efron, B., & Hinkley, D.V. 1978. Assessing the accuracy of the maximum likelihood estimator: Observed versus expected Fisher information. *Biometrika*, **65**, 457–483.

Emery, X. 2009. The kriging update equations and their application to the selection of neighboring data. *Computational Geosciences*, **13**, 269–280.

Emery, X., Arroyo, D., & Mery, N. 2022. Twenty-two families of multivariate covariance kernels on spheres, with their spectral representations and sufficient validity conditions. *Stochastic Environmental Research and Risk Assessment*, **36**, 1447–1467.

Eubank, R.L., & Speckman, P. 1990. Curve fitting by polynomial-trigonometric regression. *Biometrika*, **77**, 1–9.

Evangelou, E., Zhu, Z., & Smith, R.L. 2011. Estimation and prediction for spatial generalized linear mixed models using high order Laplace approximation. *Journal of Statistical Planning and Inference*, **141**, 3564–3577.

Falk, M.G., McGree, J.M., & Pettitt, A.N. 2014. Sampling designs on stream networks using the pseudo-Bayesian approach. *Environmental and Ecological Statistics*, **21**, 751–773.

Fanshawe, T.R., & Diggle, P.J. 2011. Spatial prediction in the presence of positional error. *Environmetrics*, **22**, 109–122.

Federov, V.V. 1972. *Theory of Optimal Experiments*. Academic Press, New York.

Fonseca, T.C.O., & Steel, M.F.J. 2011. A general class of nonseparable space-time covariance models. *Environmetrics*, **22**, 224–242.

Fouedjio, F. 2017. Second-order non-stationary modeling approaches for univariate geostatistical data. *Stochastic Environmental Research and Risk Assessment*, **31**, 1887–1906.

Fuentes, M. 2005. A formal test for nonstationarity of spatial stochastic processes. *Journal of Multivariate Analysis*, **96**, 30–54.

Fuentes, M. 2007. Approximate likelihood for large irregularly spaced spatial data. *Journal of the American Statistical Association*, **102**, 321–331.

Fuentes, M., Chen, L., & Davis, J.M. 2008. A class of nonseparable and nonstationary spatial temporal covariance functions. *Environmetrics*, **19**, 487–507.

Gabrosek, J., & Cressie, N. 2002. The effect on attribute prediction of location uncertainty in spatial data. *Geographical Analysis*, **34**, 262–285.

Geary, R.C. 1954. The contiguity ratio and statistical mapping. *The Incorporated Statistician*, **5**, 115–145.

Gelfand, A.E. 2021. Multivariate spatial process models. *Pages 1985–2016 of:* Fischer, M.M., & Nijkamp, P. (eds), *Handbook of Regional Science*. Springer, Berlin.

Gelfand, A.E., & Vounatsou, P. 2003. Proper multivariate conditional autoregressive models for spatial data analysis. *Biostatistics*, **4**, 11–25.

Gelfand, A.E., Zhu, L., & Carlin, B.P. 2001. On the change of support problem for spatio-temporal data. *Biostatistics*, **2**, 31–45.

Gelfand, A.E., Sahu, S.K., & Holland, D.M. 2012. On the effect of preferential sampling in spatial prediction. *Environmetrics*, **23**, 565–578.

Genton, M.G., & Kleiber, W. 2015. Cross-covariance functions for multivariate geostatistics. *Statistical Science*, **30**, 147–163.

Gill, P.S., & Shukla, G.K. 1985a. Efficiency of nearest neighbour balanced block designs for correlated observations. *Biometrika*, **72**, 539–544.

Gill, P.S., & Shukla, G.K. 1985b. Experimental designs and their efficiencies for spatially correlated observations in two dimensions. *Communications in Statistics - Theory and Methods*, **14**, 2181–2197.

Gneiting, T. 1999. Correlation functions for atmospheric data analysis. *Quarterly Journal of the Royal Meteorological Society*, **125**, 2449–2464.

Gneiting, T. 2002a. Compactly supported correlation functions. *Journal of Multivariate Analysis*, **83**, 493–508.

Gneiting, T. 2002b. Nonseparable, stationary covariance functions for space-time data. *Journal of the American Statistical Association*, **97**, 590–600.

Gneiting, T. 2013. Strictly and non-strictly positive definite functions on spheres. *Bernoulli*, **19**(4), 1327–1349.

Gneiting, T., Sasvári, Z., & Schlather, M. 2001. Analogies and correspondences between variograms and covariance functions. *Advances in Applied Probability*, **33**(3), 617–630.

Gneiting, T., Kleiber, W., & Schlather, M. 2010. Matérn cross-covariance functions for multivariate random fields. *Journal of the American Statistical Association*, **105**, 1167–1177.

Goldberger, A.S. 1962. Best linear unbiased prediction in the generalized linear regression model. *Journal of the American Statistical Association*, **57**(298), 369–375.

Golub, G., & Loan, C.F. Van. 1996. *Matrix Computations, 3rd ed.* Johns Hopkins University Press, Baltimore.

Goovaerts, P. 2008. Kriging and semivariogram deconvolution in the presence of irregular geographical units. *Mathematical Geosciences*, **40**, 101–128.

Goovaerts, P. 2010. Combining areal and point data in geostatistical interpolation: Applications to soil science and medical geography. *Mathematical Geosciences*, **42**, 535–554.

Gotway, C. A., & Stroup, W. W. 1997. A generalized linear model approach to spatial data analysis and prediction. *Journal of Agricultural, Biological, and Environmental Statistics*, **2**, 157–178.

Gotway, C.A., & Young, L.J. 2002. Combining incompatible spatial data. *Journal of the American Statistical Association*, **97**, 632–648.

Green, P.J. 1984. Iteratively reweighted least squares for maximum likelihood estimation, and some robust and resistant alternatives. *Journal of the Royal Statistical Society, Series B*, **46**, 149–170.

Gribov, A., Krivoruchko, K., & Ver Hoef, J.M. 2006. Modeling the semivariogram: New approach, methods comparison, and simulation study. *Pages 45–57 of:* Coburn, T.C., Yarus, J.M, & Chambers, R.L. (eds), *Stochastic Modeling and Geostatistics: Principles, Methods, and Case Studies, volume II: AAPG Computer Applications in Geology.* AAPG Special Volumes.

Griffith, D.A. 2005. Effective geographic sample size in the presence of spatial autocorrelation. *Annals of the Association of American Geographers*, **95**, 740–760.

Grondona, M.O., & Cressie, N. 1993. Efficiency of block designs under stationary second-order autoregressive errors. *Sankhya*, **55**, 267–284.

Guan, Y., Sherman, M., & Calvin, J.A. 2004. A nonparametric test for spatial isotropy using subsampling. *Journal of the American Statistical Association*, **99**, 810–821.

Guinness, J., & Fuentes, M. 2016. Isotropic covariance functions on spheres: Some properties and modeling considerations. *Journal of Multivariate Analysis*, **143**, 143–152.

Gurka, M.J. 2006. Selecting the best linear mixed model under REML. *The American Statistician*, **60**, 19–26.

Haas, T.C. 1990. Lognormal and moving window methods of estimating acid deposition. *Journal of the American Statistical Association*, **85**, 950–963.

Hanks, E.M., Schliep, E.M., Hooten, M.B., & Hoeting, J.A. 2015. Restricted spatial regression in practice: Geostatistical models, confounding, and robustness under model misspecification. *Environmetrics*, **26**, 243–254.

Hardisty, F., & Klippel, A. 2010. Analysing spatio-temporal autocorrelation with LISTA-Viz. *International Journal of Geographical Information Science*, **24**, 1510–1526.

Harris, P., Charlton, M., & Fotheringham, A.S. 2010. Moving window kriging with geographically weighted variograms. *Stochastic Environmental Research and Risk Assessment*, **24**, 1193–1209.

Hartman, L., & Hössjer, O. 2008. Fast kriging of large data sets with Gaussian Markov random fields. *Computational Statistics and Data Analysis*, **52**, 2331–2349.

Harville, D.A. 1977. Maximum likelihood approaches to variance component estimation and related problems. *Journal of the American Statistical Association*, **72**, 320–340.

Harville, D.A. 1997. *Matrix Algebra from a Statistician's Perspective.* Springer, New York.

Harville, D.A. 2001. *Matrix Algebra: Exercises and Solutions.* Springer, New York.

Harville, D.A. 2018. *Linear Models and the Relevant Distributions and Matrix Algebra.* Chapman and Hall/CRC, Boca Raton, FL.

Harville, D.A., & Jeske, D.R. 1992. Mean squared error of estimation or prediction under a general model. *Journal of the American Statistical Association*, **87**, 724–731.

Haslett, J. 1989. Geostatistical neighbourhoods and subset selection. *Pages 569–577 of:* Armstrong, M. *(ed.), Geostatistics.* Kluwer Academic Publishers, Dordrecht.

Hasselbach, L., Ver Hoef, J.M., Ford, J., Neitlich, P., Crecelius, E., Berryman, S., Wolk, B., & Bohle, T. 2005. Spatial patterns of cadmium and lead deposition on and adjacent to National Park Service lands in the vicinity of Red Dog Mine, Alaska. *Science of the Total Environment*, **348**(1–3), 211–230.

Hastie, T.J., & Tibshirani, R.J. 1990. *Generalized Additive Models.* Chapman and Hall/CRC Press, Boca Raton, FL.

Hawkins, B.A., Diniz-Filho, J.A.F., Bini, L.M., Marco, P. De, & Blackburn, T.M. 2007. Red herrings revisited: Spatial autocorrelation and parameter estimation in geographical ecology. *Ecography*, **30**, 375–384.

Hawkins, D.M., & Cressie, N. 1984. Robust kriging — a proposal. *Journal of the International Association for Mathematical Geology*, **16**, 3–18.

Heaton, M.J., Datta, A., Finley, A.O., Furrer, R., Guiness, J., Guhaniyogi, R., Gerber, F., Gramacy, R.B., Hammerling, D., Katzfuss, M., Lindgren, F., Nychka, D.W., Sun, F., & Zammit-Mangion, A. 2019. A case study competition among methods for analyzing large spatial data. *Journal of Agricultural, Biological, and Environmental Statistics*, **24**, 398–425.

Higdon, D. 1998. A process-convolution approach to modelling temperatures in the North Atlantic Ocean. *Environmental and Ecological Statistics*, **5**(2), 173–190.

Higdon, D., Swall, J., & Kern, J. 1999. Non-stationary spatial modeling. *Bayesian Statistics*, **6**(1), 761–768.

Hinkelmann, K., & Kempthorne, O. 2007. *Design and Analysis of Experiments, Volume 1: Introduction to Experimental Design.* John Wiley & Sons, New York.

Hodges, J.S., & Reich, B.J. 2010. Adding spatially-correlated errors can mess up the fixed effect you love. *The American Statistician*, **64**, 325–334.

Hoeting, J.A., Madigan, D., Raftery, A.E., & Volinsky, C.T. 1999. Bayesian model averaging: A tutorial. *Statistical Science*, **14**, 382–417.

Hoeting, J.A., Davis, R.A., Merton, A.A., & Thompson, S.E. 2006. Model selection for geostatistical models. *Ecological Applications*, **16**, 87–98.

Horn, R.A., & Johnson, C.R. 1985. *Matrix Analysis*. Cambridge University Press, Cambridge.

Horvitz, D.G., & Thompson, D.J. 1952. A generalization of sampling without replacement from a finite universe. *Journal of the American Statistical Association*, **47**, 663–685.

Huang, C., Yao, Y., Cressie, N., & Hsing, T. 2009. Multivariate intrinsic random functions for cokriging. *Mathematical Geosciences*, **41**, 887–904.

Huang, C., Zhang, H., & Robeson, S. 2011. On the validity of commonly used covariance and variogram functions on the sphere. *Mathematical Geosciences*, **43**, 721–733.

Huang, H., & Sun, Y. 2019. Visualization and assessment of spatio-temporal covariance properties. *Spatial Statistics*, **34**, 100272.

Huang, H.-C., & Chen, C.-S. 2007. Optimal geostatistical model selection. *Journal of the American Statistical Association*, **102**, 1009–1024.

Hughes, J., & Haran, M. 2013. Dimension reduction and alleviationn of confounding for spatial generalized linear mixed models. *Journal of the Royal Statistical Society, Series B*, **75**, 139–159.

Hughes-Oliver, J.M., & González-Farias, G. 1999. Parametric covariance models for shock-induced stochastic processes. *Journal of Statistical Planning and Inference*, **77**, 51–72.

Hughes-Oliver, J.M., González-Farıas, G., Lu, J.-C., & Chen, D. 1998. Parametric nonstationary correlation models. *Statistics & Probability Letters*, **40**, 267–278.

Ip, R.H., & Li, W. 2017. On some Matérn covariance functions for spatio-temporal random fields. *Statistica Sinica*, **27**, 805–822.

Irvine, K.M., Gitelman, A.I., & Hoeting, J.A. 2007. Spatial designs and properties of spatial correlation: Effects on covariance estimation. *Journal of Agricultural, Biological, and Environmental Statistics*, **12**, 450–469.

Jennrich, R.I., & Sampson, P.F. 1976. Newton-Raphson and related algorithms for maximum likelihood variance component estimation. *Technometrics*, **18**, 11–17.

Jeong, J., & Jun, M. 2015. A class of Matérn-like covariance functions for smooth processes on a sphere. *Spatial Statistics*, **11**, 1–18.

Jeong, J., Jun, M., & Genton, M.G. 2017. Spherical process models for global spatial statistics. *Statistical Science*, **32**, 501–513.

Johnson, D.E., & Graybill, F.A. 1972. An analysis of a two-way model with interaction and no replication. *Journal of the American Statistical Association*, **67**, 862–868.

Johnson, M.E., Moore, L.M., & Ylvisaker, D. 1990. Minimax and maximin distance designs. *Journal of Statistical Planning and Inference*, **26**, 131–148.

Jones, R.H. 1963. Stochastic processes on a sphere. *Annals of Mathematical Statistics*, **34**, 213–218.

Journel, A.G., & Huijbregts, C.J. 1978. *Mining Geostatistics*. Academic Press, London.

Jun, M. 2011. Non-stationary cross-covariance models for multivariate processes on a globe. *Scandinavian Journal of Statistics*, **38**, 726–747.

Jun, M. 2014. Matérn-based nonstationary cross-covariance models for global processes. *Journal of Multivariate Analysis*, **128**, 134–146.

Jun, M., & Stein, M.L. 2007. An approach to producing space–time covariance functions on spheres. *Technometrics*, **49**, 468–479.

Jun, M., & Stein, M.L. 2008. Nonstationary covariance models for global data. *Annals of Applied Statistics*, **2**, 1271–1289.

Kackar, R.N., & Harville, D.A. 1981. Unbiasedness of two-stage estimation and prediction procedures for mixed linear models. *Communications in Statistics-Theory and Methods*, **10**, 1249–1261.

Kass, R.E., & Raftery, A.E. 1995. Bayes factors. *Journal of the American Statistical Association*, **90**, 773–795.

Kaufman, C.G., Schervish, M.J., & Nychka, D.W. 2008. Covariance tapering for likelihood-based estimation in large spatial datasets. *Journal of the American Statistical Association*, **103**, 1545–1555.

Kenward, M.G., & Roger, J.H. 1997. Small sample inference for fixed effects from restricted maximum likelihood. *Biometrics*, **53**, 983–997.

Kiefer, J. 1953. Sequential minimax search for a maximum. *Proceedings of the American Mathematical Society*, **4**, 502–506.

Kiefer, J. 1975. Construction and optimality of generalized Youden designs. *Pages 333– 353 of:* Srivastava, J.N. (ed), *A Survey of Statistical Design and Linear Models*. North-Holland, Amsterdam.

Kissling, W.D., & Carl, G. 2008. Spatial autocorrelation and the selection of simultaneous autoregressive models. *Global Ecology and Biogeography*, **17**, 59–71.

Kitanidis, P.K. 1983. Statistical estimation of polynomial generalized covariance functions and hydrologic applications. *Water Resources Research*, **19**, 909–921.

Kleiber, W., Nychka, D., & Bandyopadhyay, S. 2019. A model for large multivariate spatial data sets. *Statistica Sinica*, **29**, 1085–1104.

Kleinschmidt, I., Sharp, B. L., Clarke, G. P. Y., Curtis, B., & Fraser, C. 2001. Use of generalized linear mixed models in the spatial analysis of small-area malaria incidence rates in KwaZulu Natal, South Africa. *American Journal of Epidemiology*, **153**(12), 1213–1221.

Koch, D., Lele, S., & Lewis, M.A. 2020. Computationally simple anisotropic lattice covari-ograms. *Environmental and Ecological Statistics*, **27**, 665–688.

Kühn, I. 2007. Incorporating spatial autocorrelation may invert observed patterns. *Diversity and Distributions*, **13**, 66–69.

Kunert, J. 1987. Neighbour balanced block designs for correlated errors. *Biometrika*, **74**, 717–724.

Künsch, H.R., Papritz, A., & Bassi, F. 1997. Generalized cross-covariances and their esti-mation. *Mathematical Geology*, **29**, 779–799.

Kupper, L.L. 1972. Fourier series and spherical harmonics regression. *Applied Statistics*, **21**, 121–130.

Kyriakidis, P.C. 2004. A geostatistical framework for area-to-point spatial interpolation. *Geographical Analysis*, **36**, 259–289.

Laaha, G., Skøien, J.O., & Blöschl, G. 2012. Comparing geostatistical models for river networks. *Pages 543–553 of:* Abrahamsen, P., Hauge, R., & Kolbjornsen, O. (eds), *Geostatistics Oslo 2012, Quantitative Geology and Geostatistics*, vol. 17. Springer, Dordrecht.

Langford, I.H., & Lewis, T. 1998. Outliers in multilevel data. *Journal of the Royal Statistical Society, Series A*, **161**, 121–160.

Lark, R.M. 2000. Estimating variograms of soil properties by method-of-moments and maximum likelihood. *European Journal of Soil Science*, **51**, 717–728.

Laslett, G.M., McBratney, A.B., Pahl, P.J., & Hutchinson, M.F. 1987. Comparison of several spatial prediction methods for soil pH. *Journal of Soil Science*, **38**, 325–341.

Le, N.D., & Zidek, J.V. 2006. *Statistical Analysis of Environmental Space-Time Processes*. Springer, New York.

Lee, L.-F. 2004. Asymptotic distributions of quasi-maximum likelihood estimators for spatial autoregressive models. *Econometrica*, **72**, 1899–1925.

Lele, S. 1995. Inner product matrices, kriging, and nonparametric estimation of variogram. *Mathematical Geology*, **27**, 673–692.

Lenart, E.A., Bowyer, R.T., Ver Hoef, J.M., & Ruess, R.W. 2002. Climate change and caribou: Effects of summer weather on forage. *Canadian Journal of Zoology*, **80**, 664–678.

Lennon, J.J. 2000. Red-shifts and red herrings in geographical ecology. *Ecography*, **23**, 101–113.

Lesage, J.P., & Pace, K.R. 2009. *Introduction to Spatial Econometrics*. CRC Press, Boca Raton, FL.

Leung, S.H., Loh, J.M., Yau, C.Y., & Zhu, Z. 2021. Spatial sampling design using generalized Neyman-Scott process. *Journal of Agricultural, Biological, and Environmental Statistics*, **26**, 105–127.

Levene, H. 1960. Robust tests for equality of variances. *Pages 278–292 of:* Olkin, I. (ed), *Contributions to Probability and Statistics: Essays in Honor of Harold Hotelling*. Stanford University Press, Palo Alto, CA.

Li, B., & Zhang, H. 2011. An approach to modeling asymmetric multivariate covariance functions. *Journal of Multivariate Analysis*, **102**, 1445–1453.

Li, B., Genton, M.G., & Sherman, M. 2007. A nonparametric assessment of properties of space-time covariance functions. *Journal of the American Statistical Association*, **102**, 736–744.

Li, B., Genton, M.G., & Sherman, M. 2008. Testing the covariance structure of multivariate random fields. *Biometrika*, **95**, 813–829.

Li, J., & Zimmerman, D.L. 2015. Model-based sampling design for multivariate geostatistics. *Technometrics*, **25**, 75–86.

Li, L., Brumback, B.A., Weppelmann, T.A., Morris, J.G., & Ali, A. 2016. Adjusting for unmeasured confounding due to either of two crossed factors with a logistic regression model. *Statistics in Medicine*, **35**, 3179–3188.

Liebetrau, A.M. 1979. Water quality sampling: Some statistical considerations. *Water Resources Research*, **15**, 1717–1725.

Lindgren, F., Rue, H., & Lindström, J. 2011. An explicit link between Gaussian fields and Gaussian Markov random fields: The stochastic partial differential equation approach. *Journal of the Royal Statistical Society Series B*, **73**, 423–498.

Lindsay, B.G. 1988. Composite likelihood methods. *Contemporary Mathematics*, **80**, 221–239.

Liu, Y. 2018. *Bayesian hierarchical normal intrinsic conditional autoregressive model for stream networks*. Ph.D. thesis, Department of Statistics & Actuarial Science, University of Iowa.

Lu, H.-C. 1994. *On the distributions of the sample covariogram and semivariogram and their use in testing for isotropy*. Ph.D. thesis, Department of Statistics & Actuarial Science, University of Iowa.

Lu, H.-C., & Zimmerman, D.L. 2001. Testing for isotropy and other directional symmetry properties of spatial correlation. *Preprint, Department of Statistics, University of Iowa*.

Lu, L., Anderson-Cook, C.M., & Robinson, T.J. 2011. Optimization of designed experiments based on multiple criteria utilizing a Pareto frontier. *Technometrics*, **53**, 353–365.

Lu, N., & Zimmerman, D.L. 2005. Testing for directional symmetry in spatial dependence using the periodogram. *Journal of Statistical Planning and Inference*, **129**, 369–385.

Lunn, D.J., Thomas, A., Best, N., & Spiegelhalter, D. 2000. WinBUGS - a Bayesian modelling framework: Concepts, structure, and extensibility. *Statistics and Computing*, **10**, 325–337.

Luo, Q., Griffith, D., & Wu, H. 2017. The Moran coefficient and the Geary ratio: Some mathematical and numerical comparisons. *Pages 253–269 of:* Griffith, D., Chun, Y., & Dean, D. (eds), *Advances in Geocomputation. Advances in Geographic Information Science*. Springer, Cham.

Ma, C. 2002. Spatio-temporal covariance functions generated by mixtures. *Mathematical Geology*, **34**, 965–975.

Ma, C. 2003a. Families of spatio-temporal stationary covariance functions. *Journal of Statistical Planning and Inference*, **116**, 489–501.

Ma, C. 2003b. Spatio-temporal stationary covariance models. *Journal of Multivariate Analysis*, **86**, 97–107.

Machuca-Mory, D.F., & Deutsch, C.V. 2013. Non-stationary geostatistical modeling based on distance weighted statistics and distributions. *Mathematical Geosciences*, **45**, 31–48.

Maity, A., & Sherman, M. 2012. Testing for spatial isotropy under general designs. *Journal of Statistical Planning and Inference*, **142**, 1081–1091.

Majumdar, A., & Gelfand, A.E. 2007. Multivariate spatial modeling using convolved covariance functions. *Mathematical Geology*, **39**, 225–245.

Mantoglou, A., & Wilson, J.L. 1982. The turning bands method for simulation of random fields using line generation by a spectral method. *Water Resources Research*, **18**, 1379–1394.

Mardia, K.V., & Marshall, R.J. 1984. Maximum likelihood estimation of models for residual covariance in spatial regression. *Biometrika*, **71**, 135–146.

Mardia, K.V., & Watkins, A.J. 1989. On multimodality of the likelihood in the spatial linear model. *Biometrika*, **76**, 289–295.

Mariella, L., & Tarantino, M. 2010. Spatial temporal conditional auto-regressive model: A new autoregressive matrix. *Austrian Journal of Statistics*, **39**, 223–244.

Martin, R.J. 1992. Leverage, influence, and residuals in regression models when observations are correlated. *Communications in Statistics - Theory and Methods*, **21**, 1183–1212.

Matérn, B. 1960. Spatial variation. *Meddelanden fran States Skogsforskningsinstitut*, **45**.

Matérn, B. 1960. *Spatial Variation*. Springer Verlag, New York.

Matheron, G. 1973. The intrinsic random functions and their applications. *Advances in Applied Probability*, **5**, 439–468.

Matheron, G. 1982. *Pour une analyse krigeante des données regionaliées*. Tech. rept. Ecole Nationale Supérieure des Mines de Paris.

McCullagh, P. 2002. What is a statistical model? *Annals of Statistics*, **30**, 1225–1310.

McCullagh, P., & Nelder, J.A. 1989. *Generalized Linear Models, 2nd Edition*. Chapman & Hall, Boca Raton, FL.

McDonald, T. 2020. SDraw: Spatially balanced sample draws for spatial objects. *R package version 2.1.13. https://CRAN.R-project.org/package=SDraw*.

McDonald, T.L. 2003. Review of environmental monitoring methods: Survey designs. *Environmental Monitoring and Assessment*, **85**, 277–292.

McKay, M., Beckman, R., & Conover, W. 1979. A comparison of three methods for selecting values of input variables in the analysis of output from a computer code. *Technometrics*, **21**, 239–245.

Minasny, B., McBratney, A.B., & Walvoort, D.J.J. 2007. The variance quadtree algorithm: Use for spatial sampling design. *Computers & Geosciences*, **33**, 383–392.

Montero, J.-M., Fernández-Avilés, G., & Mateu, J. 2015. *Spatial and Spatio-Temporal Geostatistical Modeling and Kriging*. John Wiley & Sons, Chichester.

Moran, P.A.P. 1950. Notes on continuous stochastic phenomena. *Biometrika*, **37**, 17–23.

Mugglin, A.S., Carlin, B.P., & Gelfand, A.E. 2000. Fully model-based approaches for spatially misaligned data. *Journal of the American Statistical Association*, **95**, 877–887.

Müller, W.G. 1999. Least-squares fitting from the variogram cloud. *Statistics and Probability Letters*, **43**, 93–98.

Müller, W.G. 2005. A comparison of spatial design methods for correlated observations. *Environmetrics*, **16**, 495–505.

Müller, W.G., & Stehlík, M. 2010. Compound optimal spatial designs. *Environmetrics*, **21**, 354–364.

Müller, W.G., & Zimmerman, D.L. 1999. Optimal designs for variogram estimation. *Environmetrics*, **10**, 23–37.

Müller, W.G., Pronzato, L., Rendas, J., & Waldl, H. 2015. Efficient prediction designs for random fields (with Discussion). *Applied Stochastic Models in Business and Industry*, **31**, 178–203.

Myers, D.E. 1982. Matrix formulation of kriging. *Journal of International Association of Mathematical Geology*, **14**, 249–257.

Myers, D.E. 1991. Pseudo-cross variograms, positive-definiteness, and cokriging. *Mathematical Geology*, **23**, 805–816.

Myers, R.H. 1990. *Classical and Modern Regression with Applications, 2nd ed.* PWS-Kent, Boston.

Neath, A.A., & Cavanaugh, J.E. 2012. The Bayesian information criterion: Background, derivation, and applications. *Wiley Interdisciplinary Reviews: Computational Statistics*, **4**, 199–203.

Neitlich, P.N., Ver Hoef, J.M., Berryman, S.D., Mines, A., Geiser, L.H., Hasselbach, L.M., & Shiel, A.E. 2017. Trends in spatial patterns of heavy metal deposition on National Park Service lands along the Red Dog Mine haul road, Alaska, 2001–2006. *PLoS ONE*, **12**, e0177936.

Nelder, J.A., & Wedderburn, R.W.M. 1972. Generalized linear models. *Journal of the Royal Statistical Society, Series A: General*, **135**, 370–384.

Nelder, John A, & Mead, Roger. 1965. A simplex method for function minimization. *The Computer Journal*, **7**(4), 308–313.

Network, RESSTE, *et al.* 2017. Analyzing spatio-temporal data with R: Everything you always wanted to know–but were afraid to ask. *Journal de la Société Française de Statistique*, **158**, 124–158.

Nirel, R., Mugglestone, M.A., & Barnett, V. 1998. Outlier-robust spectral estimation for spatial lattice processes. *Communications in Statistics - Theory and Methods*, **27**, 3095–3111.

Nychka, D., Bandyopadhyay, S., Hammerling, D., Lindgren, F., & Sain, S. 2015. A multiresolution Gaussian process model for the analysis of large spatial datasets. *Journal of Computational and Graphical Statistics*, **24**, 579–599.

Okabe, A., & Sugihara, K. 2012. *Spatial Analysis along Networks: Statistical and Computational Methods.* John Wiley & Sons, Chichester.

Olea, R.A. 1984. Sampling design optimization for spatial functions. *Mathematical Geology*, **16**, 369–392.

Ord, K. 1975. Estimation methods for models of spatial interaction. *Journal of the American Statistical Association*, **70**, 120–126.

Overton, W.S., & Stehman, S.V. 1993. Properties of designs for sampling continuous spatial resources from a triangular grid. *Communications in Statistics - Theory and Methods*, **22**, 2641–2660.

Paciorek, C.J. 2003. *Nonstationary Gaussian processes for regression and spatial modelling*. Ph.D. thesis, Department of Statistics, Carnegie Mellon University.

Paciorek, C.J. 2010. The importance of scale for spatial-confounding bias and precision of spatial regression estimators. *Statistical Science*, **25**, 107–125.

Paciorek, C.J., & Schervish, M.J. 2006. Spatial modelling using a new class of nonstationary covariance functions. *Environmetrics*, **17**, 483–506.

Papritz, A., Künsch, H.R., & Webster, R. 1993. On the pseudo cross-variogram. *Mathematical Geology*, **25**, 1015–1026.

Park, J.J., Shin, K.I., Lee, J.H., Lee, S.E., Lee, W.K., & Cho, K. 2012. Detecting and cleaning outliers for robust estimation of variogram models in insect count data. *Ecological Research*, **27**, 1–13.

Patterson, H.D., & Thompson, R. 1971. Recovery of inter-block information when block sizes are unequal. *Biometrika*, **58**, 545–554.

Patterson, H.D., & Thompson, R. 1974. Maximum likelihood estimation of components of variance. *Pages 197–207 of: Proceedings of the 8th International Biometrics Conference*. Alexandria, VA The Biometric Society.

Pebesma, E. 2012. spacetime: Spatio-temporal data in R. *Journal of Statistical Software*, **51**, 1–30.

Pebesma, E., & Gräler, B. 2023. Introduction to spatio-temporal variography, https://CRAN.R-project.org/package=st.

Pettitt, A.N., Weir, I.S., & Hart, A.G. 2002. A conditional autoregressive Gaussian process for irregularly spaced multivariate data with application to modelling large sets of binary data. *Statistical Computing*, **12**, 353–367.

Pistone, G., & Vicario, G. 2010. Comparing and generating Latin hypercube designs in kriging models. *Advances in Statistical Analysis*, **94**, 353–366.

Porcu, E., Gregori, P., & Mateu, J. 2006. Nonseparable stationary anisotropic space-time covariance functions. *Stochastic Environmental Research and Risk Assessment*, **21**, 113–122.

Porcu, E., Mateu, J., & Bevilacqua, M. 2007a. Covariance functions that are stationary or nonstationary in space and stationary in time. *Statistica Neerlandica*, **61**, 358–382.

Porcu, E., Mateu, J., Zini, A., & Pini, R. 2007b. Modelling spatio-temporal data: A new variogram and covariance structure proposal. *Statistics & Probability Letters*, **77**, 83–89.

Porcu, E., Mateu, J., & Saura, F. 2008. New classes of covariance and spectral density functions for spatio-temporal modelling. *Stochastic Environmental Research and Risk Assessment*, **22**, 65–79.

Porcu, E., Daley, D.J., Buhmann, M., & Bevilacqua, M. 2013. Radial basis functions with compact support for multivariate geostatistics. *Stochastic Environmental Research and Risk Assessment*, **27**, 909–922.

Porcu, E., Bevilacqua, M., & Genton, M.G. 2016. Spatio-temporal covariance and cross-covariance functions of the great circle distance on a sphere. *Journal of the American Statistical Association*, **111**, 888–898.

Porcu, E., Alegria, A., & Furrer, R. 2018. Modeling temporally evolving and spatially globally dependent data. *International Statistical Review*, **86**, 344–377.

Porcu, E., Furrer, R., & Nychka, D. 2020a. 30 years of space-time covariance functions. *WIREs Computational Statistics*, **13**, e1512.

Porcu, E., Zastavnyi, V., Bevilacqua, M., & Emery, X. 2020b. Stein hypothesis and screening effect for covariances with compact support. *Electronic Journal of Statistics*, **14**, 2510–2528.

Porcu, E., White, P.A. & Genton, M.G., 2023. Stationary nonseparable space-time covariance functions on networks. *Journal of the Royal Statistical Society Series B: Statistical Methodology*, **85**, 1417-1440.

Pregibon, D. 1981. Logistic regression diagnostics. *The Annals of Statistics*, **9**, 705–724.

Reich, B.J., Hodges, J.S., & Zadnik, V. 2006. Effects of residual smoothing on the posterior of the fixed effects in disease-mapping models. *Biometrics*, **62**, 1197–1206.

Ripley, B.D. 1981. *Spatial Statistics*. Wiley, New York.

Robertson, B., McDonald, T., Price, C., & Brown, J. 2018. Halton iterative partitioning: Spatially balanced sampling via partitioning. *Environmental and Ecological Statistics*, **25**, 305–323.

Robertson, B.L., Brown, J.A., McDonald, T., & Jaksons, P. 2013. BAS: Balanced acceptance sampling of natural resources. *Biometrics*, **69**, 776–784.

Robertson, B.L., McDonald, T., Price, C.J., & Brown, J.A. 2017. A modification of balanced acceptance sampling. *Statistics & Probability Letters*, **129**, 107–112.

Rodrigues, A., & Diggle, P.J. 2010. A class of convolution-based models for spatio-temporal processes with non-separable covariance structure. *Scandinavian Journal of Statistics*, **37**, 553–567.

Rouhani, S., & Myers, D.E. 1990. Problems in space-time kriging of geohydrological data. *Mathematical Geology*, **22**, 611–623.

Royle, J.A., & Nychka, D. 1998. An algorithm for the construction of spatial coverage designs with implementation in S-Plus. *Computers & Geosciences*, **24**, 479–488.

Rue, H., & Tjelmeland, H. 2002. Fitting Gaussian Markov random fields to Gaussian fields. *Scandinavian Journal of Statistics*, **29**, 31–50.

Russell, K.G., & Eccleston, J.A. 1987. The construction of optimal balanced incomplete block designs when adjacent observations are correlated. *Australian Journal of Statistics*, **29**, 84–90.

Russo, D. 1984. Design of an optimal spatial sampling network for estimating the variogram. *Soil Science Society of America Journal*, **48**, 708–716.

Sahoo, I., Guinness, J., & Reich, B.J. 2019. A test for isotropy on a sphere using spherical harmonic functions. *Statistica Sinica*, **29**, 1253–1276.

Sain, S.R., & Cressie, N. 2007. A spatial model for multivariate lattice data. *Journal of Econometrics*, **140**, 226–259.

Salvaña, M.L.O., & Genton, M.G. 2020. Nonstationary cross-covariance functions for multivariate spatio-temporal random fields. *Spatial Statistics*, **37**, 100411.

Scaccia, L., & Martin, R.J. 2005. Testing axial symmetry and separability of lattice processes. *Journal of Statistical Planning and Inference*, **131**, 19–39.

Schlather, M., Malinowski, A., Menck, P.J., Oesting, M., & Strokorb, K. 2015. Analysis, simulation and prediction of multivariate random fields with package RandomFields. *Journal of Statistical Software*, **63**, 1–25.

Schliep, E.M., Wikle, C.K., & Daw, R. 2023. Correcting for informative sampling in spatial covariance estimation and kriging predictions. *Journal of Geographical Systems*, **25**, 1–27.

Schoenberg, I.J. 1938. Metric spaces and completely monotone functions. *Annals of Mathematics*, **39**, 811–841.

Schoenberg, I.J. 1942. Positive definite functions on spheres. *Duke Mathematical Journal*, **9**, 96–108.

Schott, J.R. 2016. *Matrix Analysis for Statistics*. John Wiley & Sons, Hoboken, NJ.

Shand, L., & Li, B. 2017. Modeling nonstationarity in space and time. *Biometrics*, **73**, 759–768.

Shapiro, A., & Botha, J.D. 1991. Variogram fitting with a general class of conditionally nonnegative definite functions. *Computational Statistics and Data Analysis*, **11**, 87–96.

Shi, L., & Chen, G. 2009. Influence measures for general linear models with correlated errors. *The American Statistician*, **63**, 40–42.

Silvey, S.D. 1980. *Optimal Design*. Chapman and Hall, London.

Sjöstedt de Luna, S., & Young, A. 2003. The bootstrap and kriging prediction intervals. *Scandinavian Journal of Statistics*, **30**, 175–192.

Skoien, J.O., Merz, R., & Blöschl, G. 2006. Top-kriging — geostatistics on stream networks. *Hydrology and Earth System Sciences*, **10**, 277–287.

Snyder, J.P. 1987. *Map Projections – A Working Manual*. Tech. rept. US Geological Survey Professional Paper 1395, United States Government Printing Office, Washington, DC, 100 p.

Som, N. A., Monestiez, P., Ver Hoef, J.M., Zimmerman, D.L., & Peterson, E.E. 2014. Spatial sampling on streams: Principles for inference on aquatic networks. *Environmetrics*, **25**, 306–323.

Song, R., & Zimmerman, D.L. 2021. Modeling spatial correlation that grows on trees, with a stream network application. *Spatial Statistics*, **45**, 100536.

Stein, M.L. 1988. Asymptotically efficient prediction of a random field with a misspecified covariance function. *The Annals of Statistics*, **16**, 55–63.

Stein, M.L. 1999. *Interpolation of Spatial Data: Some Theory for Kriging*. Springer-Verlag, New York.

Stein, M.L. 2002. The screening effect in kriging. *The Annals of Statistics*, **30**, 298–323.

Stein, M.L. 2005a. Nonstationary spatial covariance functions. *Unpublished technical report.*

Stein, M.L. 2005b. Space-time covariance functions. *Journal of the American Statistical Association*, **100**, 310–321.

Stein, M.L. 2007. Spatial variation of total column ozone on a global scale. *Annals of Applied Statistics*, **1**, 191–210.

Stein, M.L. 2011. 2010 Rietz lecture: When does the screening effect hold? *The Annals of Statistics*, **39**, 2795–2819.

Stein, M.L. 2014. Limitations on low rank approximations for covariance matrices of spatial data. *Spatial Statistics*, **8**, 1–19.

Stein, M.L. 2015. When does the screening effect not hold? *Spatial Statistics*, **11**, 65–80.

Stein, M.L., & Handcock, M.S. 1989. Some asymptotic properties of kriging when the covariance function is misspecified. *Mathematical Geology*, **21**, 171–190.

Stein, M.L., Chi, Z., & Welty, L.J. 2004. Approximating likelihoods for large spatial data sets. *Journal of the Royal Statistical Society, Series B*, **66**, 275–296.

Stevens, D.L. 1997. Variable density grid-based sampling designs for continuous spatial populations. *Environmetrics*, **8**, 167–195.

Stevens, D.L., & Olsen, A.R. 2003. Variance estimation for spatially balanced samples of environmental resources. *Environmetrics*, **14**, 593–610.

Stevens, D.L., & Olsen, A.R. 2004. Spatially balanced sampling of natural resources. *Journal of the American Statistical Association*, **99**, 262–278.

Szidarovszky, F., Baafi, E.Y., & Kim, Y.C. 1987. Kriging without negative weights. *Mathematical Geology*, **19**, 549–559.

Tang, J., & Zimmerman, D. 2024. Space-time covariance models on networks. *Electronic Journal of Statistics,* 18, 490–514.

Tiefelsdorf, M., Griffith, D.A., & Boots, B. 1999. A variance-stabilizing coding scheme for spatial link matrices. *Environment and Planning A*, **31**, 165–180.

Tobler, W.R. 1970. A computer movie simulating urban growth in the Detroit region. *Economic Geography*, **46**(sup1), 234–240.

Tognelli, M.F., & Kelt, D.A. 2004. Analysis of determinants of mammalian species richness in South America using spatial autoregressive models. *Ecography*, **27**, 427–436.

Tukey, J.W. 1949. One degree of freedom for non-additivity. *Biometrics*, **5**, 232–242.

Urquhart, N.S., & Kincaid, T.M. 1999. Designs for detecting trend from repeated surveys of ecological response. *Journal of Agricultural, Biological, and Environmental Statistics*, **4**, 404–414.

Urquhart, N.S., Overton, W.S., & Birkes, D.S. 1993. Comparing sampling designs for monitoring ecological status and trends: Impact of temporal patterns. *Pages 71–85 of:* Barnett, V., & Turkman, K.F. (eds), *Statistics for the Environment.* Wiley, New York.

Vallejos, R., & Osorio, F. 2014. Effective sample size of spatial process models. *Spatial Statistics*, **9**, 66–92.

Van Marcke, E. 2010. *Random Fields: Analysis and Synthesis*. Princeton University Press, Princeton, NJ.

Vecchia, A.V. 1988. Estimation and model identification for continuous spatial processes. *Journal of the Royal Statistical Society, Series B*, **50**, 297–312.

Vega, S. Halleck, & Elhorst, J.P. 2015. The SLX model. *Journal of Regional Science*, **55**, 339–363.

Ver Hoef, J.M. 2008. Spatial methods for plot-based sampling of wildlife populations. *Environmental and Ecological Statistics*, **15**, 3–13.

Ver Hoef, J.M. 2012a. Practical considerations for experimental designs of spatially auto-correlated data using computer intensive methods. *Statistical Methodology*, **9**, 172–184.

Ver Hoef, J.M. 2012b. Who invented the delta method? *The American Statistician*, **66**, 124–127.

Ver Hoef, J.M., & Barry, R.P. 1998. Constructing and fitting models for cokriging and multivariable spatial prediction. *Journal of Statistical Planning and Inference*, **69**, 275–294.

Ver Hoef, J.M., & Boveng, P.L. 2015. Iterating on a single model is a viable alternative to multimodel inference. *Journal of Wildlife Management*, **79**, 719–729.

Ver Hoef, J.M., & Cressie, N. 1993. Multivariable spatial prediction. *Mathematical Geology*, **25**, 219–240.

Ver Hoef, J.M., & Peterson, E.E. 2010. A moving average approach for spatial statistical models of stream networks. *Journal of the American Statistical Association*, **105**, 6–18.

Ver Hoef, J.M., Cressie, N., & Barry, R.P. 2004. Flexible spatial models for kriging and cokriging using moving averages and the fast Fourier transform (FFT). *Journal of Computational and Graphical Statistics*, **13**, 265–282.

Ver Hoef, J.M., Peterson, E.E., & Theobald, D. 2006. Spatial statistical models that use flow and stream distance. *Environmental and Ecological Statistics*, **13**, 449–464.

Ver Hoef, J.M., Peterson, E.E., Clifford, D., & Shah, R. 2014. SSN: An R package for spatial statistical modeling on stream networks. *Journal of Statistical Software*, **56**, 1–45.

Ver Hoef, J.M., Hanks, E.M., & Hooten, M.B. 2018a. On the relationship between conditional (CAR) and simultaneous (SAR) autoregressive models. *Spatial Statistics*, **25**, 68–85.

Ver Hoef, J.M., Peterson, E.E., Hooten, M.B., Hanks, E.M., & Fortin, M.J. 2018b. Spatial autoregressive models for statistical inference from ecological data. *Ecological Monographs*, **88**, 36–59.

Ver Hoef, J.M., Peterson, E.E., & Isaak, D.J. 2019. Spatial statistical models for stream networks. *Pages 421–444 of:* Gelfand, A., Fuentes, M., Hoeting, J., and Smith, R. (eds), *Handbook of Environmental and Ecological Statistics*. Chapman and Hall/CRC, Boca Raton, FL.

Ver Hoef, J.M., Dumelle, M., Higham, M., Peterson, E.E., & Isaak, D.J. 2023. Indexing and partitioning the spatial linear model for large data sets. *PLoS ONE*, 18:e0291906.

Wackernagel, H. 1998. *Multivariate Geostatistics: An Introduction with Applications, 2nd ed.* Springer-Verlag, New York.

Wall, M.M. 2004. A close look at the spatial structure implied by the CAR and SAR models. *Journal of Statistical Planning and Inference*, **121**, 311–324.

Waller, L.A., & Gotway, C.A. 2004. *Applied Spatial Statistics for Public Health Data.* John Wiley & Sons, Hoboken, NJ.

Walvoort, D.J.J., Brus, D.J., & de Gruijter, J.J. 2010. An R package for spatial coverage sampling and random sampling from compact geographical strata by k-means. *Computers & Geosciences*, **36**, 1261–1267.

Wang, Z., & Zhu, Z. 2019. Spatiotemporal balanced sampling design for longitudinal area surveys. *Journal of Agricultural, Biological, and Environmental Statistics*, **24**, 245–263.

Warnes, J.J., & Ripley, B.D. 1987. Problems with likelihood estimation of covariance functions of spatial Gaussian processes. *Biometrika*, **74**, 640–642.

Warren, J.L. 2020. A nonstationary spatial covariance model for processes driven by point sources. *Journal of Agricultural, Biological and Environmental Statistics*, **25**, 415–430.

Wartenberg, D. 1985. Multivariate spatial correlation: A method for exploratory geographical analysis. *Geographical Analysis*, **17**, 263–283.

Weisberg, S. 2005. *Applied Linear Regression, 3rd ed.* Wiley, Hoboken, NJ.

Weller, Z.D., & Hoeting, J.A. 2016. A review of nonparametric hypothesis tests of isotropy properties in spatial data. *Statistical Science*, **31**, 305–324.

White, P., & Porcu, E. 2017. Towards a complete picture of covariance functions on spheres cross time. *Biometrika*, **103**, 1–23.

White, P., & Porcu, E. 2019. Towards a complete picture of stationary covariance functions on spheres cross time. *Electronic Journal of Statistics*, **13**, 2566–2594.

Whittaker, J. 1990. *Graphical Models in Applied Multivariate Statistics.* Wiley, Chichester.

Whittle, P. 1954. On stationary processes in the plane. *Biometrika*, **41**, 434–439.

Wikle, C.W., & Royle, J.A. 1999. Space-time dynamic design of environmental monitoring networks. *Journal of Agricultural, Biological, and Environmental Statistics*, **4**, 489–507.

Wikle, C.W., & Royle, J.A. 2005. Dynamic design of ecological monitoring networks for non-Gaussian spatio-temporal data. *Environmetrics*, **16**, 507–522.

Williams, R.M. 1952. Experimental designs for serially correlated observations. *Biometrika*, **39**, 151–167.

Wolfinger, R.D., & O'Connell, M. 1993. Generalized linear mixed models: A pseudo-likelihood approach. *Journal of Statistical Computation and Simulation*, **48**, 233–243.

Yadrenko, M.I. 1983. *Spectral Theory of Random Fields: Translation Series in Mathematics and Engineering.* Springer, New York.

Yaglom, A.M. 1987. *Correlation Theory of Stationary and Related Random Functions. Volume I: Basic Results.* Springer-Verlag, New York.

Yfantis, E.A., Flatman, G.T., & Behar, J.V. 1987. Efficiency of kriging estimation for square, triangular, and hexagonal grids. *Mathematical Geology*, **19**, 183–205.

Yildirim, V., & Kantar, Y. Mert. 2020. Robust estimation approach for spatial error model. *Journal of Statistical Computation and Simulation*, **90**, 1618–1638.

Ying, Z. 1991. Asymptotic properties of a maximum likelihood estimator with data from a Gaussian process. *Journal of Multivariate Analysis*, **36**, 280–296.

Zhang, H. 2004. Inconsistent estimation and asymptotically equal interpolations in model-based geostatistics. *Journal of the American Statistical Association*, **99**, 250–261.

Zhang, H., & Cai, W. 2015. When doesn't cokriging outperform kriging? *Statistical Science*, **30**, 176–180.

Zhang, H., & Zimmerman, D.L. 2005. Toward reconciling two asymptotic frameworks in spatial statistics. *Biometrika*, **92**, 921–936.

Zhu, Z., & Stein, M.L. 2005. Spatial sampling design for parameter estimation of the covariance function. *Journal of Statistical Planning and Inference*, **134**, 583–603.

Zhu, Z., & Stein, M.L. 2006. Spatial sampling design for prediction with estimated parameters. *Journal of Agricultural, Biological, and Environmental Statistics*, **11**, 24–44.

Zimmerman, D., Pavlik, C., Ruggles, A., & Armstrong, M.P. 1999. An experimental comparison of ordinary and universal kriging and inverse distance weighting. *Mathematical Geology*, **31**, 375–390.

Zimmerman, D.L. 1986. *A random field approach to spatial experiments.* Ph.D. thesis, Department of Statistics, Iowa State University.

Zimmerman, D.L. 1989a. Computationally efficient restricted maximum likelihood estimation of generalized covariance functions. *Mathematical Geology*, **21**, 655–672.

Zimmerman, D.L. 1989b. Computationally exploitable structure of covariance matrices and generalized covariance matrices in spatial models. *Journal of Statistical Computation and Simulation*, **32**, 1–15.

Zimmerman, D.L. 1993. Another look at anisotropy in geostatistics. *Mathematical Geology*, **25**, 453–470.

Zimmerman, D.L. 2006. Optimal network design for spatial prediction, covariance parameter estimation, and empirical prediction. *Environmetrics*, **17**, 635–652.

Zimmerman, D.L. 2020. *Linear Model Theory — With Examples and Exercises.* Springer, New York.

Zimmerman, D.L., & Cressie, N. 1992. Mean squared prediction error in the spatial linear model with estimated covariance parameters. *Annals of the Institute of Statistical Mathematics*, **44**, 27–43.

Zimmerman, D.L., & Holland, D.M. 2005. Complementary co-kriging: Spatial prediction using data combined from several environmental monitoring networks. *Environmetrics*, **16**, 219–234.

Zimmerman, D.L., & Li, J. 2013. Model-based frequentist design for univariate and multi-variate geostatistics. *Pages 37–53 of:* Mateu, J., & Müller, W.G. (eds), *Spatio-Temporal Design*. Wiley, Chichester.

Zimmerman, D.L., & Núñez-Antón, V.N. 2010. *Antedependence Models for Longitudinal Data*. CRC Press, Boca Raton, FL.

Zimmerman, D.L., & Stein, M. 2010. Classical geostatistical methods. *Pages 29–44 of:* Gelfand, A.E., Diggle, P.J., Fuentes, M., & Guttorp, P. (eds), *Handbook of Spatial Statistics*. Chapman and Hall/CRC Press, Boca Raton, FL.

Zimmerman, D.L., & Ver Hoef, J.M. 2017. The Torgegram for fluvial variography: Characterizing spatial dependence on stream networks. *Journal of Computational and Graphical Statistics*, **26**, 253–264.

Zimmerman, D.L., & Ver Hoef, J.M. 2022. On deconfounding spatial confounding in linear models. *The American Statistician*, **76**, 159–167.

Zimmerman, D.L., & Zimmerman, M.B. 1991. A comparison of spatial semivariogram estimators and corresponding ordinary kriging predictors. *Technometrics*, **33**, 77–99.

Zimmerman, D.L., Liu, Z., & Song, R. 2024. Nonparametric assessment of spatial dependence properties on Euclidean trees, with an application to stream networks. *Unpublished technical report*.

Index